Stem Cell Biology and Regenerative Medicine

Series Editor
Kursad Turksen, Ph.D.
kursadturksen@gmail.com

More information about this series at http://www.springer.com/series/7896

Amy Firth • Jason X.-J. Yuan
Editors

Lung Stem Cells in the Epithelium and Vasculature

Humana Press

Editors
Amy Firth
Laboratory of Genetics
Salk Institute for Biological Studies
La Jolla, CA, USA

Jason X.-J. Yuan
Department of Medicine
The Univsersity of Arizona
Tucson, AZ, USA

ISSN 2196-8985 ISSN 2196-8993 (electronic)
Stem Cell Biology and Regenerative Medicine
ISBN 978-3-319-35338-8 ISBN 978-3-319-16232-4 (eBook)
DOI 10.1007/978-3-319-16232-4

Springer Cham Heidelberg New York Dordrecht London

Softcover reprint of the hardcover 1st edition 2015

Printed on acid-free paper

Humana Press is a brand of Springer
Springer International Publishing AG Switzerland is part of Springer Science+Business Media
(www.springer.com)

Preface

Unfortunately lung disease remains a devastating cause of morbidity and mortality around the world. With an ever-increasing prevalence, the situation is not improving; indeed by 2020 COPD is expected to become the third leading cause of death worldwide. Most of the current therapeutic approaches for lung disease serve only to manage the symptoms. Lung transplantation remains one of few curative options and itself is associated with a host of other complications, particularly those due to lifelong immunosuppression.

Like for many other diseases, there is a significant appeal for using stem cells in the treatment of lung disease. Before such approaches will come to fruition, it is necessary to establish in-depth knowledge of the role of stem cells in the human lung. The identity and function of stem cells in the developing and, in particular, the adult lung has been increasingly studied over the past decade. We are thus starting to get a clearer insight into the endogenous stem cell populations of the lung, their roles in the maintenance of the lung epithelium and vasculature, and their dysregulation in the pathogenesis of lung disease. As the fields of pluripotent and adult stem cells evolve, we will develop a greater understanding of the therapeutic potential of stem cells with the hope of eventual clinical intervention in lung disease.

This book aims to provide an up-to-date review of endogenous stem cells in both the airways and vasculature of the lung in addition to discussing the rapidly evolving field of pluripotent stem cells and regenerative medicine approaches to study and treat lung disease.

La Jolla, CA, USA — Amy Firth
Tucson, AZ, USA — Jason X.-J. Yuan

Acknowledgements

A special thank you to two of the best colleagues and friends anyone could wish for, Tushar Menon Ph.D. and Amy Rommel Ph.D., for spending countless hours listening to and supporting me in the editing of this book. Your patience and tolerance were enormously appreciated.

La Jolla, CA, USA Amy Firth

Contents

Part II Pulmonary Vasculature Stem Cells

Contributors

Xingbin Ai Division of Pulmonary, Allergy, Sleep and Critical Care Medicine, Boston University School of Medicine, Boston, MA, USA

Kewal Asosingh Department of Pathobiology, Lerner Research Institute, Cleveland Clinic Lerner College of Medicine of Case Western Reserve University, Cleveland, OH, USA

Eric D. Austin Department of Pediatrics, Vanderbilt University, Nashville, TN, USA

Ena Ray Banerjee Department of Zoology, Immunology and Regenerative Medicine Research Unit, University of Calcutta, Kolkata, India

Rubin Baskir Department of Medicine, Division of Allergy, Pulmonary and Critical Care Medicine, Vanderbilt Center for Stem Cell Biology, Cell and Developmental Biology, Vanderbilt University, Nashville, TN, USA

Kelley L. Colvin Department of Pediatrics-Critical Care, University of Colorado Denver, Aurora, CO, USA

Cardiovascular Pulmonary Research, University of Colorado Denver, Aurora, CO, USA

Department of Bioengineering, University of Colorado Denver, Aurora, CO, USA

Linda Crnic Institute for Down Syndrome, University of Colorado Denver, Aurora, CO, USA

Fernanda F. Cruz Federal University of Rio de Janeiro, Rio de Janeiro, Brazil

Radhika Dixit Division of Pulmonary, Allergy, Sleep and Critical Care Medicine, Boston University School of Medicine, Boston, MA, USA

Barbara Driscoll Developmental Biology and Regenerative Medicine Program, Department of Surgery, Children's Hospital Los Angeles, University of Southern California, Los Angeles, CA, USA

John F. Engelhardt Department of Anatomy and Cell Biology, Carver College of Medicine, University of Iowa, Iowa City, IA, USA

Serpil Erzurum Department of Pathobiology, Lerner Research Institute and Respiratory Institute, Cleveland Clinic Lerner College of Medicine of Case Western Reserve University, Cleveland, OH, USA

Quentin Felty Department of Environmental and Occupational Health, Florida International University, Miami, FL, USA

Alan Fine Division of Pulmonary, Allergy, Sleep and Critical Care Medicine, Boston University School of Medicine, Boston, MA, USA

Amy L. Firth Laboratory of Genetics, The Salk Institute for Biological Studies, La Jolla, CA, USA

Orquidea Garcia Developmental Biology and Regenerative Medicine Program, Department of Surgery, Children's Hospital Los Angeles, University of Southern California, Los Angeles, CA, USA

Christophe Guignabert INSERM UMR_S 999, LabEx LERMIT, Centre Chirurgical Marie Lannelongue, Le Plessis-Robinson, France

Univ Paris-Sud, School of medicine, Kremlin-Bicêtre, France

Lauren Hartman Department of Physiology and Cell Biology, University of South Alabama, Mobile, AL, USA

Center for Lung Biology, College of Medicine, University of South Alabama, Mobile, AL, USA

Anna R. Hemnes Department of Medicine, Division of Allergy, Pulmonary and Critical Care Medicine, Vanderbilt University, Nashville, TN, USA

William Reed Henderson Jr. Department of Medicine, Center for Allergy and Inflammation, UW Medicine at South Lake Union, University of Washington, Seattle, WA, USA

Michael Hiatt Developmental Biology and Regenerative Medicine Program, Department of Surgery, Children's Hospital Los Angeles, University of Southern California, Los Angeles, CA, USA

Alice Huertas INSERM UMR_S 999, LabEx LERMIT, Centre Chirurgical Marie Lannelongue, Le Plessis-Robinson, France

Univ Paris-Sud, School of medicine, Kremlin-Bicêtre, France

AP-HP, Service de Pneumologie, Centre de Référence de l'Hypertension Pulmonaire Sévère, DHU Thorax Innovation, Hôpital de Bicêtre, France

Marc Humbert INSERM UMR_S 999, LabEx LERMIT, Centre Chirurgical Marie Lannelongue, Le Plessis-Robinson, France

Univ Paris-Sud, School of medicine, Kremlin-Bicêtre, France

AP-HP, Service de Pneumologie, Centre de Référence de l'Hypertension Pulmonaire Sévère, DHU Thorax Innovation, Hôpital de Bicêtre, France

Melanie Königshoff Comprehensive Pneumology Center, Munich, Germany

Yuru Liu Department of Pharmacology, College of Medicine, University of Illinois, Chicago, IL, USA

Xiaoming Liu Institute of Human Stem Cell Research, General Hospital of Ningxia Medical University, Yinchuan, Ningxia, IA, China

Department of Anatomy and Cell Biology, Carver College of Medicine, University of Iowa, Iowa City, IA, USA

Ozus Lohani Department of Bioengineering, University of Colorado Denver, Aurora, CO, USA

Amber Lundin Developmental Biology and Regenerative Medicine Program, Department of Surgery, Children's Hospital Los Angeles, University of Southern California, Los Angeles, CA, USA

Thomas J. Lynch Department of Anatomy and Cell Biology, Carver College of Medicine, University of Iowa, Iowa City, IA, USA

Susan Majka Department of Medicine, Division of Allergy, Pulmonary and Critical Care Medicine, Vanderbilt Center for Stem Cell Biology, Cell and Developmental Biology, Vanderbilt University, Nashville, TN, USA

Asrar B. Malik Department of Pharmacology, College of Medicine, University of Illinois, Chicago, IL, USA

Center for Lung and Vascular Biology, University of Illinois, Chicago, IL, USA

Glenn Marsboom Department of Pharmacology, College of Medicine, University of Illinois, Chicago, IL, USA

Center for Lung and Vascular Biology, University of Illinois, Chicago, IL, USA

Tushar Menon Laboratory of Genetics, The Salk Institute for Biological Studies, La Jolla, CA, USA

Samriddha Ray Cedars-Sinai Medical Center, Lung and Regenerative Medicine Institutes, Los Angeles, CA, USA

Jalees Rehman Department of Pharmacology, College of Medicine, University of Illinois, Chicago, IL, USA

Center for Lung and Vascular Biology, University of Illinois, Chicago, IL, USA

Department of Medicine, Section of Cardiology, University of Illinois, Chicago, IL, USA

Patricia R.M. Rocco Federal University of Rio de Janeiro, Rio de Janeiro, Brazil

Jonathan Rose Department of Pathobiology, Lerner Research Institute, Cleveland Clinic Lerner College of Medicine of Case Western Reserve University, Cleveland, OH, USA

Seiijiro Sakao Department of Respirology, Chiba University, Chiba, Japan

Troy Stevens Department of Physiology and Cell Biology, University of South Alabama, Mobile, AL, USA

Department of Medicine, University of South Alabama, Mobile, AL, USA

Center for Lung Biology, College of Medicine, University of South Alabama, Mobile, AL, USA

Barry R. Stripp Professor of Medicine and Biomedical Sciences, Cedars-Sinai Medical Center, Lung and Regenerative Medicine Institutes, Los Angeles, CA, USA

Franziska E. Uhl Comprehensive Pneumology Center, Munich, Germany

Norbert F. Voelkel Department of Biochemistry and Molecular Biology, Virginia Commonwealth University, Richmond, VA, USA

Darcy E. Wagner Comprehensive Pneumology Center, Munich, Germany

Jun Wei Institute of Human Stem Cell Research, General Hospital of Ningxia Medical University, Yinchuan, Ningxia, IA, China

Daniel J. Weiss Department of Medicine, Pulmonary and Critical Care, University of Vermont, Burlington, VT, USA

Michael E. Yeager Department of Pediatrics-Critical Care, University of Colorado Denver, Aurora, CO, USA

Cardiovascular Pulmonary Research, University of Colorado Denver, Aurora, CO, USA

Department of Bioengineering, University of Colorado Denver, Aurora, CO, USA

Linda Crnic Institute for Down Syndrome, University of Colorado Denver, Aurora, CO, USA

Jason X.-J. Yuan Department of Medicine, Arizona Health Sciences Center, University of Arizona, Tucson, AZ, USA

Min Zhang Department of Pharmacology, College of Medicine, University of Illinois, Chicago, IL, USA

Center for Lung and Vascular Biology, University of Illinois, Chicago, IL, USA

Part I
Airway Stem Cells

Chapter 1
Does a Lung Stem Cell Exist?

Samriddha Ray and Barry R. Stripp

1.1 Introduction

The mammalian respiratory tract can be subdivided into three functionally distinct anatomic regions: proximal conducting airways termed bronchi, distal conducting airways termed bronchioles, and the terminal gas exchange region, termed alveoli. Distribution of the constituent cells varies along this proximal-distal axis and reflects differences in function and physiologic requirement at each zone (Fig. 1.1). The different cell types of the lung airway are best defined within the epithelium that lines airspaces. Some of the major epithelial cell types include ciliated, basal, non-ciliated secretory, and alveolar cells and differ in their morphology, behavior, and contribution to maintenance of tissue homeostasis and repair (Rackley and Stripp 2012; Rock and Hogan 2011). In addition to tissue turn over, airway undergoes tissue injury and cell loss due to constant exposure to inhaled environmental pollutants and infectious agents.

The current model of adult tissue regeneration and maintenance supports the idea that regional pools of epithelial stem cells ensure maintenance and timely repair of damaged tissue (Leeman et al. 2014; Rawlins and Hogan 2006). Precedence for this model comes from studies of developing and adult murine lung epithelium. First, the lung epithelium is derived from the anterior foregut endoderm while the mesenchyme is derived from the splanchnic mesoderm. There is no evidence so far for the

S. Ray
Cedars-Sinai Medical Center, Lung and Regenerative Medicine Institutes,
Los Angeles, CA USA

B.R. Stripp, Ph.D. (✉)
Professor of Medicine and Biomedical Sciences, Cedars-Sinai Medical Center,
Lung and Regenerative Medicine Institutes, 8700 Beverly Blvd.,
AHSP Room A9401, Los Angeles, CA 90048, USA
e-mail: Barry.Stripp@cshs.org

A. Firth, J.X.-J. Yuan (eds.), *Lung Stem Cells in the Epithelium and Vasculature*,
Stem Cell Biology and Regenerative Medicine, DOI 10.1007/978-3-319-16232-4_1

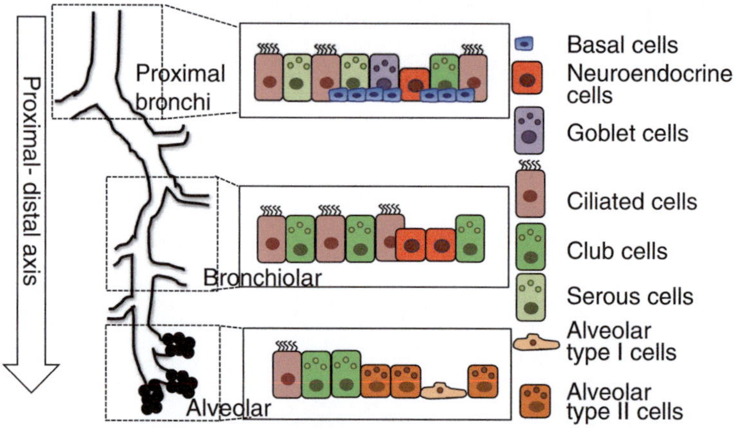

Fig. 1.1 Schematic representation of epithelial cell types found in airways of the adult mouse lung. The adult airway exhibits compartmentalization of constituent cells along the proximal—distal axis. The proximal bronchial epithelium is pseudostratified due to the presence of basal as well as luminal cells. The latter includes ciliated cells and non-ciliated secretory cells including goblet, serous, and club cells. In contrast to bronchial epithelium, distal bronchiolar airway in adult laboratory mouse strains lack basal cells and are predominantly populated by secretory club and ciliated cells, resulting in structural demarcation between the proximal and distal airways. This is unlike adult human distal airways where basal cells are maintained within distal bronchiolar epithelium. As described in the text, in vivo and in vitro assays have identified multiple stem and progenitor cell pools that include bronchial basal cells, Scgb1a1-expressing club cells and alveolar type II cells in gas exchange zone

existence of a common developmental progenitor that gives rise to epithelial as well as mesenchymal cell types of the adult respiratory system, which precludes the idea of a common stem cell that gives rise to epithelial, as well as mesenchymal, cells in mice (Perl and Whitsett 1999; Morrisey and Hogan 2010; Cardoso and Lu 2006). Secondly, during the process of lung development, the epithelial lining of airways is spatially segregated into distinct proximal and distal zones by non-overlapping lineages of cells that are specified very early in gestation (Perl et al. 2002). This results in the spatially diverse distribution of the adult lung epithelium. Lastly, as outlined below, multiple pools of regionally restricted adult stem cells have been identified in the airway epithelium that contribute to repair. One theme that emerges from these studies is that progenitor and stem cell behavior is significantly altered between steady state and following injuries that remodel the tissue microenvironment, highlighting the interdependence of cell phenotypes and behaviors within different regional microenvironments of the lung. Additionally, different progenitor/stem cell pools respond differently to the injury type and pathological condition. For instance, following bleomycin treatment of laboratory strains of mice, resident alveolar progenitors in distal airway namely surfactant protein C expressing type 2 and integrin $\alpha6$ $\beta4$ expressing stem cells are activated to promote tissue regeneration, whereas following naphthalene exposure that ablates most airway epithelial cells,

Scgb1a1-expressing naphthalene-resistant variant club progenitors are essential for airway regeneration (Chapman et al. 2011; Barkauskas et al. 2013).

These and related data underscore the existence of distinct stem cell pools to combat tissue damage from a myriad of environmental and pathological insults during the lifetime of terrestrial beings. A related but less explored question is whether there is cross talk between the different lung stem cell pools similar to what has been observed in other organ systems such as rapid and slow cycling stem cells of small intestinal crypts or between luminal and myoepithelial cells of the mammary epithelium (Van Keymeulen et al. 2011; van Es et al. 2012). In that context, recent findings that luminal Scgb1a1-expressing secretory cells can dedifferentiate into basal-like stem cells following ectopic ablation of tracheal basal cells suggest that different stem cell pools in lung can indeed cross talk and influence each other's stem cell behavior (Tata et al. 2013).

Identification of human lung stem cells has primarily relied upon in vitro culture assays of cells isolated based on expression of putative stem cell markers (Barkauskas et al. 2013; Rock et al. 2009). However, as mentioned later, strategies to study human lung stem cell behavior in vivo have been reported and advances in such approaches will be critical to better understand their roles. A putative c-kit positive pool of human stem cells has been proposed to generate both epithelial and endothelial cells in culture and promote airway and pulmonary vessel repair following cryo-injury in mice (Kajstura et al. 2011). Such studies suggest existence of a single multipotent stem cell pool, but its biological significance and roles in vivo remain unclear (Lung stem cells: looking beyond the hype 2011). Functions of lung mesenchymal cells in postnatal tissue maintenance, repair, and epithelial stem cell function are just beginning to be understood and precise knowledge of their role in epithelial repair and maintenance will be paramount (Sinclair et al. 2013).

1.2 Epithelial Cells of the Conducting Airway and Alveolar Zones

The epithelium lining proximal airways is pseudostratified in composition and composed of basal cells that contact the basement membrane and luminal cells that extend from basolateral contacts with the basement membrane to the airway lumen. Basal cells are attached to the basement membrane by hemi desmosomes and to luminal columnar cells by desmosomes (Evans et al. 1990; Evans and Plopper 1988); these cytoskeletal associations promote airway structural integrity. As discussed later, basal cells behave as multipotent stem cells to promote adult tissue regeneration.

In mice, basal cell distribution tapers along the tracheal-bronchial airway axis and are absent from bronchiolar airways. This is in contrast to adult human airway epithelium that remains pseudostratified from trachea down to small bronchioles of distal lung (Nakajima et al. 1998). The proximal airway luminal cells include ciliated cells and non-ciliated secretory cells that are specialized for mucus clearance and host defense response, respectively, as described later. Additionally, there are neuroendocrine cells that have been particularly implicated in human small-cell lung cancer (SCLC). Their function in steady state tissue maintenance remains unclear (Song et al. 2012).

The bronchiolar airway epithelium in mice is largely maintained by differentiated ciliated and non-ciliated secretory cells. The abundant cilia protruding from the apical membrane of the ciliated cells are each held to the cytoplasmic basal body through an axoneme (Jeffery and Reid 1975; Rhodin 1966). Axonemes are microtubule-based structures that are associated with dynein and kinesin motors to allow synchronous beating of cilia and mobilization of mucus and adsorbed materials. The non-ciliated secretory cells of the bronchiolar airway include morphologically and functionally distinct cells such as the goblet, serous, and club cells. Goblet and serous cells are primarily involved in secretion of mucous glycoproteins, enzymes, and antimicrobial peptides that collectively promote inactivation and clearance of inhaled agents. Defects in mucus production, clearance, and antimicrobial defense can contribute to serious chronic respiratory conditions such as the lung complications associated with cystic fibrosis (Rogers 1994). Club cells are involved in diverse functions including modulation of local inflammatory responses, host defense, xenobiotic metabolism, and tissue maintenance and regeneration (discussed later).

The distal lung gas exchange component is composed of alveolar type 1 (AT1) and type 2 (AT2) cells. Surfactant protein secreting type 2 cells maintain surface tension and prevent cell collapse following exhalation. The thin and lacy morphology of alveolar type 1 cells and their apposition to blood capillaries makes them well suited for gas exchange (Bertalanffy and Leblond 1955).

The diversity in mature epithelial cell types and their functions along the airway axis underscores the need for region-specific mechanisms to maintain and regenerate the epithelium in homeostasis and repair.

1.3 Region-Specific Stem and Progenitor Cells Maintain the Adult Airway

Identification of regional stem and progenitor cell pools has been accomplished using a combination of techniques including in vivo lineage tracing of defined epithelial cell types, in vitro 3-dimensional culture assays to investigate the behavior of flow sorted epithelial cells, and ex vivo transplant assays following isolation of putative stem cells (Barkauskas et al. 2013; Chen et al. 2012; Rock et al. 2009; Flodby et al. 2010; Kretzschmar and Watt 2012; Randell et al. 1991). Technical caveats and limitations notwithstanding these approaches have proved valuable to advance understanding of lung stem cell functions in maintenance of epithelial diversity along the airway tree.

1.3.1 Basal Cells Function as Stem Cells for Maintenance and Repair of the Pseudostratified Epithelium of Tracheal and Bronchial Airways

Identification of stem cell properties in basal cells is obtained from several lines of evidence including lineage analyses of Keratin 14 and 5 expressing cells under steady state and during tissue regeneration following injury. Lineage tracing experiments demonstrated that mouse tracheal basal cells expressing the intermediate filament protein cytokeratin 5 (Krt5) have long-term self-renewal capacity and can generate differentiating progeny that replenish both ciliated and non-ciliated luminal epithelial cells types. These properties held true under conditions of normal epithelial turnover and during renewal following exposure to toxic chemicals like sulfur dioxide and naphthalene. However, there is evidence that not all basal cells of mouse pseudostratified airways share the same molecular characteristics or have the same capacity for multipotent differentiation (Hong et al. 2004; Rock et al. 2009). It is not clear whether these differences between airway basal cells are stochastic or evidence of functionally distinct subsets. In addition to intrinsic differences, cell extrinsic factors such as changes in the microenvironment from injury significantly alter their proliferation dynamics and differentiation fate (Ghosh et al. 2011; Rock et al. 2009). In vivo data are substantiated by in vitro experiments in which isolated basal stem cells can undergo proliferation and differentiation to generate pseudostratified trachea-like structures reminiscent of in vivo tissue architecture. Additional evidence for their stem cell roles comes from their ability to promote regeneration of denuded tissue in xenograft assays using rodent and human tracheal cells (Rock et al. 2010).

Although the current model suggests basal cells as the stem cells of trachea and proximal bronchi, ectopic ablation of these cells in the trachea could trigger dedifferentiation of Scgb1a1-expressing secretory cells into basal-like stem cells to promote regeneration of pseudostratified airway. Such studies substantiate the role of non-cell autonomous factors in the regulation of airway progenitor cell behavior (Tata et al. 2013).

Evidence for their role in human airways is derived from in vivo clonal patch analyses and in vitro growth assays of isolated putative basal cells. In humans, basal cell marker expression was initially reported in stem cell-derived clonal patches from human upper airway epithelium, and their stem cell function supported by in vitro 3D culture assays (Rock et al. 2010). In vivo clonal patch analyses suggest that proliferating basal cells are maintained by a neutral drift model where loss of progenitor cells due to differentiation is compensated by duplication of neighbor cells (Teixeira et al. 2013).

1.3.2 Non-ciliated Club Cells Behave as Bronchiolar Stem Cells

Non-ciliated cells of mouse bronchiolar airways have been defined based upon their expression of the abundant secretory club cell-specific protein (CCSP or Scgb1a1). The Scgb1a1-expressing cells represent a highly heterogeneous cell type whose diversity develops early in embryonic development (Guha et al. 2012, 2014). In vivo lineage tracing experiments under steady state conditions have identified club cells as the progenitor cells of mouse bronchiolar airways, but not of tracheal airways. Thus, the current model supports the idea that basal cells are stem cells of the tracheobronchial airway in mice while Scgb1a1-expressing cells constitute the bronchiolar progenitor pool. However, as mentioned earlier, in the event of a loss of tracheal basal cells, Scgb1a1 cells undergo self-renewal and differentiation to regenerate pseudostratified tracheal airway epithelium (Rawlins et al. 2009; Tata et al. 2013). Their role in bronchiolar regeneration has been studied in the context of various injury models including those triggered by exposure to naphthalene, bleomycin, and viral infection. Whether distinct subsets of Scgb1a1-expressing cells respond to the different kinds of injury remains unexplored. One subset that is resistant to naphthalene exposure due to their lack of Cy2f2 (termed as variant club cells) promotes regeneration following naphthalene-induced epithelial damage and are localized around the neuroendocrine bodies (NEB) of mouse airway and terminal bronchoalveolar duct junctions (BADJ) (Giangreco et al. 2002; Reynolds et al. 2000a, b). Intriguingly, NEB is also associated with precursors of secretory cells such as Scgb3a2 and Upk3a (Guha et al. 2012), and thus resembles a niche for secretory cell development and progenitor function in adults. The mechanism by which these cells promote parenchymal tissue regeneration remains less clear. But in vivo Scgb1a1-expressing club cells have been shown to generate alveolar type 2 and 1 cells following tissue injury such as those from bleomycin treatment and PR 8 influenza virus infection (Rock et al. 2011; Zheng et al. 2012, 2013).

Scgb1a1-expressing cells also constitute the pool of bronchioalveolar stem cells (BASCs) that were identified by their coexpression of Scgb1a1 and alveolar type 2 protein SPC and were located at regions where terminal airway abuts with alveolar epithelial cells (Kim et al. 2005). In vivo lineage tracing analyses support their role in alveolar type 2 cell regeneration following alveolar cell-specific injury (Tropea et al. 2012). In vitro culture assays of BASCs have not only highlighted their ability to undergo differentiation into multiple epithelial lineages including bronchial, alveolar, as well as bronchioalveolar, cells but also underscored the role of surrounding endothelial cell microenvironment in determination of their differentiation fate (Lee et al. 2014).

Additional pools of multipotent adult airway progenitor cells expressing Epcam, CD49f, CD104, and CD24 (low) have been identified by in vitro matrigel-based clonal assays and have been shown to generate airway as well as alveolar cell lineages (Bertoncello and McQualter 2011; McQualter et al. 2010). In humans, Scgb1a1 or CC10 (club cell)-expressing cells represent significant proportion of proliferating cells in distal conducting airways under steady state and are reduced following cigarette smoke injuries. Their contributions to tissue repair and regeneration, however, remain unknown (Boers et al. 1999).

1.3.3 The Bronchioalveolar and Alveolar Stem Cells

Classical BrdU labeling experiments suggest that adult alveolar epithelium undergo slow renewal (Messier and Leblond 1960). In vivo lineage tracing analyses have shown that adult alveolar type 2 cells are progenitor cells that undergo self-renewal, clonal expansion, and differentiation into alveolar type 1 cells (Barkauskas et al. 2013; Desai et al. 2014). Changes in the endogenous microenvironment caused by targeted loss of type 2 cells or widespread alveolar cell loss by bleomycin treatment significantly alter stem cell behavior namely in their proliferation and differentiation rate, implicating role of cell extrinsic factors in regulation of stem cell behavior. In vitro three dimensional alveolosphere formation assays using isolated surfactant type 2 cells substantiate their self-renewal and differentiation properties (Barkauskas et al. 2013). In vitro culture assays of isolated alveolar type 1 cells implicate their role as putative stem cells although in vivo evidence warrants further research (Dobbs et al. 2010; Gonzalez et al. 2009).

Additional pools of distal lung stem cells include laminin receptor α6 β4 integrin expressing facultative stem cells that promote alveolar cell regeneration after parenchymal injury such as those induced following bleomycin treatment (Chapman et al. 2011).

1.4 Summary

Identification and function of stem cells in a slowly regenerating organ like the lung is highly context dependent. The current body of evidence does not support the existence of a single lung stem cell, rather the existence of multiple region-specific stem and associated progenitor cells that can generate all specialized cell types to fulfill adult airway function. It is, however, safe to infer that several pools of stem cells maintain the different compartments of adult respiratory tract, and that their stem cell behavior is intricately regulated by changes in their microenvironment. Existence of multiple adult stem cell pools is therefore serving multiple functions including: (1) maintenance of spatially diverse epithelial cell types of the adult airway to promote tissue homeostasis as well as (2) timely and efficient regeneration of the airway epithelium to combat the diverse injury types to which the adult airway is exposed during its lifetime.

Acknowledgement We thank members of our lab for useful comments and discussion on the chapter.

References

Barkauskas CE, Cronce MJ, Rackley CR, Bowie EJ, Keene DR, Stripp BR, Randell SH, Noble PW, Hogan BL (2013) Type 2 alveolar cells are stem cells in adult lung. J Clin Invest 123(7):3025–3036. doi:10.1172/JCI68782

Bertalanffy FD, Leblond CP (1955) Structure of respiratory tissue. Lancet 269(6905):1365–1368

Bertoncello I, McQualter J (2011) Isolation and clonal assay of adult lung epithelial stem/progenitor cells. Curr Protoc Stem Cell Biol; Chapter 2: Unit 2G. 1. doi:10.1002/9780470151808.sc02g01s16

Boers JE, Ambergen AW, Thunnissen FB (1999) Number and proliferation of clara cells in normal human airway epithelium. Am J Respir Crit Care Med 159(5 Pt 1):1585–1591. doi:10.1164/ajrccm.159.5.9806044

Cardoso WV, Lu J (2006) Regulation of early lung morphogenesis: questions, facts and controversies. Development 133(9):1611–1624. doi:10.1242/dev.02310

Chapman HA, Li X, Alexander JP, Brumwell A, Lorizio W, Tan K, Sonnenberg A, Wei Y, Vu TH (2011) Integrin α6β4 identifies an adult distal lung epithelial population with regenerative potential in mice. J Clin Invest 121(7):2855–2862. doi:10.1172/JCI57673

Chen H, Matsumoto K, Brockway BL, Rackley CR, Liang J, Lee JH, Jiang D, Noble PW, Randell SH, Kim CF, Stripp BR (2012) Airway epithelial progenitors are region specific and show differential responses to bleomycin-induced lung injury. Stem Cells 30(9):1948–1960. doi:10.1002/stem.1150

Desai TJ, Brownfield DG, Krasnow MA (2014) Alveolar progenitor and stem cells in lung development, renewal and cancer. Nature 507(7491):190–194. doi:10.1038/nature12930

Dobbs LG, Johnson MD, Vanderbilt J, Allen L, Gonzalez R (2010) The great big alveolar TI cell: evolving concepts and paradigms. Cell Physiol Biochem 25(1):55–62. doi:10.1159/000272063

Evans MJ, Plopper CG (1988) The role of basal cells in adhesion of columnar epithelium to airway basement membrane. Am Rev Respir Dis 138(2):481–483

Evans MJ, Cox RA, Shami SG, Plopper CG (1990) Junctional adhesion mechanisms in airway basal cells. Am J Respir Cell Mol Biol 3(4):341–347. doi:10.1165/ajrcmb/3.4.341

Flodby P, Borok Z, Banfalvi A, Zhou B, Gao D, Minoo P, Ann DK, Morrisey EE, Crandall ED (2010) Directed expression of Cre in alveolar epithelial type 1 cells. Am J Respir Cell Mol Biol 43(2):173–178. doi:10.1165/rcmb.2009-0226OC

Ghosh M, Brechbuhl HM, Smith RW, Li B, Hicks DA, Titchner T, Runkle CM, Reynolds SD (2011) Context-dependent differentiation of multipotential keratin 14-expressing tracheal basal cells. Am J Respir Cell Mol Biol 45(2):403–410. doi:10.1165/rcmb.2010-0283OC

Giangreco A, Reynolds SD, Stripp BR (2002) Terminal bronchioles harbor a unique airway stem cell population that localizes to the bronchoalveolar duct junction. Am J Pathol 161(1):173–182. doi:10.1016/S0002-9440(10)64169-7

Gonzalez RF, Allen L, Dobbs LG (2009) Rat alveolar type I cells proliferate, express OCT-4, and exhibit phenotypic plasticity in vitro. Am J Physiol Lung Cell Mol Physiol 297(6):L1045–L1055. doi:10.1152/ajplung.90389.2008

Guha A, Vasconcelos M, Cai Y, Yoneda M, Hinds A, Qian J, Li G, Dickel L, Johnson JE, Kimura S, Guo J, McMahon J, McMahon AP, Cardoso WV (2012) Neuroepithelial body microenvironment is a niche for a distinct subset of Clara-like precursors in the developing airways. Proc Natl Acad Sci U S A 109(31):12592–12597. doi:10.1073/pnas.1204710109

Guha A, Vasconcelos M, Zhao R, Gower AC, Rajagopal J, Cardoso WV (2014) Analysis of Notch signaling-dependent gene expression in developing airways reveals diversity of Clara cells. PLoS One 9(2):e88848. doi:10.1371/journal.pone.0088848

Hong KU, Reynolds SD, Watkins S, Fuchs E, Stripp BR (2004) Basal cells are a multipotent progenitor capable of renewing the bronchial epithelium. Am J Pathol 164(2):577–588. doi:10.1016/S0002-9440(10)63147-1

Jeffery PK, Reid L (1975) New observations of rat airway epithelium: a quantitative and electron microscopic study. J Anat 120(Pt 2):295–320

Kajstura J, Rota M, Hall SR, Hosoda T, D'Amario D, Sanada F, Zheng H, Ogorek B, Rondon-Clavo C, Ferreira-Martins J, Matsuda A, Arranto C, Goichberg P, Giordano G, Haley KJ, Bardelli S, Rayatzadeh H, Liu X, Quaini F, Liao R, Leri A, Perrella MA, Loscalzo J, Anversa P (2011) Evidence for human lung stem cells. N Engl J Med 364(19):1795–1806. doi:10.1056/NEJMoa1101324

Kim CF, Jackson EL, Woolfenden AE, Lawrence S, Babar I, Vogel S, Crowley D, Bronson RT, Jacks T (2005) Identification of bronchioalveolar stem cells in normal lung and lung cancer. Cell 121(6):823–835. doi:10.1016/j.cell.2005.03.032

Kretzschmar K, Watt FM (2012) Lineage tracing. Cell 148(1–2):33–45. doi:10.1016/j.cell.2012.01.002

Lee JH, Bhang DH, Beede A, Huang TL, Stripp BR, Bloch KD, Wagers AJ, Tseng YH, Ryeom S, Kim CF (2014) Lung stem cell differentiation in mice directed by endothelial cells via a BMP4-NFATc1-thrombospondin-1 axis. Cell 156(3):440–455. doi:10.1016/j.cell.2013.12.039

Leeman KT, Fillmore CM, Kim CF (2014) Lung stem and progenitor cells in tissue homeostasis and disease. Curr Top Dev Biol 107:207–233. doi:10.1016/B978-0-12-416022-4.00008-1

Lung stem cells: looking beyond the hype (2011). Nat Med 17(7): 788–789. doi:10.1038/nm0711-788

McQualter JL, Yuen K, Williams B, Bertoncello I (2010) Evidence of an epithelial stem/progenitor cell hierarchy in the adult mouse lung. Proc Natl Acad Sci U S A 107(4):1414–1419. doi:10.1073/pnas.0909207107

Messier B, Leblond CP (1960) Cell proliferation and migration as revealed by radioautography after injection of thymidine-H3 into male rats and mice. Am J Anat 106:247–285. doi:10.1002/aja.1001060305

Morrisey EE, Hogan BL (2010) Preparing for the first breath: genetic and cellular mechanisms in lung development. Dev Cell 18(1):8–23. doi:10.1016/j.devcel.2009.12.010

Nakajima M, Kawanami O, Jin E, Ghazizadeh M, Honda M, Asano G, Horiba K, Ferrans VJ (1998) Immunohistochemical and ultrastructural studies of basal cells, Clara cells and bronchiolar cuboidal cells in normal human airways. Pathol Int 48(12):944–953

Perl AK, Whitsett JA (1999) Molecular mechanisms controlling lung morphogenesis. Clin Genet 56(1):14–27

Perl AK, Wert SE, Nagy A, Lobe CG, Whitsett JA (2002) Early restriction of peripheral and proximal cell lineages during formation of the lung. Proc Natl Acad Sci U S A 99(16):10482–10487. doi:10.1073/pnas.152238499

Rackley CR, Stripp BR (2012) Building and maintaining the epithelium of the lung. J Clin Invest 122(8):2724–2730. doi:10.1172/JCI60519

Randell SH, Comment CE, Ramaekers FC, Nettesheim P (1991) Properties of rat tracheal epithelial cells separated based on expression of cell surface alpha-galactosyl end groups. Am J Respir Cell Mol Biol 4(6):544–554. doi:10.1165/ajrcmb/4.6.544

Rawlins EL, Hogan BL (2006) Epithelial stem cells of the lung: privileged few or opportunities for many? Development 133(13):2455–2465. doi:10.1242/dev.02407

Rawlins EL, Okubo T, Xue Y, Brass DM, Auten RL, Hasegawa H, Wang F, Hogan BL (2009) The role of Scgb1a1+ Clara cells in the long-term maintenance and repair of lung airway, but not alveolar, epithelium. Cell Stem Cell 4(6):525–534. doi:10.1016/j.stem.2009.04.002

Reynolds SD, Giangreco A, Power JH, Stripp BR (2000a) Neuroepithelial bodies of pulmonary airways serve as a reservoir of progenitor cells capable of epithelial regeneration. Am J Pathol 156(1):269–278. doi:10.1016/S0002-9440(10)64727-X

Reynolds SD, Hong KU, Giangreco A, Mango GW, Guron C, Morimoto Y, Stripp BR (2000b) Conditional clara cell ablation reveals a self-renewing progenitor function of pulmonary neuroendocrine cells. Am J Physiol Lung Cell Mol Physiol 278(6):L1256–L1263

Rhodin JA (1966) The ciliated cell. Ultrastructure and function of the human tracheal mucosa. Am Rev Respir Dis 93(3):1–15

Rock JR, Hogan BL (2011) Epithelial progenitor cells in lung development, maintenance, repair, and disease. Annu Rev Cell Dev Biol 27:493–512. doi:10.1146/annurev-cellbio-100109-104040

Rock JR, Onaitis MW, Rawlins EL, Lu Y, Clark CP, Xue Y, Randell SH, Hogan BL (2009) Basal cells as stem cells of the mouse trachea and human airway epithelium. Proc Natl Acad Sci U S A 106(31):12771–12775. doi:10.1073/pnas.0906850106

Rock JR, Randell SH, Hogan BL (2010) Airway basal stem cells: a perspective on their roles in epithelial homeostasis and remodeling. Dis Model Mech 3(9–10):545–556. doi:10.1242/dmm.006031

Rock JR, Barkauskas CE, Cronce MJ, Xue Y, Harris JR, Liang J, Noble PW, Hogan BL (2011) Multiple stromal populations contribute to pulmonary fibrosis without evidence for epithelial to mesenchymal transition. Proc Natl Acad Sci U S A 108(52):E1475–E1483. doi:10.1073/pnas.1117988108

Rogers DF (1994) Airway goblet cells: responsive and adaptable front-line defenders. Eur Respir J 7(9):1690–1706

Sinclair K, Yerkovich ST, Chambers DC (2013) Mesenchymal stem cells and the lung. Respirology 18(3):397–411. doi:10.1111/resp.12050

Song H, Yao E, Lin C, Gacayan R, Chen MH, Chuang PT (2012) Functional characterization of pulmonary neuroendocrine cells in lung development, injury, and tumorigenesis. Proc Natl Acad Sci U S A 109(43):17531–17536. doi:10.1073/pnas.1207238109

Tata PR, Mou H, Pardo-Saganta A, Zhao R, Prabhu M, Law BM, Vinarsky V, Cho JL, Breton S, Sahay A, Medoff BD, Rajagopal J (2013) Dedifferentiation of committed epithelial cells into stem cells in vivo. Nature 503(7475):218–223. doi:10.1038/nature12777

Teixeira VH, Nadarajan P, Graham TA, Pipinikas CP, Brown JM, Falzon M, Nye E, Poulsom R, Lawrence D, Wright NA, McDonald S, Giangreco A, Simons BD, Janes SM (2013) Stochastic homeostasis in human airway epithelium is achieved by neutral competition of basal cell progenitors. Elife 2:e00966. doi:10.7554/eLife.00966

Tropea KA, Leder E, Aslam M, Lau AN, Raiser DM, Lee JH, Balasubramaniam V, Fredenburgh LE, Alex Mitsialis S, Kourembanas S, Kim CF (2012) Bronchioalveolar stem cells increase after mesenchymal stromal cell treatment in a mouse model of bronchopulmonary dysplasia. Am J Physiol Lung Cell Mol Physiol 302(9):L829–L837. doi:10.1152/ajplung.00347.2011

van Es JH, Sato T, van de Wetering M, Lyubimova A, Nee AN, Gregorieff A, Sasaki N, Zeinstra L, van den Born M, Korving J, Martens AC, Barker N, van Oudenaarden A, Clevers H (2012) Dll1+ secretory progenitor cells revert to stem cells upon crypt damage. Nat Cell Biol 14(10):1099–1104. doi:10.1038/ncb2581

Van Keymeulen A, Rocha AS, Ousset M, Beck B, Bouvencourt G, Rock J, Sharma N, Dekoninck S, Blanpain C (2011) Distinct stem cells contribute to mammary gland development and maintenance. Nature 479(7372):189–193. doi:10.1038/nature10573

Zheng D, Limmon GV, Yin L, Leung NH, Yu H, Chow VT, Chen J (2012) Regeneration of alveolar type I and II cells from Scgb1a1-expressing cells following severe pulmonary damage induced by bleomycin and influenza. PLoS One 7(10):e48451. doi:10.1371/journal.pone.0048451

Zheng D, Limmon GV, Yin L, Leung NH, Yu H, Chow VT, Chen J (2013) A cellular pathway involved in Clara cell to alveolar type II cell differentiation after severe lung injury. PLoS One 8(8):e71028. doi:10.1371/journal.pone.0071028

Chapter 2
Type II Cells as Progenitors in Alveolar Repair

Yuru Liu

Abbreviations

BASC	Bronchioalveolar stem cells
BMP4	Bone morphogenic protein 4
CC10	Club Cell 10 kDa protein
EGFR	Epidermal growth factor receptor
Elf5	E74-like factor 5
Erm	Ets-related molecule
FGF	Fibroblast growth factor
FoxM1	Forkhead box protein M1
GATA6	GATA binding protein 6
GFP	Green fluorescent protein
HGF	Hepatocyte growth factor
Id2	Inhibitor of DNA binding 2
IGF	Insulin-like growth factor
IL-6/8/10	Interleukin 6/8/10
iPS	Induced pluripotent stem cells
KGF	Keratinocyte growth factor
MEF	Mouse embryonic fibroblast
MMP14	Matrix metalloproteinase 14
PDGF	Platelet-derived growth factor
SPA/B/C/D	Surfactant protein A/B/C/D
TGFβ	Transforming growth factor beta
Wnt	Wingless protein

Y. Liu, Ph.D. (✉)
Department of Pharmacology, College of Medicine, University of Illinois, Chicago, IL, USA
e-mail: yuruliu@uic.edu

A. Firth, J.X.-J. Yuan (eds.), *Lung Stem Cells in the Epithelium and Vasculature*, Stem Cell Biology and Regenerative Medicine, DOI 10.1007/978-3-319-16232-4_2

2.1 Introduction

Lung alveoli are lined by a continuous epithelial layer consisting of type I and type II cells. Type I cells constitute approximately 8 % of the total parenchymal lung cells but form more than 90 % of the alveolar surface area (Weibel 2009; Schneeberger 1997). Type I cells have a thin, flat, squamous shape with multiple branches spread over a large area; this shape facilitates their close contact with the basement membrane and capillary endothelial cells, allowing for efficient gas exchange (Schneeberger 1997; Weibel 2009). Type II cells usually reside at the corners of alveoli and are sometime referred to as "corner cells" (Mason and Shannon 1997; Weibel 2009) (Fig. 2.1a–b). Type II cells constitute approximately 15 % of all lung cells, but because of their cuboidal shape, they occupy only approximately 5 % of the alveolar surface area (Mason and Shannon 1997; Weibel 2009).

Alveolar type II cells have multiple functions. One of the most important functions of these cells is to synthesize and secrete surfactant (Mason and Shannon 1997; Mason 2006). Surfactants are synthesized in type II cell-specific organelles called "lamellar bodies." Each type II cell contains approximately 150 lamellar bodies. This structure has a mean diameter of 1 μm, contains multiple phospholipid bilayers, and gives type II cells a unique morphology (Mason and Shannon 1997; Mason 2006; Mason and Crystal 1998). In this organelle, the phospholipids are packed with the surfactant proteins Sp-A, Sp-B, Sp-C, and Sp-D, which constitute surfactant (Andreeva et al. 2007; Mason and Shannon 1997). When secreted into the alveolar space, surfactant maintains alveolar surface tension and prevents lung collapse (Mason and Shannon 1997; Mason and Crystal 1998). While Sp-A, Sp-B, and Sp-D are also produced in some bronchiolar cells (Mason and Shannon 1997; Walker et al. 1986; Fehrenbach 2001), Sp-C is produced only in type II cells and is considered a type II cell marker (Kalina et al. 1992; Mason and Shannon 1997; Fehrenbach 2001).

Other than producing surfactant, type II cells also play roles in the transepithelial transport of ions and fluids (Mason and Shannon 1997; Mason 2006), modulation of lung inflammatory responses (Mason 2006), and maintenance of epithelial homeostasis and integrity by acting as progenitor cells through proliferation and differentiating into alveolar type I cells (Mason and Shannon 1997; Mason 2006; Stripp 2008; Rock and Hogan 2011). The function of type II cells as alveoli epithelial progenitor cells during injury repair is discussed in this chapter.

Type II cell hyper-proliferation is an important feature in the resolution and repair phase following lung injury (Matthay et al. 2012; Ware and Matthay 2000; Shimabukuro et al. 2003). Evidence indicates that type II cells function as progenitor cells for the re-epithelialization of injured alveoli by converting into type I cells (Evans et al. 1973, 1975) (Fig. 2.1c). In the 1970s, Evans et al. performed classical experiments using NO_2 to damage the alveolar surface and showed that ^{3}H-TdR was first incorporated into proliferating type II cells and later labeled type I cells (Evans et al. 1973, 1975). In addition, it has been shown that isolated type II cells can differentiate into type I cells in culture (Dobbs 1990). However, knowledge of the detailed cellular and molecular mechanisms of the progenitor cell properties of type II cells has only recently started to emerge.

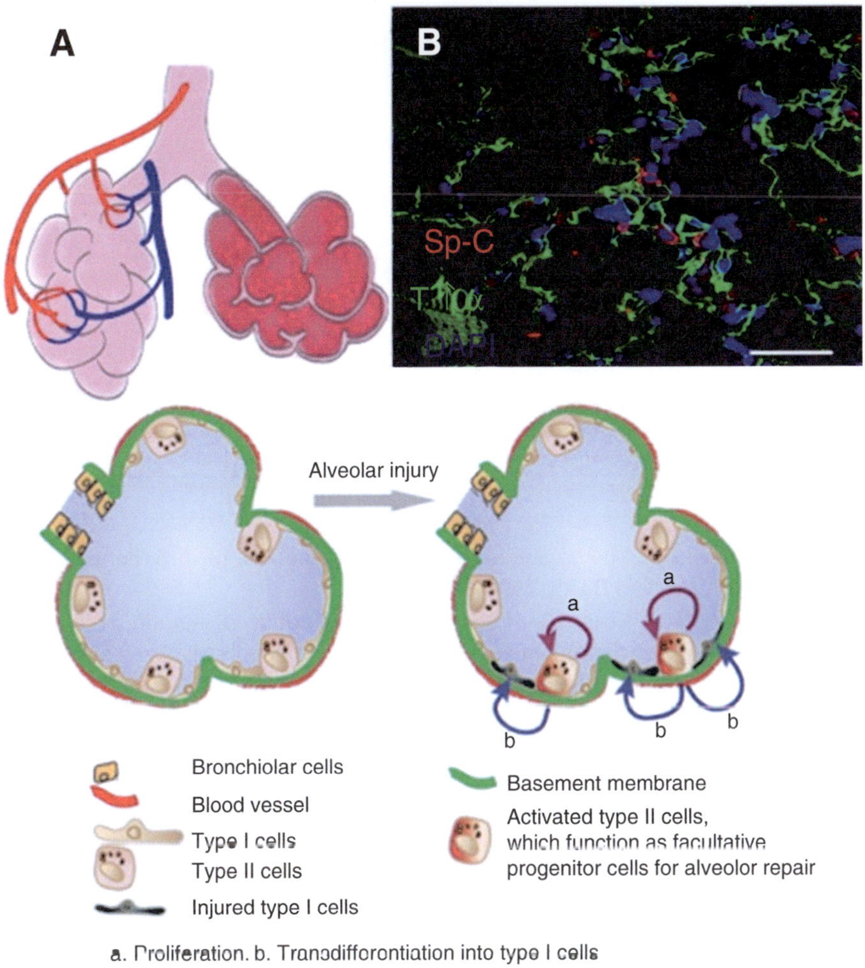

Fig. 2.1 A model showing type II cell progenitor functions following alveoli injury. (**a**) Schematic representation of alveoli. (**b**) Section of mouse lung alveoli region stained by type II cell marker Sp-C (*red*) and type I cell marker T1α (*green*). Scale bar = 50 μm. (**c**) Upon alveolar epithelial injury, some type II cells are activated, undergo proliferation and transition into type I cells to restore alveolar barrier integrity

2.2 Type II Cells Exhibit Progenitor Cell Phenotypes In Vitro

No reported cell line exhibits all the major properties of type II cells. However, techniques have been developed to isolate type II cells from human, rat, and mouse lungs (Kikkawa and Yoneda 1974; Dobbs 1990; Corti and Brody 1996). Basically, lung samples are separated into single cell suspensions using enzymes, such as dispase, and the type II cells are enriched using differential sedimentation. The type II cells

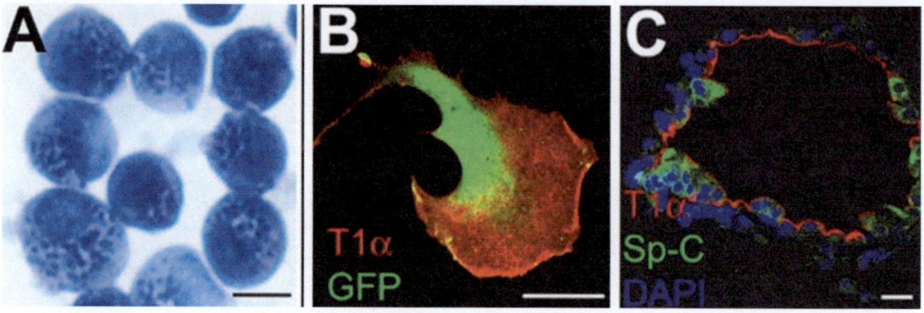

Fig. 2.2 Type II cells are able to differentiate into type I cells in culture. (**a**) Freshly isolated mouse type II cells stained with modified Papanicolaou method (Dobbs 1990). Dark blue dots are the lamellar bodies (Dobbs 1990). (**b**) After culturing on plastic surface, some type II cells underwent transition into type I-like cell and expressed type I cell marker T1α. This cell also expressed GFP in cytoplasm. (**c**) When co-cultured with lung fibroblasts in 3D matrigel, some type II cells formed alveoli-like structures that contain Sp-C expressing type II cells and T1α expressing type I cells. Scale bar = 10 μm

can be further purified using antibodies to surface antigens of different cell types (Dobbs 1990; Corti and Brody 1996) (Fig. 2.2a).

Isolated type II cells can be cultured on mixture of collagen, laminin, or fibronectin substrata on which they maintain their type II cell phenotypes (such as forming lamellar bodies and secreting surfactant) for approximately 1 week; however, the cells have a tendency to differentiate into flat type I-like cells (Dobbs 1990; Rice et al. 2002; Gobran and Rooney 2004; Rannels et al. 1987; Paine and Simon 1996). If isolated type II cells are cultured on plastic in the presence of 5–10 % fetal bovine serum and without exogenous extracellular matrix, they undergo a morphological change, lose their lamellar bodies, and differentiate into flat type I-like cells within 2–7 days (Dobbs 1990). Some of these type II cell-derived flat cells express type I cell markers such as aquaporin 5 and T1α (Williams 2003) (Fig. 2.2b). Thus, in vitro studies have suggested that type II cells may be precursors of type I cells.

Recently, a new 3D culturing technique of culturing type II cells has been developed (Barkauskas et al. 2013; McQualter et al. 2010; Chen et al. 2012a). When isolated type II cells are cultured in Matrigel (a mixture of extracellular matrix components secreted by Engelbreth-Holm-Swarm (EHS) mouse sarcoma cells) (Kleinman and Martin 2005), they form cysts consisting of polarized monolayers of type II cells, and the cells secrete surfactant into the lumen (Dodelet and Pasquale 2000). However, when type II cells are co-cultured with lung mesenchymal cells in Matrigel (Barkauskas et al. 2013; McQualter et al. 2010; Chen et al. 2012a), they form 3D alveoli-like structures with flat type I-like cells lining the lumen and cuboidal type II-like cells facing the matrix (Fig. 2.2c). By mixing cells from different lineage-labeled lines, it appears that each of the alveolar-like spheres can be derived from a single cell (Barkauskas et al. 2013). These results show that at least some subgroups of type II cells have stem cell characteristics that include the ability to expand clonally and differentiate into multiple lineages.

From the above studies, the behavior of type II cells varies depending on the culture conditions. Developing new culture techniques will lead to novel discoveries of the progenitor cell aspects of type II cells. Some semi-in vivo culture methods, such as growing cells in subcutaneous Matrigel plugs or in a renal capsule (Chapman et al. 2011; Lee et al. 2014), may provide additional insight into type II cell progenitor functions.

2.3 Type II Cells Act as Progenitor Cells During Lung Repair in Animal Models

To study the progenitor cell behavior of type II cells during alveolar repair in vivo, it is important to have proper animal models to mimic various alveolar injuries; these studies also provide a bridge between patients and laboratory research. Because of their size and the availability of genetic approaches, mouse and rat models are widely used. Typically, three types of agents are used to target the alveolar epithelium in experimental animals: (1) the inhalation of gasses, such as toxic NO_2 (Evans et al. 1973, 1975); because high concentrations of O_2 can cause alveolar damage in rodents, hyperoxia is a frequently used alveolar injury model (Pogach et al. 2007); (2) the administration of chemicals or antibiotics, such as acid (Matute-Bello et al. 2008) or bleomycin (Flozak et al. 2010); and (3) the introduction of pathogens such as the intratracheal injection of *Pseudomonas aeruginosa* bacteria (Sadikot et al. 2005; Liu et al. 2011) or the H1N1 influenza virus (Kumar et al. 2011) (see Table 2.1).

The mechanisms that cause alveolar damage vary in different injury models. NO_2 causes cell death by nitrating oxidants and the subsequent inflammatory responses (Persinger et al. 2001). In the hyperoxia model, high oxygen concentrations (usually 80–100 %) release reactive oxygen species or free radicals derived from O_2, causing necrosis or apoptosis of alveolar epithelia cells (Pagano and Barazzone-Argiroffo 2003). Hyperoxia also leads to the activation of NFκB signaling and inflammatory responses that further enhance alveolar damage (Matute-Bello et al. 2008). One limitation of the hyperoxia model is that, in humans with normal lungs, exposure to 100 % oxygen does not cause clinical or pathological lung injury (Matute-Bello et al. 2008). The acid aspiration model is usually accomplished by the instillation of HCl into the trachea. The acid causes alveolar cell death, increased alveolar barrier permeability and an acute inflammatory response (Matute-Bello et al. 2008). Bleomycin is an antineoplastic antibiotic (Matute-Bello et al. 2008; Wansleeben et al. 2013) that, when administered into the lung, forms a complex with oxygen and metals, such as Fe, leading to the production of oxygen radicals that cause alveolar cell death (Matute-Bello et al. 2008; Wansleeben et al. 2013). At the same time, the bleomycin-treated lung undergoes an inflammatory response that increases damage (Wansleeben et al. 2013; Matute-Bello et al. 2008). The intratracheal instillation of bacteria causes pneumonia. Some bacteria such as *Pseudomonas aeruginosa* produce toxins that penetrate the cell membrane, causing cell death (Sadikot et al. 2005). The H1N1 influenza virus causes cell death and an inflammatory response (Wansleeben et al. 2013; Hendrickson and Matthay 2013; Kumar et al. 2011) (Table 2.1).

Table 2.1 Selected animal models of alveolar epithelial injury

Model	Mechanism of injury	Lung features (emphasizing type II cells)	References
Inhalation of toxic gas NO_2	Generation of nitrating oxidants; subsequent inflammation	Acute injury; type II cell proliferation and transition into type I cells	Evans et al. (1973, 1975), Persinger et al. (2001)
Hyperoxia	Generation of reactive oxygen species; inflammation; cell necrosis or apoptosis; degradation of extracellular matrix	Acute lung injury; neutrophilic infiltration followed by type II cell proliferation; fibroblast proliferation and scarring	Pagano and Barazzone-Argiroffo (2003), Lee et al. (2006), Matute-Bello et al. (2008)
Acid aspiration	Cell death; disruption of the alveolar/capillary barrier; neutrophilic infiltration	Acute lung injury and inflammation; type II cell proliferation, heal with fibrosis	Matute-Bello et al. (2008), Pogach et al. (2007)
Bleomycin	Formation of complex with metal and oxygen; production of oxygen radicals; inflammation	Inflammation and acute alveoli injury; type II cell proliferation and conversion to type I cells; later lead to reversible fibrosis	Zhao et al. (2002), Matute-Bello et al. (2008), Flozak et al. (2010), Chapman et al. (2011), Rock et al. (2011), Barkauskas et al. (2013), Wansleeben et al. (2013)
Intrapulmonary bacteria, e.g., intratracheal inject live *P. aeruginosa*	Bacteria and the toxin damage cell; neutrophil infiltration; inflammatory response	Acute lung injury and inflammatory response; increased alveoli barrier permeability; type II cell proliferation and differentiation into type I cells; lung repaired after 7 days	Sadikot et al. (2005), Liu et al. (2011)
Viral infection, e.g., H1N1	Destruction of alveoli cells; excessive inflammatory responses	Trp63+ cells proliferate and produce pod-like structures in alveoli	Kumar et al. (2011), Wansleeben et al. (2013), Hendrickson and Matthay (2013)
Diptheria toxin	Toxin kill cells; no apparent inflammation	Remaining type II cells proliferate and transit into type I cells	Barkauskas et al. (2013)
Unilateral pneumonectomy	Mechanical injury	Compensatory growth of existing lobes; type II cell proliferation	Nolen-Walston et al. (2008)

Excessive inflammation is a factor common to most of these models and plays an important role in causing cellular damage (Matute-Bello et al. 2011). The only exception listed in Table 2.1 is a model of diphtheria toxin-induced lung injury (Barkauskas et al. 2013). Using a knock-in mouse in which the expression of diphtheria toxin in type II cells is induced following tamoxifen injection, there does not appear to be a clear inflammatory response (Barkauskas et al. 2013) (Table 2.1).

Some types of injuries are reversible; for example, *Pseudomonas aeruginosa-induced* lung injury is usually repaired in approximately 7–10 days (Liu et al. 2011). By contrast, some injuries lead to chronic disease; for example, bleomycin injection can induce acute lung injury in the early phase; later, the lung frequently develops fibrosis (Flozak et al. 2010; Chapman et al. 2011; Zhao et al. 2002). Recent findings indicate that the lung alveoli may utilize different repair mechanisms in response to different types of damage. For example, a group of putative progenitor cells called BASC (for bronchioalveolar stem cells) appear to migrate and proliferate in response to bleomycin-induced injury (Barkauskas et al. 2013), but they do not respond to hyperoxia (Rawlins et al. 2009). This result is not surprising because the pathological mechanism as well as the extent of injury varies in different models. Most agents cause damage in multiple cell types in the alveoli, some injuries may be localized to certain foci, and others affect a more extended region. One might hypothesize that certain local facultative progenitor cells may be sufficient to repair a local mild injury, while putative stem cells reside in certain niches in the alveoli or airway may be mobilized in response to a more severe injury (Vaughan and Chapman 2013). Therefore, it is important to compare several injury models to study the mechanisms of alveolar repair.

Type II cells have been shown to be involved in repair in different injury models (Table 2.1). In response to NO_2, type II cells proliferate, as revealed by ^3H labeling (Evans et al. 1973, 1975). In addition, in pulse labeling experiments, the ^3H label appeared first in type II cells and later in type I cells, suggesting that type II cells give rise to type I cells (Evans et al. 1973, 1975). Following hyperoxia or acid aspiration, type II cells engage in hyper-proliferation (Lee et al. 2006; Pogach et al. 2007; Desai et al. 2014). Following bleomycin-induced injury, type II cells proliferate, and lineage tracing experiments using this injury model showed that type II cells also differentiate into type I cells (Barkauskas et al. 2013; Rock et al. 2011). In a *Pseudomonas aeruginosa*-induced injury model (Liu et al. 2011), type II cells undergo proliferation in the early phase and differentiate into type I cells in the later repair phase (Liu et al. 2011). Finally, in the unilateral pneumonectomy (PNX) model, type II cell proliferation also plays an important role in the compensatory growth of the remaining lung (Nolen-Walston et al. 2008). Therefore, the involvement of type II cells is a common repair mechanism following alveolar injury. However, the exact molecular mechanisms and the subgroups of type II cells that are involved in the repair of different injuries may differ.

2.4 Lineage Analysis of Type II Cells Involved in Alveoli Repair

Even though the historical experiments by Evans et al. discussed above (Evans et al. 1973, 1975) strongly suggested that type II alveolar cells behave as progenitor cells by proliferating and differentiating into type I cells following injury, it is only with the recent development of new genetic approaches that direct evidence has shown that type II cells give rise to type I cells during alveolar homeostasis and repair (Rock et al. 2011; Barkauskas et al. 2013; Desai et al. 2014).

By adopting the yeast Cre DNA recombination enzyme and the *cis*-element *loxP* site (Soriano 1999), new genetic lineage tracing techniques have been developed in mouse models by coupling three elements together: an inducible Cre recombinase-expressing system; a type II cell-specific promoter; and a reporter allele tagged with a fluorescent protein (Kretzschmar and Watt 2012). Figure 2.3 shows an example of this system. With the administration of certain chemicals, in this case, tamoxifen, into mice, Cre recombinase is expressed in Sp-C-expressing type II cells, and Cre activates the expression of a fluorescent protein, such as GFP, which will permanently label the type II cells in adult mice; the fate of these labeled cells can be traced any time thereafter (Rock et al. 2011; Desai et al. 2014). Therefore, even if these type II cells later differentiate into another cell type, the cells can still be visualized by the fluorescent signals.

Using this lineage tracing mouse line and low doses of tamoxifen, Barkauskas et al. showed that only a small number of type II cells were labeled with the fluorescent proteins. When traced after 12–30 weeks, it was found that these labeled cells formed small clusters, indicating that type II cells undergo a slow self-renewal (Barkauskas et al. 2013). The same studies also showed that the labeled type II cells slowly differentiated into type I cells during steady-state tissue maintenance. Immediately after the mice were injected with tamoxifen, only type II cells were labeled by fluorescent proteins. However, after a 24-week chase, some flat cells expressing type I cell markers were also labeled with the same fluorescent signals, indicating that the lineage-labeled type II cells had converted into type I cells. These results are consistent with another lineage tracing study using a different mouse line (Desai et al. 2014) and support the conclusions from earlier cell tracing studies using ^{3}H-labeled thymidine showing that the steady state turnover time of type II cells is on the order of 4–6 weeks (Spencer and Shorter 1962; Fehrenbach 2001).

From these studies, type II cells have been shown to be relatively quiescent in the absence of injury. However, in response to alveolar injury, type II cells are activated and enter the repairing program (Mason and Shannon 1997; Barkauskas et al. 2013; Desai et al. 2014). Using the above lineage tracing mouse lines, it is shown that after injury, the conversion of type II cells into type I cells was greatly accelerated (Barkauskas et al. 2013).

Furthermore, in another study that traced the type II cells following *Pseudomonas aeruginosa*-induced injury, it is found that some lineage-labeled type II cells lost the type II cell marker Sp-C at 5–7 days post-injury, and these cells did not yet express

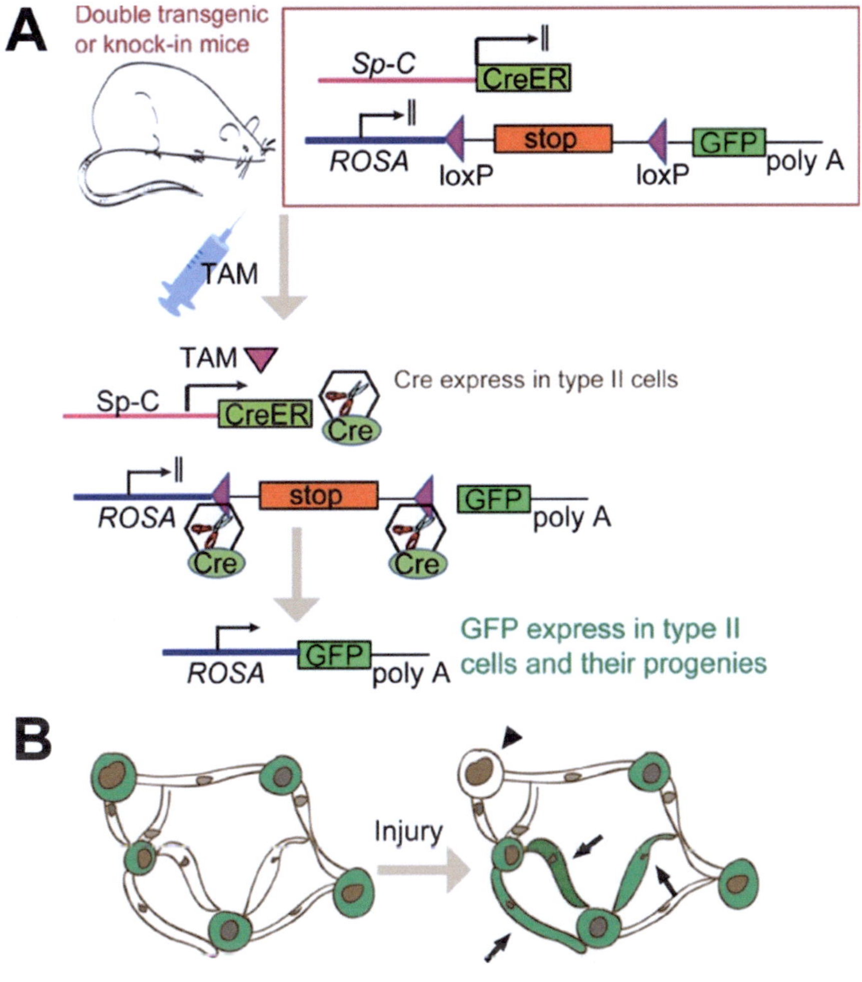

Fig. 2.3 A typical lineage tracing study shows progenitor function of type II cells. (**a**) A double transgenic (or knock-in) mouse harbors (1) an inducible Cre-expressing system, in this case a *Cre-ER* in which the expression of Cre recombinase is dependent on tamoxifen (TAM). The Cre expression is also under the control of a type II cell-specific promoter *Sp-C*; (2) a reporter tagged with a fluorescent protein (in this case GFP). But GFP is normally not expressed because a stop sequence flanked by two *loxP* sites reside upstream of it. With the administration of TAM into this mouse, Cre is expressed in Sp-C-expressing type II cells, and Cre deletes the stop sequence to activate the GFP expression in type II cells through the constitutive *ROSA* promoter, and then GFP will permanently label the type II cells as well as their progenies. (**b**) Schematic representation of a typical result of the lineage tracing studies. Mice were injected with TAM before injury, so that the type II cells were permanently labeled with GFP while type I cells were unlabeled. Following alveoli injury, some type II cells underwent transition into type I cells; therefore, some GFP-labeled type I cells appeared (*arrow*). In addition, some type II cells without GFP labeling appeared in the lung (*arrowhead*), suggesting that some Sp-C^{-} cells gave rise to type II cells

or only expressed faint level of type I cell markers. These may represent the intermediate cells undergoing type II to I transition (Liu et al. 2011).

Another question related to the progenitor function of type II cells in repair is how injured type II cells are replenished. Because type II cells undergo hyper-proliferation following injury (Mason and Shannon 1997), one might hypothesize that type II cells are able to repair themselves solely by self-renewal. However, recent lineage tracing studies have shown that type II cells may not be the only source for self-replenishment (Barkauskas et al. 2013; Chapman et al. 2011). It has been shown using lineage tracing mouse lines (Fig. 2.2), that, following bleomycin-induced injury, the percentage of lineage-labeled type II cells versus the total number of type II cells decreased in the injured areas. These results indicate that some non-lineage-labeled cells, i.e., cells that did not express Sp-C before the injury, gave rise to type II cells during repair (Barkauskas et al. 2013; Chapman et al. 2011).

2.5 Subgroups of Type II Cells Display Progenitor Features

Type II cells have long been considered a homogeneous population. However, recent studies have indicated that type II cells are heterogeneous and consist of subpopulations. For example, studies using transgenic mice that express GFP under the control of the *Sp-C* promoter have revealed that certain type II cells express higher levels of GFP than others (Lee et al. 2013; Roper et al. 2003). It would, therefore, be informative to separate type II cells into subpopulations because some of these subgroups may be particularly important in the repair of injured alveoli.

It has been found that approximately 5–10 % of type II cells express low levels of CC10. CC10, also known as CCSP or Scgbb1a1, is a secretory protein expressed at high levels in Club cells located in the airways. A number of Sp-C$^+$CC10$^+$ type II cells are located in the alveolar region; the function of these cells is unknown. A rare population of SP-C$^+$CC10$^+$ cells located at terminal bronchioles is known as BASC. BASCs show certain stem cell properties in vitro including forming colonies when co-cultured with mouse embryonic fibroblasts (MEFs) (Kim et al. 2005a, b) as well as multi-lineage differentiation when cultured in Matrigel (Kim et al. 2005a, b; Lee et al. 2014).

Some subpopulations of type II cells appear following injury. Studies by Reddy et al. showed that type II cells isolated from hyperoxia-treated rats can be separated into two populations after a brief culture; some cells expressed lower levels of E-cadherin than others (Driscoll et al. 2000; Reddy et al. 2004). Furthermore, the E-cadherin^{high} population appeared to be relatively quiescent, while the E-cadherin^{low} population was more proliferative and expressed higher levels of telomerase activity (Driscoll et al. 2000). Therefore, certain type II cells appear to show more progenitor cell properties than their counterparts.

A subgroup of type II cells expressing the progenitor cell marker Sca-1 appear in the early repair phase following *Pseudomonas aeruginosa*-induced injury (Liu et al. 2011). These cells are mostly derived from Sp-C-expressing type II cells and can account for more than 40 % of the total type II cells following injury (Liu et al. 2011). This Sca-1$^+$ subpopulation of type II cells exhibits a higher proliferation rate and also has a higher potential to convert into type I cells compared with the Sca-1$^-$ counterpart (Liu et al. 2011, 2014). These cells express higher levels of the transcription factor FoxM1 than the Sca-1$^-$ type II cells, and FoxM1 is required for the alveolar repair process (Liu et al. 2011). Thus, the Sca-1$^+$ subgroup of type II cells appear to be responsible for the repair.

The results from the above studies indicate that certain subpopulations of type II cells manifesting progenitor cell features appear at the repair phase following injury. However, it is unclear whether these putative regenerative type II cells appearing post-injury are derived from the expansion of existing stem cell pools located in an undefined niche or whether they are derived from quiescent, terminally differentiated type II cells. It would also be interesting to identify which signals or factors induce the formation of progenitor-like type II cell subgroups following injury.

Another point is that, in most studies, Sp-C is considered a type II cell-specific marker. In other words, type II cells are defined as Sp-C$^+$ cells. However, it has been observed that some type II cells express lower levels of Sp-C than others (Chapman et al. 2011; Liu et al. 2011; Lee et al. 2013). In addition, it has been reported that a small number of epithelial cells located in alveoli do not show type I cell phenotypes and do not express detectable levels of Sp-C (Chapman et al. 2011). It remains to be determined whether these cells can be regarded as a subtype of type II cells or type II-like cells. Among the Sp-C$^-$ non-type I alveolar epithelial cells is a group of cells expressing the α6β4 integrin (Chapman et al. 2011); α6β4 integrin double-positive cells have progenitor cell features and may participate in repair following bleomycin-induced injury (Chapman et al. 2011).

2.6 Cellular and Molecular Mechanisms Regulating Type II Cell Progenitor Properties

The cellular and molecular mechanisms underlying how type II cells engage in the repair of a damaged alveolar barrier is unclear. It is known that type II cells can convert into type I cells; however, the detailed cellular processes involved in this transition are unknown. One possibility is that certain relatively undifferentiated subgroups of type II cells located at unknown locations differentiate into type I cells. Another possibility is that differentiated, mature type II cells undergo trans-differentiation and give rise to type I cells. It is also possible that mature type II cells first undergo dedifferentiation into intermediate precursors and then differentiate into type I cells. Electron microscopy studies suggest that there are intermediate cell types appearing in NO_2-injured lung that show morphological characters of both type II and type I cells (Evans et al. 1975). It has also been reported that alveolar cells co-expressing type II and type I cell surface markers appear following *Staphylococcus aureus*-induced lung injury (Clegg et al. 2005). The existence of intermediate cells supports the transition of type II to type I cells.

A possible mechanism to initiate type II cell activation into the repair process may be signals related to the injury. Recent studies have suggested that the inflammatory *milieu* that forms following most types of alveolar injury can create alveolar regenerative signals (Pociask et al. 2013; Buckley et al. 2011). In injuries induced by hyperoxia, the generation of oxidants may also signal the initiation of the repair process (Pogach et al. 2007). Type II cells, in response to unknown signals related to specific injuries, can start proliferation and a type I cell transition, and they may also migrate to the injured region and engage in repair.

Little is known regarding the molecules that regulate the type II cell activation that leads to alveolar repair. One would expect that the transcriptional programs involved in embryonic lung development, such as those under the control of the FGF and Wnt pathways (Cardoso 2001; Morrisey and Hogan 2010; Warburton et al. 2008), may be activated post-injury and may contribute to repair. However, even though FGF and Wnt signaling appear to be involved in alveolar repair (Tanjore et al. 2013; Flozak et al. 2010; Ghosh et al. 2013), several transcription factors (Id2, Erm, Gata6, and Elf5) that are involved in alveolar development have not been shown to have elevated expression in type II cells following *Pseudomonas aeruginosa*-induced injury (Liu et al. 2011; Liu and Hogan 2002). Therefore, even if there may be some correlations, the repair process does not completely recapitulate the developmental process in embryogenesis.

Several growth factors appear to be able to regulate certain aspects of the progenitor properties of type II cells (Table 2.2). IGF expression is enhanced following hyperoxia-induced injury, and exogenous IGF promotes the type II to type I transition in vitro (Narasaraju et al. 2006). FGF7 (KGF) induces cultured type II cells to proliferate, while preventing them from type I cell differentiation (Qiao et al. 2008; Portnoy et al. 2004; Borok et al. 1998). EGF and HGF can also stimulate cultured type II cells to proliferate (Mason et al. 1994; Portnoy et al. 2004; Desai et al. 2014). The intratracheal injection of HGF or KGF stimulates type II cell proliferation in the lung (Panos et al. 1996; Yano et al. 2000; Fehrenbach et al. 1999, 2003; Ware and Matthay 2002). By contrast, TGFβ expression in the bleomycin-injured lung suggests that it plays a negative regulatory role in the proliferation of type II cells during early repair phase (Khalil et al. 1994). In culture, TGFβ inhibits type II cell proliferation (Khalil et al. 1994) but promotes the conversion of cultured type II cells into type I cells (Zhao et al. 2013), while BMP4, another member of the TGF superfamily, antagonizes this differentiation (Zhao et al. 2013).

The Wnt/β-Catenin signaling pathway plays an important role in the differentiation of lung epithelial cells during development (Li et al. 2005; Shu et al. 2005), and recent studies have shown that Wnt/β-Catenin signaling activity is elevated following bleomycin-induced injury (Tanjore et al. 2013; Flozak et al. 2010). Inhibition of this signaling in cultured type II cells prevented their conversion into type I cells (Tanjore et al. 2013; Flozak et al. 2010; Ghosh et al. 2013). In addition, the microRNA miR375 appears to play a role in the type II cell proliferation/differentiation process through the regulation of the Wnt/β-Catenin signaling pathway (Wang et al. 2013) (Table 2.2).

Table 2.2 Some molecules involved in type II cell-mediated alveoli repair

Molecules	Function in type II cell-mediated repair	References
IGF	Expressions of IGF and receptors are enhanced in type II cells following hyperoxia-induced injury, and exogenous IGF promotes the type II to type I transition in vitro	Narasaraju et al. (2006)
FGF7	Induces cultured type II cells to proliferate, while preventing them from type I cell differentiation; intratracheal injection of KGF stimulates type II cell proliferation in the lung	Borok et al. (1998), Fehrenbach et al. (1999), Yano et al. (2000), Ware and Matthay (2002), Fehrenbach et al. (2003), Portnoy et al. (2004), Qiao et al (2008)
EGF	Stimulate cultured type II cells to proliferate	Portnoy et al. (2004), Desai et al. (2014)
HGF	Intra tracheal injection of HGF stimulates type II cell proliferation in the lung	Mason et al. (1994), Panos et al. (1996), Ware and Matthay (2002)
TGFβ	Expression decrease at early repair phase and increase at later phase following bleomycin injury. In culture, TGFβ inhibits type II cell proliferation but promotes the conversion of cultured type II cells into type I cells	Khalil et al. (1994), Zhao et al. (2013)
BMP4	Antagonizes cultured type II cells from differentiation into type I cells	Zhao et al. (2013)
Wnt/β--Catenin	Activity was elevated following bleomycin-induced injury. Inhibition of this signaling in cultured type II cells prevented their conversion into type I cells	Tanjore et al. (2013), Flozak et al. (2010), Ghosh et al. (2013)
miR375	Inhibit type II to type I cell differentiation. Act through the regulation of the Wnt/β-Catenin signaling pathway	Wang et al. (2013)
FoxM1	Expressed in the Sca-1$^+$ subpopulation of type II cells that appear at the early repair phase following *Pseudomonas*-induced injury. Studies using mouse model to disrupt FoxM1 in type II cells showed that FoxM1 is required in these cells for their proliferation and transition into type I cells	Liu et al. (2011)

Thus, FGF, TGF, and Wnt are signaling molecules that are possibly released in response to alveolar damage that may initiate the type II cell-dependent repair process. However, many of these results were generated by expression analysis and in vitro studies; it remains to be determined whether all these factors function in the in vivo context of alveolar repair.

It is also important to understand the transcriptional mechanisms that reprogram type II cells to acquire the progenitor property and initiate the repair process. Recently, it has been found that the forkhead transcription factor FoxM1 plays an essential role in the type II cell-mediated repair of alveolar injury induced by *Pseudomonas aeruginosa* infection (Liu et al. 2011). FoxM1 expression is elevated in the Sca-1$^+$ subpopulation of type II cells that appear at the early repair phase following alveolar injury (Liu et al. 2011), and genetic studies using mouse models that disrupt FoxM1 in type II cells showed that FoxM1 is required in these cells for their proliferation and transition into type I cells that lead to repair (Liu et al. 2011) (Table 2.2).

Finally, epigenetic modifications may also play important roles in the progenitor cell function of type II cells (Marconett et al. 2013).

2.7 Regulation of Type II Cell Progenitor Properties by the Microenvironment

As discussed above, in vitro culture experiments showed that the progenitor cell features of type II cells differ in different extracellular matrices, and they show distinct behaviors when co-cultured with other cell types (Rice et al. 2002; Dobbs 1990; Demaio et al. 2009; Shannon et al. 1987; Sannes 1991; Leiner et al. 2006; Olsen et al. 2005). These results indicate that the progenitor cell properties of type II cells are tightly controlled by the cell–matrix and cell–cell interactions related to their particular environments before or after injury. In response to most acute lung injuries, many changes occur, including increases in microvascular permeability, edema and the proliferation of fibroblasts, deposit of collagen matrix and leukocyte recruitment (Ware and Matthay 2000; Matthay et al. 2012). All these factors contribute to an altered microenvironment, and it is likely that the initiation of type II cell activation is triggered by a combination of these factors.

Type II cells are normally in close contact with lung stromal cells in the basal layer. The interaction of type II cells with stromal cells, including fibroblasts, following injury has been studied extensively (Fehrenbach 2001; Thannickal et al. 2004; Mason and Crystal 1998). Fibroblast-derived soluble factors can promote the proliferation of type II cells (Fehrenbach 2001). Recently, a population of PDGFα-expressing stromal cells that include fibroblasts and lipofibroblasts have been identified that normally reside in close proximity with type II cells, and they might form a niche to maintain type II cell stemness (Barkauskas et al. 2013; Chen et al. 2012b). When co-cultured with type II cells, these cells promote type II cell self-renewal as well as differentiation toward type I cell (Barkauskas et al. 2013). Thus, fibroblasts and lipofibroblasts located adjacent to type II cells could play a role in regulating type II cell progenitor behavior through direct cell–cell contact as well as by secreting soluble factors such as growth factors or by remodeling the extracellular matrix (El Ghalbzouri and Ponec 2004).

One important feature of the lung is that alveoli are highly vascularized to facilitate gas exchange, and capillary endothelial cells reside in close proximity to alveolar epithelial cells (Komarova and Malik 2010; Bhattacharya 2005). Recent studies have revealed that alveolar capillary endothelial cells play a role in promoting alveolar regeneration following unilateral pneumonectomy (Ding et al. 2011). Controlling MMP14-dependent release of EGFR ligands appears to be one of the mechanisms that endothelial cells use to instruct type II cells to initiate the repair process (Ding et al. 2011).

Inflammatory cells, such as macrophages and neutrophils, are recruited to the alveolar space, and these cells release various inflammatory factors (Matthay et al. 2012; Ware and Matthay 2000). Among these factors are TNF, IL-1, IL-6, IL-8, and IL-10 (Fehrenbach 2001; Matthay et al. 2012). Many of these factors stimulate the proliferation, migration, and changing of cell fate of various target cells through NFκB signaling (dos Santos et al. 2012; Chen and Greene 2004; Zhang et al. 2009), and it is highly likely that these inflammatory mediators also play a role in directing type II cells to acquire progenitor phenotypes. In fact, it has been shown that IL-1β was able to stimulate migration and proliferation of cultured type II cells (Yang et al. 1999; Geiser et al. 2000). It is also possible that inflammatory cells send regeneration signals to type II cells through cell–cell contact. It has been reported that neutrophil transmigration through cultured lung alveolar type II cells activates Wnt/β-Catenin signaling in type II cells, and this is a possible mechanism for repair initiation (Zemans et al. 2011).

2.8 Potential Therapy of Lung Disease Utilizing the Progenitor Cell Properties of Type II Cells

Studies of the progenitor properties of type II cells could lead to the discovery of novel therapeutic approaches for the treatment of chronic and acute lung diseases. Discovering the distinct molecular elements of the transition from a quiescent to a regenerative type II cell could result in the identification of new pharmaceutical targets to accelerate lung repair. If we can dissect type II cell-mediated alveolar repair into a sequential process, it may be possible to intervene pharmacologically at each step to promote recovery and prevent or reverse chronic lung disease related to injuries.

Furthermore, because type II cells can be obtained through in vitro differentiation from iPS cells or lung biopsies, a potential therapeutic approach would be transplanting type II cells into injured lungs (Huang et al. 2014; Longmire et al. 2012). In fact, there is one report showing that by transplanting normal type II cells into bleomycin-injured rat lungs, the fibrosis formation induced by bleomycin is delayed or even reversed (Serrano-Mollar et al. 2007). The identification of subgroups of type II cells that function in regeneration will substantially improve such cell transplantation therapies by using distinct populations of type II cells.

Finally, recent lineage tracing studies showed that some lung adenocarcinoma cells are derived from type II cells (Desai et al. 2014; Xu et al. 2012). Therefore, understanding the regulation of type II cell progenitor properties may shed light on the mechanisms of lung cancer initiation.

Acknowledgements I am grateful to the grant support from National Institutes of Health (HL105947-01 to YL), and I thank Dr. Varsha Suresh Kumar for providing images.

References

Andreeva AV, Kutuzov MA, Voyno-Yasenetskaya TA (2007) Regulation of surfactant secretion in alveolar type II cells. Am J Physiol Lung Cell Mol Physiol 293(2):L259–L271

Barkauskas CE, Michael J, Cronce MJ, Rackley CR, Bowie EJ, Keene DR, Stripp BR, Randell SH, Noble PW, Hogan BLM (2013) Type 2 alveolar cells are stem cells in adult lung. J Clin Invest 123(7):3025–3036

Bhattacharya J (2005) Alveolocapillary cross-talk: Giles F. Filley lecture. Chest 128(6 Suppl): 553S–555S

Borok Z, Lubman RL, Danto SI, Zhang XL, Zabski SM, King LS, Lee DM, Agre P, Crandall ED (1998) Keratinocyte growth factor modulates alveolar epithelial cell phenotype in vitro: expression of aquaporin 5. Am J Respir Cell Mol Biol 18(4):554–561. doi:10.1165/rcmb.18.4.2838

Buckley S, Shi W, Carraro G, Sedrakyan S, Da Sacco S, Driscoll BA, Perin L, De Filippo RE, Warburton D (2011) The milieu of damaged alveolar epithelial type 2 cells stimulates alveolar wound repair by endogenous and exogenous progenitors. Am J Respir Cell Mol Biol 45(6):1212–1221

Cardoso W (2001) Molecular regulation of lung development. Annu Rev Physiol 63:471–494

Chapman HA, Li X, Alexander JP, Brumwell A, Lorizio W, Tan K, Sonnenberg A, Wei Y, Vu TH (2011) Integrin α6β4 identifies an adult distal lung epithelial population with regenerative potential in mice. J Clin Invest 121(7):2855–2862

Chen LF, Greene WC (2004) Shaping the nuclear action of NF-kappaB. Nat Rev Mol Cell Biol 5(5):392–401

Chen H, Matsumoto K, Brockway BL, Rackley CR, Liang J, Lee JH, Jiang D, Noble PW, Randell SH, Kim CF, Stripp BR (2012a) Airway epithelial progenitors are region specific and show differential responses to bleomycin-induced lung injury. Stem Cells 30(9):1948–1960. doi:10.1002/stem.1150

Chen L, Acciani T, Le Cras T, Lutzko C, Perl AK (2012b) Dynamic regulation of platelet-derived growth factor receptor alpha expression in alveolar fibroblasts during realveolarization. Am J Respir Cell Mol Biol 47(4):517–527. doi:10.1165/rcmb.2012-0030OC

Clegg GR, Tyrrell C, McKechnie SR, Beers MF, Harrison D, McElroy MC (2005) Coexpression of RTI40 with alveolar epithelial type II cell proteins in lungs following injury: identification of alveolar intermediate cell types. Am J Physiol Lung Cell Mol Physiol 289(3):L382–L390

Corti M, Brody AR, Harrison JH (1996) Isolation and primary culture of murine alveolar type II cells. Am J Respir Cell Mol Biol 14(4):309–315

Demaio L, Tseng W, Balverde Z, Alvarez JR, Kim KJ, Kelley DG, Senior RM, Crandall ED, Borok Z (2009) Characterization of mouse alveolar epithelial cell monolayers. Am J Physiol Lung Cell Mol Physiol 296(6):L1051–L1058. doi:10.1152/ajplung.00021.2009

Desai TJ, Brownfield DG, Krasnow MA (2014) Alveolar progenitor and stem cells in lung development, renewal and cancer. Nature 507(7491):190–194. doi:10.1038/nature12930

Ding BS, Nolan DJ, Guo P, Babazadeh AO, Cao Z, Rosenwaks Z, Crystal RG, Simons M, Sato TN, Worgall S, Shido K, Rabbany SY, Rafii S (2011) Endothelial-derived angiocrine signals induce and sustain regenerative lung alveolarization. Cell 147(3):539–553. doi:10.1016/j.cell.2011.10.003

Dobbs LG (1990) Isolation and culture of alveolar type II cells. Am J Physiol 258(4 Pt 1): L134–L147

Dodelet VC, Pasquale EB (2000) Eph receptors and ephrin ligands: embryogenesis to tumorigenesis. Oncogene 19(49):5614–5619

dos Santos G, Kutuzov MA, Ridge KM (2012) The inflammasome in lung diseases. Am J Physiol Lung Cell Mol Physiol 303(8):L627–L633

Driscoll B, Buckley S, Bui KC, Anderson KD, Warburton D (2000) Telomerase in alveolar epithelial development and repair. Am J Physiol Lung Cell Mol Physiol 279(6):L1191–L1198

El Ghalbzouri A, Ponec M (2004) Diffusible factors released by fibroblasts support epidermal morphogenesis and deposition of basement membrane components. Wound Repair Regen 12(3):359–367

Evans MJ, Cabral LJ, Stephens RJ, Freeman G (1973) Renewal of alveolar epithelium in the rat following exposure to NO2. Am J Pathol 70(2):175–198

Evans MJ, Cabral LJ, Stephens RJ, Freeman G (1975) Transformation of alveolar type 2 cells to type 1 cells following exposure to NO2. Exp Mol Pathol 22(1):142–150

Fehrenbach H (2001) Alveolar epithelial type II cell: defender of the alveolus revisited. Respir Res 2(1):33–46

Fehrenbach H, Kasper M, Tschernig T, Pan T, Schuh D, Shannon JM, Muller M, Mason RJ (1999) Keratinocyte growth factor-induced hyperplasia of rat alveolar type II cells in vivo is resolved by differentiation into type I cells and by apoptosis. Eur Respir J 14(3):534–544

Fehrenbach A, Bube C, Hohlfeld JM, Stevens P, Tschernig T, Hoymann HG, Krug N, Fehrenbach H (2003) Surfactant homeostasis is maintained in vivo during keratinocyte growth factor-induced rat lung type II cell hyperplasia. Am J Respir Crit Care Med 167(9):1264–1270

Flozak AS, Lam AP, Russell S, Jain M, Peled ON, Sheppard KA, Beri R, Mutlu GM, Budinger GS, Gottardi CJ (2010) {beta}-catenin/TCF signaling is activated during lung injury and promotes the survival and migration of alveolar epithelial cells. J Biol Chem 285(5):3157–3167

Geiser T, Jarreau PH, Atabai K, Matthay MA (2000) Interleukin-1beta augments in vitro alveolar epithelial repair. Am J Physiol Lung Cell Mol Physiol 279(6):L1184–L1190

Ghosh MC, Gorantla V, Makena PS, Luellen C, Sinclair SE, Schwingshackl A, Waters CM (2013) Insulin-like growth factor-I stimulates differentiation of ATII cells to ATI-like cells through activation of Wnt5a. Am J Physiol Lung Cell Mol Physiol 305(3):L222–L228. doi:10.1152/ajplung.00014.2013

Gobran LI, Rooney SA (2004) Pulmonary surfactant secretion in briefly cultured mouse type II cells. Am J Physiol Lung Cell Mol Physiol 286(2):L331–L336

Hendrickson CM, Matthay MA (2013) Viral pathogens and acute lung injury: investigations inspired by the SARS epidemic and the 2009 H1N1 influenza pandemic. Semin Respir Crit Care Med 34(4):475–486. doi:10.1055/s-0033-1351122

Huang SX, Islam MN, O'Neill J, Hu Z, Yang YG, Chen YW, Mumau M, Green MD, Vunjak-Novakovic G, Bhattacharya J, Snoeck HW (2014) Efficient generation of lung and airway epithelial cells from human pluripotent stem cells. Nat Biotechnol 32(1):84–91. doi:10.1038/nbt.2754

Kalina M, Mason RJ, Shannon JM (1992) Surfactant protein C is expressed in alveolar type II cells but not in Clara cells of rat lung. Am J Respir Cell Mol Biol 6(6):594–600

Khalil N, O'Connor RN, Flanders KC, Shing W, Whitman CI (1994) Regulation of type II alveolar epithelial cell proliferation by TGF-beta during bleomycin-induced lung injury in rats. Am J Physiol 267(5 Pt 1):L498–L507

Kikkawa Y, Yoneda K (1974) The type II epithelial cell of the lung. I. Method of isolation. Lab Invest 30(1):76–84

Kim CF, Jackson EL, Woolfenden AE, Lawrence S, Babar I, Vogel SM, Crowley D, Bronson RT, Jacks T (2005a) Identification of bronchioalveolar stem cells in normal lung and lung cancer. Cell 121(6):823–835

Kim IM, Ramakrishna S, Gusarova GA, Yoder HM, Costa RH, Kalinichenko VV (2005b) The forkhead box m1 transcription factor is essential for embryonic development of pulmonary vasculature. J Biol Chem 280(23):22278–22286

Kleinman HK, Martin GR (2005) Matrigel: basement membrane matrix with biological activity. Semin Cancer Biol 15(5):378–386

Komarova Y, Malik AB (2010) Regulation of endothelial permeability via paracellular and transcellular transport pathways. Annu Rev Physiol 72:463–493. doi:10.1146/annurev-physiol-021909-135833

Kretzschmar K, Watt FM (2012) Lineage tracing. Cell 148(1–2):33–45. doi:10.1016/j.cell.2012.01.002

Kumar PA, Hu Y, Yamamoto Y, Hoe NB, Wei TS, Mu D, Sun Y, Joo LS, Dagher R, Zielonka EM, de Wang Y, Lim B, Chow VT, Crum CP, Xian W, McKeon F (2011) Distal airway stem cells yield alveoli in vitro and during lung regeneration following H1N1 influenza infection. Cell 147(3):525–538

Lee JC, Reddy R, Barsky L, Weinberg K, Driscoll B (2006) Contribution of proliferation and DNA damage repair to alveolar epithelial type 2 cell recovery from hyperoxia. Am J Physiol Lung Cell Mol Physiol 290(4):L685–L694

Lee JH, Kim J, Gludish D, Roach RR, Saunders AH, Barrios J, Woo AJ, Chen H, Conner DA, Fujiwara Y, Stripp BR, Kim CF (2013) Surfactant protein-C chromatin-bound green fluorescence protein reporter mice reveal heterogeneity of surfactant protein C-expressing lung cells. Am J Respir Cell Mol Biol 48(3):288–298. doi:10.1165/rcmb.2011-0403OC

Lee JH, Bhang DH, Beede A, Huang TL, Stripp BR, Bloch KD, Wagers AJ, Tseng YH, Ryeom S, Kim CF (2014) Lung stem cell differentiation in mice directed by endothelial cells via a BMP4-NFATc1-thrombospondin-1 axis. Cell 156(3):440–455. doi:10.1016/j.cell.2013.12.039

Leiner KA, Newman D, Li CM, Walsh E, Khosla J, Sannes PL (2006) Heparin and fibroblast growth factors affect surfactant protein gene expression in type II cells. Am J Respir Cell Mol Biol 35(5):611–618. doi:10.1165/rcmb.2006-0159OC

Li C, Hu L, Xiao J, Chen H, Li JT, Bellusci S, Delanghe S, Minoo P (2005) Wnt5a regulates Shh and Fgf10 signaling during lung development. Dev Biol 287(1):86–97

Liu Y, Hogan BL (2002) Differential gene expression in the distal tip endoderm of the embryonic mouse lung. Gene Expr Patterns 2:229–233

Liu Y, Suresh Kumar V, Zhang W, Rehman J, Malik AB (2014) Activation of type II cells into regenerative Sca-1+ cells during alveolar repair. Am J Respir Cell Mol Biol. [Epub ahead of print]

Liu Y, Sadikot RT, Adami GR, Kalinichenko VV, Pendyala S, Natarajan V, Zhao YY, Malik AB (2011) FoxM1 mediates the progenitor function of type II epithelial cells in repairing alveolar injury induced by Pseudomonas aeruginosa. J Exp Med 208(7):1473–1484

Longmire TA, Ikonomou L, Hawkins F, Christodoulou C, Cao Y, Jean JC, Kwok LW, Mou H, Rajagopal J, Shen SS, Dowton AA, Serra M, Weiss DJ, Green MD, Snoeck HW, Ramirez MI, Kotton DN (2012) Efficient derivation of purified lung and thyroid progenitors from embryonic stem cells. Cell Stem Cell 10(4):398–411. doi:10.1016/j.stem.2012.01.019

Marconett CN, Zhou B, Rieger ME, Selamat SA, Dubourd M, Fang X, Lynch SK, Stueve TR, Siegmund KD, Berman BP, Borok Z, Laird-Offringa IA (2013) Integrated transcriptomic and epigenomic analysis of primary human lung epithelial cell differentiation. PLoS Genet 9(6):e1003513. doi:10.1371/journal.pgen.1003513

Mason RJ (2006) Biology of alveolar type II cells. Respirology 11:S12–S15

Mason RJ, Crystal RG (1998) Pulmonary cell biology. Am J Respir Crit Care Med 157(4 Pt 2): S72–S81

Mason RJ, Shannon JM (1997) Alveolar type II cells. In: Crystal RG, West JB, Weibel ER, Barnes PJ (eds) The lung: scientific foundations, vol 1, 3, 2nd edn. Lippincott Williams & Wilkins, Philadelphia, pp 543–556

Mason RJ, Leslie CC, McCormick-Shannon K, Deterding RR, Nakamura T, Rubin JS, Shannon JM (1994) Hepatocyte growth factor is a growth factor for rat alveolar type II cells. Am J Respir Cell Mol Biol 11(5):561–567

Matthay MA, Ware LB, Zimmerman GA (2012) The acute respiratory distress syndrome. J Clin Invest 122(8):2731–2740

Matute-Bello G, Frevert CW, Martin TR (2008) Animal models of acute lung injury. Am J Physiol Lung Cell Mol Physiol 295(3):L379–L399. doi:10.1152/ajplung.00010.2008

Matute-Bello G, Downey G, Moore BB, Groshong SD, Matthay MA, Slutsky AS, Kuebler WM (2011) An official American Thoracic Society workshop report: features and measurements of experimental acute lung injury in animals. Am J Respir Cell Mol Biol 44(5):725–738. doi:10.1165/rcmb.2009-0210ST

McQualter JL, Yuen K, Williams B, Bertoncello I (2010) Evidence of an epithelial stem/progenitor cell hierarchy in the adult mouse lung. Proc Natl Acad Sci U S A 107(4):1414–1419. doi:10.1073/pnas.0909207107

Morrisey EE, Hogan BL (2010) Preparing for the first breath: genetic and cellular mechanisms in lung development. Dev Cell 18(1):8–23

Narasaraju TA, Chen H, Weng T, Bhaskaran M, Jin N, Chen J, Chen Z, Chinoy MR, Liu L (2006) Expression profile of IGF system during lung injury and recovery in rats exposed to hyperoxia: a possible role of IGF-1 in alveolar epithelial cell proliferation and differentiation. J Cell Biochem 97(5):984–998

Nolen-Walston RD, Kim CF, Mazan MR, Ingenito EP, Gruntman AM, Tsai L, Boston R, Woolfenden AE, Jacks T, Hoffman AM (2008) Cellular kinetics and modeling of bronchioalveolar stem cell response during lung regeneration. Am J Physiol Lung Cell Mol Physiol 294(6):L1158–L1165. doi:10.1152/ajplung.00298.2007

Olsen CO, Isakson BE, Seedorf GJ, Lubman RL, Boitano S (2005) Extracellular matrix-driven alveolar epithelial cell differentiation in vitro. Exp Lung Res 31(5):461–482

Pagano A, Barazzone-Argiroffo C (2003) Alveolar cell death in hyperoxia-induced lung injury. Ann N Y Acad Sci 1010:405–416

Paine R, Simon RH (1996) Expanding the frontiers of lung biology through the creative use of alveolar epithelial cells in culture. Am J Physiol 270(4 Pt 1):L484–L486

Panos RJ, Patel R, Bak PM (1996) Intratracheal administration of hepatocyte growth factor/scatter factor stimulates rat alveolar type II cell proliferation in vivo. Am J Respir Cell Mol Biol 15(5):574–581

Persinger RL, Blay WM, Heintz NH, Hemenway DR, Janssen-Heininger YM (2001) Nitrogen dioxide induces death in lung epithelial cells in a density-dependent manner. Am J Respir Cell Mol Biol 24(5):583–590

Pociask DA, Scheller EV, Mandalapu S, McHugh KJ, Enelow RI, Fattman CL, Kolls JK, Alcorn JF (2013) IL-22 is essential for lung epithelial repair following influenza infection. Am J Pathol 182(4):1286–1296

Pogach MS, Cao Y, Millien G, Ramirez MI, Williams MC (2007) Key developmental regulators change during hyperoxia-induced injury and recovery in adult mouse lung. J Cell Biochem 100(6):1415–1429

Portnoy J, Curran-Everett D, Mason RJ (2004) Keratinocyte growth factor stimulates alveolar type II cell proliferation through the extracellular signal-regulated kinase and phosphatidylinositol 3-OH kinase pathways. Am J Respir Cell Mol Biol 30(6):901–907

Qiao R, Yan W, Clavijo C, Mehrian-Shai R, Zhong Q, Kim KJ, Ann D, Crandall ED, Borok Z (2008) Effects of KGF on alveolar epithelial cell transdifferentiation are mediated by JNK signaling. Am J Respir Cell Mol Biol 38(2):239–246

Rannels SR, Yarnell JA, Fisher CS, Fabisiak JP, Rannels DE (1987) Role of laminin in maintenance of type II pneumocyte morphology and function. Am J Physiol 253(6 Pt 1):C835–C845

Rawlins EL, Okubo T, Xue Y, Brass DM, Auten RL, Hasegawa H, Wang F, Hogan BL (2009) The role of Scgb1a1+ Clara cells in the long-term maintenance and repair of lung airway, but not alveolar, epithelium. Cell Stem Cell 4(6):525–534

Reddy R, Buckle YS, Doerken M, Barsky L, Weinberg K, Anderson KD, Warburton D, Driscoll B (2004) Isolation of a putative progenitor subpopulation of alveolar epithelial type 2 cells. Am J Physiol Lung Cell Mol Physiol 286(4):L658–L667

Rice WR, Conkright JJ, Na CL, Ikegami M, Shannon JM, Weaver TE (2002) Maintenance of the mouse type II cell phenotype in vitro. Am J Physiol Lung Cell Mol Physiol 283(2):L256–L264

Rock JR, Hogan BL (2011) Epithelial progenitor cells in lung development, maintenance, repair, and disease. Annu Rev Cell Dev Biol 27:493–512

Rock JR, Barkauskas CE, Cronce MJ, Xue Y, Harris JR, Liang J, Noble PW, Hogan BL (2011) Multiple stromal populations contribute to pulmonary fibrosis without evidence for epithelial to mesenchymal transition. Proc Natl Acad Sci U S A 108(52):E1475–E1483

Roper JM, Staversky RJ, Finkelstein JN, Keng PC, O'Reilly MA (2003) Identification and isolation of mouse type II cells on the basis of intrinsic expression of enhanced green fluorescent protein. Am J Physiol Lung Cell Mol Physiol 285(3):L691–L700

Sadikot RT, Blackwell TS, Christman JW, Prince AS (2005) Pathogen-host interactions in Pseudomonas aeruginosa pneumonia. Am J Respir Crit Care Med 171(11):1209–1223

Sannes PL (1991) Structural and functional relationships between type II pneumocytes and components of extracellular matrices. Exp Lung Res 17(4):639–659

Schneeberger EE (1997) Alveolar type I cells. In: Crystal RG, West JB, Weibel ER, Barnes PJ (eds) The lung: scientific foundations. Lippincott Williams & Wilkins, Philadelphia, pp 535–542

Serrano-Mollar A, Nacher M, Gay-Jordi G, Closa D, Xaubet A, Bulbena O (2007) Intratracheal transplantation of alveolar type II cells reverses bleomycin-induced lung fibrosis. Am J Respir Crit Care Med 176(12):1261–1268

Shannon JM, Mason RJ, Jennings SD (1987) Functional differentiation of alveolar type II epithelial cells in vitro: effects of cell shape, cell-matrix interactions and cell-cell interactions. Biochim Biophys Acta 931(2):143–156

Shimabukuro DW, Sawa T, Gropper MA (2003) Injury and repair in lung and airways. Crit Care Med S524–531

Shu W, Guttentag S, Wang Z, Andl T, Ballard P, Lu MM, Piccolo S, Birchmeier W, Whitsett JA, Millar SE, Morrisey EE (2005) Wnt/beta-catenin signaling acts upstream of N-myc, BMP4, and FGF signaling to regulate proximal-distal patterning in the lung. Dev Biol 283(1):226–239

Soriano P (1999) Generalized lacZ expression with the ROSA26 Cre reporter strain. Nat Genet 21(1):70–71

Spencer H, Shorter RG (1962) Cell turnover in pulmonary tissues. Nature 194:880

Stripp BR (2008) Hierarchical organization of lung progenitor cells: is there an adult lung tissue stem cell? Proc Am Thorac Soc 5(6):695–698

Tanjore H, Degryse AL, Crossno PF, Xu XC, McConaha ME, Jones BR, Polosukhin VV, Bryant AJ, Cheng DS, Newcomb DC, McMahon FB, Gleaves LA, Blackwell TS, Lawson WE (2013) beta-catenin in the alveolar epithelium protects from lung fibrosis after intratracheal bleomycin. Am J Respir Crit Care Med 187(6):630–639

Thannickal VJ, Toews GB, White ES, Lynch JP, Martinez FJ (2004) Mechanisms of pulmonary fibrosis. Annu Rev Med 55:395–417

Vaughan AE, Chapman HA (2013) Regenerative activity of the lung after epithelial injury. Biochim Biophys Acta 1832(7):922–930. doi:10.1016/j.bbadis.2012.11.020

Walker SR, Williams MC, Benson B (1986) Immunocytochemical localization of the major surfactant apoproteins in type II cells, Clara cells, and alveolar macrophages of rat lung. J Histochem Cytochem 34(9):1137–1148

Wang Y, Huang C, Reddy Chintagari N, Bhaskaran M, Weng T, Guo Y, Xiao X, Liu L (2013) miR-375 regulates rat alveolar epithelial cell trans-differentiation by inhibiting Wnt/beta-catenin pathway. Nucleic Acids Res 41(6):3833–3844. doi:10.1093/nar/gks1460

Wansleeben C, Barkauskas CE, Rock JR, Hogan BL (2013) Stem cells of the adult lung: their development and role in homeostasis, regeneration, and disease. Wiley Interdiscip Rev Dev Biol 2(1):131–148

Warburton D, Perin L, Defilippo R, Bellusci S, Shi W, Driscoll B (2008) Stem/progenitor cells in lung development, injury repair, and regeneration. Proc Am Thorac Soc 5(6):703–706

Ware LB, Matthay MA (2000) The acute respiratory distress syndrome. N Engl J Med 342(18):1334–1349

Ware LB, Matthay MA (2002) Keratinocyte and hepatocyte growth factors in the lung: roles in lung development, inflammation, and repair. Am J Physiol Lung Cell Mol Physiol 282(5):L924–L940

Weibel ER (2009) What makes a good lung? Swiss Med Wkly 139(27–28):375–386

Williams MC (2003) Alveolar type I cells: molecular phenotype and development. Annu Rev Physiol 65:669–695

Xu X, Rock JR, Lu Y, Futtner C, Schwab B, Guinney J, Hogan BL, Onaitis MW (2012) Evidence for type II cells as cells of origin of K-Ras-induced distal lung adenocarcinoma. Proc Natl Acad Sci U S A 109(13):4910–4915. doi:10.1073/pnas.1112499109

Yang GH, Osanai K, Takahashi K (1999) Effects of interleukin-1 beta on DNA synthesis in rat alveolar type II cells in primary culture. Respirology 4(2):139–145

Yano T, Mason RJ, Pan T, Deterding RR, Nielsen LD, Shannon JM (2000) KGF regulates pulmonary epithelial proliferation and surfactant protein gene expression in adult rat lung. Am J Physiol Lung Cell Mol Physiol 279(6):L1146–L1158

Zemans RL, Briones N, Campbell M, McClendon J, Young SK, Suzuki T, Yang IV, De Langhe S, Reynolds SD, Mason RJ, Kahn M, Henson PM, Colgan SP, Downey GP (2011) Neutrophil transmigration triggers repair of the lung epithelium via beta-catenin signaling. Proc Natl Acad Sci U S A 108(38):15990–15995. doi:10.1073/pnas.1110144108

Zhang Y, Tomann P, Andl T, Gallant NM, Huelsken J, Jerchow B, Birchmeier W, Paus R, Piccolo S, Mikkola ML, Morrisey EE, Overbeek PA, Scheidereit C, Millar SE, Schmidt-Ullrich R (2009) Reciprocal requirements for EDA/EDAR/NF-kappaB and Wnt/beta-catenin signaling pathways in hair follicle induction. Dev Cell 17(1):49–61

Zhao J, Shi W, Wang YL, Chen H, Bringas PJ, Datto MB, Frederick JP, Wang XF, Warburton D (2002) Smad3 deficiency attenuates bleomycin-induced pulmonary fibrosis in mice. Am J Physiol Lung Cell Mol Physiol 282(3):L585–L593

Zhao L, Yee M, O'Reilly MA (2013) Transdifferentiation of alveolar epithelial type II to type I cells is controlled by opposing TGF-beta and BMP signaling. Am J Physiol Lung Cell Mol Physiol 305(6):L409–L418. doi:10.1152/ajplung.00032.2013

Chapter 3
Stem Cell Niches in the Lung

Thomas J. Lynch, Xiaoming Liu, Jun Wei, and John F. Engelhardt

Abbreviations

AEC I/II	Alveolar epithelial cell I/II
AF	Autofluorescence
ALDH1	Aldehyde dehydrogenase 1 family
AQP3	Aquaporin 3
BADJ	Bronchioalveolar duct junction
BASC	Bronchioalveolar stem cell
CCSP	Club cell secretory protein
CD16	Fc receptor, IgG, low affinity III
CD24	CD24 molecule
CD31	Platelet/Endothelial cell adhesion molecule 1 (aka PECAM1)
CD32	Fc receptor, IgG, low affinity IIb
CD34	CD34 antigen
CD45	Protein tyrosine phosphatase, receptor type, C (aka Ptprc)
CD73	5′ Nucleotidase, Ecto (aka Nt5e)

T.J. Lynch • J.F. Engelhardt (✉)
Department of Anatomy and Cell Biology, Carver College of Medicine, University of Iowa, 51 Newton Road, Room 1-111 BSB, Iowa City, IA 52242, USA
e-mail: john-engelhardt@uiowa.edu

X. Liu
Institute of Human Stem Cell Research, General Hospital of Ningxia Medical University, Yinchuan, Ningxia, IA 750004, China

Department of Anatomy and Cell Biology, Carver College of Medicine, University of Iowa, 51 Newton Road, Room 1-111 BSB, Iowa City, IA 52242, USA

J. Wei
Institute of Human Stem Cell Research, General Hospital of Ningxia Medical University, Yinchuan, Ningxia, IA 750004, China

A. Firth, J.X.-J. Yuan (eds.), *Lung Stem Cells in the Epithelium and Vasculature*, Stem Cell Biology and Regenerative Medicine, DOI 10.1007/978-3-319-16232-4_3

CD151	CD151 antigen
CDH1	Cadherin 1, type 1, E-cadherin (aka E-Cad)
CFTR	Cystic fibrosis transmembrane conductance regulator
CGRP	Calcitonin gene-related peptide
CK	Cytokeratin
CSC	Cancer stem cell
EpCAM	Epithelial cell adhesion molecule
F3	Tissue factor (aka TF)
FGF	Fibroblast growth factor
GSI-A_3B	*Griffonia simplicifolia* isolectin A_3B
H33342	Hoechst 33342
ITGA6	Integrin alpha 6 (aka CD49f)
ITGB4	Integrin beta 4 (aka CD104)
LGR6	Leucine-rich repeat containing g protein-coupled receptor 6
LRC	Label retaining cell
Ly6a	Lymphocyte antigen 6 complex, locus A (aka Sca-1)
Ly76	Lymphocyte antigen 76 (aka TER119)
lrMSC	Lung-resident mesenchymal stromal cell
MSC	Mesenchymal stromal cell
NEB	Neuroendocrine body
NGFR	Nerve growth factor receptor
NSCLC	Non-small cell lung cancer
PDGFRα	Platelet-derived growth factor receptor alpha
PNEC	Pulmonary neuroendocrine cell
SAE	Surface airway epithelium
SCLC	Small cell lung cancer
SMG	Submucosal gland
SPC	Surfactant protein C (aka Sftpc)
TGFβ	Transforming growth factor beta
TROP2	Tumor-associated calcium signal transducer 2 (aka Tacstd2)
TTF1	Thyroid transcription factor 1 (aka NKx2.1)

3.1 Introduction

The epithelium of the conducting and respiratory airways in the adult lung is composed of numerous phenotypically distinct epithelial cell types tailored to perform region-specific functions. Because the lung is exposed to the external environment and to inhaled pathogens, its airways must have a rapid capacity to regenerate if injured; this is essential to preserving an epithelial barrier and normal lung functions. Both during injury repair and in the context of homeostatic turnover, cell regeneration depends on various types of stem/progenitor cells that are positioned throughout the pulmonary tree (Borthwick et al. 2001; Hong et al. 2001; Kim et al. 2005; Liu et al. 2006; Liu and Engelhardt 2008; Rawlins et al. 2009b; Reynolds and Malkinson 2010; Rock and Hogan 2011).

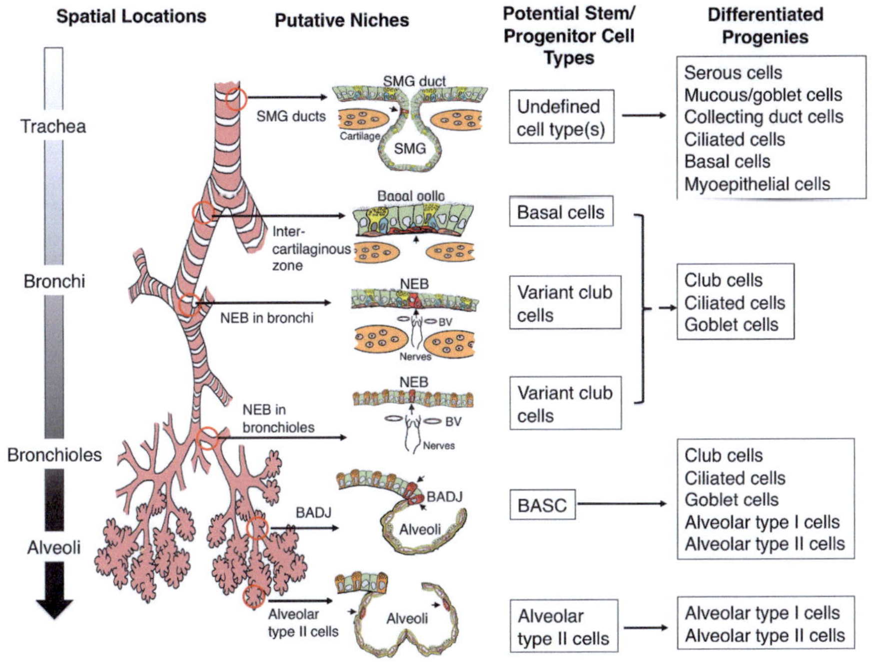

Fig. 3.1 Illustration of potential stem/progenitor cell niches in the lung of the adult mouse. The lung can be divided into three major levels of conducting airways (the trachea, bronchi, and bronchioles) plus the gas-exchanging alveoli. Distinct region-specific stem/progenitor cell niches are thought to exist along the proximal-distal axis of the airway. These include: SMG ducts in the proximal trachea, basal cells within intercartilaginous zones of the trachea and primary bronchi, NEBs in the intralobar bronchi and bronchioles, and the BADJ and alveolar spaces within the alveoli. Progenitor/stem cells (marked in *red* and *listed*) reside in their respective local niches and these environments enable them to maintain their stem/progenitor properties and control their ability to differentiate into various progeny cell types. *SMG* submucosal gland, *NEB* neuroepithelial body, *BADJ* bronchioalveolar duct junction, *BV* blood vessel

As in other adult tissues and organs, the stem/progenitor cells of the adult lung are undifferentiated cells and have the capacity to remain multipotent, self-renew, and produce differentiated progeny present in the physiological domain in which they reside. Throughout the airway tree, several distinct cell types carry out local repair in response to injury (Fig. 3.1). In mice, such cells include a subset of basal cells (within the proximal airway) (Rock et al. 2009, 2010; Cole et al. 2010; Hajj et al. 2007; Hong et al. 2004a), basal-like cells within the ducts of SMGs (Borthwick et al. 2001; Engelhardt 2001; Engelhardt et al. 1995; Hegab et al. 2011, 2012b; Xie et al. 2011; Lynch and Engelhardt 2014), a subset of naphthalene-resistant variant club cells (within the NEBs of the bronchi and bronchioles) (Guha et al. 2012; Hong et al. 2001; Reynolds et al. 2000a; Xing et al. 2012; Reynolds and Malkinson 2010), a subset of SPC expressing club cells at the BADJ (Giangreco et al. 2002; Kim et al. 2005; Rawlins et al. 2009b; Zheng et al. 2013), and a subset of alveolar type II cells (Barkauskas et al. 2013; Fujino et al. 2011).

Studies using murine models have revealed several region-specific stem cell niches along the proximal-distal axis of the airway that maintain distinct subpopulations of progenitors. Stem/progenitor cells are mobilized from these epithelial niches to maintain tissue homeostasis during injury repair and normal cellular turnover. The coordination of molecular and cellular events in the microenvironment of stem cell niches plays a pivotal role in maintaining the balance of stem/progenitors and differentiated cells that are needed for regeneration in the lung (Fig. 3.1). In this chapter, we review the diversity of cell types, including potential stem/progenitor cells, that have been identified in the adult lung, and discuss advances in our understanding of stem/progenitor cell niches and their roles in injury repair and lung cancer.

3.2 Cellular Diversity in the Adult Lung

Based on its anatomical and functional features, the lung epithelium can be divided into three domains: the proximal cartilaginous airways (trachea and bronchi), the bronchioles (bronchioles, terminal bronchioles, and respiratory bronchioles), and the alveoli. The epithelial cell types in each of these domains are distinguished by their morphology, cellular phenotype (i.e., proteins they express), and function. The proximal airway of the mouse is lined with a pseudostratified columnar epithelium composed mainly of basal, club, goblet, and ciliated cells; the secretory SMGs reside beneath this surface airway epithelium (SAE) and are limited to the proximal trachea in mice (Hansell and Moretti 1969; Pack et al. 1980; Widdicombe et al. 2001; Jeffery 1983; Liu et al. 2006). The major cell types in the human proximal airway differ slightly from those in mice and include basal, intermediate, goblet, non-ciliated columnar, and ciliated cells (Jeffery 1983; Liu et al. 2006; Mercer et al. 1994). Furthermore, in humans the SMGs are present throughout the cartilaginous airways, including the trachea and bronchi. These glands are composed of an interconnecting network of serous acini and mucus tubules, which secrete antibacterial factors, mucous, and fluid into the airway lumen (Wine and Joo 2004). In the distal mouse and human airways (i.e., bronchioles), club, ciliated, neuroendocrine, and goblet cells are the major cell types, and neuroendocrine cells are found both individually and in clusters within NEBs (Mercer et al. 1994; Van Lommel et al. 1999; Plopper et al. 1980; Liu et al. 2006). However, the bronchioles of human lungs have also been shown to contain basal cells, albeit at lower abundance than in the proximal regions (Tamai 1983; Rock et al. 2010). The alveolar epithelium is lined by surfactant-producing cuboidal alveolar type II epithelial cells (AECII) and squamous gas-exchanging alveolar type I epithelial cells (AECI) (Liu et al. 2006). The major epithelial cell types that are present at various locations throughout the airway are listed in Table 3.1.

Table 3.1 Major epithelial cell types in the lungs of adult mice and humans

Cell types	Cell-type markers	Stem/ Progenitor cells	Lineage cell type(s)	Candidate niches	Reference(s)
Basal cells	Cytokeratin 5, Cytokeratin 14, P63, NGFR, ITGA6	Yes	Basal, club, goblet, and ciliated cells	Intercartilaginous zones and SMGs	Hong et al. (2004b), Hajj et al. (2007), Hong et al. (2004a), Rock et al. (2009), Schoch et al. (2004), Cole et al. (2010), Engelhardt et al. (1995), Xie et al. (2011), Borthwick et al. (2001), Hegab et al. (2011, 2012b)
Intermediate cells	Cytokeratin 5, Cytokeratin 14, P63	Yes	Goblet and ciliated cells	Undefined (transient amplifying progenitor)	Engelhardt et al. (1995), Evans et al. (1986)
Club cells (conducting airways)	CCSP (Scgb1a1)	Yes	Club, basal[a], goblet, and ciliated cells	NEB	Hong et al. (2001), Boers et al. (1999), Rawlins et al. (2009b), Reynolds et al. (2000b), Reynolds et al. (2000a), Tata et al. (2013)
Club cells (BADJ)	CCSP (Scgb1a1), SPC	Yes	Club, AECII, and AEC1 cells	BADJ	Zacharek et al. (2011), Kim et al. (2005), Regala et al. (2009), Tropea et al. (2012), Zheng et al. (2013), Rock et al. (2011)
Ciliated cells	FoxJ1, Tubulin IV	No	Not applicable	Not applicable	Pardo-Saganta et al. (2013), Rawlins and Hogan (2008)
Goblet cells	Mucin 5AC, Mucin 5B, Spdef	No	Not applicable	Not applicable	Engelhardt et al. (1995), Shimizu et al. (1996)
Non-ciliated columnar cells (non-goblet or non-club)	Undefined	Yes	Undefined	Undefined	Evans et al. (1986)
AECI cells	Aquaporin 5, Podoplanin RAGE	No	Not applicable	Not applicable	Barkauskas et al. (2013), Kinnard et al. (1994)

(continued)

Table 3.1 (continued)

Cell types	Cell-type markers	Stem/ Progenitor cells	Lineage cell type(s)	Candidate niches	Reference(s)
AECII cells	SPB, SPC, Lamp3, Abca3, and integrin α6β4	Yes	ATI and ATII cells	Alveoli space	Reddy et al. (2004), Barkauskas et al. (2013), Fujino et al. (2011), Chapman et al. (2011)
PNEC	CGRP, Ascl1, Pgp9.5	Yes	PNEC, club[b], and ciliated cells[b]	NEB	Song et al. (2012), Hong et al. (2001), Peake et al. (2000), Boers et al. (1996), Guha et al. (2012)
MSC	CD73/90/105, Vimentin, Prolys-4-hydroxylase	Yes	Adipocyte, chondrocyte, osteocyte, myofibroblasts, and fibroblasts	Bone marrow, circulation system, and pulmonary interstitium	Lama et al. (2007), Sabatini et al. (2005), Sinclair et al. (2013), Martin et al. (2008)

NEB neuroepithelial body, *BADJ* bronchioalveolar duct junction, *PNEC* pulmonary neuroendocrine cell, *MSC* mesenchymal stem cell, *AECI* alveolar type I epithelial cells, *AECII* alveolar type II epithelial cells

[a]Occurs only in the setting of basal cell ablation

[b]Findings from studies using different transgenic models conflict with respect to club and ciliated lineages derived cells from PNECs

3.3 Potential Stem Cells in the Adult Lung

Stem/progenitor cells are crucial for development, tissue homeostasis, and injury repair in the lung. Studies using epithelial reconstitution assays, murine injury models, and lineage tracing approaches have identified several region-specific stem/progenitor cell populations in the adult lung of mice and humans. Basal cells in the proximal airways, variant club cells in bronchioles, bronchoalveolar stem cells (BASCs) in BADJs, and a subset of AECII in alveolar spaces have all been identified as stem/progenitor cells (Table 3.1).

In the trachea and main-stem bronchi, basal cells are the principal stem cells involved in homeostasis and injury repair and have the capacity to generate all the major cell types found in the proximal airway, including basal, ciliated, goblet, and granular secretory cells (including club cells) (Hong et al. 2004a, b; Hajj et al. 2007; Rock et al. 2009; Schoch et al. 2004; Cole et al. 2010; Engelhardt et al. 1995). The intermediate cells in the human proximal airway are so named because they are generally thought to represent an intermediate state of differentiation from basal cells and to serve as a transient amplifying cell population with the capacity to differentiate into ciliated and goblet cells (Engelhardt et al. 1995; Mercer et al. 1994). Intermediate cells do not exist in the mouse proximal airway, potentially because of the less pseudostratified nature of their smaller diameter airways. Of note, studies of murine lung injury involving BrdU labeling demonstrated that label-retaining cells (LRCs) reside predominantly in the ducts of SMGs, suggesting that these glands serve as a stem cell niche in the proximal airway (Xie et al. 2011; Borthwick et al. 2001; Engelhardt et al. 1995; Engelhardt 2001; Rock et al. 2009). Importantly, the SMG-localized LRCs have the capacity to undergo sequential rounds of cell division despite their slowly cycling phenotype (Xie et al. 2011; Lynch and Engelhardt 2014). Nevertheless, because lineage tracing of glandular LRCs has not yet been possible, the ability of these stem cells to produce specific airway cell types remains unclear. Several cellular markers have been utilized to identify and isolate basal cells. These include cytokeratin 5 (CK5), cytokeratin 14 (CK14), and aquaporin 3 (Rock et al. 2009, 2010; Schoch et al. 2004). Using a CK5-CreER^{T2} transgenic mouse line, Rock et al. further demonstrated that basal cells are capable of differentiating into club and ciliated cells, both at steady state and during injury repair (Rock et al. 2009). In addition, they identified nerve growth factor receptor (NGFR) and integrin α6 (ITGA6, also called CD49f) as markers on the surfaces of isolated human basal stem cells (Rock et al. 2009). Similarly, Ghosh et al. identified a CD49f^{bright}/Sca-1^{+}/ALDH1^{+} (Aldehyde dehydrogenase 1) subset of tracheal basal cells as region-specific stem cells, and demonstrated that these cells could generate niches in vitro and contribute to tracheal epithelial maintenance and injury repair (Ghosh et al. 2011). These studies suggested that basal cells play key roles in both homeostasis and injury repair of the proximal airway.

In the intralobar bronchiolar airways, a subset of the variant club cells that express club cell secretory protein (CCSP, also called Scgb1a1) but not CyP450-2F2 (CCSP^{+}, CyP450-2F2^{-}) can self-renew and produce both club cells and ciliated cells (Hong et al. 2001; Rawlins et al. 2009b; Reynolds and Malkinson 2010; Xing et al. 2012; Guha et al. 2012). This CCSP$^+$/CyP450-2F2$^-$ subset was also found at the BADJ of distal bronchioles, where it contributed to airway epithelial regeneration following naphthalene-mediated depletion of CyP450-2F2$^+$ club cells (Giangreco et al. 2002). Kim et al. subsequently identified BASCs as a subpopulation of cells that express CCSP and pro-surfactant protein C (SPC) and serve as region-specific stem cell at the BADJ (Kim et al. 2005). Using naphthalene- and bleomycin-induced murine models of lung injury repair, Kim et al. further demonstrated that these cells possessed the capacity to self-renew and to produce differentiated epithelial cells in vivo, and that BASCs could differentiate into club cells and alveolar epithelial cells in an ex vivo clonogenic assay (Kim et al. 2005). Conversely, an in vivo lineage tracing experiment using a CCSP(Scgb1a1)-CreER™ knock-in mouse line revealed that club cells generated daughter club cells and ciliated cells but not alveolar cells following hypoxia-induced lung injury (Rawlins et al. 2009b). However, subsequent lineage tracing studies following alveolar injury by influenza infection and bleomycin exposure support the finding that CCSP-expressing stem/progenitors can give rise to AECI and AECII cells (Zheng et al. 2013; Rock et al. 2011). These injury-dependent influences on BASC-derived lineages suggest that either specific injury signals may invoke different responses and/or that multiple subsets of BASCs exist with different capacities for differentiation. The later hypothesis is consistent with findings suggesting that BASCs in the distal airways might include a heterogeneous population of progenitor cells (Teisanu et al. 2009, 2011; Chen et al. 2012). A study by Teisanu et al. classified club cells with the surface antigen profile CD45$^-$/CD31$^-$/CD34$^-$/EpCAM$^+$/Sca-1^{low} into two subgroups based on their autofluorescence (AF) profiles and suggested that club cells in the AF^{low} population are naphthalene resistant, whereas their AF^{high} counterparts were not (Teisanu et al. 2011). Indeed, mice that were exposed to naphthalene showed significantly greater proliferation in AF^{low} club cells compared to AF^{high} club cells, and conversely, mice exposed to ozone showed significantly greater proliferation in the AF^{high} club cell fraction compared to AF^{low} club cells (Teisanu et al. 2011). McQualter et al. demonstrated that an EpCAM$^+$/Sca-1^{low}/Integrin $\alpha6\beta4^+$/CD24^{low} fraction of epithelial stem/progenitor cells was capable of self-renewing and differentiating into a variety of airway epithelial lineages, including alveolar epithelial cells (McQualter and Bertoncello 2012). These studies provide evidence that BASCs play key roles in the repair of injury to both bronchiolar and alveolar cells, as well as in homeostasis.

In the pulmonary alveolus, surfactant-producing AECII cells have long been recognized as stem/progenitor cells for the squamous AECI cells in the adult lung (Adamson and Bowden 1974; Evans et al. 1975). In vitro assays of cell proliferation and clonogenicity, as well as in vivo analyses following epithelial injury and lineage tracing, have produced mounting evidence that a subset of AECIIs have the capacity to proliferate and restore the alveolar epithelium by producing either new AECII cells or their squamous AECI counterparts (Reddy et al. 2004; Barkauskas et al. 2013; Fujino et al. 2011). Equally noteworthy were findings suggesting that integrin $\alpha6\beta4$ is a biomarker for a subset of stem/progenitor cells in alveolar epithelia;

SPC$^-$/integrin α6β4$^+$ cells were found resident in the alveolar epithelia, where they were able to regenerate SPC$^+$ AECII cells (Chapman et al. 2011). The notion that AECII cells are region-specific stem/progenitors was recently confirmed by work from the Hogan laboratory, which employed a genetic SPC-labeled lineage tracing assay and an in vitro 3D culture model. This study produced convincing evidence that AECII cells were able to maintain the homeostasis of alveolar epithelia during both steady-state turnover and injury repair (Barkauskas et al. 2013). Several strategies that rely on biomarkers to identify and isolate adult lung stem/progenitor cells are listed in Table 3.2.

3.4 Stem Cell Niches in the Adult Lung

As discussed above, a vast body of evidence has demonstrated that distinct stem/progenitor cell populations reside in specific anatomical niches (Fig. 3.1), where diverse cell types and signals coordinate the behavior of stem cells during homeostasis and following injury. Stem cell niches are discrete microenvironmental units within a tissue that can provide one or more of the following features important for stem cell control: a unique extracellular matrix; supporting cell types; unique innervation and nearby vasculature; and diffusible factors that allow stem cells to maintain a capacity to self-renewal and control their proliferation and differentiation in the setting of injury (Fuchs et al. 2004). The anatomical sites of airway stem cell niches are typically epithelial structures associated with these unique features described above (e.g., innervation, support cells). Although much remains to be learned about how components of airway niches coordinate stem/progenitor cell behavior and phenotype, data from organ systems that have been studied more extensively suggest that they are likely important in the lung as well.

In the following discussion of stem cell niches in the airway, we focus on the unique anatomic and biologic properties of each niche within a particular region of the lung, and on how these features may contribute to repair following injury. In particular, we concentrate on studies of slowly cycling stem/progenitor cells in the mouse lung, since nucleotide label retention has been one of the most commonly used methods for tracking the anatomic locations of stem cell niches in the lung.

3.4.1 The Tracheal Surface Airway Epithelium

Within the tracheal SAE, subsets of basal cells are thought to be the major stem/progenitor cells. Following injury, LRCs tend to cluster within intercartilaginous zones of the distal trachea and larger bronchi along the basal lamina of the surface epithelium (Borthwick et al. 2001). These intercartilaginous zones tend to be sites of high blood vessel concentration and nerve penetration to the epithelium (Baker et al. 1986; McDonald 1988). These features are likely important biologic

Table 3.2 Experimental strategies used to isolate and characterize progenitor cells based on phenotypic characteristics

Input	Phenotypic markers[a,b]	Assay	Differentiation/Proliferation potential	References
Human nasal polyps	$CD151^+/F3^+$	In vitro ALI, ex vivo xenograft assays	Fully differentiated and functional airway epithelium including basal, columnar, goblet, and ciliated cells	Hajj et al. (2007)
Human fetal airway xenografts	$AQP3^-$	Ex vivo xenograft assay	Mature epithelia including ciliated, secretory, and basal cells within 4–6 weeks (compared to 6–20 weeks in $AQP3^{hi}$ cells) and by 20 weeks both $AQP3^-$ and $AQP3^{hi}$ cells generated mature epithelia including glands containing mucous and secretory cells	Avril-Delplanque et al. (2005)
Human trachea	$NGFR^+/ITGA6^+$	In vitro 3D colony-forming assay	Colonies containing basal and ciliated cells	Rock et al. (2009)
Human trachea and bronchi	$CD45^-/$ $H33342^{Side\ Pop}$	In vitro limiting dilution assay, ALI culture	Cobblestone colonies of basal cells on plastic and basal, columnar, and goblet cells in ALI cultures	Hackett et al. (2008)
Human lung	$CD45^-/CD31^-/$ $CD73^-/CD34^-/$ $CDH1^+/LGR6^+$	In vitro limiting dilution assay, in vivo kidney organoid assay	Colonies containing club, AECII, and AECI cells and single cells from both freshly isolated and clonally expanded cells generated clones in mouse kidney capsule grafts	Oeztuerk-Winder et al. (2012)
Human lung	$CD45^-/CD31^-/$ $EpCAM^+/HTII\text{-}280^+$	In vitro 3D colony-forming assay	Colonies containing AECII cells	Barkauskas et al. (2013)
Mouse trachea	$GSI\text{-}A_3B^+/NGFR^+$	In vitro 3D colony-forming assay	Colonies containing basal and ciliated cells	Rock et al. (2009)
Mouse trachea	$CD45^-/CD31^-/$ $Ly76^-/Ly6a^+/Itga6^{hi}/$ $Aldefluor^+$	In vitro colony forming assay(s) in vivo airway injury model	Rimmed clones in a co-culture system with irradiated feeder cells; basal, ciliated, and club cells in a mixed ALI culture system; proliferated in vivo following naphthalene injury	Ghosh et al. (2011)
Mouse tracheal surface epithelium	$Trop2^+/Itga6^+/$ $Aldefluor^+$	In vitro 3D colony-forming assay	Luminal colonies containing basal, ciliated, columnar, serous, and mucous cells in 3D cultures	Hegab et al. (2011, 2014)
Mouse tracheal glandular epithelium	$Trop2^+/Itga6^+/$ $Aldefluor^+$	In vitro 3D colony-forming assay, in vivo fat pad assay	Dense colonies containing basal, columnar, serous, and mucous cells in 3D cultures, and basal, serous, mucous, and myoepithelial cells in an in vivo fat pad assay	Hegab et al. (2011, 2014)

Mouse lung	CD45^{-}/CD31^{-}/EpCAM^{+}/Itgb4^{+}/LysoTracker^{+}	In vitro 3D colony-forming assay	Colonies containing AECII cells	Van der Velden et al. (2013)
Mouse lung	CD45^{-}/CD16^{-}/CD32^{-}/EpCAM^{hi}/Itgb4^{+}	In vitro 3D colony-forming assay, in vivo kidney organoid assay	Colonies containing AECII, club, and cell expressing both SPC and CCSP in a 3D colony-forming assay and within a mixed kidney capsule organiod assay	Chapman et al. (2011)
Mouse lung	CD45^{-}/CD31^{-}/Ly6a^{+}/CD34^{+}	In vitro limiting dilution assay	Mesenchymal colonies with osteogenic and chondrogenic potential (data not shown)	McQualter et al. (2009)
Mouse lung	CD45^{-}/CD31^{-}/EpCAM^{hi}/Itga6^{+}/Itgb4^{+}/CD24^{lo}	in vitro 3D colony-forming assay	Cystic colonies containing ciliated, club, goblet, basal, and secretory cells, saccular colonies containing club, AECII cells, and mixed colonies containing mucous, ciliated, and AECII cells	McQualter et al. (2010)
Transgenic mouse trachea	CK5-GFP^{+}/GSI-A_3B^{+}	In vitro 3D colony-forming assay, in vivo lineage tracing	Colonies containing ciliated and basal cells; lineage-labeled CK5-expressing cells generated basal, club, and ciliated cells both at steady state and following SO_2 injury	Rock et al. (2009)
Transgenic mouse lung	CD45^{-}/CD31^{-}/CD34^{-}/EpCAM^{+}/SFTPC-GFP$^{-/lo/hi}$	In vitro 3D colony-forming assay	GFP^{neg} cells generated spheroid colonies containing AECII, club, serous, ciliated, and basal cells, GFP^{lo} cells generated irregularly shaped colonies containing AECII, club, and ciliated cells, and GFP^{hi} cells generated dense colonies containing AECII cells	Chen et al. (2012)
Transgenic mouse lung	CD45^{-}/CD31^{-}/EpCAM^{+}/Sftpc-CreER; Rosa-Tm^{+}	In vitro 3D colony-forming assay, in vivo lineage tracing	Colonies containing AECII and AECI cells; lineage-labeled Sftpc-expressing cells proliferate at steady state and following injury	Barkauskas et al. (2013)

[a]Official gene symbols are substituted for CD49f (Itga6), Sca-1 (Ly6a), TER119 (Ly76), E-Cad (CDH1), and Tissue Factor (F3)

[b]GSI-A_3B *Griffonia simplicifolia* isolectin A_3B has been shown to bind to airway basal cells, Aldefluor substrate reports enzymatic aldehyde dehydrogenase activity, LysoTracker (LysoTracker Green DND-26 (Invitrogen)) is a dye that stains acidic intracellular compartments

components and may function to localize stem/progenitor cells within this niche at homeostasis; they could also mediate injury responses that direct changes in stem/progenitor cell behavior. It has been suggested that these intercartilaginous zones enable a subset of surface basal cells to maintain multipotency within the mouse proximal airway (Borthwick et al. 2001; Engelhardt 2001; Liu and Engelhardt 2008; Rock et al. 2009, 2010; Hong et al. 2004b). Subpopulations of basal cells with the capacity for self-renewal and differentiation have also been described by others, based on clonogenic assays and lineage tracing studies (Cole et al. 2010; Hajj et al. 2007; Hong et al. 2004a; Schoch et al. 2004; Rock et al. 2009). For example, Rock et al. recently identified a subset of basal cells that were marked with $p63^+/NGFR^+/CK5^+$ that were able to self-renew and to generate luminal daughter cells within an in vitro 3D tracheosphere assay (Rock et al. 2009). Lineage tracing studies in mice expressing a CK5-promoter driven CreER transgene further demonstrated that CK5-expressing basal cells could give rise to ciliated and club cells in the tracheobronchial airways, both at steady state and following injury (Rock and Hogan 2011; Rock et al. 2010).

3.4.2 The Tracheal Submucosal Glands

A link between SMGs and stem/progenitor cells in the SAE was first discovered through retroviral lineage tracing experiments using human airway epithelial cells and a rat trachea xenograft model (Engelhardt et al. 1995). In these studies, retrovirally tagged human tracheobronchial epithelial cells were expanded in a denuded rat trachea that had been subcutaneously implanted into nu/nu mice. These cultures contained diverse populations of airway cells that were capable of clonal expansion within the xenografted airway. Phenotypic analysis of clones established a working model for progenitor/progeny relationships in the adult human proximal airway. Although seven clonal classes were discovered, the most abundant clone phenotype was multipotent and contained basal, intermediate, ciliated, and goblet cells. These multipotent clones were also the largest in size, supporting the hypothesis that they were derived from stem/progenitors with the largest capacity for expansion. Notably, SMGs also formed within these xenografts, and lineage tracing revealed that they were always associated with multipotent clones on the SAE. Expansion of basal cell progenitors in vitro prior to seeding into xenografts reduced the complexity of possible outcomes in clone phenotypes observed, giving rise to multipotent clones almost exclusively. These findings suggested that a small subset of basal cells are multipotent for SAE cell types and also have the capacity to form SMGs (Engelhardt et al. 1995). Additionally, these studies demonstrated with early passage primary human airway epithelial cells that a diverse range of progenitors exist in the human proximal airway with unipotent and bipotent capacities for differentiation. Later, clonal analysis in mice expressing a CK14-CreER transgene confirmed these findings and demonstrated that at least two subsets of basal cells exist with either unipotent or multipotent capacity for differentiation (Hong et al. 2004b). Consistent with the finding that a subset of adult airway stem cells have the capacity to generate SMGs, LRCs localized within glands or glandular ducts following tracheal epithelial regeneration following injury (in response to both SO_2 and detergent treatment), suggesting that a subset of slowly cycling glandular epithelial cells are tissue-specific stem/progenitor cells and are capable of regenerating the airway epithelium after injury (Borthwick et al. 2001; Engelhardt 2001). Based on dual nucleotide sequential labeling experiments, these glandular LRCs retain the capacity to divide following repeated injury but remain slowly cycling (Xie et al. 2011; Lynch and Engelhardt 2014). Glandular LRCs make up a small fraction of total glandular cells (0.39 % ± 0.03 %) at 90 days after injury and only about 10 % of glandular LRCs reenter the cell cycle following a second injury and remain slowly cycling (Xie et al. 2011). Thus, if slowly cycling glandular LRCs are a stem cell, they represent approximately 0.04 % of total glandular cells.

Similar observations on glandular-derived stem cells were made in a murine model of hypoxic-ischemic injury (Hegab et al. 2011, 2012b). In these studies, Hegab et al. found that the SMG duct cell population included stem/progenitor cells that shared phenotypic features with surface airway basal cells and were resistant to epithelial injury in the context of tracheal hypoxic-ischemic injury. In vitro and ex vivo assays carried out with epithelial stem/progenitor cells isolated from the SMG duct have demonstrated that these cells are capable of self-renew and can generate several cell types found in the SAE and SMGs (Hegab et al. 2011, 2012a, 2014). Furthermore, in vitro colony forming assays using epithelia isolated from the gland-rich proximal region of the mouse trachea have revealed that these cells have a higher potential for proliferation than their counterparts from the gland-free distal trachea (Xie et al. 2011).

Cumulatively, these studies provide convincing evidence that the SMGs serve as a stem/progenitor cell niche for the proximal airway. The positioning of stem/progenitor cell niches within SMGs likely has biologic significance beyond simply the maintenance of glandular cell types. For example, SMGs are less exposed to the external environment and pathogens that threaten the lung, thus glandular stem cell niches are more protected. Additionally, SMGs are highly innervated (Nadel 1983; Wine 2007) and their secretions are regulated (i.e., enhanced) in response to injury of the SAE (Xie et al. 2011). Given that SMGs play an important role in airway innate immunity by producing secretions that regulate the composition of fluid, electrolytes, mucus, and antibacterial factors at the airway surface (Wine and Joo 2004; Wang et al. 2001; Dajani et al. 2005), it is not surprising that the regulation of glandular secretions following airway insults might be coordinated with the mobilization of glandular stem/progenitor cells that regenerate the airway surface. Interestingly, studies of cystic fibrosis suggest that defects in glandular secretions caused by the lack of the cystic fibrosis transmembrane conductance regulator (CFTR) chloride channel alter the SMG stem/progenitor cell niche by dysregulating the calcitonin gene-related peptide (CGRP) neuropeptide (Xie et al. 2011). Such studies have demonstrated that, in mice deficient for CFTR, slowly cycling LRCs relocate from SMGs to the SAE following naphthalene injury and that this is accompanied by a redistribution of highly proliferative stem/progenitor cells from proximal gland-rich regions of the trachea to regions of the SAE that lack glands. CGRP is induced in SMGs following airway injury and leads to the induction of gland secretions by activating CFTR, however, CGRP is constitutively upregulated in SMGs of cystic fibrosis humans, ferrets, pigs, and mice and this altered neuroendocrine signaling is thought to be the basis of stem cell niche dysfunction (Xie et al. 2011). Such findings emphasize the plastic nature of airway stem/progenitor cell niches.

3.4.3 The Neuroepithelial Bodies in the Intralobar Airways

Pulmonary NEBs are found within the epithelia of intrapulmonary airways (bronchi and bronchioles) and contain specialized CGRP-expressing pulmonary neuroendocrine cells (PNECs) (Cutz et al. 2013). The NEBs are extensively innervated and intrapulmonary bronchial capillaries are fenestrated at these sites (Lauweryns et al. 1972, 1974; Cutz et al. 2013). Both innervation and NEBs have been shown to be highest at sites of bifurcation in the airway (Elftman 1943; Cutz et al. 2013). Given these unique anatomic characteristics of NEBs, it is not surprising they appear to be stem/progenitor cell niches for the intralobar airways during both normal cell turnover and injury repair (Hong et al. 2001; Reynolds et al. 2000a).

At least two distinct cell types exist within NEBs—CCSP$^+$/CyP450$^-$ variant club cells and the above-mentioned CGRP$^+$ PNECs (Reynolds et al. 2000a, b). Lineage tracing studies using murine models have demonstrated that both the PNECs, and variant club cells associated with NEBs, have the capacity to self-renew and to differentiate into club and/or ciliated cells following naphthalene injury (Song et al. 2012; Hong et al. 2001; Xing et al. 2012; Guha et al. 2012). A subpopulation of naphthalene-resistant CCSP$^+$/CyP450$^-$ variant club cells was identified as stem/progenitor cells in distal airways (Reynolds et al. 2000a, b; Hong et al. 2001; Rawlins et al. 2009b). Guha et al. also recently identified a distinct subset of CCSP^{low}/CyP450$^-$/Scgb3a2$^+$-expressing club cells resident in NEBs for which Notch signals and the transcription factor TTF1 (Nkx2.1) played a crucial role in determining the secretory cell fate in developing murine airways, supporting the idea that the NEB microenvironment is a stem cell niche for variant club-like stem cell precursors (Guha et al. 2012). Notch signaling in club cells was also found in the adult lung, in which Notch1 was required for repopulating lost club cells following airway epithelial injury (Xing et al. 2012). Previously, Hong et al. ablated CCSP-expressing cells—including club and variant club cells—by treating transgenic mice that expressed thymidine kinase from a CCSP promoter with ganciclovir, and then studied the lineage potential of the CGRP-expressing PNEC progenitors (Hong et al. 2001). The group found that, although PNECs replicated following club cell ablation, they were unable to regenerate CCSP-expressing club cells or ciliated cells, suggesting that PNECs are not competent to regenerate the mouse bronchiolar epithelium (Hong et al. 2001). However, Song et al. obtained different results using another approach to tag the PNEC lineage (Song et al. 2012). Specifically, these investigators introduced a CreER transgene into the CGRP locus and used this transgene to lineage trace or ablate CGRP-expressing PNECs. They found that fate mapped CGRP-expressing PNECs could generate both club and ciliated cells following naphthalene injury, but that when PNECs were ablated using Cre-activated diphtheria toxin (DTA), ciliate cells were not regenerated (Song et al. 2012). The apparent discrepancies between the outcomes in Hong et al. and Song et al., which suggest that PNECs are either unipotent or multipotent, respectively, are likely related to the methods of airway injury used in these studies and may reflect a high level of plasticity in distal airway progenitors.

3.4.4 The Bronchioalveolar Duct Junctions

Mounting evidence suggests that within the terminal bronchioles the BADJ is a niche for BASCs that are capable of regenerating both bronchiolar and alveolar epithelial cell lineages following injury (Zacharek et al. 2011; Kim et al. 2005; Regala et al. 2009; Tropea et al. 2012; Zheng et al. 2013; Rock et al. 2011). In vitro studies suggested that BASCs have the capacity to differentiate into club, AECII, and AECI cells. However, lineage tracing studies using CCSP-CreER knock-in mice did not substantiate these findings in vivo, at least in the cases of naphthalene- and hyperoxia-induced acute injury to the lung; the CCSP-expressing progenitors did not give rise to the alveolar epithelium in these contexts (Rawlins et al. 2009a). Nevertheless, when other models of alveolar injury (influenza infection or bleomycin exposure) were tested, lineage traced CCSP-expressing progenitors gave rise to labeled AECI and AECII cells (Zheng et al. 2013; Rock et al. 2011; Tropea et al. 2012). Thus, the contribution of BASCs to alveolar injury repair may depend on injury-specific regulatory factors within the BADJ microenvironment.

3.4.5 The Alveoli

The terminal end of the respiratory tree is composed of alveolar sacs, whose cellular composition includes AECI and AECII cells, capillaries, and lung-resident mesenchymal stromal cells (lrMSCs). AECII cells, which produce surfactant protein C (SPC), have been suggested to serve as a stem/progenitor cells from which AECI and AECII cells are regenerated after alveolar injury (Adamson and Bowden 1974; Barkauskas et al. 2013). However, a recent study that used an SPC-CreER mouse model to map the fates of AEC cells following bleomycin injury found that the majority of AECII cells in fibrotic areas did not arise from preexisting SPC-expressing AECII cells (Chapman et al. 2011), but rather from a subset of previously unrecognized AECs. These cells expressed the laminin receptor integrin $\alpha6\beta4$ but not CCSP or SPC and expanded to form a differentiated alveolar-like epithelium containing CCSP-expressing cells and SPC-expressing AECII cells in an ex vivo kidney capsule model (Chapman et al. 2011). By contrast, in a more recent study by Barkauskas et al., SPC-expressing AECII cells were found to self-renew and differentiate into AECI cells, both at steady state and following alveolar injury (Barkauskas et al. 2013). These investigators went on to show, using an in vitro differentiation 3D culture model, that individual AECII cells produced self-renewing "alveolospheres" that comprised both AECI and AECII cells. Of note, co-culturing AECII cells with a PDGFRα-expressing subpopulation of lung mesenchymal cells significantly increased the efficiency of formation of self-renewing alveolospheres. Thus, these PDGFRα-expressing lung stromal cells, which include alveolar fibroblasts and lipofibroblasts, appear to be components of the AECII stem/progenitor cell niche within the alveolus (Barkauskas et al. 2013). Taken together with studies on the BADJ, these studies suggest that multiple stem/progenitor cell niches in the distal lung may contribute to repair of the alveolus following injury, and that the active niche in the context of homeostasis resides in the alveolus.

3.5 MSCs and Stem/Progenitor Cell Niches in the Adult Lung

Increasing evidence indicates that MSCs are important components of epithelial stem/progenitor niches in the adult lung, and that they play an essential role in orchestrating epithelial regeneration during both homeostasis and injury repair (McQualter et al. 2010, 2013; Volckaert et al. 2011, 2013; Gong et al. 2014). lrMSCs can be isolated from bronchioalveolar lavage (BAL) fluid (Lama et al. 2007) and lung tissue (Ricciardi et al. 2012) using the techniques of differential plastic adherence and enzymatic dissociation, respectively. Studies evaluating in vitro co-culture models have demonstrated that lrMSCs are not only key for the proliferation and differentiation of epithelial stem cells (McQualter et al. 2010) but also are able to differentiate into AECII cells when co-cultured with AECII cells in a transwell model (Gong et al. 2014). In this context, lrMSCs can contribute to lung repair by secreting FGF-10 and TGF-β, and thereby promoting re-epithelialization (McQualter et al. 2010, 2013; Volckaert et al. 2011, 2013). In the developing lung, FGF-10 is central to regulating BMP, Wnt, and Shh signaling pathways, which are responsible for coordinating differentiation in this context (Morrisey and Hogan 2010). In the adult lung, TGF-β signaling by mesenchymal cells regulates the secretion of FGF-10 and provides a cue that is necessary for epithelial regeneration (McQualter et al. 2010, 2013). Two subpopulations of lrMSCs were found—$CD166^-$ lrMSCs, which have the capacity to differentiate into lipofibroblast and myofibroblast cell types and to support epithelial stem cell proliferation and differentiation in vitro, and $CD166^+$ lrMSCs, which are limited to producing cells of the myofibroblast lineage and fail to support epithelial stem cell proliferation and differentiation in vitro (McQualter et al. 2013). Studies by Volckaert et al., which used a naphthalene-based model of lung injury, have identified lrMSCs as important components of the bronchiolar stem/progenitor cell niche. These studies implicate parabronchial smooth muscle cells (PSMCs) in the regulation of naphthalene-resistant club cells at the BADJ and adjacent to NEBs by activating Wnt/FGF-10 signaling (Volckaert et al. 2011). In addition, the Wnt target gene c-Myc was found to be critical for both activating the PSMC niche and inducing FGF-10 expression (Volckaert et al. 2013). Mechanistically, FGF-10 secreted by PMSCs activated Notch signaling and Snail expression in naphthalene-resistant club cells, and subsequently initiated the repair process by promoting club cell proliferation and differentiation (Volckaert et al. 2011, 2013).

3.6 Stem/Progenitor Cell Niches and Cancer-Initiating Stem Cells in the Lung

Lung cancer is a heterogeneous disease in terms of its phenotypic diversity and anatomical sites of origin in the airways. Lung cancers can be subdivided into two major groups—small cell lung cancers (SCLCs) and non-small cell lung cancers (NSCLCs). SCLC is characterized by neuroendocrine cell morphology and accounts for ~15 % of lung malignancies; NSCLC accounts for the remaining cases (~85 %) and can be further subdivided into three distinct histological subtypes: squamous cell carcinoma (SCC), adenocarcinoma, and large cell carcinoma (Travis et al. 2013). The morphologies and molecular properties (e.g., activation of the Wnt, Hedgehog (Hh), and Notch signaling pathways) of each subtype have led to the hypothesis that lung cancers are derived from stem cells in the lung (Alamgeer et al. 2013; Lundin and Driscoll 2013). Currently, it is possible to isolate lung cancer stem cells (CSCs) based on the expression of several tumor markers, including aldehyde dehydrogenase (ALDH), CD133, CD44, and the ability to efflux certain dyes such as Hoechst (Alamgeer et al. 2013).

Although lung CSCs have not been as well characterized as other tumors, the current understanding of the phenotypes of region-specific airway epithelial stem/progenitor cells has led to the hypothesis that cancers initiate at specifically those anatomic locations in which stem cell niches reside. This hypothesis is supported, in part, by findings from animal models of lung cancer; the most common sites of origin for different lung cancer types correlate with distinct, region-specific airway stem/progenitor cell niches (Kitamura et al. 2009; Succony and Janes 2014; Leeman et al. 2014). Notably, mouse adenocarcinomas are characterized by the expression of the transcription factor Nkx2.1 (TTF1), CCSP, and SPC and arise from BADJs, suggesting that cancer-initiating progenitor cells arise from within club or AECII stem/progenitor-cell populations (Kim et al. 2005; Imielinski et al. 2012; Travis et al. 2013; Xu et al. 2012). Lung SCCs are characterized by differentiation into squamous cells with a basal cell phenotype, and can be subdivided based on mRNA expression levels, into classes of cells that resemble basal cell progenitors in the SAE or SMGs (Wilkerson et al. 2010), two sites at which stem cell niches exist. Similarly, SCLCs are found predominantly in the intermediate airways and are characterized by the expression of a range of neuroendocrine cell markers, including CGRP (Song et al. 2012; Kelley et al. 1994; Carraresi et al. 2006). Thus, SCLCs may originate from CGRP-expressing progenitors within the NEB stem/progenitor cell niche. This hypothesis is further supported by experiments using transgenic mice deficient for Rb1 and p53 in specifically the club cells, AECII cells, or PNECs; in these animals, SCLCs arise most frequently from NEB-resident PNECs (Sutherland et al. 2011). These findings suggest that mutations that dysregulate airway stem/progenitor cell niches play important roles in selecting lung CSCs that outcompete other progenitors and promote cancer initiation, metastasis, and chemoresistance (Takebe and Ivy 2010; Chen et al. 2014).

3.7 Perspective Summary

The results from in vitro and in vivo clonogenic assays and lineage tracing analyses in various experimental models have suggested that region-specific stem/progenitor cells reside within distinct niches in the lung. At least five unique epithelial stem/progenitor cell niches have been proposed in the lung, and the signals that induce the expansion of progenitors and specification of daughter cells from each of these niches appear to be diverse. Moreover, in the mouse models that have been studied, this often depends on the type of injury. The available data also suggest that some progenitors impart a high level of lineage plasticity to the lung, with committed differentiated cell types capable of adopting stem cell properties and reestablishing stem cell niches in the setting of severe airway injury. Given that abnormalities in lung stem/progenitor cell niches can occur in the context of genetic disease, viral infection, and lung cancer, it will be important to define the cues that are intrinsic to lung cells, as well as those that are extrinsic (i.e., present in the unique regional niches of the lung). Such knowledge is expected to provide effective new avenues for the treatment of lung diseases.

Acknowledgment This work was supported by grants from the NIH (DK047967) and the University of Iowa Center for Gene Therapy (DK054759).

Conflict of Interest Statement: None of the authors has a financial relationship with a commercial entity that has an interest in the subject of this manuscript.

References

Adamson IY, Bowden DH (1974) The type 2 cell as progenitor of alveolar epithelial regeneration. A cytodynamic study in mice after exposure to oxygen. Lab Invest 30(1):35–42

Alamgeer M, Peacock CD, Matsui W, Ganju V, Watkins DN (2013) Cancer stem cells in lung cancer: evidence and controversies. Respirology 18(5):757–764. doi:10.1111/resp.12094

Avril-Delplanque A, Casal I, Castillon N, Hinnrasky J, Puchelle E, Peault B (2005) Aquaporin-3 expression in human fetal airway epithelial progenitor cells. Stem Cells 23(7):992–1001. doi:10.1634/stemcells.2004-0197

Baker DG, McDonald DM, Basbaum CB, Mitchell RA (1986) The architecture of nerves and ganglia of the ferret trachea as revealed by acetylcholinesterase histochemistry. J Comp Neurol 246(4):513–526. doi:10.1002/cne.902460408

Barkauskas CE, Cronce MJ, Rackley CR, Bowie EJ, Keene DR, Stripp BR, Randell SH, Noble PW, Hogan BL (2013) Type 2 alveolar cells are stem cells in adult lung. J Clin Invest 123(7):3025–3036. doi:10.1172/JCI68782

Boers JE, den Brok JL, Koudstaal J, Arends JW, Thunnissen FB (1996) Number and proliferation of neuroendocrine cells in normal human airway epithelium. Am J Respir Crit Care Med 154(3 Pt 1):758–763

Boers JE, Ambergen AW, Thunnissen FB (1999) Number and proliferation of clara cells in normal human airway epithelium. Am J Respir Crit Care Med 159(5 Pt 1):1585–1591

Borthwick DW, Shahbazian M, Krantz QT, Dorin JR, Randell SH (2001) Evidence for stem-cell niches in the tracheal epithelium. Am J Respir Cell Mol Biol 24(6):662–670

Carraresi L, Martinelli R, Vannoni A, Riccio M, Dembic M, Tripodi S, Cintorino M, Santi S, Bigliardi E, Carmellini M, Rossini M (2006) Establishment and characterization of murine small cell lung carcinoma cell lines derived from HPV-16 E6/E7 transgenic mice. Cancer Lett 231(1):65–73. doi:http://dx.doi.org/10.1016/j.canlet.2005.01.027

Chapman HA, Li X, Alexander JP, Brumwell A, Lorizio W, Tan K, Sonnenberg A, Wei Y, Vu TH (2011) Integrin alpha6beta4 identifies an adult distal lung epithelial population with regenerative potential in mice. J Clin Invest 121(7):2855–2862. doi:10.1172/JCI57673

Chen H, Matsumoto K, Brockway BL, Rackley CR, Liang J, Lee JH, Jiang D, Noble PW, Randell SH, Kim CF, Stripp BR (2012) Airway epithelial progenitors are region specific and show differential responses to bleomycin-induced lung injury. Stem Cells 30(9):1948–1960. doi:10.1002/stem.1150

Chen WJ, Ho CC, Chang YL, Chen HY, Lin CA, Ling TY, Yu SL, Yuan SS, Chen YJ, Lin CY, Pan SH, Chou HY, Chen YJ, Chang GC, Chu WC, Lee YM, Lee JY, Lee PJ, Li KC, Chen HW, Yang PC (2014) Cancer-associated fibroblasts regulate the plasticity of lung cancer stemness via paracrine signalling. Nat Commun 5:3472. doi:10.1038/ncomms4472

Cole BB, Smith RW, Jenkins KM, Graham BB, Reynolds PR, Reynolds SD (2010) Tracheal Basal cells: a facultative progenitor cell pool. Am J Pathol 177(1):362–376. doi:10.2353/ajpath.2010.090870

Cutz E, Pan J, Yeger H, Domnik NJ, Fisher JT (2013) Recent advances and controversies on the role of pulmonary neuroepithelial bodies as airway sensors. Semin Cell Dev Biol 24(1):40–50. doi:10.1016/j.semcdb.2012.09.003

Dajani R, Zhang Y, Taft PJ, Travis SM, Starner TD, Olsen A, Zabner J, Welsh MJ, Engelhardt JF (2005) Lysozyme secretion by submucosal glands protects the airway from bacterial infection. Am J Respir Cell Mol Biol 32(6):548–552

Elftman AG (1943) The afferent and parasympathetic innervation of the lungs and trachea of the dog. Am J Anat 72:1–27

Engelhardt JF (2001) Stem cell niches in the mouse airway. Am J Respir Cell Mol Biol 24(6):649–652

Engelhardt JF, Schlossberg H, Yankaskas JR, Dudus L (1995) Progenitor cells of the adult human airway involved in submucosal gland development. Development 121(7):2031–2046

Evans MJ, Cabral LJ, Stephens RJ, Freeman G (1975) Transformation of alveolar type 2 cells to type 1 cells following exposure to NO2. Exp Mol Pathol 22(1):142–150

Evans MJ, Shami SG, Cabral-Anderson LJ, Dekker NP (1986) Role of nonciliated cells in renewal of the bronchial epithelium of rats exposed to NO_2. Am J Pathol 123(1):126–133

Fuchs E, Tumbar T, Guasch G (2004) Socializing with the neighbors: stem cells and their niche. Cell 116(6):769–778

Fujino N, Kubo H, Suzuki T, Ota C, Hegab AE, He M, Suzuki S, Suzuki T, Yamada M, Kondo T, Kato H, Yamaya M (2011) Isolation of alveolar epithelial type II progenitor cells from adult human lungs. Lab Invest 91(3):363–378. doi:10.1038/labinvest.2010.187

Ghosh M, Helm KM, Smith RW, Giordanengo MS, Li B, Shen H, Reynolds SD (2011) A single cell functions as a tissue-specific stem cell and the in vitro niche-forming cell. Am J Respir Cell Mol Biol 45(3):459–469. doi:10.1165/rcmb.2010-0314OC

Giangreco A, Reynolds SD, Stripp BR (2002) Terminal bronchioles harbor a unique airway stem cell population that localizes to the bronchoalveolar duct junction. Am J Pathol 161(1): 173–182

Gong X, Sun Z, Cui D, Xu X, Zhu H, Wang L, Qian W, Han X (2014) Isolation and characterization of lung resident mesenchymal stem cells capable of differentiating into alveolar epithelial type II cells. Cell Biol Int 38(4):405–411. doi:10.1002/cbin.10240

Guha A, Vasconcelos M, Cai Y, Yoneda M, Hinds A, Qian J, Li G, Dickel L, Johnson JE, Kimura S, Guo J, McMahon J, McMahon AP, Cardoso WV (2012) Neuroepithelial body microenvironment is a niche for a distinct subset of Clara-like precursors in the developing airways. Proc Natl Acad Sci U S A 109(31):12592–12597. doi:10.1073/pnas.1204710109

Hackett TL, Shaheen F, Johnson A, Wadsworth S, Pechkovsky DV, Jacoby DB, Kicic A, Stick SM, Knight DA (2008) Characterization of side population cells from human airway epithelium. Stem Cells 26(10):2576–2585. doi:10.1634/stemcells.2008-0171

Hajj R, Baranek T, Le Naour R, Lesimple P, Puchelle E, Coraux C (2007) Basal cells of the human adult airway surface epithelium retain transit-amplifying cell properties. Stem Cells 25(1):139–148. doi:10.1634/stemcells.2006-0288

Hansell MM, Moretti RL (1969) Ultrastructure of the mouse tracheal epithelium. J Morphol 128(2):159–169. doi:10.1002/jmor.1051280203

Hegab AE, Ha VL, Gilbert JL, Zhang KX, Malkoski SP, Chon AT, Darmawan DO, Bisht B, Ooi AT, Pellegrini M, Nickerson DW, Gomperts BN (2011) Novel stem/progenitor cell population from murine tracheal submucosal gland ducts with multipotent regenerative potential. Stem Cells 29(8):1283–1293. doi:10.1002/stem.680

Hegab AE, Ha VL, Darmawan DO, Gilbert JL, Ooi AT, Attiga YS, Bisht B, Nickerson DW, Gomperts BN (2012a) Isolation and in vitro characterization of basal and submucosal gland duct stem/progenitor cells from human proximal airways. Stem Cells Transl Med 1(10):719–724. doi:10.5966/sctm.2012-0056 sctm.2012-0056

Hegab AE, Nickerson DW, Ha VL, Darmawan DO, Gomperts BN (2012b) Repair and regeneration of tracheal surface epithelium and submucosal glands in a mouse model of hypoxic-ischemic injury. Respirology 17(7):1101–1113. doi:10.1111/j.1440-1843.2012.02204.x

Hegab AE, Ha VL, Bisht B, Darmawan DO, Ooi AT, Zhang KX, Paul MK, Kim YS, Gilbert JL, Attiga YS, Alva-Ornelas JA, Nickerson DW, Gomperts BN (2014) Aldehyde dehydrogenase activity enriches for proximal airway basal stem cells and promotes their proliferation. Stem Cells Dev 23(6):664–675. doi:10.1089/scd.2013.0295

Hong KU, Reynolds SD, Giangreco A, Hurley CM, Stripp BR (2001) Clara cell secretory protein-expressing cells of the airway neuroepithelial body microenvironment include a label-retaining subset and are critical for epithelial renewal after progenitor cell depletion. Am J Respir Cell Mol Biol 24(6):671–681

Hong KU, Reynolds SD, Watkins S, Fuchs E, Stripp BR (2004a) Basal cells are a multipotent progenitor capable of renewing the bronchial epithelium. Am J Pathol 164(2):577–588

Hong KU, Reynolds SD, Watkins S, Fuchs E, Stripp BR (2004b) In vivo differentiation potential of tracheal basal cells: evidence for multipotent and unipotent subpopulations. Am J Physiol Lung Cell Mol Physiol 286(4):L643–L649

Imielinski M, Berger AH, Hammerman PS, Hernandez B, Pugh TJ, Hodis E, Cho J, Suh J, Capelletti M, Sivachenko A, Sougnez C, Auclair D, Lawrence MS, Stojanov P, Cibulskis K, Choi K, de Waal L, Sharifnia T, Brooks A, Greulich H, Banerji S, Zander T, Seidel D, Leenders F, Ansen S, Ludwig C, Engel-Riedel W, Stoelben E, Wolf J, Goparju C, Thompson K, Winckler W, Kwiatkowski D, Johnson BE, Janne PA, Miller VA, Pao W, Travis WD, Pass HI, Gabriel SB, Lander ES, Thomas RK, Garraway LA, Getz G, Meyerson M (2012) Mapping the hallmarks of lung adenocarcinoma with massively parallel sequencing. Cell 150(6):1107–1120. doi:10.1016/j.cell.2012.08.029

Jeffery PK (1983) Morphologic features of airway surface epithelial cells and glands. Am Rev Respir Dis 128(2 Pt 2):S14–S20

Kelley MJ, Snider RH, Becker KL, Johnson BE (1994) Small cell lung carcinoma cell lines express mRNA for calcitonin and alpha- and beta-calcitonin gene related peptides. Cancer Lett 81(1):19–25. doi:10.1016/0304-3835(94)90159-7

Kim CF, Jackson EL, Woolfenden AE, Lawrence S, Babar I, Vogel S, Crowley D, Bronson RT, Jacks T (2005) Identification of bronchioalveolar stem cells in normal lung and lung cancer. Cell 121(6):823–835

Kinnard WV, Tuder R, Papst P, Fisher JH (1994) Regulation of alveolar type II cell differentiation and proliferation in adult rat lung explants. Am J Respir Cell Mol Biol 11(4):416–425

Kitamura H, Okudela K, Yazawa T, Sato H, Shimoyamada H (2009) Cancer stem cell: implications in cancer biology and therapy with special reference to lung cancer. Lung Cancer 66(3): 275–281. doi:10.1016/j.lungcan.2009.07.019

Lama VN, Smith L, Badri L, Flint A, Andrei AC, Murray S, Wang Z, Liao H, Toews GB, Krebsbach PH, Peters-Golden M, Pinsky DJ, Martinez FJ, Thannickal VJ (2007) Evidence for tissue-resident mesenchymal stem cells in human adult lung from studies of transplanted allografts. J Clin Invest 117(4):989–996. doi:10.1172/JCI29713

Lauweryns JM, Cokelaere M, Theunynck P (1972) Neuroepithelial bodies in the respiratory mucosa of various mammals. A light optical, histochemical and ultrastuctural investigation. Z Zellforsch Mikrosk Anat 135:569–592

Lauweryns JM, Cokelaere M, Theunynck P, Deleersnyder M (1974) Neuroepithelial bodies in mammalian respiratory mucosa: light optical, histochemical an ultrastructural studies. Chest 65(Suppl):22S–29S

Leeman KT, Fillmore CM, Kim CF (2014) Lung stem and progenitor cells in tissue homeostasis and disease. Curr Top Dev Biol 107:207–233. doi:10.1016/B978-0-12-416022-4.00008-1

Liu X, Engelhardt JF (2008) The glandular stem/progenitor cell niche in airway development and repair. Proc Am Thorac Soc 5(6):682–688. doi:10.1513/pats.200801-003AW

Liu X, Driskell RR, Engelhardt JF (2006) Stem cells in the lung. Methods Enzymol 419: 285–321

Lundin A, Driscoll B (2013) Lung cancer stem cells: progress and prospects. Cancer Lett 338(1): 89–93. doi:10.1016/j.canlet.2012.08.014

Lynch TJ, Engelhardt JF (2014) Progenitor cells in proximal airway epithelial development and regeneration. J Cell Biochem 115(10):1637–45. doi:10.1002/jcb.24834

Martin J, Helm K, Ruegg P, Varella-Garcia M, Burnham E, Majka S (2008) Adult lung side population cells have mesenchymal stem cell potential. Cytotherapy 10(2):140–151. doi:10.1080/14653240801895296

McDonald DM (1988) Neurogenic inflammation in the rat trachea. I. Changes in venules, leucocytes and epithelial cells. J Neurocytol 17(5):583–603

McQualter JL, Bertoncello I (2012) Concise review: Deconstructing the lung to reveal its regenerative potential. Stem Cells 30(5):811–816. doi:10.1002/stem.1055

McQualter JL, Brouard N, Williams B, Baird BN, Sims-Lucas S, Yuen K, Nilsson SK, Simmons PJ, Bertoncello I (2009) Endogenous fibroblastic progenitor cells in the adult mouse lung are highly enriched in the sca-1 positive cell fraction. Stem Cells 27(3):623–633. doi:10.1634/stemcells.2008-0866

McQualter JL, Yuen K, Williams B, Bertoncello I (2010) Evidence of an epithelial stem/progenitor cell hierarchy in the adult mouse lung. Proc Natl Acad Sci U S A 107(4):1414–1419. doi:10.1073/pnas.0909207107

McQualter JL, McCarty RC, Van der Velden J, O'Donoghue RJ, Asselin-Labat ML, Bozinovski S, Bertoncello I (2013) TGF-beta signaling in stromal cells acts upstream of FGF-10 to regulate epithelial stem cell growth in the adult lung. Stem Cell Res 11(3):1222–1233. doi:10.1016/j.scr.2013.08.007

Mercer RR, Russell ML, Roggli VL, Crapo JD (1994) Cell number and distribution in human and rat airways. Am J Respir Cell Mol Biol 10(6):613–624. doi:10.1165/ajrcmb.10.6.8003339

Morrisey EE, Hogan BL (2010) Preparing for the first breath: genetic and cellular mechanisms in lung development. Dev Cell 18(1):8–23. doi:10.1016/j.devcel.2009.12.010

Nadel JA (1983) Neural control of airway submucosal gland secretion. Eur J Respir Dis Suppl 128(Pt 1):322–326

Oeztuerk-Winder F, Guinot A, Ochalek A, Ventura JJ (2012) Regulation of human lung alveolar multipotent cells by a novel p38alpha MAPK/miR-17-92 axis. EMBO J 31(16):3431–3441. doi:10.1038/emboj.2012.192

Pack RJ, Al-Ugaily LH, Morris G, Widdicombe JG (1980) The distribution and structure of cells in the tracheal epithelium of the mouse. Cell Tissue Res 208(1):65–84

Pardo-Saganta A, Law BM, Gonzalez-Celeiro M, Vinarsky V, Rajagopal J (2013) Ciliated cells of pseudostratified airway epithelium do not become mucous cells after ovalbumin challenge. Am J Respir Cell Mol Biol 48(3):364–373. doi:10.1165/rcmb.2012-0146OC

Peake JL, Reynolds SD, Stripp BR, Stephens KE, Pinkerton KE (2000) Alteration of pulmonary neuroendocrine cells during epithelial repair of naphthalene-induced airway injury. Am J Pathol 156(1):279–286

Plopper CG, Hill LH, Mariassy AT (1980) Ultrastructure of the nonciliated bronchiolar epithelial (Clara) cell of mammalian lung. III. A study of man with comparison of 15 mammalian species. Exp Lung Res 1(2):171–180

Rawlins EL, Hogan BL (2008) Ciliated epithelial cell lifespan in the mouse trachea and lung. Am J Physiol Lung Cell Mol Physiol 295(1):L231–L234. doi:10.1152/ajplung.90209.2008

Rawlins EL, Clark CP, Xue Y, Hogan BL (2009a) The Id2+ distal tip lung epithelium contains individual multipotent embryonic progenitor cells. Development 136(22):3741–3745. doi:10.1242/dev.037317

Rawlins EL, Okubo T, Xue Y, Brass DM, Auten RL, Hasegawa H, Wang F, Hogan BL (2009b) The role of Scgb1a1+ Clara cells in the long-term maintenance and repair of lung airway, but not alveolar, epithelium. Cell Stem Cell 4(6):525–534. doi:10.1016/j.stem.2009.04.002

Reddy R, Buckley S, Doerken M, Barsky L, Weinberg K, Anderson KD, Warburton D, Driscoll B (2004) Isolation of a putative progenitor subpopulation of alveolar epithelial type 2 cells. Am J Physiol Lung Cell Mol Physiol 286(4):L658–L667

Regala RP, Davis RK, Kunz A, Khoor A, Leitges M, Fields AP (2009) Atypical protein kinase Ci is required for bronchioalveolar stem cell expansion and lung tumorigenesis. Cancer Res 69(19):7603–7611. doi:10.1158/0008-5472.CAN-09-2066

Reynolds SD, Malkinson AM (2010) Clara cell: progenitor for the bronchiolar epithelium. Int J Biochem Cell Biol 42(1):1–4. doi:10.1016/j.biocel.2009.09.002

Reynolds SD, Giangreco A, Power JH, Stripp BR (2000a) Neuroepithelial bodies of pulmonary airways serve as a reservoir of progenitor cells capable of epithelial regeneration. Am J Pathol 156(1):269–278

Reynolds SD, Hong KU, Giangreco A, Mango GW, Guron C, Morimoto Y, Stripp BR (2000b) Conditional clara cell ablation reveals a self-renewing progenitor function of pulmonary neuroendocrine cells. Am J Physiol Lung Cell Mol Physiol 278(6):L1256–L1263

Ricciardi M, Malpeli G, Bifari F, Bassi G, Pacelli L, Nwabo Kamdje AH, Chilosi M, Krampera M (2012) Comparison of epithelial differentiation and immune regulatory properties of mesenchymal stromal cells derived from human lung and bone marrow. PLoS One 7(5):e35639. doi:10.1371/journal.pone.0035639

Rock JR, Hogan BL (2011) Epithelial progenitor cells in lung development, maintenance, repair, and disease. Annu Rev Cell Dev Biol 27:493–512. doi:10.1146/annurev-cellbio-100109-104040

Rock JR, Onaitis MW, Rawlins EL, Lu Y, Clark CP, Xue Y, Randell SH, Hogan BL (2009) Basal cells as stem cells of the mouse trachea and human airway epithelium. Proc Natl Acad Sci U S A 106(31):12771–12775. doi:10.1073/pnas.0906850106

Rock JR, Randell SH, Hogan BL (2010) Airway basal stem cells: a perspective on their roles in epithelial homeostasis and remodeling. Dis Model Mech 3(9–10):545–556. doi:10.1242/dmm.006031

Rock JR, Barkauskas CE, Cronce MJ, Xue Y, Harris JR, Liang J, Noble PW, Hogan BL (2011) Multiple stromal populations contribute to pulmonary fibrosis without evidence for epithelial to mesenchymal transition. Proc Natl Acad Sci U S A 108(52):E1475–E1483. doi:10.1073/pnas.1117988108

Sabatini F, Petecchia L, Tavian M, Jodon de Villeroche V, Rossi GA, Brouty-Boye D (2005) Human bronchial fibroblasts exhibit a mesenchymal stem cell phenotype and multilineage differentiating potentialities. Lab Invest 85(8):962–971

Schoch KG, Lori A, Burns KA, Eldred T, Olsen JC, Randell SH (2004) A subset of mouse tracheal epithelial basal cells generates large colonies in vitro. Am J Physiol Lung Cell Mol Physiol 286(4):L631–L642

Shimizu T, Takahashi Y, Kawaguchi S, Sakakura Y (1996) Hypertrophic and metaplastic changes of goblet cells in rat nasal epithelium induced by endotoxin. Am J Respir Crit Care Med 153(4 Pt 1):1412–1418. doi:10.1164/ajrccm.153.4.8616574

Sinclair K, Yerkovich ST, Chambers DC (2013) Mesenchymal stem cells and the lung. Respirology 18(3):397–411. doi:10.1111/resp.12050

Song H, Yao E, Lin C, Gacayan R, Chen MH, Chuang PT (2012) Functional characterization of pulmonary neuroendocrine cells in lung development, injury, and tumorigenesis. Proc Natl Acad Sci U S A 109(43):17531–17536. doi:10.1073/pnas.1207238109

Succony L, Janes SM (2014) Airway stem cells and lung cancer. QJM 107(8):607–12. doi:10.1093/qjmed/hcu040

Sutherland KD, Proost N, Brouns I, Adriaensen D, Song JY, Berns A (2011) Cell of origin of small cell lung cancer: inactivation of Trp53 and Rb1 in distinct cell types of adult mouse lung. Cancer Cell 19(6):754–764. doi:10.1016/j.ccr.2011.04.019

Takebe N, Ivy SP (2010) Controversies in cancer stem cells: targeting embryonic signaling pathways. Clin Cancer Res 16(12):3106–3112. doi:10.1158/1078-0432.CCR-09-2934

Tamai S (1983) Basal cells of the human bronchiole. Acta Pathol Jpn 33(1):123–140

Tata PR, Mou H, Pardo-Saganta A, Zhao R, Prabhu M, Law BM, Vinarsky V, Cho JL, Breton S, Sahay A, Medoff BD, Rajagopal J (2013) Dedifferentiation of committed epithelial cells into stem cells in vivo. Nature 503(7475):218–223. doi:10.1038/nature12777

Teisanu RM, Lagasse E, Whitesides JF, Stripp BR (2009) Prospective isolation of bronchiolar stem cells based upon immunophenotypic and autofluorescence characteristics. Stem Cells 27(3):612–622. doi:10.1634/stemcells.2008-0838

Teisanu RM, Chen H, Matsumoto K, McQualter JL, Potts E, Foster WM, Bertoncello I, Stripp BR (2011) Functional analysis of two distinct bronchiolar progenitors during lung injury and repair. Am J Respir Cell Mol Biol 44(6):794–803. doi:10.1165/rcmb.2010-0098OC

Travis WD, Brambilla E, Riely GJ (2013) New pathologic classification of lung cancer: relevance for clinical practice and clinical trials. J Clin Oncol 31(8):992–1001. doi:10.1200/JCO.2012.46.9270

Tropea KA, Leder E, Aslam M, Lau AN, Raiser DM, Lee JH, Balasubramaniam V, Fredenburgh LE, Alex Mitsialis S, Kourembanas S, Kim CF (2012) Bronchioalveolar stem cells increase after mesenchymal stromal cell treatment in a mouse model of bronchopulmonary dysplasia. Am J Physiol Lung Cell Mol Physiol 302(9):L829–L837. doi:10.1152/ajplung.00347.2011

Van der Velden JL, Bertoncello I, McQualter JL (2013) LysoTracker is a marker of differentiated alveolar type II cells. Respir Res 14:123. doi:10.1186/1465-9921-14-123

Van Lommel A, Bolle T, Fannes W, Lauweryns JM (1999) The pulmonary neuroendocrine system: the past decade. Arch Histol Cytol 62(1):1–16

Volckaert T, Dill E, Campbell A, Tiozzo C, Majka S, Bellusci S, De Langhe SP (2011) Parabronchial smooth muscle constitutes an airway epithelial stem cell niche in the mouse lung after injury. J Clin Invest 121(11):4409–4419. doi:10.1172/JCI58097

Volckaert T, Campbell A, De Langhe S (2013) c-Myc regulates proliferation and Fgf10 expression in airway smooth muscle after airway epithelial injury in mouse. PLoS One 8(8):e71426. doi:10.1371/journal.pone.0071426

Wang X, Zhang Y, Amberson A, Engelhardt JF (2001) New models of the tracheal airway define the glandular contribution to airway surface fluid and electrolyte composition. Am J Respir Cell Mol Biol 24(2):195–202

Widdicombe JH, Chen LL, Sporer H, Choi HK, Pecson IS, Bastacky SJ (2001) Distribution of tracheal and laryngeal mucous glands in some rodents and the rabbit. J Anat 198(Pt 2):207–221

Wilkerson MD, Yin X, Hoadley KA, Liu Y, Hayward MC, Cabanski CR, Muldrew K, Miller CR, Randell SH, Socinski MA, Parsons AM, Funkhouser WK, Lee CB, Roberts PJ, Thorne L,

Bernard PS, Perou CM, Hayes DN (2010) Lung squamous cell carcinoma mRNA expression subtypes are reproducible, clinically important, and correspond to normal cell types. Clin Cancer Res 16(19):4864–4875. doi:10.1158/1078-0432.CCR-10-0199

Wine JJ (2007) Parasympathetic control of airway submucosal glands: central reflexes and the airway intrinsic nervous system. Auton Neurosci 133(1):35–54. doi:10.1016/j.autneu.2007.01.008

Wine JJ, Joo NS (2004) Submucosal glands and airway defense. Proc Am Thorac Soc 1(1):47–53. doi:10.1513/pats.2306015

Xie W, Fisher JT, Lynch TJ, Luo M, Evans TI, Neff TL, Zhou W, Zhang Y, Ou Y, Bunnett NW, Russo AF, Goodheart MJ, Parekh KR, Liu X, Engelhardt JF (2011) CGRP induction in cystic fibrosis airways alters the submucosal gland progenitor cell niche in mice. J Clin Invest 121(8):3144–3158. doi:10.1172/JCI41857

Xing Y, Li A, Borok Z, Li C, Minoo P (2012) NOTCH1 is required for regeneration of Clara cells during repair of airway injury. Stem Cells 30(5):946–955. doi:10.1002/stem.1059

Xu X, Rock JR, Lu Y, Futtner C, Schwab B, Guinney J, Hogan BL, Onaitis MW (2012) Evidence for type II cells as cells of origin of K-Ras-induced distal lung adenocarcinoma. Proc Natl Acad Sci U S A 109(13):4910–4915. doi:10.1073/pnas.1112499109

Zacharek SJ, Fillmore CM, Lau AN, Gludish DW, Chou A, Ho JW, Zamponi R, Gazit R, Bock C, Jager N, Smith ZD, Kim TM, Saunders AH, Wong J, Lee JH, Roach RR, Rossi DJ, Meissner A, Gimelbrant AA, Park PJ, Kim CF (2011) Lung stem cell self-renewal relies on BMI1-dependent control of expression at imprinted loci. Cell Stem Cell 9(3):272–281. doi:10.1016/j.stem.2011.07.007

Zheng D, Limmon GV, Yin L, Leung NH, Yu H, Chow VT, Chen J (2013) A cellular pathway involved in Clara cell to alveolar type II cell differentiation after severe lung injury. PLoS One 8(8):e71028. doi:10.1371/journal.pone.0071028

Chapter 4
Bronchioalveolar Stem Cells in Cancer

Michael Hiatt, Orquidea Garcia, Amber Lundin, and Barbara Driscoll

Abbreviations

AAH	Atypical adenomatous hyperplasia's
AEC1/1	Alveolar epithelial cell 1/2
BASC	Bronchioalveolar stem cell
CC10	Clara cell 10 kDa
CCSP	Clara cell secretory protein
COPD	Chronic obstructive pulmonary disease
CSC	Cancer stem cell
EGFR	Epithelial growth factor receptor
Hh	Hedgehog
IPF	Idiopathic pulmonary hypertension
Krt	Keratin
NE	Neuroendocrine cell
NSCLC	None small cell lung cancer
OB	Obliterative bronchitis
SCGB1A1	Secretoglobin family 1A member 1 protein
SCLC	Small cell lung cancer
SFTPC	Surfactant protein C
TTF-1	Thyroid transcription factor 1 (aka NKx2.1)
Trp63	Tumor repressor protein 63
WNT	Wingless-related integration site

M. Hiatt • O. Garcia • A. Lundin • B. Driscoll, Ph.D. (✉)
Developmental Biology and Regenerative Medicine Program, Department of Surgery, Children's Hospital Los Angeles, MS 35, University of Southern California, Los Angeles, CA 90027, USA
e-mail: bdriscoll@chla.usc.edu

A. Firth, J.X.-J. Yuan (eds.), *Lung Stem Cells in the Epithelium and Vasculature*, Stem Cell Biology and Regenerative Medicine, DOI 10.1007/978-3-319-16232-4_4

4.1 Introduction: Stem Cells and Cancer

Stem cells are defined by their expression of pluripotent and/or multipotent markers and are considered the undifferentiated source for the generation and regeneration of all tissues. The stem cell ability to self-renew via symmetrical division ensures that an undifferentiated pool of cells is maintained within each tissue and throughout the lifespan of the organism (Reya et al. 2001; He et al. 2009). The stem cell ability to differentiate via asymmetrical division drives the generation of specialized tissues during development and during regeneration and/or repair of injury in the adult organism (Morrison and Kimble 2006). The choice to divide symmetrically or asymmetrically appears to depend on a variety of factors, including the potency of the cell in question, the position of the stem cell within its niche, and the requirements of the tissue supported by the stem cell for regeneration or repair (Snippert et al. 2010; Scadden 2014). One underlying mechanism for asymmetric versus symmetric division is controlled by distribution of specific molecules within the cell that define polarity, with a major role played by the Notch-binding protein Numb (Dingli et al. 2007), though other critical, cell scaffolding molecules also contribute (Liu et al. 2014). Paracrine signaling from neighboring niche cells and the surrounding tissue can also drive distinct stem cell responses. All these aspects of stem cell biology continue to play critical roles when a normal tissue stem cell is transformed into a cancer stem cell (Dingli et al. 2007; Pardal et al. 2003).

It is hypothesized that cancers arise when the normal proliferative response to injury becomes uncontrolled due to mutations to pro-proliferative oncogenes, anti-proliferative tumor-suppressing genes, and/or alterations in the cell-supporting environment. Hyperproliferation may stay localized and relatively benign or may eventually lead to malignancy and invasion of surrounding tissues. Tumors are often comprised of a heterogeneous population of cells, within which are cells that possess stem cell-like properties and behaviors (Reya et al. 2001). Multiple studies have shown that this subpopulation does not just contribute to the growth of the tumor within its original niche, but retains the ability form de novo tumors at other locations (Pardal et al. 2003; Buzzeo et al. 2007; Glinsky 2008). Along with unrestricted self-renewal, these cells also retain the capacity for multipotent differentiation, giving rise to the heterogeneous lineages of cancer cells that comprise tumors. These distinctive properties, which are similar in many ways to the behaviors of somatic stem cells, have led them to be classified as cancer stem cells (CSCs) (Guo et al. 2006; Girouard and Murphy 2011). However, the "Cancer Stem Cell Hypothesis" extends this concept, in that the data that demonstrate stem-like properties for CSCs also support the hypothesis that they originally arise from resident tissue stem cells (Reya et al. 2001; Rahman et al. 2011). These data suggest that tumors are initiated by the uncontrolled growth of stem cells or differentiated cells that have been reprogrammed to a less differentiated state (Spike and Wahl 2011). Because each tissue in the body carries a resident stem cell population for the purposes of regeneration, it is hypothesized that each also harbors tumor-initiating cells, which produce lesions that reflect both their stem cell origin and their originating tissue (Clarke and Fuller 2006).

4.2 Role of Bronchioalveolar Stem Cells in Development, Homeostasis, and Injury Repair

In the lung, each functional compartment harbors distinct stem cell populations, each of which displays characteristic molecular signaling patterns and responses to adult tissue injury (Wansleeben et al. 2013). In developing lung, signaling via pathways critical to the embryo as a whole (WNT, Hedgehog (Hh), and Notch) are supplemented by more tissue-specific signaling patterns. Lung tissue is specified early in development in a small stem cell population originating from the laryngotracheal groove by the expression of Thyroid transcription factor 1 (Ttf-1, also termed Nkx2.1) (Cardoso and Kotton 2008; Morrisey and Hogan 2010). In murine models, this is rapidly followed by expression of a variety of signaling factors that specify the morphogenesis required to create the branched structure of the bronchial and alveolar epithelium, mesenchyme, and supporting vasculature (Metzger et al. 2008; Maeda et al. 2007; Griffiths et al. 2005; Que et al. 2009). Mature mouse and human lung diverge in morphology and corresponding epithelial stem cell populations in the upper airways, with the trachea of the mouse more closely resembling the upper airways in humans (Rock and Hogan 2011). Humans also possess respiratory bronchioles and goblet cells that are absent or rare, respectively, in murine lung. However, more distal markers and stem cell populations show some overlap (Wansleeben et al. 2013). In mice, Trp63-Krt5-positive basal stem cells of the upper airways give way to Secretoglobin family 1A, member 1 protein (SCGB1A1) positive club cells (formerly termed CC10- or CCSP-positive Clara cells) of the lower airway (Reynolds et al. 2000), and can transiently upregulate Krt14 in response to injury (Reynolds and Malkinson 2010; Cole et al. 2010; Daniely et al. 2004; Ghosh et al. 2010; Hong et al. 2004; Rock et al. 2009). Small airways also harbor scattered, injury-responsive neuroendocrine cells (NE) (Reynolds et al. 2000; Linnoila 2006). Within alveoli, the terminal breathing structure of the distal lung, the gas exchange function of large, capillary-adjacent alveolar epithelial type 1 cells (AEC1) is supported by less common alveolar epithelial type 2 cells (AEC2). AEC2 perform the differentiated function of surfactant expression, but also exhibit a stem cell-like responsiveness to injury (Griffiths et al. 2005; Evans et al. 1973, 1975) Functional facets of the pulmonary neuroendocrine system). In vitro, multiple studies have shown differentiation of AEC2 to AEC1 and recent lineage-tracing studies following injury have shown evidence that AEC2 can serve as progenitors for AEC1 in vivo (Evans et al. 1975; Marconett et al. 2013). Recent studies have also shown that the bronchioalveolar duct junction, situated between the distal and terminal airways leading to the alveolus, harbors rare, multipotent bronchiolaveolar stem cells (BASCs). BASCs typically express both SCGB1A1 and surfactant protein (as assessed by expression of surfactant protein C (SFTPC)) (Kim et al. 2005; Kim 2007), though recently a stem-like population of bronchioalveolar duct junction cells that are SFTPC-negative and express integrin α6ß4 have been identified (Chapman et al. 2011).

The lung is considered a low turnover organ, presumably meaning a decreased likelihood of propagating the mutations that are essential for the formation of malignancies. However, its critical function and constant exposure to the environment creates a strong drive toward homeostasis, resulting in discrete regions of proliferation that can be detected following environmental and experimental injury to the distal lung (Bowden et al. 1968; Giangreco et al. 2009; Snyder et al. 2009). The source of this proliferation has been assigned to the stem cells of the various lung compartments, including small airway club and NE cells, alveolar AEC2s and BASCs, the latter of which have been shown to be particularly responsive to the injury/regeneration signals unleashed by pneumonectomy (Nolen-Walston et al. 2008; Jackson et al. 2011; Eisenhauer et al. 2013). Likewise, the SFTPC-negative-integrin α6ß4-positive bronchioalveolar duct junction population is a label retaining cell that has been shown to respond to the distal lung injury caused by bleomycin treatment by proliferating, making it a candidate bronchiolaveolar long-term stem cell (Chapman et al. 2011).

4.3 Role of Stem Cells in Lung Cancer

Worldwide, the morbidity and mortality caused by lung cancer outranks that of all other cancers, with approximately 1.5 million deaths per year attributed to the disease (The Lancet 2013). Lung tumors arise in all areas of the lung and its variable nature reflects the specialized nature of the various compartments of the organ (Curtis et al. 2010). Thus, within a number of main categories of lung cancer, a large number of tumor subtypes are now recognized. In addition, even within lung tumors, a high degree of heterogeneity can be found, with a continuum of undifferentiated to well-differentiated cell phenotypes observed. Tumors composed of mixtures of cell types with little variation in differentiation status can also be found. These observations make it difficult to pinpoint the type of initiating cell most common to lung cancer, but instead indicate more complex tumor-initiating events, where a tumor-promoting environment may activate multiple cell types or, alternatively, in which initiating cells with a broad potential for differentiation are activated. However, both scenarios can be supported by a sequence of mutations that lead to tumor formation initiated by a select population of lung cancer stem cells (Sutherland and Berns 2010; Peacock and Watkins 2008; Sullivan et al. 2010; Giangreco et al. 2007).

Lung cancers are prevalent despite the slow rate of epithelial cell turnover and the propensity for human lung tissue to scar rather than regenerate. Over the past several decades, correlative observations have led to even stronger causative data showing that lung cancers arise due to both the self-inflicted insult of smoking and the passive insults due to environmental, especially atmospheric, pollution (which may include secondary cigarette smoke) (Samet 2013; Dresler 2013; Gomperts et al. 2011). Finally, recent evidence of epigenetic contributions to the initiation and progression of lung cancer has added an additional dimension to the etiology of lung cancers (Risch and Plass 2008; Scherf et al. 2013). Both cigarette smoke and airborne pollutants have been shown to contain toxins and carcinogens that drive the development of lung inflammation (Walser et al. 2008; Kundu and Surh 2012).

Both this change in the lung epithelial cell environment, as well as direct effects of mutagens on the resident cell populations, can be considered potential triggering events in the transformation of lung stem cells to cancer stem cells. The toxins and reactive oxygen species (ROS) generated by side stream (secondary) cigarette smoke are proven agents of lung cell DNA damage, impairment of normal epithelial cell function, and alterations in gene expression. In addition, cytokines activated during an inflammatory event often remain within the lung as a proximate, underlying cause of chronic pulmonary inflammation. As injury-responsive cells, lung stem cells are clearly vulnerable to mutagenic events and an environment that conceivably leads to an altered, cancer stem cell phenotype (Peebles et al. 2007; Kundu and Surh 2008; Gonda et al. 2009).

4.4 Bronchioalveolar Cells as Stem Cells for Lung Cancer: Data from Pathology

Cancer is identified by its originating cell, i.e., the cell that incurred the first oncogenic mutation. Lung cancers are classified into histological categories based on the initiating cell type (Davidson et al. 2013). Apart from a very small percentage (2 %) of cancers of non-defined origin, the two main groups of cancers that arise from lung cells are small cell lung cancers (SCLC), which comprise approximately 18 % of diagnosed cases, and non-small cell lung cancers (NSCLC), which account for the remaining 80 %. SCLC is the more aggressive and potentially lethal form of the disease, with a 5-year US survival rate of only 5 %. While NSCLC are more common, they metastasize at a slower rate. The NSCLC 5-year survival rate in the US is slightly better, at 15 %, but taken together, survival rates for all lung cancers remain exceedingly poor (The Lancet 2013).

The most common types of SCLC exhibit dense, neurosecretory granules, indicating a putative origin as small airway neuroendocrine cells. Bronchioalveolar cells more commonly appear to contribute to the three subcategories of NSCLC, with evidence based on both the epithelial nature of the tumor cell populations and the location of the initiating tumor at airway branch points, within the bronchioalveolar duct junctions or alveoli. These subcategories are characterized as adenocarcinoma, squamous cell carcinoma, and large cell lung carcinomas (Sakashita et al. 2014). In addition to tumors that arise from lung cells, the highly vascularized lung is also a preferential site for metastatic growth of tumor cells of extra-pulmonary origin, commonly including, but not limited to, breast cancers and melanomas.

While lung cancer is a common type of cancer, chronic lung diseases are even more common and are hypothesized to be manifestations of repeated lung epithelial cell injury. In chronic obstructive pulmonary disease (COPD), obliterative bronchitis (OB), and idiopathic pulmonary fibrosis (IPF), changes in the lung cellular environment are triggered by epithelial cell destruction. Changes in immune cell number and function lead to the release of inflammatory cytokines, activation of lung mesenchymal cell populations, and/or influx of circulating cells. The damage produced by repeated inflammation has a significant impact on the resident cell populations responsible for the homeostatic regulation and repair of epithelium. It is hypothesized that homeostatic repair, and the necessary increase in mitosis required for restoration of homeostatic conditions within the lung, increases the probability of carcinogenic mutations occurring within injury-responsive cell populations. These mutations can therefore accumulate to the point where carcinogenic changes occur and lung cancer develops (Peacock and Watkins 2008). Histological observations have linked both COPD and IPF to development of hyperplasias that can lead to cancer, including the formation of atypical adenomatous hyperplasia's (AAH) at the periphery of lungs where fibrotic plaque formation has also occurred (Mizushima and Kobayashi 1995; Sekine et al. 2012). Though originally hypothesized to arise from AEC2s based on pathological analysis of human tissue, later studies using experimental models showed that bronchioalveolar cells may also contribute to AAH formation (Kim et al. 2005). In fact, controversy surrounding the cell of origin of these lesions, which are believed to be a precursor of adenocarcinoma, still persists.

4.5 Bronchioalveolar Cells as Stem Cells for Lung Cancer: Data from Experimental Models

Human AAH can, under conditions favorable to tumor formation, transition to adenocarcinomas. The cell of origin for these tumors is often observed to be SFTPC-positive and has long been considered to be the surfactant secretory AEC2 of the alveolus. Early transgenic mouse models utilized constitutive overexpression of the viral tumor-suppressing binding protein SV40, which inactivates both Trp53 and pRb. Expression of this potent oncogene was driven by either murine or rabbit club cell *Scgb1a1* promoters or the human *Sftpc* promoter, and each model exhibited distinct development of alveolar or bronchiolar tumors in the form of adenocarcinomas with no marker overlap, indicating singular cells of origin (Wikenheiser et al. 1992, 1993; Sandmöller et al. 1994, 1995; Magdaleno et al. 1997; Wikenheiser and Whitsett 1997). In even earlier studies, spontaneous lung cancers in the form of adenocarcinomas were found to preferentially develop when a mutated K-Ras was utilized as a transgene (Suda et al. 1987). As oncogenically mutated K-Ras can be found in approximately 25 % of human ACs, these latter studies were followed up by development of a conditional, oncogenic K-Ras transgenic model, termed K-RasG12D (Jackson et al. 2001). The earliest data from this model confirmed that the hyperplastic cells induced when mutant K-Ras was expressed were indeed SFTPC-positive. However, evidence from human pathology that bronchiolar cells could also contribute to adenocarcinoma formation proved more difficult to model.

These conflicting observations appeared to be resolved by the identification of SCGB1A1-SFTPC co-expressing BASCs identified by their response to oncogenic K-Ras overexpression in the K-RasG12D model (Kim et al. 2005). Using this model, the rapid expansion of BASCs due to K-Ras signaling led to studies that defined BASC involvement in lung cancer tumorigenesis. Similarly, studies where the tumor suppressors p27 and PTEN were inactivated also stimulated BASC hyperproliferation and subsequent AAH and adenocarcinoma formation (Besson et al. 2007; Yanagi et al. 2007). However, as with the identity of BASCs as a stem cell progenitor for AEC2 and airway cells (Rawlins et al. 2009), the identity of BASCs as a bronchioalveolar cancer stem cell remains controversial.

A more recent study by Xu and colleagues posited that AEC2s, and not BASCs, were the initiating cell specifically for malignant adenocarcinoma formation in the lung due to mutant K-ras expression. This study used different transgenic models, where K-rasG12D was expressed via knock-in at loci for the *Sftpc* gene, the *Scgb1a1* gene or both. Comparison of these models showed formation of AEC2-like, club cell-like, or BASC-like hyperplasias in the corresponding knock-in models, but confirmed that only mice that expressed mutant K-Ras in SFTPC-positive AEC2 exhibited the ability to progress to adenocarcinoma (Xu et al. 2012). These observations were echoed by Mainardi and colleagues with data derived from a lung ubiquitous mutant K-Ras (K-RasG12V) transgenic model where malignant adenocarcinomas arose only from AEC2. Furthermore, in this model, conditional expression of K-RasG12V produced hyperplasias throughout the distal lung of both bronchiolar and bronchioalveolar origin, with only those that exhibited AEC2 markers progressing to adenocarcinoma (Mainardi et al. 2014). However, a subsequent study by Sutherland and colleagues that combined the use of conditional K-RasG12D expression, lineage tracing analyses, and knock down of Trp53 showed a more complex outcome (Sutherland et al. 2014). In these models, both SCGB1A1-positive club cells and SFTPC-positive AEC2 were capable of forming adenocarcinomas and the knock down of Trp53 was required for progression to a metastatic, invasive phenotype. These models all differ in the method used to conditionally induce K-RasG12D or K-RasG12V expression from floxed alleles (via genetically induced Cre-recombinase expression in the Xu and one portion of the Mainardi studies versus via intratracheal administration of Cre-recombinase expressing adenoviruses in the remainder of the Mainardi study and in the Sutherland approach) and it remains to be seen if underlying differences in location or transgene expression, local inflammatory response to the transducing vectors and/or hyperplastic cells, as well as penetrance and levels of mutant K-Ras expression affect the outcome in these models. However, all these results expand the pool of potential stem/progenitor-like distal lung-initiating cell types and challenge the concept that a single, bronchioalveolar cancer stem cell plays a primary role in the initiation of adenocarcinomas.

4.6 Conclusions

The stem cell-like resistance to injury and proliferative potentials of cancer-initiating cells appears to contribute to cancer growth and resistance to treatment. Further research into the validity of BASCs, club cells, and/or AEC2s as stem or progenitor sources of lung carcinogenesis is needed to properly identify these populations as lung tumor cells of origin, particularly in human lung. Identifying cancer stem cells within bronchioalveolar lung tumors will not only illuminate the mechanistic underpinnings of some of the most common and deadly types of lung cancer, but may provide a focus for a wide range of possible treatments and therapies that specifically target stem-like cells. Several possible therapeutic targets unique to these cells include the repair or correction of dysfunctional signaling cascades including altered EGFR, Wnt, Hedgehog, and Notch pathways (Alison et al. 2009). These signaling pathways play vital roles in lung development, response to injury, and the regulation of stem cell self-renewal. They may also play a role in the initiation of tumorigenesis, where a variety of mutations have been shown to cause dysregulation of the processes of stem cell renewal and directed an appropriate differentiation. Characterizations of these mutations are currently being pursued using genome-wide analyses, which are rapidly progressing to the point of becoming useful supplements to standard pathological approaches for identification of tumor-initiating cells (Sakashita et al. 2014). The possibility of customized therapeutic approaches, as well as a rapid method for addressing therapeutic resistance, could lie in the ability of those who study and treat bronchioalveolar tumors to isolate and analyze their initiating cancer stem cells.

Acknowledgment This work was partially supported by NIH grant R01 HL 065352 to B.D., by a training grant from the California Institute of Regenerative Medicine and by an endowment from the Pasadena Guild.

References

Alison MR, LeBrenne AC, Islam S (2009) Stem cells and lung cancer: future therapeutic targets? Expert Opin Biol Ther 9(9):1127–1141. doi:10.1517/14712590903103803

Besson A, Hwang HC, Cicero S, Donovan SL, Gurian-West M, Johnson D, Clurman BE, Dyer MA, Roberts JM (2007) Discovery of an oncogenic activity in p27Kip1 that causes stem cell expansion and a multiple tumor phenotype. Genes Dev 21(14):1731–1746. doi:10.1101/gad.1556607

Bowden DH, Davies E, Wyatt JP (1968) Cytodynamics of pulmonary alveolar cells in the mouse. Arch Pathol 86(6):667–670

Buzzeo MP, Scott EW, Cogle CR (2007) The hunt for cancer-initiating cells: a history stemming from leukemia. Leukemia 21(8):1619–1627. doi:10.1038/sj.leu.2404768

Cardoso WV, Kotton DN (2008) Specification and patterning of the respiratory system. In: Melton D, Girard L (eds) StemBook. Harvard Stem Cell Institute, Cambridge. doi:10.3824/stembook.1.10.1

Chapman HA, Li X, Alexander JP, Brumwell A, Lorizio W, Tan K, Sonnenberg A, Wei Y, Vu TH (2011) Integrin α6β4 identifies an adult distal lung epithelial population with regenerative potential in mice. J Clin Invest 121(7):2855–2862. doi:10.1172/JCI57673

Clarke MF, Fuller M (2006) Stem cells and cancer: two faces of Eve. Cell 124(6):1111–1115. doi:10.1016/j.cell.2006.03.011

Cole BB, Smith RW, Jenkins KM, Graham BB, Reynolds PR, Reynolds SD (2010) Tracheal basal cells: a facultative progenitor cell pool. Am J Pathol 177(1):362–376. doi:10.2353/ajpath.2010.090870

Curtis SJ, Sinkevicius KW, Li D, Lau AN, Roach RR, Zamponi R, Woolfenden AE, Kirsch DG, Wong K-K, Kim CF (2010) Primary tumor genotype is an important determinant in identification of lung cancer propagating cells. Cell Stem Cell 7(1):127–133. doi:10.1016/j.stem.2010.05.021

Daniely Y, Liao G, Dixon D, Linnoila RI, Lori A, Randell SH, Oren M, Jetten AM (2004) Critical role of p63 in the development of a normal esophageal and tracheobronchial epithelium. Am J Physiol Cell Physiol 287(1):C171–C181. doi:10.1152/ajpcell.00226.2003

Davidson MR, Gazdar AF, Clarke BE (2013) The pivotal role of pathology in the management of lung cancer. J Thorac Dis 5(Suppl 5):S463–S478. doi:10.3978/j.issn.2072-1439.2013.08.43

Dingli A, Traulsen F, Michor F (2007) (A)symmetric stem cell replication and cancer. PLoS Comput Biol 3(3):e53. doi:10.1371/journal.pcbi.0030053

Dresler C (2013) The changing epidemic of lung cancer and occupational and environmental risk factors. Thorac Surg Clin 23(2):113–122. doi:10.1016/j.thorsurg.2013.01.015

Eisenhauer P, Earle B, Loi R, Sueblinvong V, Goodwin M, Allen GB, Lundblad L, Mazan MR, Hoffman AM, Weiss DJ (2013) Endogenous distal airway progenitor cells, lung mechanics, and disproportionate lobar growth following long-term postpneumonectomy in mice. Stem Cells 31(7):1330–1339. doi:10.1002/stem.1377

Evans MJ, Cabral LJ, Stephens RJ, Freeman G (1973) Renewal of alveolar epithelium in the rat following exposure to NO2. Am J Pathol 70(2):175–198

Evans MJ, Cabral LJ, Stephens RJ, Freeman G (1975) Transformation of alveolar type 2 cells to type 1 cells following exposure to NO2. Exp Mol Pathol 22(1):142–150. doi:10.1016/0014-4800(75)90059-3

Ghosh M, Brechbuhl HM, Smith RW, Li B, Hicks DA, Titchner T, Runkle CM, Reynolds SD (2010) Context-dependent differentiation of multipotential keratin 14-expressing tracheal basal cells. Am J Respir Cell Mol Biol 45(2):403–410. doi:10.1165/rcmb.2010-0283OC

Giangreco A, Groot KR, Janes SM (2007) Lung cancer and lung stem cells strange bedfellows? Am J Respir Crit Care Med 175(6):547–553. doi:10.1164/rccm.200607-984PP

Giangreco A, Arwert EN, Rosewell IR, Snyder J, Watt FM, Stripp BR (2009) Stem cells are dispensable for lung homeostasis but restore airways after injury. Proc Natl Acad Sci U S A 106(23):9286–9291. doi:10.1073/pnas.0900668106

Girouard SD, Murphy GF (2011) Melanoma stem cells: not rare, but well done. Lab Invest 91(5):647–664. doi:10.1038/labinvest.2011.50

Glinsky GV (2008) "Stemness" genomics law governs clinical behavior of human cancer: implications for design making in disease management. J Clin Oncol 26(17):2846–2853. doi:10.1200/JCO.2008.17.0266

Gomperts BN, Spira A, Massion PP, Walser TC, Wistuba II, Minna JD, Dubinett SM (2011) Evolving concepts in lung carcinogenesis. Semin Respir Crit Care Med 32(1):32–43. doi:10.1055/s-0031-1272867

Gonda TA, Tu S, Wang TC (2009) Chronic inflammation, the tumor microenvironment and carcinogenesis. Cell Cycle 8(13):2005–2013. doi:10.4161/cc.8.13.8985

Griffiths MJ, Bonnet D, Janes SM (2005) Stem cells of the alveolar epithelium. Lancet 366:249–260. doi:10.1016/S0140-6736(05)66916-4

Guo W, Lasky JL, Wu H (2006) Cancer stem cells. Pediatr Res 59(4 Pt 2):59R–64R. doi:10.1203/01.pdr.0000203592.04530.06

He S, Nakada D, Morrison SJ (2009) Mechanisms of stem cell self-renewal. Annu Rev Cell Dev Biol 25:377–406. doi:10.1146/annurev.cellbio.042308.113248

Hong KU, Reynolds SD, Watkins S, Fuchs E, Stripp BR (2004) Basal cells are a multipotent progenitor capable of renewing the bronchial epithelium. Am J Pathol 164(2):577–588. doi:10.1016/S0002-9440(10)63147-1

Jackson EL, Willis N, Mercer K, Bronson RT, Crowley D, Montoya R, Jacks T, Tuveson DA (2001) Analysis of lung tumor initiation and progression using conditional expression of oncogenic K-ras. Genes Dev 15(24):3243–3248. doi:10.1101/gad.943001

Jackson SR, Lee J, Reddy R, Williams GN, Kikuchi A, Freiberg Y, Warburton D, Driscoll B (2011) Partial pneumonectomy of telomerase null mice carrying shortened telomeres initiates cell growth arrest resulting in a limited compensatory growth response. Am J Physiol Lung Cell Mol Physiol 300(6):L898–L909. doi:10.1152/ajplung.00409.2010

Kim C (2007) Paving the road for lung stem cell biology: bronchioalveolar stem cells and other putative distal lung stem cells. Am J Physiol Lung Cell Mol Physiol 293(5):L1092–L1098. doi:10.1152/ajplung.00015.2007

Kim CF, Jackson EL, Woolfenden AE, Lawrence S, Babar I, Vogel S, Crowley D, Bronson RT, Jacks T (2005) Identification of bronchioalveolar stem cells in normal lung and lung cancer. Cell 121:823–835. doi:10.1016/j.cell.2005.03.032

Kundu JK, Surh YJ (2008) Inflammation: gearing the journey to cancer. Mutat Res 659(1–2):15–30. doi:10.1016/j.mrrev.2008.03.002

Kundu JK, Surh YJ (2012) Emerging avenues linking inflammation and cancer. Free Radic Biol Med 52:2013–2037. doi:10.1016/j.freeradbiomed.2012.02.035

Linnoila RI (2006) Functional facets of the pulmonary neuroendocrine system. Lab Invest 86(5):425–444. doi:10.1038/labinvest.3700412

Liu JC, Lerou PH, Lahav G (2014) Stem cells: balancing resistance and sensitivity to DNA damage. Trends Cell Biol 24(5):268–274. doi:10.1016/j.tcb.2014.03.002

Maeda Y, Dave V, Whitsett JA (2007) Transcriptional control of lung morphogenesis. Physiol Rev 87(1):219–244. doi:10.1152/physrev.00028.2006

Magdaleno SM, Wang G, Mireles VL, Ray MK, Finegold MJ, DeMayo FJ (1997) Cyclin-dependent kinase inhibitor expression in pulmonary Clara cells transformed with SV4O large T antigen in transgenic mice. Cell Growth Differ 8(2):145–155

Mainardi S, Mijimolle N, Francoz S, Vicente-Dueñas C, Sánchez-García I, Barbacid M (2014) Identification of cancer initiating cells in K-Ras driven lung adenocarcinoma. Proc Natl Acad Sci U S A 111(1):255–260. doi:10.1073/pnas.1320383110

Marconett CN, Zhou B, Rieger ME, Selamat SA, Dubourd M, Fang X, Lynch SK, Stueve TR, Siegmund KD, Berman BP, Borok Z, Laird-Offringa IA (2013) Integrated transcriptomic and epigenomic analysis of primary human lung epithelial cell differentiation. PLoS Genet 9(6):e1003513. doi:10.1371/journal.pgen.1003513

Metzger RJ, Klein OD, Martin GR, Krasnow MA (2008) The branching programme of mouse lung development. Nature 453:745–750. doi:10.1038/nature07005

Mizushima Y, Kobayashi M (1995) Clinical characteristics of synchronous multiple lung cancer associated with idiopathic pulmonary fibrosis. A review of Japanese cases. Chest 108(5):1272–1277. doi:10.1378/chest.108.5.1272

Morrisey EE, Hogan BL (2010) Preparing for the first breath: genetic and cellular mechanisms in lung development. Dev Cell 18(1):8–23. doi:10.1016/j.devcel.2009.12.010

Morrison SJ, Kimble J (2006) Asymmetric and symmetric stem-cell divisions in development and cancer. Nature 441:1068–1074. doi:10.1038/nature04956

Nolen-Walston RD, Kim CF, Mazan MR, Ingenito EP, Gruntman AM, Tsai L, Boston R, Woolfenden AE, Jacks T, Hoffman AM (2008) Cellular kinetics and modeling of bronchioalveolar stem cell response during lung regeneration. Am J Physiol Lung Cell Mol Physiol 294(6):L1158–L1165. doi:10.1152/ajplung.00298.2007

Pardal R, Clarke MF, Morrison SJ (2003) Applying the principles of stem-cell biology to cancer. Nat Rev 3(12):895–902. doi:10.1038/nrc1232

Peacock CD, Watkins DN (2008) Cancer stem cells and the ontogeny of lung cancer. J Clin Oncol 26(17):2883–2889. doi:10.1200/JCO.2007.15.2702

Peebles KA, Lee JM, Mao JT, Hazra S, Reckamp KL, Krysan K, Dohadwala M, Heinrich EL, Walser TC, Cui X, Baratelli FE, Garon E, Sharma S, Dubinett SM (2007) Inflammation and lung carcinogenesis: applying findings in prevention and treatment. Expert Rev Anticancer Ther 7(10):1405–1421. doi:10.1586/14737140.7.10.1405

Que J, Luo X, Schwartz RJ, Hogan BL (2009) Multiple roles for Sox2 in the developing and adult mouse trachea. Development 136(11):1899–1907. doi:10.1242/dev.034629

Rahman M, Deleyrolle L, Vedam-Mai V, Azari H, Abd-El-Barr M, Reynolds BA (2011) The cancer stem cell hypothesis: failures and pitfalls. Neurosurgery 68(2):531–545. doi:10.1227/NEU.0b013e3181ff9eb5

Rawlins EL, Okubo T, Xue Y, Brass DM, Auten RL, Hasegawa H, Wang F, Hogan BL (2009) The role of Scgb1a1+ Clara cells in the long-term maintenance and repair of lung airway, but not alveolar, epithelium. Cell Stem Cell 4(6):525–534. doi:10.1016/j.stem.2009.04.002

Reya T, Morrison SJ, Clarke MF, Weissman IL (2001) Stem cells, cancer, and cancer stem cells. Nature 414:105–111. doi:10.1038/35102167

Reynolds SD, Malkinson AM (2010) Clara cell: progenitor for the bronchiolar epithelium. Int J Biochem Cell Biol 42(1):1–4. doi:10.1016/j.biocel.2009.09.002

Reynolds SD, Giangreco A, Power JH, Stripp BR (2000) Neuroepithelial bodies of pulmonary airways serve as a reservoir of progenitor cells capable of epithelial regeneration. Am J Pathol 156(1):269–278. doi:10.1016/S0002-9440(10)64727-X

Risch A, Plass C (2008) Lung cancer epigenetics and genetics. Int J Cancer 123(1):1–7. doi:10.1002/ijc.23605

Rock JR, Hogan BL (2011) Epithelial progenitor cells in lung development, maintenance, repair, and disease. Annu Rev Cell Dev Biol 27:493–512. doi:10.1146/annurev-cellbio-100109-104040

Rock JR, Onaitis MW, Rawlins EL, Lu Y, Clark CP, Xue Y, Randell SH, Hogan BL (2009) Basal cells as stem cells of the mouse trachea and human airway epithelium. Proc Natl Acad Sci U S A 106(31):12771–12775. doi:10.1073/pnas.0906850106

Sakashita S, Sakashita M, Sound Tsao M (2014) Genes and pathology of non-small cell lung carcinoma. Semin Oncol 41(1):28–39. doi:10.1053/j.seminoncol.2013.12.008

Samet JM (2013) Tobacco smoking: the leading cause of preventable disease worldwide. Thorac Surg Clin 23(2):103–112. doi:10.1016/j.thorsurg.2013.01.009

Sandmöller A, Halter R, Gómez-La-Hoz E, Gröne HJ, Suske G, Paul D, Beato M (1994) The uteroglobin promoter targets expression of the SV40 T antigen to a variety of secretory epithelial cells in transgenic mice. Oncogene 9(10):2805–2815

Sandmöller A, Halter R, Suske G, Paul D, Beato M (1995) A transgenic mouse model for lung adenocarcinoma. Cell Growth Diff 6(1):97–103

Scadden DT (2014) Nice neighborhood: emerging concepts of the stem cell niche. Cell 157(1):41–50. doi:10.1016/j.cell.2014.02.013

Scherf DB, Sarkisyan N, Jacobsson H, Claus R, Bermejo JL, Peil B, Gu L, Muley T, Meister M, Dienemann H, Plass C, Risch A (2013) Epigenetic screen identifies genotype-specific promoter DNA methylation and oncogenic potential of CHRNB4. Oncogene 32(8):3329–3338. doi:10.1038/onc.2012.344

Sekine Y, Katsura H, Koh E, Hiroshima K, Fujisawa T (2012) Early detection of COPD is important for lung cancer surveillance. Eur Respir J 39(5):1230–1240. doi:10.1183/09031936.00126011

Snippert HJ, van der Flier LG, Sato T, van Es JH, van den Born M, Kroon-Veenboer C, Barker N, Klein AM, van Rheenen J, Simons BD, Clevers H (2010) Intestinal crypt homeostasis results from neutral competition between symmetrically dividing Lgr5 stem cells. Cell 143:134–144. doi:10.1016/j.cell.2010.09.016

Snyder JC, Teisanu RM, Stripp BR (2009) Endogenous lung stem cells and contribution to disease. J Pathol 217(2):254–264. doi:10.1002/path.2473

Spike BT, Wahl GM (2011) p53, stem cells, and reprogramming: tumor suppression beyond guarding the genome. Genes Cancer 2(4):404–419. doi:10.1177/1947601911410224

Suda Y, Aizawa S, S-i H, Inoue T, Furuta Y, Suzuki M, Hirohashi S, Ikawa Y (1987) Driven by the same Ig enhancer and SV4O T promoter ras induced lung adenomatous tumors, myc induced pre-B cell lymphomas and SV40 large T gene a variety of tumors in transgenic mice. EMBO J 6(13):4055–4066

Sullivan JP, Minna JD, Shay JW (2010) Evidence for self-renewing lung cancer stem cells and their implications in tumor initiation, progression, and targeted therapy. Cancer Metastasis Rev 29(1):61–72. doi:10.1007/s10555-010-9216-5

Sutherland KD, Berns A (2010) Cell of origin of lung cancer. Mol Oncol 4(5):397–403. doi:10.1016/j.molonc.2010.05.002

Sutherland KD, Song JY, Kwon MC, Proost N, Zevenhoven J, Berns A (2014) Multiple cells-of-origin of mutant K-Ras-induced mouse lung adenocarcinoma. Proc Natl Acad Sci U S A 111(13):4952–4957. doi:10.1073/pnas.1319963111

The Lancet (2013) 382:659. doi:10.1016/S2213-2600(13)70149-8
Walser T, Cui X, Yanagawa J, Lee JM, Heinrich E, Lee G, Sharma S, Dubinett SM (2008) Smoking and lung cancer: the role of inflammation. Proc Am Thorac Soc 5(8):811–815. doi:10.1513/pats.200809-100TH
Wansleeben C, Barkauskas CE, Rock JR, Hogan BLM (2013) Stem cells of the adult lung: their development and role in homeostasis, regeneration, and disease. Wiley Interdiscip Rev Dev Biol 2(1):131–148. doi:10.1002/wdev.58
Wikenheiser KA, Whitsett JA (1997) Tumor progression and cellular differentiation of pulmonary adenocarcinomas in SV40 large T antigen transgenic mice. Am J Respir Cell Mol Biol 16(6):713–723. doi:10.1165/ajrcmb.16.6.9191473
Wikenheiser KA, Clark JC, Linnoila RI, Stahlman MT, Whitsett JA (1992) Simian virus 40 large T antigen directed by transcriptional elements of the human surfactant protein C gene produces pulmonary adenocarcinomas in transgenic mice. Cancer Res 52(19):5342–5352
Wikenheiser KA, Vorbroker DK, Rice WR, Clark JC, Bachurski CJ, Oie HK, Whitsett JA (1993) Production of immortalized distal respiratory epithelial cell lines from surfactant protein C/simian virus 40 large tumor antigen transgenic mice. Proc Natl Acad Sci U S A 90(23): 11029–11033
Xu X, Rock JR, Lu Y, Futtner C, Schwab B, Guinney J, Hogan BL, Onaitis MW (2012) Evidence for type II cells as cells of origin of K-Ras-induced distal lung adenocarcinoma. Proc Natl Acad Sci U S A 109(13):4910–4915. doi:10.1073/pnas.1112499109
Yanagi S, Kishimoto H, Kawahara K, Sasaki T, Sasaki M, Nishio M, Yajima N, Hamada K, Horie Y, Kubo H, Whitsett JA, Mak TW, Nakano T, Nakazato M, Suzuki A (2007) Pten controls lung morphogenesis, bronchioalveolar stem cells, and onset of lung adenocarcinomas in mice. J Clin Invest 117(10):2929–2940. doi:10.1172/JCI31854

Chapter 5
Role of Progenitors in Pulmonary Fibrosis and Asthma

Ena Ray Banerjee and William Reed Henderson Jr.

Abbreviations

αSMA	Alpha smooth muscle actin
ABC	ATP-binding cassette
ADSC	Adipose stromal cell
AFSC	Amniotic fluid stem cell
AQP-1	Aquaporin 1
AQP-5	Aquaporin 5
AT1	Alveolar type 1
AT2	Alveolar type 2
BADJ	Bronchoalveolar duct junction
BAL	Bronchoalveolar lavage
BASC	Bronchioalveolar stem cell
BEGM	Bronchial epithelial growth medium
BM-MSC	Bone marrow-derived mesenchymal stem cell
BRDU	Bromodeoxyuridine
CBP	CREB-binding protein
CC-10	Club cell-specific protein
CCL5	Chemokine (C–C motif) ligand 5

E. Ray Banerjee, Ph.D. (✉)
Department of Zoology, Immunology and Regenerative Medicine Research Unit, Laboratory Room 214, University of Calcutta, 35, Ballygunge Circular Road, Kolkata 700 019, India
e-mail: enarb1@gmail.com

W.R. Henderson Jr., M.D.
Department of Medicine, Center for Allergy and Inflammation, UW Medicine at South Lake Union, University of Washington, Brotman Building, Room 253, 850 Republican Street, Seattle, WA 98109-4725, USA
e-mail: wrhchem@uw.edu

A. Firth, J.X.-J. Yuan (eds.), *Lung Stem Cells in the Epithelium and Vasculature*, Stem Cell Biology and Regenerative Medicine, DOI 10.1007/978-3-319-16232-4_5

CCL11	Chemokine (C–C motif) ligand 11
CCL12	Chemokine (C–C motif) ligand 12
CCL17	Chemokine (C–C motif) ligand 17
CCL24	Chemokine (C–C motif) ligand 24
CCR2	Chemokine (C–C motif) receptor 2
CCSP	Club cell-specific promoter
CD31	Cluster of differentiation 31
CD34	Cluster of differentiation 34
CD45	Cluster of differentiation 45
CD147	Cluster of differentiation 147
CD206	Cluster of differentiation 206
CK5	Cytokeratin 5
CysLT	Cysteinyl leukotriene
$CysLT_1R$	Cysteinyl leukotriene 1 receptor
CXCR4	Chemokine (C-X-C motif) receptor 4
CXCL1	Chemokine (C-X-C motif) ligand 1
CXCL12	Chemokine (C-X-C motif) ligand 12
DAB	3,3′-Diaminobenzidine
ECFC	Endothelial colony-forming cell
ECM	Extracellular matrix
EGF	Epidermal growth factor
EMT	Epithelial–mesenchymal transition
ERK	Extracellular signal-regulated kinase
ESC	Embryonic stem cell
Fgf10	Fibroblast growth factor 10
FGF	Fibroblast growth factor
hESC	Human embryonic stem cell
HIF-1α	Hypoxia-inducible factor 1 alpha
HIF-2α	Hypoxia-inducible factor 2 alpha
ICAM	Intracellular adhesion molecule 1
IL-4	Interleukin 4
IL-5	Interleukin 5
IL-6	Interleukin 6
IL-12	Interleukin 12
IL-13	Interleukin 13
IL-33	Interleukin 33
IFN-γ	Interferon-γ
IPF	Idiopathic pulmonary fibrosis
iPSC	Induced pluripotent stem cell
JNK	Jun amino-terminal kinase
KDR	Vascular endothelial growth factor receptor2
KGF	Keratinocyte growth factor
Klf-4	Kruppel-like factor 4
LEF-1	Lymphoid enhancer factor-1
LRC	Label-retaining cell

MMP	Matrix metalloproteinase
MMP-2	Matrix metalloproteinase 2
MMP-3	Matrix metalloproteinase 3
MMP-7	Matrix metalloproteinase 7
MRC1	Mannose receptor (cluster of differentiation 206)
MSC	Mesenchymal stem cell
Nf-κB	Nuclear factor kappa B
OVA	Ovalbumin
PDGF	Platelet-derived growth factor
SAGM	Small airways growth medium
SDF-1	Stromal-derived factor 1
SP	Side population
SPC	Surfactant protein C
SPD	Surfactant protein D
ST2L	Interleukin 33 receptor component
SVP	Stromal vascular fraction
TCF	T cell factor
TGF-β	Transforming growth factor beta
Th1	T-helper type 1
Th2	T-helper type 2
Th17	T-helper type 17
VEGF	Vascular endothelial growth factor
VEGF-A	Vascular endothelial growth factor A
WISP-1	Wnt1-inducible-signaling pathway protein 1

5.1 Stem Cells in the Repair of Lung Injury

The recovery of the lungs from injury is dependent on the regeneration of the airway cells and repair of the damaged structure. When this repair process is aberrant, chronic debilitating lung diseases ensue. Critical for the regeneration of normal lung structure and function are stem cells that can generate differentiated lung lineage-specific cells at the same time as perpetuating in an undifferentiated state for future mobilization and differentiation (Weissman et al. 2001; Smith 2001). Stem cells are typified by their self-renewal capability and unlimited proliferative capacity (Kim et al. 2005). In the steady state, stem cells are located in specific microenvironments or niches where they proliferate infrequently. Various progenitor cells participate in the repair process including circulating progenitor epithelial cells and bone marrow-derived cells and resident progenitor cells in the lungs (Mehrad et al. 2007).

Specific chemokine ligand-receptor interactions are key for the recruitment of progenitor cells to the airways. For example, in a mouse model of tracheal transplantation, cytokeratin 5 $(CK5)^{+}$chemokine (C-X-C motif) receptor 4 $(CXCR4)^{+}$ cells are circulating progenitor epithelial cells that move to the injured airways to restore the pseudostratified epithelium in a process mediated by the chemokine receptor CXCR4 and chemokine (C-X-C motif) ligand 12 (CXCL12) (Gomperts et al. 2006). Once recruited to the airways, the stem cells can form lung mesenchymal cells as well as airway and alveolar epithelial cells. A common feature of many lung disorders is dysregulated repair of the injured lung tissues leading to airway remodeling in patients with chronic asthma and irreversible fibrosis in patients with idiopathic pulmonary fibrosis (IPF).

Progenitor cells are identified by flow cytometry as a side population (SP) based on their (1) expression of ATP-binding cassette (ABC) superfamily multidrug efflux pumps and (2) rapid efflux of the DNA-staining fluorochrome, Hoechst 33342 (Giangreco et al. 2002, 2004). By this characterization, adult progenitor cells are widely found, including in the lungs (Alison et al. 2002). Lung SP cells express the common leukocyte marker cluster of differentiation 45 (CD45). These cells may originate from bone marrow-derived circulating cells or resident pulmonary cells that differentiate into endothelial, mesenchymal, and epithelial cells in the damaged airways to interact with resident mesenchymal cells in the repair of lung injury (Abe et al. 2001; Schmidt et al. 2003; Summer et al. 2003, 2004; Gomperts et al. 2006).

5.2 Lung Fibrosis in Patients with Idiopathic Pulmonary Fibrosis

Pulmonary fibrosis results when airway epithelial cells, smooth muscle cells, and endothelial cells are injured and aberrant repair ensues with accumulation of ECM components that destroy the architecture of the lungs. In response to airway injury, IPF patients have progressive and inexorable development of lung fibrosis leading to lung failure and early death. Although the fibrogenic process in this disorder is not altered by current therapies including corticosteroids, reduced disease progress and decline in lung function have recently been reported with two new therapeutic interventions—the cytokine inhibitor pirfenidone (King et al., 2014) and the intracellular tyrosine kinase inhibitor nintedanib (Richeldi et al. 2014).

In IPF patients, lung injury results in the progressive loss of alveolar type 1 (AT1) cells and endothelial cells that form the gas exchange unit of the airways (Daly et al. 1997). The AT1 cells regulate the alveolar fluid balance. AT1 cell disruption is followed by abnormal epithelial repair in patients with IPF characterized by replacement of damaged AT1 cells by alveolar type 2 (AT2) cells that secrete surfactant and impaired regeneration of AT1 cells. The disrupted AT1 cells may also be replaced by resident non-ciliated bronchiolar epithelial progenitor cells and circulating bone marrow-derived progenitors (McElroy and Kasper 2004). Release of matrix metalloproteinase (MMP)s and growth factors such as transforming growth factor-β (TGF-β) and fibroblast growth factor (FGF)s, chemokines, and cytokines by the perturbed epithelial cells leads to activation and proliferation of mesenchymal cells and development of fibroblast/myofibroblast foci in a process termed epithelial–mesenchymal transition that is key for lung fibrogenesis (McElroy and

Kasper 2004; Henderson et al. 2010; Xi et al. 2014). Myofibroblasts are potent secretors of collagen and also interfere with the epithelial cell repair process in IPF by promoting epithelial cell apoptosis.

Although the mechanisms of the lung fibrosis are poorly understood in IPF, important roles have been attributed to TGF-β and T-helper type 2 (Th2) chemokines. For example, bleomycin-treated mice have increased expression of lung C–C chemokines, in particular chemokine (C–C motif) ligand 17 (CCL17) in airway epithelial cells (Belperio et al. 2004); neutralization of CCL17 leads to decreased lung fibrosis. In the IPF airways, the subepithelial fibroblastic foci stain extensively for the proteoglycan versicans (Bensadoun et al. 1996). Myofibroblasts located in these foci stain for procollagen type 1 consistent with collagen synthesis in these damaged areas (Bensadoun et al. 1996). This process results in destruction of the parenchyma through deposition of extracellular matrix components (Henderson et al. 2010; Xi et al. 2014). Circulating mesenchymal fibrocytes (i.e., cluster of differentiation (CD45)$^+$/collagen-1$^+$ cells; see Reilkoff et al. 2011 for review of properties used to characterize fibrocytes) that derive from monocyte precursors may contribute to the differentiation of myofibroblasts in the lungs of patients with IPF (Andersson-Sjöland et al. 2008, 2011). An increase in fibrocytes to greater than 5 % of total blood leukocytes has been associated with early mortality in these patients (Moeller et al. 2009). Endothelial damage leads to interstitial edema accompanied by deposition of collagen fibers in the alveolus in areas of AT1 cell necrosis. In healing, limited proliferation of the AT2 cells and subsequent differentiation into AT1 and Club cells are critical to restoration of normal gas exchange.

In the lung injury repair process, IPF patients have been found to have a decreased number of early endothelial progenitor cells (EPCs) (Malli et al. 2013). Also observed is an increased endogenous expression of vascular endothelia growth factor (VEGF) that may be related to the hypoxic state of these patients since hypoxia increases VEGF transcription in EPCs (Avouac et al. 2008). By flow cytometry and cell culture, an imbalance of circulating endothelial cells and progenitors has been demonstrated in a prospective, multicentric cohort of IPF patients (Smadja et al. 2013). Compared to healthy control subjects, IPF patients have a significant increase in circulating endothelial cells and decrease in circulating cluster of differentiation 34 (CD34)$^+$ progenitor cells expressing KDR (i.e., the VEGF receptor 2) and endothelial colony-forming cell (ECFC)s (Smadja et al. 2013). Patients experiencing a significant worsening of their lung function in comparison to stable IPF patients exhibit a marked increase in their circulating endothelial cells and the proliferative potential of their ECFCs (Smadja et al. 2013). A deleterious effect of this imbalance in IPF may be the differentiation of the increased endothelial cells and ECFCs into mesenchymal cells and then into collagen-secreting fibroblasts and myofibroblasts in the injured airways.

Bone marrow-derived mesenchymal stem cell (MSC)s have been studied for their potential role in airway fibrosis in IPF. In comparison to healthy controls, IPF MSCs have similar immunophenotypic cell markers, capacity to differentiate to adipogenic, osteogenic, and chondrogenic phenotypes, and gene expression of VEGF and FGF (Antoniou et al. 2010). In contrast to controls, IPF MSCs have significantly greater expression of CXCR4 and the axis stromal cell-derived factor-1 (SDF-1), suggesting that these factors could promote the migration of the MSCs to the lungs in response to injury and potentially modulate the profibrotic response.

5.2.1 *Wnt/β-Catenin Signaling in Patients with IPF*

The lungs of patients with IPF strongly express genes encoding ECM molecules (e.g., collagen and fibronectin), MMPs (e.g., matrix metalloproteinase-3 (MMP-3) and matrix metalloproteinase-7 (MMP-7)), the metalloproteinase inducer cluster of differentiation 147 (CD147)/EMMPRIN, and Wnt family members (e.g., secreted frizzled-related protein 2 and Wnt1-inducible signal pathway protein (WISP-1)) (Henderson et al. 2010; Zimmerman et al. 2012). Wnt signaling activation in the mesenchyme may impede the correct differentiation of the alveolar epithelium. The Wnt/β-catenin (i.e., canonical Wnt signaling) pathway is typified by nuclear accumulation of β-catenin to form a complex with members of the T cell factor (TCF)/lymphoid enhancer factor-1 (LEF-1) family of transcription factors (Zimmerman et al. 2012). The transcriptional coactivators, Creb-binding protein (CBP) or p300, are then recruited by β-catenin. Self-renewal by various tissue stem cells such as hematopoeitic stem cells is promoted by Wnt/β-catenin signaling. Wnt/β-catenin pathway activation can either promote or inhibit cell differentiation (Miki et al. 2011). ICG-001, a small molecule inhibitor of Wnt/β-catenin signaling, has been useful in β-catenin/CBP-driven transcription and is key for proliferation without differentiation in both cancer progenitors and normal cells, but a switch to β-catenin/p300-driven gene expression is necessary to initiate normal cellular differentiation (Teo and Kahn 2010). The wound healing process may fail to terminate when altered coactivator usage (CBP versus p300) by β-catenin occurs to result in the development and progression of fibrotic disorders.

5.2.2 *Mouse Bleomycin Model of IPF*

The primary adverse effect of the antineoplastic agent bleomycin is pulmonary fibrosis. Intratracheal administration of bleomycin into the lungs of rodents produces acute inflammation with edema, alveolar epithelial injury, and infiltration by inflammatory cells. This acute process is followed by chronic changes characterized by fibroblast infiltration and proliferation, myofibroblast differentiation, tissue remodeling, proliferation of fibroblasts and pneumocytes, and thickening of the alveolar walls similar to the histopathologic process in patients with IPF (Chua et al. 2005). In mice, bleomycin leads to ectopic expression of the AT2 cell marker surfactant protein C (SPC) in the distal bronchiolar epithelium in cells that co-express the Club cell-specific protein 10 (CC-10) marker, suggesting a unique progenitor cell population may participate in the epithelial cell repair process (Daly et al. 1997). In bleomycin-induced lung fibrosis in mice, aberrant Wnt signaling has an important role in fibrotic lung injury. Inhibition of aberrant Wnt/β-catenin/CBP-driven transcription by ICG-001 not only prevents but reverses airway fibrosis (Henderson et al. 2010). Therefore, inhibition of Wnt/β-catenin/CBP-dependent transcription may provide a novel therapeutic intervention in fibrotic airway diseases.

Amniotic fluid stem cells (AFSCs) are multipotent cells found in amniotic fluid that express Oct4 and SSEA-4 stem cell markers and reduce pro-fibrotic chemokines such as chemokine (C–C motif) ligand 12 (CCL12) in renal fibrosis models. In the mouse bleomycin model of IPF, intravenously infused AFSCs that express chemokine (C–C motif) receptor 2 (CCR2) (i.e., the high affinity receptor for CCL12) migrate to fibrotic foci in the airways where increased levels of CCL12 are found (Garcia et al. 2013). At these sites of injury, the AFSCs limit the bleomycin-induced fibrotic alveolar and parenchymal remodeling and pulmonary function loss (Garcia et al. 2013). This ameliorating effect of AFSCs may be via their release of matrix metalloproteinase-2 (MMP-2) that cleaves CCL12 to form a product with CCR2 receptor antagonist activity. Thus, transplantation of AFSCs may provide a unique therapeutic option for decreasing pulmonary fibrosis.

5.2.3 Generation of Lung Lineage-Specific Airway Cells from Stem Cells and Effect of Wnt/β-Catenin Signaling

Murine and human embryonic stem cells (ESCs) have been employed to generate AT1 cells, AT2 cells, and Club cells (Ali et al. 2002; Rippon et al. 2004; Samadikuchaksaraei et al. 2006; Wang et al. 2007; Aguilar et al. 2009; Banerjee et al. 2012; Garcia et al. 2013) and repair airway injury in fibrosis models (Banerjee et al. 2012). For example, the effect of the two primary populations of stem cells in the bone marrow, MSCs and hematopoietic stem cell (HSC)s, on bleomycin-induced lung fibrosis has been studied (Aguilar et al. 2009). The HSCs and MSCs were engineered to express keratinocyte growth factor (KGF) through an inducible lentivirus vector (Tet-on). Although collagen accumulation in the lungs was reduced by both MSCs and HSCs, only the HSC population reduced the histopathologic injury. KGF-induced proliferation of endogenous AT2 cells was a key effect of the ameliorating effect of the stem cell transplants in this IPF model (Aguilar et al. 2009).

Human (h) ESC H7 cells have been differentiated in vitro into lung epithelial lineage-specific cells (i.e., AT1 and AT2 epithelial cells and Club cells) as an initial developmental step for a cell-based strategy to repair pulmonary injury in the bleomycin mouse IPF model (Banerjee et al. 2012). Undifferentiated hESCs after culture in embryoid body formation were cultured in either a "small airways growth medium" (SAGM) or "bronchiolar epithelial growth medium" (BEGM) for differentiation into lung lineage-specific cells. Differentiation of the hESCs was skewed to a predominantly AT2 phenotype (i.e., 68 % AT2 cells, 12 % AT1 cells, and 4 % Club cells) by culture in SAGM and to a Club cell phenotype (i.e., 33 % Club cells, 12 % AT2, and 2 % AT1 cells) by culture in BEGM by immunostaining (aquaporin-5 (AQP-5), caveolin, and intracellular adhesion molecule-1 (ICAM-1) for AT1 cells, surfactant protein C (SPC) and surfactant protein D (SPD) and aquaporin-1 (AQP-1) for AT2 cells, and CC-10 for Club cells) and by electron microscopy. mRNA expression for the AT2 marker SPC increased 15-fold in the hESCs cultured in SAGM, whereas expression of the Club cell marker CC-10 increased sixfold in the cells cultured in BEGM. It was noted that incubation of the hESCs after differentiation into alveolar and bronchiolar non-ciliated epithelial cells with ICG-001, the small molecule inhibitor of Wnt/β-catenin/CBP transcription, changed the cells from an AT2 phenotype to a predominantly AT1 phenotype as demonstrated by flow cytometry and immunocytochemistry (Banerjee et al. 2012). Thus, inhibition of Wnt/β-catenin pathway in a CBP-dependent process promotes transdifferentiation of type 2 alveolar epithelial cells to type 1 cells.

5.2.4 Transplantation of Lung Lineage-Specific Airway Cells from Stem Cells and Effect of Wnt/β-Catenin Signaling for Resolution of Pulmonary Fibrosis

Transplantation of lung lineage-differentiated hESCs into mice treated with bleomycin has been performed to detect their homing to injured lung and determine whether they could reduce lung fibrosis. Since the probability of mice rejecting the human stem cell xenografts was increased, Rag2γC$^{-/-}$ immunodeficient mice that had undergone sublethal irradiation were employed in these transplant studies (Banerjee et al. 2012). The hESC cells (1×10^5/transplant) consisting of a heterogeneous cell population (i.e., non-ciliated cells with AT2, AT1, and Club cell marker phenotype) that were differentiated in SAGM in the absence or presence of ICG-001 were administered intratracheally 1 week after bleomycin treatment. Human cells engrafted in the airways (Fig. 5.1), and the bleomycin-induced interstitial and alveolar fibrosis, were significantly decreased in receiving the stem cell transplants but not in controls without human cell transplants (Fig. 5.2) (Banerjee et al. 2012). Whereas total lung collagen content increased ~4-fold in bleomycin-treated Rag2γC$^{-/-}$ mice, total collagen was reduced to baseline levels of saline-treated controls (Fig. 5.2). In a similar manner, the increased expression of collagen, and VEGF genes seen in bleomycin-treated controls, was significantly decreased in animals receiving lung lineage-specific hESCs differentiated in the absence or presence of ICG-001 (Fig. 5.2). It is unknown whether the engrafted human cells resolve the fibrosis directly by replicating and replacing the injured lung or indirectly by induction of resident cell proliferation or by modulation of ECM stromal components or growth factors via a paracrine effect.

The resolution of bleomycin-induced airway fibrosis in the mice receiving the human stem cell transplants was accompanied by an expansion in progenitor numbers in the bone marrow, blood, spleen, bronchoalveolar (BAL) fluid, and lung and

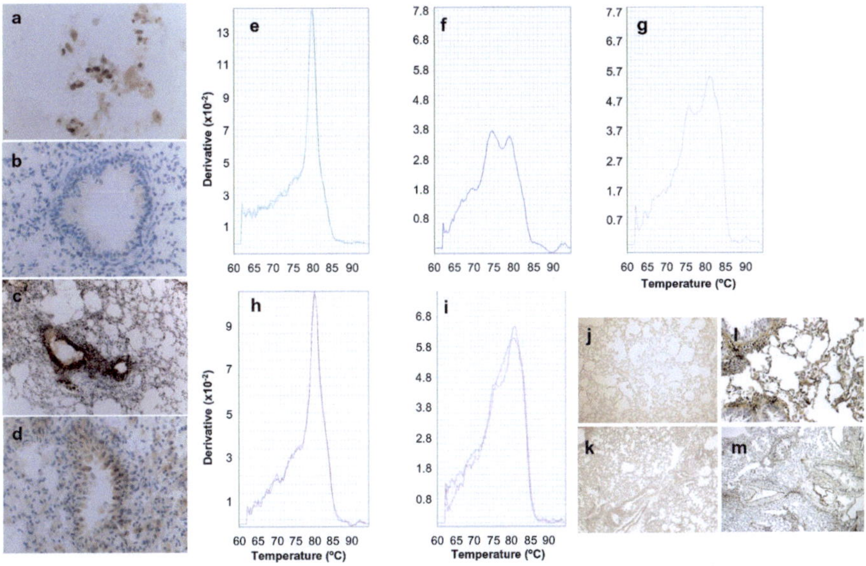

Fig. 5.1 Human cells are found in the lungs of bleomycin-treated mice after transplantation with hES cells differentiated into lung lineage-specific cells. 3,3′-Diaminobenzidine (DAB) immunocytochemistry using anti-human nuclear factor antibody was employed to detect human cells in the lungs of bleomycin-treated Rag2γC$^{-/-}$mice transplanted with lung lineage-specific differentiated H7 hES cells (**a–d**). The transfer of the human cells was confirmed by qPCR of human *Alu* element sequence (**e–i**), and in situ hybridization with human pan-centromeric probe (**j–m**). The treatment groups consisted of mice given 10^5 H7 hES cells differentiated in SAGM alone (SAGM group; **c**, **h**, **l**) or with 5 μM ICG-001 (SAGM+ICG-001 group; **d**, **i**, **m**). The control groups consisted of either untreated mice (**f**, **j**) or bleomycin-treated mice (Bleo/Saline group; **b**, **g**, **k**). The positive control of H7 hES cells (**a**, **e**) is also shown. Human nuclear-specific antibody staining (**a–d**) and DAB-positive pan-centromeric probe reactions (**l**, **m**) are indicated by the brown staining; dissociation curves are shown for the qPCR reaction using the Alu-specific primer (**e–i**). Figure and legend reprinted with permission by PLoS One (Banerjee ER, Laflamme MA, Papayannopoulou T, Kahn M, Murry CE, Henderson WR Jr (2012). Human embryonic stem cells differentiated to lung lineage-specific cells ameliorate pulmonary fibrosis in a xenograft transplant mouse model. PLoS One 7 (3):e33165. doi: 10.1371/journal.pone.0033165. Epub 2012 Mar 28)

increase in both AT1 and AT2 cells that were not observed in control mice. ICG-001 treatment of the cells differentiated in SAGM in vitro expanded the Club cell population after engraftment and caused a further marked increase in the number of lung progenitors, suggesting that modification of the cells in vitro by this specific inhibitor of Wnt/β-catenin/CBP-signaling resulted in a persistent effect on progenitor number in the lungs (Banerjee et al. 2012). These data in a mouse IPF model indicate that progenitor cells from both hematopoetic (i.e., bone marrow) and non-hematopoetic (i.e., lung, splenic lymphoid tissue) niches circulate to the lungs after local injury where they contribute to the resolution of the ensuing pulmonary fibrosis.

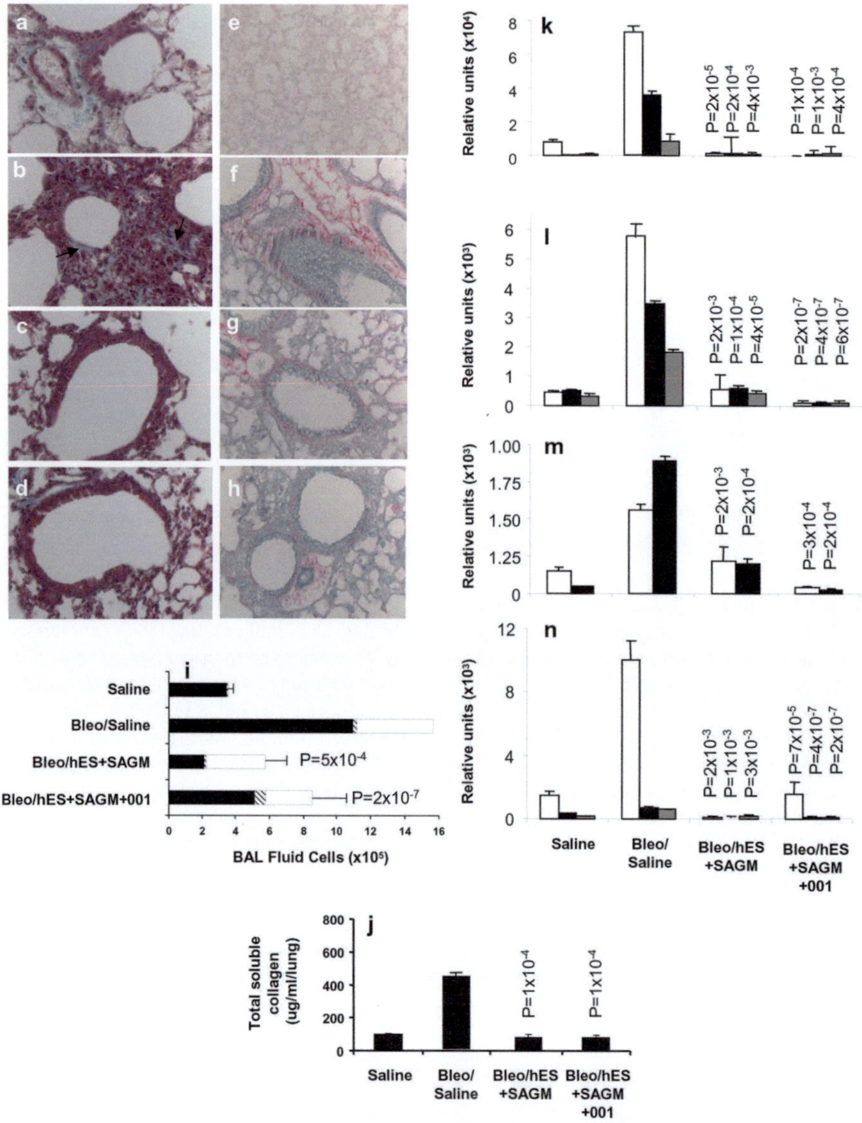

Fig. 5.2 Transplantation of lung lineage-specific differentiated hES cells reverses bleomycin-induced lung inflammation and fibrosis. Transplant groups consisted of Bleomycin-treated Rag2γC$^{-/-}$ mice transplanted intratracheally with 10^5 H7 hES cells differentiated in SAGM alone (Bleo/hES + SAGM group; **c**, **g**, **i-n**), or presence of 5 μM ICG-001 (Bleo/hES + SAGM + ICG-001 group; **d**, **h**, **i–n**). Control groups consisted of saline-treated control Rag2γC$^{-/-}$ mice (Saline group; **a**, **e**) and bleomycin-treated Rag2γC$^{-/-}$ mice given saline intratracheally (Bleo/Saline group; **c**, **g**, **i–n**). Collagen was detected by Masson's trichrome (**a–d**) and Picro Sirius red (**e–h**) stains. BAL fluid cell counts (**i**) are shown for macrophages (*solid bars*), lymphocytes (*hatched bars*), and neutrophils (*open bars*). Total soluble collagen/lung was measured by the Sircol™ assay (**j**). qPCR was used to determine differential expression in the lungs of the following genes: collagen 1α2 (*white bars*), collagen 3α1 (*black bars*), and collagen 6α1 (*gray bars*) (**k**); transforming growth

5.2.5 Stem Cell Transplantation in Patients with IPF

Clinical research studies to investigate the safety and potential efficacy in ameliorating lung fibrosis in IPF patients have recently been initiated (Kørbling and Estrov 2003). The stromal vascular fraction (SVP) within the adipose tissue contains adipose stromal cells (ADSCs) that, in culture, acquire immunophenotypic characteristics of bone marrow-derived MSCs and reduce lung fibrosis in animal models. In a Phase 1b, non-randomized, non-placebo-controlled prospective unicentric study to assess safety, autologous ADSCs-SVF were administered endobronchially in 14 IPF patients of mild-moderate severity (Tzouvelekis et al. 2013). The stem cell infusions (three at monthly intervals; 0.5×10^6 cells/kg body weight/infusion) were well-tolerated with no clinically significant adverse reactions observed. Of note, no ectopic tissue was observed by whole body CT scan 12 months after the first endobronchial infusion in the study group. Retention of hexametazine ^{99m}Tc-HMPAO radiolabeled cells within the lungs for 24 h after the endobronchial delivery was observed; further tracking of the cells was not possible due to the inability to produce signal greater than 24 h after infusion (Tzouvelekis et al. 2013).

MSCs derived from term placenta have been employed in a Phase 1b non-randomized, dose escalation study in patients with IPF of moderate severity to test the safety of this potential therapeutic approach (Chambers et al. 2014). In this single-center study, eight patients received either 1×10^6 MSCs/kg ($n=4$) or 2×10^6 MSCs/kg ($n=4$) intravenously and were followed for a 6-month period for adverse events and lung function. Short-term safety of the infusions was found with one subject having possible embolization of the MSCs to an impaired pulmonary vascular bed, but rapid resolution of the resulting minimal hemodynamic and gas exchange decrements. At 6 months after MSC infusion, lung function and high-resolution chest CT scores were unchanged from baseline. Pulmonary disease worsening was observed in two of the eight subjects, but occurred approximately at 176 days and 6 months after the stem cell transplants. It is unknown whether this decline in lung function was related to the MSC transplants since some studies have suggested that MSCs contribute to the profibrotic process in lung injury (Walker et al. 2011) or rather consistent with the natural history of the rapid decline in lung function observed in this uniformly fatal lung disorder that has a median survival period of less than 4 years. Thus, initial steps in stem cell transplantation have not exhibited significant toxicity and set the stage for prospective,

Fig. 5.2 (continued) factor-β_1 (*white bars*), transforming growth factor-β_2 (*black bars*), and transforming growth factor-β_3 (*grey bars*) (**l**); fibroblast growth factor-1 (*white bars*) and fibroblast growth factor-2 (*black bars*) (**m**); and vascular endothelial growth factor-A (*white bars*), vascular endothelial growth factor-B (*black bars*), and vascular endothelial growth factor-C (*gray bars*) (**n**). $P<0.05$ values compared to bleomycin-treated control group administered saline are shown. Figure and legend reprinted with permission by PLoS One (Banerjee ER, Laflamme MA, Papayannopoulou T, Kahn M, Murry CE, Henderson WR Jr (2012) Human embryonic stem cells differentiated to lung lineage-specific cells ameliorate pulmonary fibrosis in a xenograft transplant mouse model. PLoS One 7 (3):e33165. doi: 10.1371/journal.pone.0033165. Epub 2012 Mar 28)

controlled multi-center studies to confirm the safety of this therapeutic intervention and assess its efficacy in resolving lung fibrosis and reversing or preventing further decline in lung function in IPF patients.

5.2.6 Characterization of Lung Stem Cell Niches

Stem cell niches are protected tissue sites where stem cells reside and mobilize in response to local injury (Watt and Hogan 2000; Spradling et al. 2001; Hong et al. 2001; Lynch and Engelhardt 2014). The tissue stem cells have an undifferentiated phenotype and rarely proliferate. The lung stem cell niches have been examined using rodent models of epithelial denudation and recovery (Hong et al. 2001, 2004; Londhe et al. 2011). Bronchiolar Club cell progenitors are critical for the restoration of alveolar epithelial cells, bronchiolar, and bronchial cells after airway damage. For example, deletion of Club cell progenitors by ganciclovir treatment of transgenic mice expressing HSV-TK driven by a Club cell-specific promoter (CCSP) prevents reconstitution of the normal alveolar and bronchiolar epithelium (Hong et al. 2001; Londhe et al. 2011). Ciliated cells in the airways that survive injury induced by naphthalene and other toxic agents release Wnt7b that induces in the parabronchial smooth muscle cell niche expression of fibroblast growth factor 10 (Fgf10). Fgf10 is important for the maintenance of the distal airway epithelial progenitor cells (Volckaert et al. 2011; Volckaert and De Langhe 2014). The Fgf10-mediated activation of the parabronchial smooth muscle cell niche is regulated by the Wnt target *c-Myc*; ablation of *c-Myc* impairs regeneration of the airway epithelium after injury (Volckaert et al. 2013; Volckaert and De Langhe 2014).

A basal cell population expressing cytokeratin 14 has a multipotent differentiation capacity for repair of airway epithelium (Hong et al. 2001). Stem cell niches in the tracheal and proximal conducting airways have been identified by their resistance to injury with sulfur dioxide and the detergent polidocanol (Borthwick et al. 2001). Bronchioalveolar stem cells (BASCs) are naphthalene-resistant cells located at the junction between the conducting and respiratory epithelium in the bronchoalveolar duct junction (BADJ) in terminal bronchioles (Giangreco et al. 2002; Kim et al. 2005; Summer et al. 2003). Club cells can be restored by naphthalene-resistant stem cells associated with neuroepithelial bodies in proximal airways (Kim et al. 2005). The BADJ-associated Sca1$^+$ CD34$^+$ cluster of differentiation 31 (CD31)$^-$ CD45$^-$ cell population is dual-positive for both pro-SPC and CCSP (Kim et al. 2005).

The bleomycin mouse model of pulmonary fibrosis has also been employed to characterize lung stem cell niches. In these studies, the DNA analogue bromodeoxyuridine (BrdU) was administered intraperitoneally after intratracheal delivery of bleomycin in pulse chase experiments to identify label-retaining cell (LRC) BrdU$^+$ stem cells (Banerjee and Henderson 2012b). BrdU$^+$ LRCs were found in both luminal and basal locations throughout the trachea within the first week of lung injury induced by bleomycin and were later observed in submucosal gland ducts in the proximal trachea and also near the cartilage–intercartilage junction of the distal trachea. Contributing to regeneration of bleomycin-damaged lung in this mouse model were progenitors from the hematopoietic pool in the lungs (Banerjee and Henderson 2012b).

5.3 Allergen-Induced Airway Remodeling in Asthma

Airway remodeling is a characteristic feature in patients with asthma with structural changes beginning in early childhood (Lazaar and Panettieri 2003; Holgate et al. 2003; Payne et al. 2003). Key structural changes include airway wall thickening, goblet cell metaplasia with mucus hypersecretion, smooth muscle hyperplasia, angiogenesis, and subepithelial fibrosis (Henderson et al. 2002, 2006; Banerjee and Henderson 2013). Th2 cytokines (i.e., interleukin-4 (IL-4), interleukin 5 (Il-5), interleukin 13 (IL-13)), and cysteinyl leukotrienes (CysLTs) C_4, D_4, and E_4 play an important role in the immunopathogenesis of the remodeling (Holgate et al. 2003; Henderson et al. 2002, 2006). IL-13 and the growth factor TGF-β are most closely associated with the development of the airway fibrosis and altered lung function and hyperresponsiveness. TGF-β and IL-13 increase expression of the cysteinyl$_1$ receptor ($CysLT_1R$) for cysteinyl leukotriene (CysLT)s C_4, D_4, and E_4 on human airway smooth muscle cells, which is likely important in the thickening of the airway smooth muscle layer since CysLTs augment epidermal growth factor (EGF)-induced proliferation of human airway smooth muscle cells (Mehrotra and Henderson 2009). An excess decline in lung function is observed in both children and adult patients with these features of chronic asthma and who have frequent recurrent severe exacerbations that are refractory to corticosteroid therapy (The Childhood Asthma Management Program Research Group 2000; Covar et al., 2004; Guilbert et al. 2006; Bai et al. 2007).

Thickening of the airway wall occurs from deposition of ECM proteins such as collagen, tenascin, fibronectin, and laminin and proteoglycans such as lumican, biglycan, and versican in the subbasement membrane lamina reticularis (Huang et al. 1999; Christie et al. 2004). Spatial differences have been observed in asthmatic airways in the production of ECM components with greater versican production in distal compared to central airway fibroblasts (Nihlberg et al. 2010). In patients with mild asthma, the thickening of the basement membrane correlates the number of tissue fibrocytes suggesting that fibroblast progenitor cells may play a major role in the airway remodeling process (Nihlberg et al. 2006). After bronchial allergen challenge, fibrocytes localize to the bronchial submucosa in patients with allergic asthma (Schmidt et al. 2003). The circulating fibrocytes from allergic asthmatics have increased levels of the interleukin-33 (IL-33) receptor component ST2L compared to non-asthmatic subjects (Bianchetti et al. 2012). Both fibrocyte chemotaxis and proliferation are stimulated in vitro by the epithelial-derived cytokine IL-33 (Bianchetti et al. 2012), providing a link between allergen-induced airway injury with fibrocyte infiltration in asthma.

Increased circulating $CD34^{+}CD45^{+}$/collagen-1^{+} fibrocytes are found in asthmatic patients who have chronic airflow obstruction compared to those without airflow obstruction (Wang et al. 2008). In vitro, fibrocytes from these subjects with chronic airflow obstruction can be transformed by TGF-β_1 into myofibroblasts, critical cells for the remodeling process (Wang et al. 2008). Proteomic analyses of BAL fluid from patients with mild asthma have demonstrated that increased expression levels of the acute phase glycoprotein haptoglobin occur in conjunction with differentiation of fibrocytes into fibroblast-like cells (Larsen et al. 2006). These data suggest a potential role for haptoglobin in trafficking fibroblast progenitor cells to the airways and promoting their differentiation into activated fibroblasts, key for the increased production of ECM proteins in the thickening airways. In patients with mild-to-refractory asthma, there is a marked increase in the number of fibrocytes in their airway smooth muscle bundles compared to healthy controls (Saunders et al. 2009). Fibrocyte chemotaxis and chemokinesis are stimulated in part by platelet-derived growth factor (PDGF) released by airway smooth muscle cells (Saunders et al. 2009). Induced sputum from patients with severe asthma compared to patients with less-severe and treatment-responsive asthma contain increased levels of the chemokine (C–C motif) ligand 5 (CCL5), chemokine (C–C motif) ligand 11 (CCL11), and chemokine (C–C motif) ligand 24 (CCL24) that promote fibrocyte chemotaxis (Isgrò et al. 2013).

Accompanying the thickening of the subendothelial basement membrane and airway wall is an increase in the vascular bed and arteriolar fibrosis (Salvato 2001). The increased number of vessels occurs in both the medium and small airways of asthmatics with a marked increase in the number of angiogenic factor-positive cells compared to normal subjects (Hashimoto et al. 2005). The elevated levels of the angiogenic cytokine VEGF found in the BAL fluid of asthmatics (as compared to non-asthmatic subjects) correlate with the increased expression of VEGF and hypoxia-inducible factor-1α (HIF-1α) and hypoxia-inducible factor-2α (HIF-2α) in the airway submucosa in these patients with asthma (Lee et al. 2006). HIF-1α that induces the transcriptional response to hypoxia regulates progenitor cell trafficking to injured tissues through induction of the chemokine SDF-1 (Ceradini et al. 2004; Ceradini and Gurtner 2005).

The interaction of allergens that contain innate immune-activating components with mannose receptor (MRC1; cluster of differentiation 206 (CD206)), a C-type lectin receptor found on fibrocytes, has recently been explored. Epidemiologic studies in New York City have found that the cockroach allergen Bla g 2 was more often found in home bed dust in neighborhoods with a high asthma prevalence than those with a low asthma prevalence (Olmedo et al. 2011). Bla g 2 binds to CD206 and is taken up by human fibrocytes that express CD206 in a process blocked by anti-human mannose receptor antibody (Tsai et al. 2013). Bla g 2-induced cytokine (i.e., TNF-α and interleukin 6 (Il-6)) secretion and nuclear factor kappa-B (Nf-κB), extracellular signal-regulated kinase (ERK), and Jun amino-terminal kinase (JNK) activation in the fibrocytes were also mediated via CD206 (Tsai et al. 2013). These data suggest that innate pattern recognition interaction between cockroach allergens and CD206 in circulating fibrocytes can modulate allergic responses in the airways.

Because airway remodeling and the loss of lung function in asthmatics are not reversed by corticosteroids or other therapies, there is intense interest in discovery of new approaches to reversing this process; one such approach is stem cell-based intervention. Mouse asthma models that replicate key features of Th2 cytokine/TGF-β-driven airway remodeling have recently been used to further understand the role of progenitor cells in this process. In these models, there is allergen-induced recruitment of T cell, macrophage, neutrophil, and basophil progenitors as well as stem cells to the lungs and BAL fluid (Banerjee and Henderson 2012a, 2013; Gao et al. 2014).

Induced pluripotent stem cells (iPSCs) may serve as a source for multilineage differentiation and proliferation. iPSCs reprogrammed from adult somatic cells that were transfected by Oct-4/Sox-2/Klf-4 but not by *c-Myc* were investigated in an ovalbumin (OVA) allergen-driven mouse asthma model with eosinophilia for their modulatory effect on airway inflammatory responses (Wang et al. 2013). In this Th2 cytokine-mediated model, Th2 antibody responses, IL-5 levels in the BAL fluid, and airway hyperresponsiveness of the OVA-treated mice were significantly decreased by the administration of iPSCs. Intranasal administration of miRNA-reprogrammed iPSCs that do not express oncogenic transcription factors such as *c-Myc* and Kruppel-like factor-4 (Klf-4) has also been demonstrated to enhance T regulatory (Treg) cell expansion and IL-10 cytokine release in the lungs and reduce airway remodeling in mice with Th2 cytokine-driven airway inflammation (Ogulur et al. 2014).

Infusion of mouse (Nemeth et al. 2010) and human (Bonfield et al. 2010) MSCs inhibits allergen-induced levels of IgE and the Th2 cytokines IL-4, IL-5, and IL-13 in BAL cells consistent with a dampening of the Th2 asthma phenotype. MSCs derived from adipose tissue infiltrate the lungs after intravenous administration in a mouse model of airway remodeling induced by house dust mite stimulation (Mariñas-Pardo et al. 2014). In this model, reduction in airway inflammation by the MSCs was associated with an initial increase in T-helper type 1 (Th1) chemokines interferon-γ (IFN-γ) and interleukin 12 (IL-12), suggesting a skewing toward a Th1 immune response and away from a pro-allergic Th2 response (Mariñas-Pardo et al. 2014). In a cockroach allergen-induced mouse asthma model, TGF-β promotes migration of MSCs to the airways in a process blocked by TGF-β-neutralizing antibodies (Gao et al. 2014). Similarly, tumor-derived MSCs administered intravenously in an OVA-driven model of allergic asthma inhibit the Th2 phenotype by increasing lung production of TGF-β_1 (Song et al., 2014).

In a mouse model of T-helper type 17 (Th17) cytokine-mediated neutrophilic allergic airway inflammation employing *Aspergilllus fumigatus* as the allergen, administration of syngeneic MSCs decreased airway hyperresponsiveness and Th17-induced inflammation in the lungs (Lathrop et al. 2014). In dust mite-driven airway remodeling in mice, HIF-1α inhibition reduces eosinophilic airway inflammation and recruitment of EPCs to the lungs by decreasing vascular endothelial growth factor-A (VEGF-A) and chemokine (C-X-C motif) ligand 1 (CXCL1) levels in the lungs (Byrne et al. 2013). Dust mite-induced proliferation of peripheral blood mononuclear cells from patients with allergic asthma, but not from allergic individuals without asthma, is inhibited by MSCs (Kapoor et al. 2012). These collective data suggest that Th2 cytokine-induced airway eosinophilic and neutrophilic inflammatory and remodeling responses may be modulated by stem cell therapies.

5.4 Conclusions

There has been an explosion in our knowledge of the role of progenitors in the restoration of lung cells and architecture after airway injury. Circulating progenitors of both hematopoietic and non-hematopoietic origin and local tissue stem cells mobilize and move to the sites of injury in the airways in response to specific signals such as chemokines. Animal models have demonstrated the great potential for administration of multipotent stem cells to reduce the fibrogenic and other airway remodeling responses observed in patients with of IPF and chronic asthma who have aberrant repair. Great excitement exists for future clinical research studies to explore the therapeutic potential of stem cell transplants in patients with progressive lung disorders.

References

Abe R, Donnelly SC, Peng T, Bucala R, Metz CN (2001) Peripheral blood fibrocytes: differentiation pathway and migration to wound sites. J Immunol 166(12):7556–7562

Aguilar S, Scotton CJ, McNulty K, Nye E, Stamp G, Laurent G, Bonnet D, Janes SM (2009) Bone marrow stem cells expressing keratinocyte growth factor via an inducible lentivirus protects against bleomycin-induced pulmonary fibrosis. PLoS One 4(11):e8013. doi:10.1371/journal.pone.0008013

Ali NN, Edgar AJ, Samadikuchaksaraei A, Timson CM, Romanska HM, Polak JM, Bishop AE (2002) Derivation of type II alveolar epithelial cells from murine embryonic stem cells. Tissue Eng 8(4):541–550

Alison MR, Poulsom R, Forbes S, Wright NA (2002) An introduction to stem cells. J Pathol 197(4):419–423

Andersson-Sjöland A, de Alba CG, Nihlberg K, Becerril C, Ramírez R, Pardo A, Westergren-Thorsson G, Selman M (2008) Fibrocytes are a potential source of lung fibroblasts in idiopathic pulmonary fibrosis. Int J Biochem Cell Biol 40(10):2129–2140. doi:10.1016/j.biocel.2008.02.012, Epub 2008 Mar 11

Andersson-Sjöland A, Nihlberg K, Eriksson L, Bjermer L, Westergren-Thorsson G (2011) Fibrocytes and the tissue niche in lung repair. Respir Res 12:76. doi:10.1186/1465-9921-12-76

Antoniou KM, Papadaki HA, Soufla G, Kastrinaki MC, Damianaki A, Koutala H, Spandidos DA, Siafakas NM (2010) Investigation of bone marrow mesenchymal stem cells (BM MSCs) involvement in idiopathic pulmonary fibrosis (IPF). Respir Med 104(10):1535–1542. doi:10.1016/j.rmed.2010.04.015, Epub 2010 May 18

Avouac J, Wipff J, Goldman O, Ruiz B, Couraud PO, Chiocchia G, Kahan A, Boileau C, Uzan G, Allanore Y (2008) Angiogenesis in systemic sclerosis impaired expression of vascular endothelial growth factor receptor 1 in endothelial progenitor-derived cells under hypoxic conditions. Arthritis Rheum 58(11):3550–3561. doi:10.1002/art.23968

Bai TR, Vonk JM, Postma DS, Boezen HM (2007) Severe exacerbations predict excess lung function decline in asthma. Eur Respir J 30(3):452–456, Epub 2007 May 30

Banerjee ER, Henderson WR Jr (2012a) Defining the molecular role of gp91phox in the immune manifestation of acute allergic asthma using a preclinical murine model. Clin Mol Allergy 10(1):2. doi:10.1186/1476-7961-10-2

Banerjee ER, Henderson WR Jr (2012b) Characterization of lung stem cell niches in a mouse model of bleomycin-induced fibrosis. Stem Cell Res Ther 3(3):21. doi:10.1186/scrt112

Banerjee ER, Henderson WR Jr (2013) Role of T cells in a gp91 phox knockout murine model of acute allergic asthma. Allergy Asthma Clin Immunol 9(1):6. doi:10.1186/1710-1492-9-6

Banerjee ER, Laflamme MA, Papayannopoulou T, Kahn M, Murry CE, Henderson WR Jr (2012) Human embryonic stem cells differentiated to lung lineage-specific cells ameliorate pulmonary fibrosis in a xenograft transplant mouse model. PLoS One 7(3):e33165. doi:10.1371/journal.pone.0033165, Epub 2012 Mar 28

Belperio JA, Dy M, Murray L, Burdick MD, Xue SRM, Keane MP (2004) The role of the Th2 CC chemokine ligand CCL17 in pulmonary fibrosis. J Immunol 173(7):4692–4698

Bensadoun ES, Burke AK, Hogg JC, Roberts CR (1996) Proteoglycan deposition in pulmonary fibrosis. Am J Respir Crit Care Med 154(6 Pt 1):1819–1828

Bianchetti L, Marini MA, Isgrò M, Bellini A, Schmidt M, Mattoli S (2012) IL-33 promotes the migration and proliferation of circulating fibrocytes from patients with allergen-exacerbated asthma. Biochem Biophys Res Commun 426(1):116–121. doi:10.1016/j.bbrc.2012.08.047, Epub 2012 Aug 17

Bonfield TL, Koloze M, Lennon DP, Zuchowski B, Yang SE, Caplan AI (2010) Human mesenchymal stem cells suppress chronic airway inflammation in the murine ovalbumin asthma model. Am J Physiol Lung Cell Mol Physiol 299(6):L760–L770. doi:10.1152/ajplung.00182.2009, Epub 2010 Sep 3

Borthwick DW, Shahbazian M, Krantz QT, Dorin JR, Randell SH (2001) Evidence for stem-cell niches in the tracheal epithelium. Am J Respir Cell Mol Biol 24(6):662–670

Byrne AJ, Jones CP, Gowers K, Rankin SM, Lloyd CM (2013) Lung macrophages contribute to house dust mite driven airway remodeling via HIF 1α. PLoS One 8(7):e69246. doi:10.1371/journal.pone.0069246

Ceradini DJ, Gurtner GC (2005) Homing to hypoxia: HIF-1 as a mediator of progenitor cell recruitment to injured tissue. Trends Cardiovasc Med 15(2):57–63

Ceradini DJ, Kulkarni AR, Callaghan MJ, Tepper OM, Bastidas N, Kleinman ME, Capla JM, Galiano RD, Levine JP, Gurtner GC (2004) Progenitor cell trafficking is regulated by hypoxic gradients through HIF-1 induction of SDF-1. Nat Med 10(8):858–864, Epub 2004 Jul 4

Chambers DC, Enever D, Ilic N, Sparks L, Whitelaw K, Ayres J, Yerkovich ST, Khalil D, Atkinson KM, Hopkins PM (2014) A phase 1b study of placenta-derived mesenchymal stromal cells in patients with idiopathic pulmonary fibrosis. Respirology. doi:10.1111/resp.12343, Epub ahead of print

Christie PE, Jonas M, Tsai CH, Chi EY, Henderson WR Jr (2004) Increase in laminin expression in allergic airway remodelling and decrease by dexamethasone. Eur Respir J 24(1): 107–115

Chua F, Gauldie J, Laurent GJ (2005) Pulmonary fibrosis, searching for model answers. Am J Respir Cell Mol Biol 33(1):9–13

Covar RA, Spahn JD, Murphy JR, Szefler SJ, Childhood Asthma Management Program Research Group (2004) Progression of asthma measured by lung function in the childhood asthma management program. Am J Respir Crit Care Med 170(3):234–241, Epub 2004 Mar 17

Daly HE, Baecher-Allan CM, Barth RK, D'Angio CT, Finkelstein JN (1997) Bleomycin induces strain-dependent alterations in the pattern of epithelial cell-specific marker expression in mouse lung. Toxicol Appl Pharmacol 142(2):303–310

Gao P, Zhou Y, Xian L, Li C, Xu T, Plunkett B, Huang SK, Wan M, Cao X (2014) Functional effects of TGF-β1 on mesenchymal stem cell mobilization in cockroach allergen-induced asthma. J Immunol 192(10):4560–4570. doi:10.4049/jimmunol.1303461, Epub 2014 Apr 7

Garcia O, Carraro G, Turcatel G, Hall M, Sedrakyan S, Roche T, Buckley S, Driscoll B, Perin L, Warburton D (2013) Amniotic fluid stem cells inhibit the progression of bleomycin-induced pulmonary fibrosis via CCL2 modulation in bronchoalveolar lavage. PLoS One 8(8):e71679. doi:10.1371/journal.pone.0071679

Giangreco A, Reynolds SD, Stripp BR (2002) Terminal bronchioles harbor a unique airway stem cell population that localizes to the bronchoalveolar duct junction. Am J Pathol 161(1):173–182, Epub 2003 Aug 8

Giangreco A, Shen H, Reynolds SD, Stripp BR (2004) Molecular phenotype of airway side population cells. Am J Physiol Lung Cell Mol Physiol 286(4):L624–L630

Gomperts BN, Belperio JA, Rao PN, Randell SH, Fishbein MC, Burdick MD, Strieter RM (2006) Circulating progenitor epithelial cells traffic via CXCR4/CXCL12 in response to airway injury. J Immunol 176(3):1916–1927

Guilbert TW, Morgan WJ, Zeiger RS, Mauger DT, Boehmer SJ, Szefler SJ, Bacharier LB, Lemanske RF Jr, Strunk RC, Allen DB, Bloomberg GR, Heldt G, Krawiec M, Larsen G, Liu AH, Chinchilli VM, Sorkness CA, Taussig LM, Martinez FD (2006) Long-term inhaled corticosteroids in preschool children at high risk for asthma. N Engl J Med 354(19):1985–1997

Hashimoto M, Tanaka H, Abe S (2005) Quantitative analysis of bronchial wall vascularity in the medium and small airways of patients with asthma and COPD. Chest 127(3):965–972

Henderson WR Jr, Tang LO, Chu SJ, Tsao SM, Chiang GK, Jones F, Jonas M, Pae C, Wang H, Chi EY (2002) A role for cysteinyl leukotrienes in airway remodeling in a mouse asthma model. Am J Respir Crit Care Med 165(1):108–116

Henderson WR Jr, Chiang GK, Tien YT, Chi EY (2006) Reversal of allergen-induced airway remodeling by $CysLT_1$ receptor blockade. Am J Respir Crit Care Med 173(7):718–728, Epub 2005 Dec 30

Henderson WR Jr, Chi EY, Ye X, Nguyen C, Tien YT, Zhou B, Borok Z, Knight DA, Kahn M (2010) Inhibition of Wnt/β-catenin/CREB binding protein (CBP) signaling reverses pulmonary fibrosis. (2010). Proc Natl Acad Sci U S A 107(32):14309–14314. doi:10.1073/pnas.1001520107, Epub 2010 Jul 21

Holgate ST, Peters-Golden M, Panettieri RA, Henderson WR Jr (2003) Roles of cysteinyl leukotrienes in airway inflammation, smooth muscle function, and remodeling. J Allergy Clin Immunol 111(1 Suppl):S18–S34; discussion S34-6

Hong KU, Reynolds SD, Giangreco A, Hurley CM, Stripp BR (2001) Club cell secretory protein-expressing cells of the airway neuroepithelial body microenvironment include a label-retaining subset and are critical for epithelial renewal after progenitor cell depletion. Am J Respir Cell Mol Biol 24(6):671–681

Hong KU, Reynolds SD, Watkins S, Fuchs E, Stripp BR (2004) Basal cells are a multipotent progenitor capable of renewing the bronchial epithelium. Am J Pathol 164(2):577–588

Huang J, Olivenstein R, Taha R, Hamid Q, Ludwig M (1999) Enhanced proteoglycan deposition in the airway wall of atopic asthmatics. Am J Respir Crit Care Med 160(2):725–729

Isgrò M, Bianchetti L, Marini MA, Bellini A, Schmidt M, Mattoli S (2013) The C-C motif chemokine ligands CCL5, CCL11, and CCL24 induce the migration of circulating fibrocytes from patients with severe asthma. Mucosal Immunol 6(4):718–727. doi:10.1038/mi.2012.109, Epub 2012 Nov 14

Kapoor S, Patel SA, Kartan S, Axelrod D, Capitle E, Rameshwar P (2012) Tolerance-like mediated suppression by mesenchymal stem cells in patients with dust mite allergy-induced asthma. J Allergy Clin Immunol 129(4):1094–1101. doi:10.1016/j.jaci.2011.10.048, Epub 2011 Dec 22

Kim CF, Jackson EL, Woolfenden AE, Lawrence S, Babar I, Vogel S, Crowley D, Bronson RT, Jacks T (2005) Identification of bronchioalveolar stem cells in normal lung and lung cancer. Cell 121(6):823–835

King TE Jr, Bradford WZ, Castro-Bernardini S, Fagan EA, Glaspole I, Glassberg MK, Gorina E, Hopkins PM, Kardatzke D, Lancaster L, Lederer DJ, Nathan SD, Pereira CA, Sahn SA, Sussman R, Swigris JJ, Noble PW, ASCEND Study Group (2014) A phase 3 trial of pirfenidone in patients with idiopathic pulmonary fibrosis. N Engl J Med 370(22):2083–2092. doi:10.1056/NEJMoa1402582, Epub 2014 May 18

Kørbling M, Estrov Z (2003) Adult stem cells for tissue repair—a new therapeutic concept? N Engl J Med 349(6):570–582

Larsen K, Macleod D, Nihlberg K, Gürcan E, Bjermer L, Marko-Varga G, Westergren-Thorsson G (2006) Specific haptoglobin expression in bronchoalveolar lavage during differentiation of circulating fibroblast progenitor cells in mild asthma. J Proteome Res 5(6):1479–1483

Lathrop MJ, Brooks EM, Bonenfant NR, Sokocevic D, Borg ZD, Goodwin M, Loi R, Cruz F, Dunaway CW, Steele C, Weiss DJ (2014) Mesenchymal stromal cells mediate *Aspergillus* hyphal extract-induced allergic airway inflammation by inhibition of the Th17 signaling pathway. Stem Cells Transl Med 3(2):194–205. doi:10.5966/sctm.2013-0061, Epub 2014 Jan 16

Lazaar AL, Panettieri RA Jr (2003) Is airway remodeling clinically relevant in asthma? Am J Med 115(8):652–659

Lee SY, Kwon S, Kim KH, Moon HS, Song JS, Park SH, Kim YK (2006) Expression of vascular endothelial growth factor and hypoxia-inducible factor in the airway of asthmatic patients. Ann Allergy Asthma Immunol 97(6):794–799

Londhe VA, Maisonet TM, Lopez B, Jeng JM, Li C, Minoo P (2011) A subset of epithelial cells with CCSP promoter activity participates in alveolar development. Am J Respir Cell Mol Biol 44(6):804–812. doi:10.1165/rcmb.2009-0429OC, Epub 2010 Aug 6

Lynch TJ, Engelhardt JF (2014) Progenitor cells in proximal airway epithelial development and regeneration. J Cell Biochem 115(10):1637–1645. doi:10.1002/jcb.24834, Epub ahead of print

Malli F, Koutsokera A, Paraskeva E, Zakynthinos E, Papagianni M, Makris D, Tsilioni I, Molyvdas PA, Gourgoulianis KI, Daniil Z (2013) Endothelial progenitor cells in the pathogenesis of idiopathic pulmonary fibrosis: an evolving concept. PLoS One 8(1):e53658. doi:10.1371/journal.pone.0053658, Epub 2013 Jan 14

Mariñas-Pardo L, Mirones I, Amor-Carro O, Fraga-Iriso R, Lema-Costa B, Cubillo I, Rodríguez Milla MA, García-Castro J, Ramos-Barbón D (2014) Mesenchymal stem cells regulate airway contractile tissue remodeling in murine experimental asthma. Allergy 69(6):730–740. doi:10.1111/all.12392, Epub 2014 Apr 21

McElroy MC, Kasper M (2004) The use of alveolar epithelial type I cell-selective markers to investigate lung injury and repair. Eur Respir J 24(4):664–673

Mehrad B, Keane MP, Gomperts BN, Strieter RM (2007) Circulating progenitor cells in chronic lung disease. Expert Rev Respir Med 1(1):157–165. doi:10.1586/17476348.1.1.157

Mehrotra AK, Henderson WR Jr (2009) The role of leukotrienes in airway remodeling. Curr Mol Med 9(3):383–391

Miki T, Yasuda SY, Kahn M (2011) Wnt/β-catenin signaling in embryonic stem cell self-renewal and somatic cell reprogramming. Stem Cell Rev 7(4):836–846. doi:10.1007/s12015-011-9275-1

Moeller A, Gilpin SE, Ask K, Cox G, Cook D, Gauldie J, Margetts PJ, Farkas L, Dobranowski J, Boylan C, O'Byrne PM, Strieter RM, Kolb M (2009) Circulating fibrocytes are an indicator of poor prognosis in idiopathic pulmonary fibrosis. Am J Respir Crit Care Med 179(7):588–594. doi:10.1164/rccm.200810-1534OC, Epub 2009 Jan 16

Nemeth K, Keane-Myers A, Brown JM, Metcalfe DD, Gorham JD, Bundoc VG, Hodges MG, Jelinek I, Madala S, Karpati S, Mezey E (2010) Bone marrow stromal cells use TGF-β to suppress allergic responses in a mouse model of ragweed-induced asthma. Proc Natl Acad Sci U S A 107(12):5652–5657. doi:10.1073/pnas.0910720107, Epub 2010 Mar 15

Nihlberg K, Larsen K, Hultgardh-Nilsson A, Malmstrom A, Bjermer L, Westergren-Thorsson G (2006) Tissue fibrocytes in patients with mild asthma: a possible link to thickness of reticular basement membrane? Respir Res 7:50

Nihlberg K, Andersson-Sjoland A, Tufvesson E, Erjefalt JS, Bjermer L, Westergren-Thorsson G (2010) Altered matrix production in the distal airways of individuals with asthma. Thorax 65(8):670–676. doi:10.1136/thx.2009.129320

Ogulur I, Gurhan G, Aksoy A, Duruksu G, Inci C, Filinte D, Kombak FE, Karaoz E, Akkoc T (2014) Suppressive effect of compact bone-derived mesenchymal stem cells on chronic airway remodeling in murine model of asthma. Int Immunopharmacol 20(1):101–109. doi:10.1016/j.intimp.2014.02.028, Epub 2014 Mar 6

Olmedo O, Goldstein IF, Acosta L, Divjan A, Rundle AG, Chew GL, Mellins RB, Hoepner L, Andrews H, Lopez-Pintado S, Quinn JW, Perera FP, Miller RL, Jacobson JS, Perzanowski MS (2011) Neighborhood differences in exposure and sensitization to cockroach, mouse, dust mite, cat, and dog allergens in New York City. J Allergy Clin Immunol 128(2):284–292.e7. doi:10.1016/j.jaci.2011.02.044, Epub 2011 May 4

Payne DN, Rogers AV, Adelroth E, Bandi V, Guntupalli KK, Bush A, Jeffery PK (2003) Early thickening of the reticular basement membrane in children with difficult asthma. Am J Respir Crit Care Med 167(1):78–82

Reilkoff RA, Bucala R, Herzog EL (2011) Fibrocytes: emerging effector cells in chronic inflammation. Nat Rev Immunol 11(6):427–435. doi:10.1038/nri2990, Epub 2011 May 20

Richeldi L, du Bois RM, Raghu G, Azuma A, Brown KK, Costabel U, Cottin V, Flaherty KR, Hansell DM, Inoue Y, Kim DS, Kolb M, Nicholson AG, Noble PW, Selman M, Taniguchi H, Brun M, Le Maulf F, Girard M, Stowasser S, Schlenker-Herceg R, Disse B, Collard HR, Trial Investigators INPULSIS (2014) Efficacy and safety of nintedanib in idiopathic pulmonary fibrosis. N Engl J Med 370(22):2071–2082. doi:10.1056/NEJMoa1402584

Rippon HJ, Ali NN, Polak JM, Bishop AE (2004) Initial observations on the effect of medium composition on the differentiation of murine embryonic stem cells to alveolar type II cells. Cloning Stem Cells 6(2):49–56

Salvato G (2001) Quantitative and morphological analysis of the vascular bed in bronchial biopsy specimens from asthmatic and non-asthmatic subjects. Thorax 56(12):902–906

Samadikuchaksaraei A, Cohen S, Isaac K, Rippon HJ, Polak JM, Bielby RC, Bishop AE (2006) Derivation of distal airway epithelium from human embryonic stem cells. Tissue Eng 12(4):867–875

Saunders R, Sutcliffe A, Kaur D, Siddiqui S, Hollins F, Wardlaw A, Bradding P, Brightling C (2009) Airway smooth muscle chemokine receptor expression and function in asthma. Clin Exp Allergy 39(11):1684–1692. doi:10.1111/j.1365-2222.2009.03310.x, Epub 2009 Sep 7

Schmidt M, Sun G, Stacey MA, Mori L, Mattoli S (2003) Identification of circulating fibrocytes as precursors of bronchial myofibroblasts in asthma. J Immunol 171(1):380–389

Smadja DM, Mauge L, Nunes H, d'Audigier C, Juvin K, Borie R, Carton Z, Bertil S, Blanchard A, Crestani B, Valeyre D, Gaussem P, Israel-Biet D (2013) Imbalance of circulating endothelial cells and progenitors in idiopathic pulmonary fibrosis. Angiogenesis 16(1):147–157. doi:10.1007/s10456-012-9306-9, Epub 2012 Sep 16

Smith AG (2001) Embryo-derived stem cells: of mice and men. Annu Rev Cell Dev Biol 17:435–462

Song C, Yuan Y, Wang XM, Li D, Zhang GM, Huang B, Feng ZH (2014) Passive transfer of tumour-derived MDSCs inhibits asthma-related airway inflammation. Scand J Immunol 79(2):98–104. doi:10.1111/sji.12140

Spradling A, Drummond-Barbosa D, Kai T (2001) Stem cells find their niche. Nature 414(6859): 98–104

Summer R, Kotton DN, Sun X, Ma B, Fitzsimmons K, Fine A (2003) Side population cells and Bcrp1 expression in lung. Am J Physiol Lung Cell Mol Physiol 285(1):L97–L104, Epub 2003 Mar 7

Summer R, Kotton DN, Sun X, Fitzsimmons K, Fine A (2004) Translational physiology: origin and phenotype of lung side population cells. Am J Physiol Lung Cell Mol Physiol 287(3):L477–L483, Epub 2004 Mar 26

Teo JL, Kahn M (2010) The Wnt signaling pathway in cellular proliferation and differentiation: a tale of two coactivators. Adv Drug Deliv Rev 62(12):1149–1155. doi:10.1016/j.addr.2010.09.012, Epub 2010 Oct 21

The Childhood Asthma Management Program Research Group (2000) Long-term effects of budesonide or nedocromil in children with asthma. N Engl J Med 343(15):1054–1063

Tsai YM, Hsu SC, Zhang J, Zhou YF, Plunkett B, Huang SK, Gao PS (2013) Functional interaction of cockroach allergens and mannose receptor (CD206) in human circulating fibrocytes. PLoS One 8(5):e64105. doi:10.1371/journal.pone.0064105

Tzouvelekis A, Paspaliaris V, Koliakos G, Ntolios P, Bouros E, Oikonomou A, Zissimopoulos A, Boussios N, Dardzinski B, Gritzalis D, Antoniadis A, Froudarakis M, Kolios G, Bouros D (2013) A prospective, non-randomized, no placebo-controlled, phase Ib clinical trial to study the safety of the adipose derived stromal cells-stromal vascular fraction in idiopathic pulmonary fibrosis. J Transl Med 11:171. doi:10.1186/1479-5876-11-171

Volckaert T, De Langhe S (2014) Lung epithelial stem cells and their niches: Fgf10 takes center stage. Fibrogen Tiss Rep 7:8. doi:10.1186/1755-1536-7-8

Volckaert T, Dill E, Campbell A, Tiozzo C, Majka S, Bellusci S, De Langhe SP (2011) Parabronchial smooth muscle constitutes an airway epithelial stem cell niche in the mouse lung after injury. J Clin Invest 121(11):4409–4419. doi:10.1172/JCI58097

Volckaert T, Campbell A, De Langhe S (2013) c-Myc regulates proliferation and Fgf10 expression in airway smooth muscle after airway epithelial injury in mouse. PLoS One 8:e71426. doi:10.1371/journal.pone.0071426

Walker N, Badri L, Wettlaufer S, Flint A, Sajjan U, Krebsbach PH, Keshamouni VG, Peters-Golden M, Lama VN (2011) Resident tissue-specific specific mesenchymal progenitor cells contribute to fibrogenesis in human lung allografts. Am J Pathol 178(6):2461–2469. doi:10.1016/j.ajpath.2011.01.058

Wang D, Haviland D, Burns A, Zsigmond E, Wetsel R (2007) A pure population of lung alveolar epithelial type II cells derived from human embryonic stem cells. Proc Natl Acad Sci U S A 104(11):4449–4454, Epub 2007 Mar 2

Wang CH, Huang CD, Lin HC, Lee KY, Lin SM, Liu CY, Huang KH, Ko YS, Chung KF, Kuo HP (2008) Increased circulating fibrocytes in asthma with chronic airflow obstruction. Am J Respir Crit Care Med 178(6):583–591. doi:10.1164/rccm.200710-1557OC, Epub 2008 Jun 26

Wang CY, Chiou GY, Chien Y, Wu CC, Wu TC, Lo WT, Chen SJ, Chiou SH, Peng HJ, Huang CF (2013) Induced pluripotent stem cells without c-Myc reduce airway responsiveness and allergic reaction in sensitized mice. Transplantation 96(11):958–965. doi:10.1097/TP.0b013e3182a53ef7

Watt FM, Hogan BL (2000) Out of Eden: stem cells and their niches. Science 287(5457):1427–1430

Weissman IL, Anderson DJ, Gage F (2001) Stem and progenitor cells: origins, phenotypes, lineage commitments, and transdifferentiations. Annu Rev Cell Dev Biol 17:387–403

Xi Y, Tan K, Brumwell AN, Chen SC, Kim YH, Kim TJ, Wei Y, Chapman HA (2014) Inhibition of epithelial-to-mesenchymal transition and pulmonary fibrosis by methacycline. Am J Respir Cell Mol Biol 50(1):51–60. doi:10.1165/rcmb.2013-0099OC

Zimmerman ZF, Moon RT, Chien AJ (2012) Targeting Wnt pathways in disease. Cold Spring Harb Perspect Biol 4(11):a008086. doi:10.1101/cshperspect.a008086

Chapter 6
Challenges of Cell Therapy for Lung Diseases and Critical Illnesses

Fernanda F. Cruz, Patricia R.M. Rocco, and Daniel J. Weiss

Abbreviations

ARDS	Acute respiratory distress syndrome
CCL2	Chemokine ligand 2
COPD	Chronic obstructive pulmonary disease
eNOS	Endothelial nitric oxide synthase
EPC	Endothelial progenitor cell
GMP	Good manufacturing process
HGF	Hepatocyte growth factor
IDO	Indoleamine 2,3-dioxygenase
IFN-β	Interferon beta
IL-10	Interleukin 10
IL-6	Interleukin 6
IND	Investigational new drug
IPF	Idiopathic pulmonary fibrosis
ISCT	International Society for Cellular Therapy
KGF	Keratinocyte growth factor
MIP-2	Macrophage inflammatory protein 2-alpha
miRNA	Microribonucleic acid
MSC	Mesenchymal stromal (stem) cells
NHBLI	National Heart Blood and Lung Institute

F.F. Cruz • P.R.M. Rocco
Federal University of Rio de Janeiro, Rio de Janeiro, Brazil

D.J. Weiss, M.D., Ph.D. (✉)
Department of Medicine, Pulmonary and Critical Care, University of Vermont, 149 Beaumont Avenue, Room 226 Health Sciences Research Facility, Burlington, VT 05405, USA
e-mail: daniel.weiss@med.uvm.edu

A. Firth, J.X.-J. Yuan (eds.), *Lung Stem Cells in the Epithelium and Vasculature*, Stem Cell Biology and Regenerative Medicine, DOI 10.1007/978-3-319-16232-4_6

NO	Nitric oxide
PACT	Production Assistance in Cellular Therapies
TGF-β	Transforming growth factor beta
TNF-α	Tumour necrosis factor alpha
TRAIL	Tumour necrosis factor-related apoptosis-inducing ligand

6.1 Introduction

Pulmonary diseases, including acute respiratory distress syndrome (ARDS), asthma, chronic obstructive pulmonary disease (COPD), idiopathic pulmonary fibrosis (IPF), bronchopulmonary dysplasia, and occupational diseases such as silicosis, remain important causes of morbidity and mortality worldwide. Some of these, such as COPD and asthma, in contrast to many other major diseases, are increasing in prevalence. COPD, the fourth leading cause of disease mortality worldwide, is expected to be the third by 2020, and thus remains a major public health concern. Current available treatments for lung diseases may lessen the severity of symptoms, but there is still a pressing need for new therapeutic approaches, since no existing treatment has been shown to reduce disease progression, reverse the pathological changes, and restore the organ functionality. Lung transplantation is considered the only curative approach for end-stage chronic diseases; however, there is a significant shortage of suitable donor lungs and many on waiting lists die before a lung becomes available. Further, lung transplantation requires lifelong immunosuppression and five-year mortality after transplantation is approximately 50 %. Lung transplantation is also not a realistic option for patients in many parts of the world. New therapeutic approaches are thus desperately needed (Weiss 2014).

Approaches utilizing cell-based therapies for lung diseases have progressed rapidly in recent years. Systemic or local (intratracheal) administration of different stem and progenitor cell types has been demonstrated to have efficacy in different pre-clinical models of lung diseases (Weiss et al. 2013a, b; Kotton 2012; Lau et al. 2012). The different cell types have included endothelial progenitor cells (EPCs), bone marrow-derived mononuclear cells, amniotic fluid cells, and mesenchymal stromal (stem) cells (MSCs) (Weiss 2014; Weiss et al. 2013a, b). However, the majority of available pre-clinical data have focused on investigation of MSCs.

6.2 Mesenchymal Stromal (Stem) Cells

MSCs were first described in 1968, and since then, have been widely investigated for their applications in stem cell-based regeneration studies. The nomenclature has evolved over time as MSCs were initially named fibroblastic colony-forming units, subsequently as marrow stromal cells, mesenchymal stem cells, mesenchymal stromal cells, or as multipotent mesenchymal stromal cells. Nowadays, the application of the more commonly currently utilized terms, mesenchymal stem cell or mesenchymal stromal cell, is inconsistent in the literature. This nomenclature loosely depends on whether MSCs are being used for their ability to differentiate into lineages potentially useful in regenerative medicine efforts and structural repair, or utilizing their immunomodulatory properties in the absence of structural engraftment (Lotfinegad et al. 2014).

MSCs are described as self-renewal, fibroblastoid, non-phagocytic, adherent cells which are able to differentiate in vitro into some cell lineages, in particular, culture systems (Mafi et al. 2011; Paunescu et al. 2007; Oswald et al. 2004; Tran et al. 2011). In addition to the bone marrow (Li and Ikehara 2013; Kern et al. 2006), MSCs have been found in other sources, including liver (Najimi et al. 2007), lung (Sabatini et al. 2005; Lama et al. 2007), brain (Kang et al. 2010), adipose tissue (Kern et al. 2006; Pawitan 2009; Zuk et al. 2002; Zannettino et al. 2008), peripheral blood (Chong et al. 2012), cornea (Choong et al. 2007), synovium (Jones et al. 2010), thymus (Krampera et al. 2007), dental pulp (Gronthos 2011; Gronthos et al. 2000), periosteum (Nakahara et al. 1990), tendon (Bi et al. 2007), fallopian tube (Jazedje et al. 2009), placenta (Sabapathy et al. 2012; Igura et al. 2004), amniotic fluid (You et al. 2008), Wharton's jelly (Wang et al. 2004), umbilical cord (Capelli et al. 2011; Romanov et al. 2003), and umbilical cord blood (Kern et al. 2006).

However, definition and investigation of MSCs continue to be confounded by several issues. For instance, there can be important differences in MSC properties, such as cell surface epitopes, secretome, immunomodulatory properties, lineage tendencies, and genomic stability, according to the tissue, strain, and species that MSCs are derived from (Keating 2012; Prockop and Oh 2012a, b; Romieu-Mourez et al. 2012; Baer and Geiger 2012). Further, there is growing evidence that MSCs are heterogeneous and that different MSC subtypes exist, even in cells isolated from the same source. Thus, delineating functional differences between MSCs isolated from different sources is an area of current intense investigation (Viswanathan et al. 2014).

To foster a more uniform characterization of MSCs and facilitate the exchange of data among investigators, the Mesenchymal and Tissue Stem Cell Committee of the International Society for Cellular Therapy (ISCT) have proposed minimal criteria to define human MSCs, which are listed as: (1) MSCs must be plastic-adherent when maintained in standard culture conditions; (2) MSCs must express CD105, CD73, and CD90 in at least 95 % of cell population, and lack expression of CD45, CD34, CD14 or CD11b, CD79 alpha or CD19, and HLA-DR surface molecules as measured by flow cytometry; (3) MSCs must differentiate into osteoblasts, adipocytes, and chondroblasts in vitro (Dominici et al. 2006; Horwitz et al. 2005). These criteria are currently being updated given the continued advances in understanding MSC biology with particular focus on developing potency assays applicable to clinical applications (Viswanathan et al. 2014). To address some of the variations in properties of cultured MSCs, an NCRR/NIH-sponsored Center for Preparation and Distribution of Adult Stem Cells (MSCs) serves as a pre-clinical resource for standardized preparations of mouse, rat, and human MSCs (http://medicine.tamhsc.edu/irm/msc-distribution.html). The NHBLI also sponsors the Production Assistance in Cellular Therapies (PACT) program, a training and GMP manufacturing resource that supports pre-clinical, IND preparation, and clinical investigations with MSCs and other cell therapy (https://secure.emmes.com/pactweb/Facilities).

6.3 Mechanisms of Action

While therapeutic interest in MSCs initially focused on exploring their capacity for multilineage differentiation to directly regenerate tissues and organs (Pittenger et al. 1999; Caplan and Bruder 2001), they are now also viewed as potent immunomodulators of disease-associated tissue microenvironments (Caplan 2009). Thus, the current translational landscape for MSCs includes therapeutic models involving direct tissue regeneration as well as indirect, through their anti-inflammatory and immunomodulatory effects on damaged and diseased tissues (Bianco et al. 2013; Griffin et al. 2013; Le Blanc and Mougiakakos 2012; Prockop and Oh 2012a, b) (Fig. 6.1). The capacity of MSCs to broadly modify the activity of most major components of the innate and adaptive immune system is now seen, along with their pro-angiogenic and cytoprotective effects, as an essential component of their therapeutic potential for many disease targets (Caplan and Bruder 2001; Bianco et al. 2013; Griffin et al. 2013; Le Blanc and Mougiakakos 2012).

The mechanisms by which MSCs might alleviate inflammation and injury are not completely understood and, as in other organ systems, likely involve multiple pathways including release of soluble mediators and/or microsomal particles as well as cell–cell contact. Importantly, the mechanisms of MSCs actions are different in different lung diseases and reflect the ability of the MSCs to sense and respond differently to different inflammatory environments (Weiss 2014; Weiss et al. 2013a, b). Much current interest in MSCs has focused on soluble factors due to their ability to secrete multiple paracrine factors such as growth factors, factors regulating

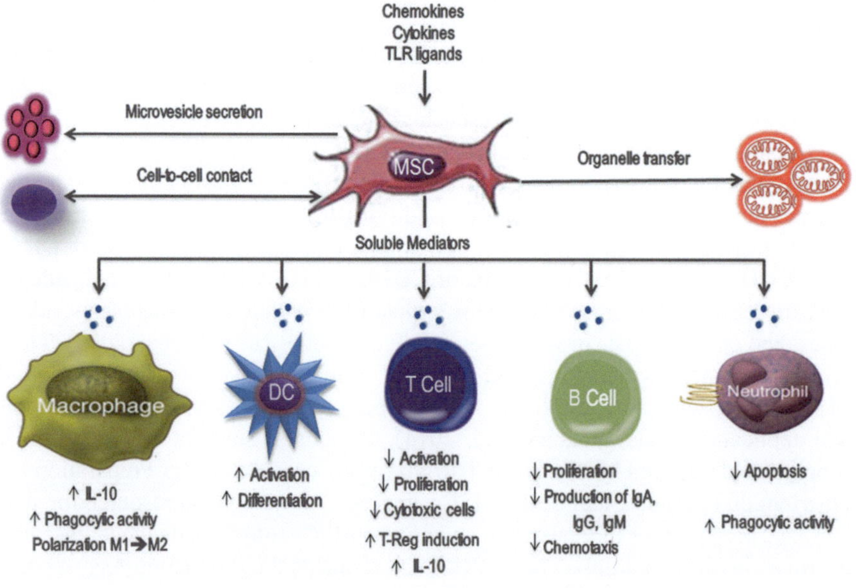

Fig. 6.1 Mechanisms of action of MSCs. MSCs promote benefitial effects through cell-to-cell interactions and through secretion of soluble mediators, microvesicles and whole organelles, that can directly affect many cells, regulating, for example, the innate and adaptative immune system

endothelial and epithelial permeability, factors regulating innate and adaptive immunity, anti-inflammatory cytokines, and more recently, antimicrobial peptides. Some of the soluble mediators implicated in the different model systems include IL-6, IL-10, indoleamine 2,3-dioxygenase (IDO), Nitrous Oxide (NO), hepatocyte growth factor (HGF), and transforming growth factor (TGF)-β (Lotfinegad et al. 2014). Transduction or transfection of the MSCs to over-express secreted mediators including angiopoietin-1 or keratinocyte growth factor (KGF) further decreases endotoxin-mediated lung injury presumably through abrogation of endotoxin-mediated endothelial injury (Xu et al. 2008; Chen et al. 2013). While native MSCs are effective, MSCs transduced to over-express eNOS, IL-10, KGF, or a CCL2 inhibitor were found to be more effective in preventing monocrotaline-induced pulmonary hypertension, ischemia-reperfusion-induced lung injury, or bleomycin-induced pulmonary inflammation and subsequent fibrosis, respectively (Prockop and Oh 2012a, b; Keating 2012; Weiss 2013; Antunes et al. 2014; Kanki-Horimoto et al. 2006). MSCs appear also to act in part by decreasing the increased endothelial permeability found in acute lung injury, by secreting antibacterial peptides, by promoting an anti-inflammatory M2 phenotype in alveolar macrophages, by increasing monocyte phagocytic activity, and by reducing collagen fiber content associated with increased metalloproteinase-8 expression and decreased expression of tissue inhibitor of metalloproteinase-1 (Weiss et al. 2013a, b). However, MSCs may not always ameliorate lung injury with some pre-clinical data suggesting that MSCs may contribute to established lung fibrosis (Epperly et al. 2003; Yan et al. 2007; Weiss and Ortiz 2013).

In addition, the ability to secrete microparticles that contain not only proteins but RNA or miRNA species which can modulate the expression of multiple genes make these packaging vesicles an attractive and quite plausible means for MSCs to regulate multiple pathways and produce a robust therapeutic effect in different lung injury models (Lee et al. 2012; Aliotta et al. 2012; Zhang et al. 2012a, b; Thebaud and Stewart 2012; Islam et al. 2012). Besides this, direct mitochondrial transfer from MSCs to ATII cells through connexin 43-mediated cell–cell bridges has been demonstrated to replenish endotoxin-depleted ATP stores and restore surfactant secretion (Islam et al. 2012).

Importantly, MSCs can also exert effects on lung inflammation and injury through primary interactions with the immune system rather than through direct actions in lung. For example, available information demonstrates that MSCs alleviate endotoxin-induced acute lung injury in mouse models inhibiting Th1 response through release of soluble anti-inflammatory, anti-bacterial, and angiogenic substances, including IL-10, angiopoietin 1, KGF, and others (Mei et al. 2007; Lee et al. 2009a, b; Danchuk et al. 2011; Gupta et al. 2012; Ionescu et al. 2012a, b). In contrast, MSC administration in mouse models of asthma (allergic airways inflammation) ameliorates airways hyper-responsiveness by reducing Th2/Th17-mediated inflammation through effects on antigen-specific T lymphocytes and by up-regulating T-regulatory cells (Cho et al. 2009; Park et al. 2010; Nemeth et al. 2010; Firinci et al. 2011; Goodwin et al. 2011). As such, as MSC-based therapies are developed for lung diseases, the specific disease pathogenesis in the context of the known actions of the MSCs must be carefully considered (Fig. 6.2).

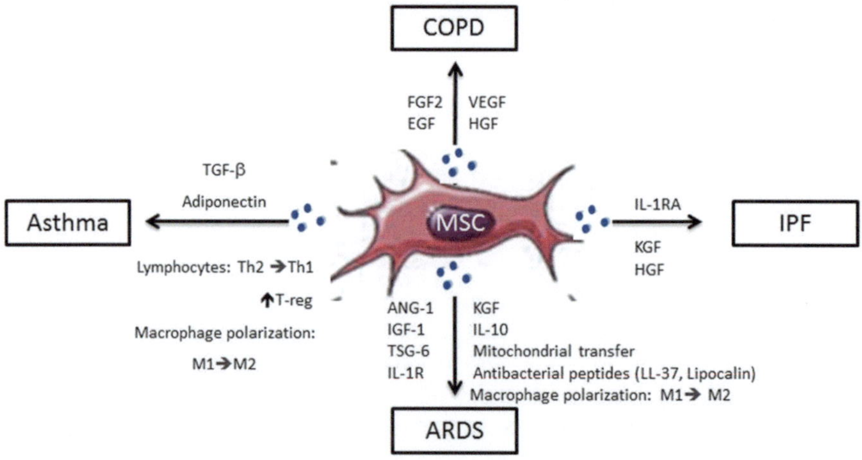

Fig. 6.2 MSCs trigger different responses according to each lung pathological environment. MSCs can secrete different soluble mediators depending on the lung disease microenvironment they get exposed to

6.4 Localization of MSCs in Lung After Systemic MSC Administration

Following systemic administration of MSCs isolated from bone marrow, adipose, placenta, or cord blood, a number of studies demonstrate that the cells initially localize in the lung vascular bed and that lung injury results in increased localization and/or retention of marrow-derived cells in lung (reviewed by Weiss 2013; Antunes et al. 2014). Whether this represents formation of cell emboli in the lung vasculature or specific adherence to pulmonary vascular adhesion or other molecules remains unclear. Further, the source of the MSCs may influence retention in the lung. For example, MSCs derived from human umbilical cord blood are cleared more rapidly from the lungs than are human bone marrow-derived MSCs (Nystedt et al. 2013). This reflects both differences in size of the MSCs from different sources as well as differential expression of specific integrin and proteoglycan patterns. Retention in the lung may also trigger the MSCs to have functional effects. For example, embolization of systemically administered MSCs in lung was felt to result in secretion of an anti-inflammatory protein, TSG-6 (Lee et al. 2009a, b). However, although bone marrow- or adipose-derived MSCs can be induced in vitro to express phenotypic markers of alveolar or airway epithelial cells, retention of MSCs in the lung is generally transient. Structural engraftment of MSCs as lung epithelium is a rare event of uncertain physiologic significance in lung (Loi et al. 2006; Sueblinvong et al. 2008; Ma et al. 2011; Maria and Tran 2011; Li et al. 2012; Yan et al. 2012; Baer 2011). However, some available data suggests that systemically administered MSCs can engraft as fibroblasts or myofibroblasts under certain fibrosing injury conditions, further discussed below (Antunes et al. 2014; Kanki-Horimoto et al. 2006). This is a potential undesirable effect of the MSCs.

6.5 Use of MSCs in Lung Diseases

A steadily increasing number of articles demonstrate efficacy of either systemic or intratracheal administration of MSCs obtained from bone marrow, adipose, cord blood, or placenta in a growing spectrum of lung injury models in mice and in a slowly growing number of clinical investigations in lung diseases (reviewed by Weiss 2013; Antunes et al. 2014). This includes mouse models of acute lung injury and bacterial lung infection (Gupta et al. 2012; Ionescu et al. 2012a, b; Danchuk et al. 2011; Kim et al. 2011; Sun et al. 2011a, b; Xu et al. 2012; Zhang et al. 2013), asthma (Firinci et al. 2011; Goodwin et al. 2011; Kapoor et al. 2011; Kavanagh and Mahon 2011; Lee et al. 2011; Ou-Yang et al. 2011; Ionescu et al. 2012a, b; Lathrop et al. 2014), bronchopulmonary dysplasia (Chang et al. 2011; Pierro et al. 2013; Zhang et al. 2010, 2012a, b; Tropea et al. 2012; Sutsko et al. 2013), COPD (Hoffman et al. 2011; Katsha et al. 2011; Schweitzer et al. 2011; Ingenito et al. 2012; Kim et al. 2012), ischemia re-perfusion injury (Yang et al. 2009; Manning et al. 2010; Sun et al. 2011a, b), post-inflammatory lung fibrosis (Ortiz et al. 2003, 2007; Rojas et al. 2005; Zhao et al. 2008; Aguilar et al. 2009; Kumamoto et al. 2009; Moodley et al. 2009; Cargnoni et al. 2010; Cabral et al. 2011; Lee et al. 2010; Saito et al. 2011), pulmonary hypertension (Lee et al. 2012; Baber et al. 2007; Umar et al. 2009; Kanki-Horimoto et al. 2006; Hansmann et al. 2012; Liang et al. 2011), sepsis and burns (Gonzalez-Rey et al. 2009; Nemeth et al. 2009; Iyer et al. 2010; Mei et al. 2010; Yagi et al. 2010a, b; Krasnodembskaya et al. 2012), and other critical illness or autoimmune-related lung injuries including hemorrhagic shock, lupus, pancreatitis, silicosis, and ventilator-induced lung injury (Shi et al. 2012; Pati et al. 2011; Wang et al. 2012; Lassance et al. 2009; Chimenti et al. 2012; Curley et al. 2012). Systemically administered MSCs can also home to tumors, through as yet unclear chemotactic mechanisms, and have been utilized for delivery of chemotherapeutic and other anti-tumor agents in mouse lung tumor models. This may provide a viable therapy for lung cancers, particularly with MSCs engineered to express anti-tumor compounds such as tumor necrosis factor-related apoptosis-inducing ligand (TRAIL) or Interferon beta (IFN-β) (Kanehira et al. 2007; Rachakatla et al. 2007; Stoff-Khalili et al. 2007; Xin et al. 2007; Zhang et al. 2008; Matsuzuka et al. 2010; Loebinger et al. 2009a, b; Heo et al. 2011; Hu et al. 2012). MSC administration has also been demonstrated to alleviate inflammation and injury produced by intratracheal instillation of either endotoxin or bacterial in human lung explants (Lee et al. 2009a, b, 2013).

In parallel with robust pre-clinical data, a slowly growing number of clinical investigations of MSC-based therapy in different lung diseases including ARDS, COPD, IPF, and silicosis are occurring (Table 6.1). In the following sections, the rationale for potential MSC effects, available pre-clinical data, and considerations of clinical trials of MSCs in ARDS, COPD, IPF, and silicosis will be considered.

Table 6.1 Worldwide stem cell clinical trials on lung diseases. Brief description of study location, number of enrolled patients, source and dose of MSCs, frequency of injections, route of administration, follow-up length, study status and clinical trials registration number in clinicaltrials.gov

Disease	Location	Patients	Cell type	Dose	Frequency	Delivery	Follow-up	Status	ClinicalTrials.gov
ARDS	USA	69	BM-MSC	$1–10 \times 10^6$/kg	Single dose	Intravenous	12 months	Recruiting	NCT01775774
	USA	60	BM-MSC	1×10^7/kg	Single dose	Intravenous	12 months	Recruiting	NCT02097641
	China	12	AD-MSC	1×10^6/kg	Single dose	Intravenous	28 days	Recruiting	NCT01902082
	China	20	MEN-MSC	1×10^7/kg	Twice weekly for 2 weeks	Intravenous	14 days	Recruiting	NCT02095444
	Sweden	10	BM-MSC	[a]	[a]	[a]	12 months	Recruiting	NCT02215811
COPD	Brazil	4	BMDMC	1×10^8/ml	Single dose	Intravenous	12 months	Completed	NCT01110252
	USA	62	BM-MSC	1×10^8	Four monthly	Intravenous	2 years	Completed	NCT00683722
	Brazil	10	BM-MSC	[a]	Single dose	Endobronchial	4 months	Recruiting	NCT01872624
	Netherlands	10	BM-MSC	[a]	Twice weekly	Intravenous	8 weeks	Completed	NCT01306513
	Russia	30	BM-MSC	2×10^8	Every 2 months for 1 year	Intravenous	2 years	Recruiting	NCT01849159
	Iran	12	BM-MSC	6×10^7	Single dose	Endobronchial	1 year	Not recruiting	NCT01758055
	Mexico	30	AD-MSC	[a]	Single dose	Intravenous	6 months	Recruiting	NCT01559051
IPF	USA	25	BM-MSC	2×10^7	Single dose	Intravenous	60 weeks	Recruiting	NCT02013700
	Spain	18	BM-MSC	Escalating doses	[a]	Endobronchial	12 months	Recruiting	NCT01919827
	Australia	8	PL-MSC	$1–2 \times 10^6$/kg	Single dose	Intravenous	6 months	Not recruiting	NCT01385644
Silicosis	Brazil	5	BM-MSC	2×10^7/kg	Single dose	Endobronchial	6 months	Recruiting	NCT01239862

ARDS acute respiratory distress syndrome, *COPD* chronic obstructive pulmonary disease, *IPF* idiopathic pulmonary fibrosis, *AD-MSC* adipose-derived mesenchymal stem cells, *BM-MSC* bone marrow-derived mesenchymal stem cells, *MEN-MSC* menstrual blood-derived mesenchymal stem cells, *PL-MSC* placental-derived mesenchymal stem cells

[a]Data not available

6.6 Acute Respiratory Distress Syndrome

The immunomodulatory and reparative potential of MSCs makes them potential therapeutic tools for the acute inflammatory response to infection and pulmonary injury seen in ARDS. Several pre-clinical studies on ARDS have demonstrated that MSCs may improve the pulmonary and systemic inflammation characteristic of the disease (Rojas et al. 2005; Nemeth et al. 2009; Mei et al. 2010; Gupta et al. 2007). In models of endotoxin- or bacterial-induced ARDS mice and in explanted human lungs, MSC administration not only attenuates inflammation by decreasing several inflammatory mediators, including tumor necrosis factor-alpha (TNF-α), macrophage inflammatory protein 2-alpha (MIP-2), IFN-γ, IL-1β, MIP-1α, IL-6, IL-8, and keratinocyte-derived cytokine in plasma and bronchoalveolar lavage fluid, but it is also able to rescue epithelial cells with mitochondrial dysfunction by mitochondria transfer (Islam et al. 2012; Spees et al. 2006). In addition, MSCs favorably influence the host response to bacterial infections, the commonest and most severe cause of ARDS. MSC therapy can reduce bacterial counts via a number of mechanisms, including increased antimicrobial peptide secretion, such as lipocalin-2 (Gupta et al. 2012), and enhanced macrophage phagocytosis (Krasnodembskaya et al. 2012; Nemeth et al. 2009). MSCs also enhance repair following lung injury, as evidenced by the findings that both intravenous (Curley et al. 2012) and intratracheal (Curley et al. 2013) MSC therapy restore lung function following ventilator-induced lung injury via a KGF-dependent mechanism. Based on these promising preclinical findings, a number of early-phase clinical trials have begun to investigate the potential of MSC therapy for severe ARDS.

Currently, five studies of MSC therapy safety in patients with ARDS are listed in ClinicalTrials.gov. At the University of California, San Francisco, a phase I, multicenter, open-label dose escalation clinical trial is in progress to assess the safety of intravenous infusion of allogeneic bone marrow-derived human MSCs in ARDS (NCT01775774) and a phase II, multicenter study was initiated in March, 2014, to assess the safety and efficacy of a single dose of allogeneic bone marrow-derived human MSCs infusion in patients with ARDS. In Sweden, a phase I, multi-center, open-label, non-randomized controlled trial is also testing the safety of bone-marrow-derived MSCs in ARDS (NCT02215811). Two phase I, randomized, double-blind, placebo-controlled trials are also taking place in China to test the safety of systemic infusion of allogeneic human adipose MSCs (NCT01902082) and of MSCs derived from menstrual blood (NCT02095444) in ARDS patients.

6.7 Chronic Obstructive Pulmonary Disease

In several preclinical studies, MSC administration has been demonstrated to attenuate inflammation by decreasing levels of inflammatory mediators, such as IL-1β, TNF-α, IL-8, as well as decrease apoptosis (Huh et al. 2011; Zhen et al. 2010), improve parenchymal repair (increased levels of KGF, HGF, and epidermal growth factor), and increase lung perfusion (Huh et al. 2011; Shigemura et al. 2006; Guan et al. 2013). Based on these preclinical findings, several groups are investigating the therapeutic potential of MSC therapy in COPD patients.

The first safety trial registered in ClinicalTrials.gov (NCT01110252) assessed systemic administration of autologous bone marrow mononuclear cells in four Brazilian patients/volunteers with advanced COPD (stage IV dyspnea) and found no obvious adverse effects after 1 year (Ribeiro-paes et al. 2011). In a recent trial carried out in the United States (NCT00683722), using non-HLA matched allogeneic bone marrow-derived MSCs obtained from healthy volunteers (Prochymal®; Osiris Therapeutics Inc), sixty-two patients were randomized to double-blinded intravenous infusions of either allogeneic MSCs or vehicle control. Patients received four monthly infusions (100×10^6 cells/infusion) and were subsequently followed for 2 years after the first infusion (Weiss et al. 2013a, b). This trial demonstrated that use of MSCs in COPD patients may be considered safe, as there were no infusion reactions and no deaths or serious adverse events deemed related to MSC administration. However, no significant differences were observed in the overall number of adverse events, frequency of COPD exacerbations, or severity of disease in patients treated with MSCs. A significant decrease was observed in circulating C-reactive protein in MSC-treated patients giving a potential mechanistic clue of MSC actions.

A phase I, non-randomized, open-label study in Brazil is currently recruiting patients diagnosed with severe heterogeneous emphysema to evaluate the safety of one-way endobronchial valves combined with bone-marrow MSCs (NCT01872624). Another phase I, non-randomized, non-blinded, prospective study to test the safety and feasibility of administration of bone-marrow MSCs before and after lung volume reduction surgery for severe pulmonary COPD has been concluded in the Netherlands (NCT01306513). Results for this study are pending. An open-label, non-randomized, multicenter study is currently underway in Mexico to evaluate the safety and efficacy of autologous adipose-derived stem cell transplantation in GOLD moderate-severe patients (NCT01559051).

6.8 Idiopathic Pulmonary Fibrosis

When administered early after injury is instituted, MSCs attenuate inflammation and prevent development of bleomycin-induced lung fibrosis in mice, the most commonly utilized experimental model. However, administration of MSCs at time intervals longer than 7 days after bleomycin administration had no effect on established fibrotic changes in either mouse or pig lungs (Ortiz et al. 2003). Further, using a different model of lung fibrosis induced by radiation exposure in rodents, MSCs administered at time points at which established fibrotic changes were present, MSCs were detected in the interstitium as myofibroblasts suggesting that fibroblastic differentiation of MSC occurred in response to mediators produced in the injured tissue (Epperly et al. 2003; Yan et al. 2007). These data suggest that MSC administration in the setting of an established or ongoing fibrotic response may worsen the disease process and augment scarring in injured tissue rather than reversing it. As such, available data only supports a potential ameliorating effect of MSC administration in fibrotic lung diseases if administered early in the disease course during active inflammation (Weiss et al. 2013a, b). At present, there is no data to support an ameliorating effect of MSCs on established lung fibrosis. Thus, careful consideration must be given to clinical investigations of MSCs in fibrotic lung diseases.

Despite these concerns, there are three trials listed in ClinicalTrials.gov that are taking place to evaluate the safety and feasibility of MSC therapy in IPF patients. In the United States, a phase I/II, randomized, blinded, and placebo-controlled trial is recruiting 25 IPF patients to investigate the safety, tolerability, and potential efficacy of intravenous infusion of allogeneic human MSCs (NCT02013700). Another phase I, open-label, multicenter, non-randomized study will evaluate the safety and feasibility of the endobronchial infusion of autologous bone-marrow MSCs at escalating doses in patients with mild-to-moderate IPF at Navarra University in Spain (NCT01919827). A third phase I, open-label, single-center, non-randomized dose-escalation study in Australia evaluated the safety and feasibility of placental-derived MSC infusion in IPF patients (NCT01385644). Initial results from this trial demonstrate no adverse effects over the 6-month follow-up period. A fourth trial, not listed in clinicaltrials.gov, reported no adverse effects of endobronchial administration of autologous adipose-derived MSCs over a 1-year follow-up period (Tzouvelekis et al. 2013).

6.9 Silicosis

Preclinical studies using an experimental model of silicosis demonstrated that both systemic and intratracheal administration of autologous BMMCs reduce inflammation and fibrosis (Lassance et al. 2009; Lopes-Pacheco et al. 2013). These positive effects encouraged a non-randomized, phase I trial of endobronchial administration of autologous BMMCs in patients with chronic and accelerated silicosis in Brazil (NCT01239862). In this study, three patients each received 2×10^7 bone marrow-derived cells labeled with 99mTc. The cellular infusion procedure was well tolerated by the patients, and no respiratory, cardiovascular, or hematological complications were observed. Scintigraphy showed an increase in lung perfusion in the basal region up to day 180 after the infusion, while the apex and midzone areas presented reduced perfusion at day 180 (Loivos et al. 2010; Souza et al. 2012). However, no subsequent clinical study of MSCs in silicosis has occurred.

6.10 Conclusions and Future Directions

Cell therapy approaches for lung diseases and critical illnesses including ARDS, COPD, IPF, and silicosis continue to evolve at a rapid pace. Pre-clinical studies with MSCs have generated a great amount of enthusiasm as a beneficial therapy for lung diseases and critical illnesses. Initial clinical trials have demonstrated that MSC administration is safe, with few adverse effects, but substantial challenges still have to be overcome before MSCs can be used for clinical practice. As such, further studies focusing on understanding the mechanisms of action of MSCs must be more investigated in order to continue to develop rational approaches for clinical trials. Nonetheless, cell-based therapies with MSCs and other cell types offer potential hope for these devastating and incurable pulmonary diseases.

References

Aguilar S, Scotton CJ, McNulty K et al (2009) Bone marrow stem cells expressing keratinocyte growth factor via an inducible lentivirus protects against bleomycin-induced pulmonary fibrosis. PLoS One 4:e8013

Aliotta JM, Lee D, Puente N, Faradyan S, Sears E, Amaral A, Goldberg L, Dooner MS, Pereira M, Quesenberry PJ (2012) Progenitor/stem cell fate determination: interactive dynamics of cell cycle and microvesicles. Stem Cells Dev 21(10):1627–1638

Antunes MA, Laffey JG, Pelosi P, Rocco PR (2014) Mesenchymal stem cell trials for pulmonary diseases. J Cell Biochem. doi:10.1002/jcb.24783

Baber SR, Deng W, Master RG, Bunnell BA, Taylor BK, Murthy SN, Hyman AL, Kadowitz PJ (2007) Intratracheal mesenchymal stem cell administration attenuates monocrotaline-induced pulmonary hypertension and endothelial dysfunction. Am J Physiol Heart Circ Physiol 292:H1120–H1128

Baer PC (2011) Adipose-derived stem cells and their potential to differentiate into the epithelial lineage. Stem Cells Dev 20(10):1805–1816

Baer PC, Geiger H (2012) Adipose-derived mesenchymal stromal/stem cells: tissue localization, characterization, and heterogeneity. Stem Cells Int 2012:812693

Bi Y, Ehirchiou D, Kilts TM, Inkson CA, Embree MC, Sonoyama W et al (2007) Identification of tendon stem/progenitor cells and the role of the extracellular matrix in their niche. Nat Med 13(10):1219–1227

Bianco P, Cao X, Frenette PS et al (2013) The meaning, the sense and the significance: translating the science of mesenchymal stem cells into medicine. Nat Med 19:35–42

Cabral RM, Branco E, Rizzo Mdos S, Ferreira GJ, Gregores GB, Samoto VY, Stopiglia AJ, Maiorka PC, Fioretto ET, Capelozzi VL, Borges JB, Gomes S, Beraldo MA, Carvalho CR, Miglino MA (2011) Cell therapy for fibrotic interstitial pulmonary disease: experimental study. Microsc Res Tech 74:957–962

Capelli C, Gotti E, Morigi M, Rota C, Weng L, Dazzi F et al (2011) Minimally manipulated whole human umbilical cord is a rich source of clinical-grade human mesenchymal stromal cells expanded in human platelet lysate. Cytotherapy 13(7):786–801

Caplan AI (2009) Why are MSCs therapeutic? New data: new insight. J Pathol 217:318–324

Caplan AI, Bruder SP (2001) Mesenchymal stem cells: building blocks for molecular medicine in the 21st century. Trends Mol Med 7:259–264

Cargnoni A, Gibelli L, Tosini A et al (2010) Transplantation of allogeneic and xenogeneic placenta-derived cells reduces bleomycin-induced lung fibrosis. Cell Transplant 18:405–422

Chang YS, Choi SJ, Sung DK, Kim SY, Oh W, Yang YS, Park WS (2011) Intratracheal transplantation of human umbilical cord blood-derived mesenchymal stem cells dose-dependently attenuates hyperoxia-induced lung injury in neonatal rats. Cell Transplant 20(11):1843–1854

Chen J, Li C, Gao X, Li C, Liang Z, Yu L, Li Y, Xiao X, Chen L (2013) Keratinocyte growth factor gene delivery via mesenchymal stem cells protects against lipopolysaccharide-induced acute lung injury in mice. PLoS One 8(12)

Chimenti L, Luque T, Bonsignore MR, Ramirez J, Navajas D, Farre R (2012) Pre-treatment with mesenchymal stem cells reduces ventilator-induced lung injury. Eur Respir J 40:939–948

Cho KS, Park HK, Park HY et al (2009) IFATS collection: immunomodulatory effects of adipose tissue-derived stem cells in an allergic rhinitis mouse model. Stem Cells 27:259–265

Chong PP, Selvaratnam L, Abbas AA, Kamarul T (2012) Human peripheral blood derived mesenchymal stem cells demonstrate similar characteristics and chondrogenic differentiation potential to bone marrow derived mesenchymal stem cells. J Orthop Res 30(4):634–642

Choong PF, Mok PL, Cheong SK, Then KY (2007) Mesenchymal stromal cell-like characteristics of corneal keratocytes. Cytotherapy 9(3):252–258

Curley GF, Hayes M, Ansari B, Shaw G, Ryan A, Barry F, O'Brien T, O'Toole D, Laffey JG (2012) Mesenchymal stem cells enhance recovery and repair following ventilator-induced lung injury in the rat. Thorax 67:496–501

Curley GF, Ansari B, Hayes M, Devaney J, Masterson C, Ryan A, Barry F, O'Brien T, Toole DO, Laffey JG (2013) Effects of intratracheal mesenchymal stromal cell therapy during recovery and resolution after ventilator-induced lung injury. Anesthesiology 118:924–932

Danchuk S, Ylostalo JH, Hossain F, Sorge R, Ramsey A, Bonvillain RW, Lasky JA, Bunnell BA, Welsh DA, Prockop DJ, Sullivan DE (2011) Human multipotent stromal cells attenuate lipopolysaccharide-induced acute lung injury in mice via secretion of tumor necrosis factor-alpha-induced protein 6. Stem Cell Res Ther 2(3):27

Dominici M, Le Blanc K, Mueller I, Slaper-Cortenbach I, Marini F, Krause D et al (2006) Minimal criteria for defining multipotent mesenchymal stromal cells. The International Society for Cellular Therapy position statement. Cytotherapy 8(4):315–317

Epperly MW, Guo H, Gretton JE, Greenberger JS (2003) Bone marrow origin of myofibroblasts in irradiation pulmonary fibrosis. Am J Respir Cell Mol Biol 29(2):213–224

Firinci F, Karaman M, Baran Y, Bagriyanik A, Ayyildiz ZA, Kiray M, Kozanoglu I, Yilmaz O, Uzuner N, Karaman O (2011) Mesenchymal stem cells ameliorate the histopathological changes in a murine model of chronic asthma. Int Immunopharmacol 11:1120–1126

Gonzalez-Rey E, Anderson P, Gonzalez MA, Rico L, Buscher D, Delgado M (2009) Human adult stem cells derived from adipose tissue protect against experimental colitis and sepsis. Gut 58:929–939

Goodwin M, Sueblinvong V, Eisenhauer P, Ziats NP, Leclair L, Poynter ME, Steele C, Rincon M, Weiss DJ (2011) Bone marrow derived mesenchymal stromal cells inhibit Th2-mediated allergic airways inflammation in mice. Stem Cells 29:1137–1148

Griffin MD, Ryan AE, Alagesan S et al (2013) Anti-donor immune responses elicited by allogeneic mesenchymal stem cells: what have we learned so far? Immunol Cell Biol 91:40–51

Gronthos S (2011) The therapeutic potential of dental pulp cells: more than pulp fiction? Cytotherapy 13(10):1162–1163

Gronthos S, Mankani M, Brahim J, Robey PG, Shi S (2000) Postnatal human dental pulp stem cells (DPSCs) in vitro and in vivo. Proc Natl Acad Sci U S A 97(25):13625–13630

Guan XJ, Song L, Han FF, Cui ZL, Chen X, Guo XJ, Xu WG (2013) Mesenchymal stem cells protect cigarette smoke-damaged lung and pulmonary function partly via VEGF-VEGF receptors. J Cell Biochem 114:323–335

Gupta N, Su X, Popov B, Lee JW, Serikov V, Matthay MA (2007) Intrapulmonary delivery of bone marrow-derived mesenchymal stem cells improves survival and attenuates endotoxin-induced acute lung injury in mice. J Immunol 179:1855–1863

Gupta N, Krasnodembskaya A, Kapetanaki M, Mouded M, Tan X, Serikov V, Matthay MA (2012) Mesenchymal stem cells enhance survival and bacterial clearance in murine Escherichia coli pneumonia. Thorax 67:533–539

Hansmann G, Fernandez-Gonzalez A, Aslam M, Vitali SH, Martin T, Mitsialis SA, Kourembanas S (2012) Mesenchymal stem cell-mediated reversal of bronchopulmonary dysplasia and associated pulmonary hypertension. Pulm Circ 2:170–181

Heo SC, Lee KO, Shin SH, Kwon YW, Kim YM, Lee CH, Kim YD, Lee MK, Yoon MS, Kim JH (2011) Periostin mediates human adipose tissue-derived mesenchymal stem cell-stimulated tumor growth in a xenograft lung adenocarcinoma model. Biochim Biophys Acta 1813(12):2061–2070

Hoffman AM, Paxson JA, Mazan MR, Davis AM, Tyagi S, Murthy S, Ingenito EP (2011) Lung-derived mesenchymal stromal cell post-transplantation survival, persistence, paracrine expression, and repair of elastase-injured lung. Stem Cells Dev 20(10):1779–1792

Horwitz EM, Le Blanc K, Dominici M, Mueller I, Slaper-Cortenbach I, Marini FC et al (2005) Clarification of the nomenclature for MSC: The International Society for Cellular Therapy position statement. Cytotherapy 7(5):393–395

Hu YL, Huang B, Zhang TY, Miao PH, Tang GP, Tabata Y, Gao J (2012) Mesenchymal stem cells as a novel carrier for targeted delivery of gene in cancer therapy based on nonviral transfection. Mol Pharm 9:2698–2709

Huh JW, Kim SY, Lee JH, Lee JS, Van Ta Q, Kim M, Oh YM, Lee YS, Lee SD (2011) Bone marrow cells repair cigarette smoke-induced emphysema in rats. Am J Physiol Lung Cell Mol Physiol 301(3):L255–L266

Igura K, Zhang X, Takahashi K, Mitsuru A, Yamaguchi S, Takashi TA (2004) Isolation and characterization of mesenchymal progenitor cells from chorionic villi of human placenta. Cytotherapy 6(6):543–553

Ingenito EP, Tsai L, Murthy S, Tyagi S, Mazan M, Hoffman A (2012) Autologous lung-derived mesenchymal stem cell transplantation in experimental emphysema. Cell Transplant 21(1): 175–189

Ionescu L, Byrne RN, van Haaftern T, Vadivel A, Alphonse RS, Rey-Parra GJ, Weissmann G, Hall A, Eaton F, Thebaud B (2012a) Stem cell conditioned medium improves acute lung injury in mice: in vivo evidence for stem cell paracrine action. Am J Physiol Lung Cell Mol Physiol 303:L967–L977

Ionescu LI, Alphonse RS, Arizmendi N, Morgan B, Abel M, Eaton F, Duszyk M, Vliagoftis H, Aprahamian TR, Walsh K, Thebaud B (2012b) Airway delivery of soluble factors from plastic-adherent bone marrow cells prevents murine asthma. Am J Respir Cell Mol Biol 46:207–216

Islam MN, Das SR, Emin MT, Wei M, Sun L, Westphalen K, Rowlands DJ, Quadri SK, Bhattacharya S, Bhattacharya J (2012) Mitochondrial transfer from bone-marrow-derived stromal cells to pulmonary alveoli protects against acute lung injury. Nat Med 8(5):759–765

Iyer SS, Torres-Gonzalez E, Neujahr DC, Kwon M, Brigham KL, Jones DP, Mora AL, Rojas M (2010) Effect of bone marrow-derived mesenchymal stem cells on endotoxin-induced oxidation of plasma cysteine and glutathione in mice. Stem Cells Int 2010:868076

Jazedje T, Perin PM, Czeresnia CE, Maluf M, Halpern S, Secco M et al (2009) Human fallopian tube: a new source of multipotent adult mesenchymal stem cells discarded in surgical procedures. J Transl Med 7:46

Jones E, Churchman SM, English A, Buch MH, Horner EA, Burgoyne CH et al (2010) Mesenchymal stem cells in rheumatoid synovium: enumeration and functional assessment in relation to synovial inflammation level. Ann Rheum Dis 69(2):450–457

Kanehira M, Xin H, Hoshino K, Maemondo M, Mizuguchi H, Hayakawa T, Matsumoto K, Nakamura T, Nukiwa T, Saijo Y (2007) Targeted delivery of NK4 to multiple lung tumors by bone marrow-derived mesenchymal stem cells. Cancer Gene Ther 14:894–903

Kang SG, Shinojima N, Hossain A, Gumin J, Yong RL, Colman H et al (2010) Isolation and perivascular localization of mesenchymal stem cells from mouse brain. Neurosurgery 67(3): 711–720

Kanki-Horimoto S, Horimoto H, Mieno S et al (2006) Implantation of mesenchymal stem cells overexpressing endothelial nitric oxide synthase improves right ventricular impairments caused by pulmonary hypertension. Circulation 114(1 Suppl):I181–I185

Kapoor S, Patel SA, Kartan S, Axelrod D, Capitle E, Rameshwar P (2011) Tolerance-like mediated suppression by mesenchymal stem cells in patients with dust mite allergy-induced asthma. J Allergy Clin Immunol 129(4):1094–1101

Katsha AM, Ohkouchi S, Xin H, Kanehira M, Sun R, Nukiwa T, Saijo Y (2011) Paracrine factors of multipotent stromal cells ameliorate lung injury in an elastase-induced emphysema model. Mol Ther 19:196–203

Kavanagh H, Mahon BP (2011) Allogeneic mesenchymal stem cells prevent allergic airway inflammation by inducing murine regulatory T cells. Allergy 66(4):523–531

Keating A (2012) Mesenchymal stromal cells: new directions. Cell Stem Cell 10(6):709–716

Kern S, Eichler H, Stoeve J, Kluter H, Bieback K (2006) Comparative analysis of mesenchymal stem cells from bone marrow, umbilical cord blood, or adipose tissue. Stem Cells 24(5):1294–1301

Kim ES, Chang YS, Choi SJ, Kim JK, Yoo HS, Ahn SY, Sung DK, Kim SY, Park YR, Park WS (2011) Intratracheal transplantation of human umbilical cord blood-derived mesenchymal stem cells attenuates Escherichia coli-induced acute lung injury in mice. Respir Res 12:108

Kim SY, Lee JH, Kim HJ, Park MK, Huh JW, Ro JY, Oh YM, Lee SD, Lee YS (2012) Mesenchymal stem cell-conditioned media recovers lung fibroblasts from cigarette smoke-induced damage. Am J Physiol Lung Cell Mol Physiol 302:L891–L908

Kotton DN (2012) Next generation regeneration: the hope and hype of lung stem cell research. Am J Respir Crit Care Med 12:1255–1260

Krampera M, Sartoris S, Liotta F, Pasini A, Angeli R, Cosmi L et al (2007) Immune regulation by mesenchymal stem cells derived from adult spleen and thymus. Stem Cells Dev 16(5): 797–810

Krasnodembskaya A, Samarani G, Song Y, Zhuo H, Su X, Lee JW, Gupta N, Petrini M, Matthay MA (2012) Human mesenchymal stem cells reduce mortality and bacteremia in gram-negative sepsis in mice in part by enhancing the phagocytic activity of blood monocytes. Am J Physiol Lung Cell Mol Physiol 302:L1003–L1013

Kumamoto M, Nishiwaki T, Matsuo N et al (2009) Minimally cultured bone marrow mesenchymal stem cells ameliorate fibrotic lung injury. Eur Respir J 34:740–748

Lama VN, Smith L, Badri L, Flint A, Andrei AC, Murray S et al (2007) Evidence for tissue-resident mesenchymal stem cells in human adult lung from studies of transplanted allografts. J Clin Invest 117(4):989–996

Lassance RM, Prota LF, Maron-Gutierrez T, Garcia CS, Abreu SC, Passaro CP, Xisto DG, Castiglione RC, Carreira H Jr, Ornellas DS, Santana MC, Souza SA, Gutfilen B, Fonseca LM, Rocco PR, Morales MM (2009) Intratracheal instillation of bone marrow-derived cell in an experimental model of silicosis. Respir Physiol Neurobiol 169:227–233

Lathrop MJ, Brooks EM, Bonenfant NR, Sokocevic D, Borg ZD, Goodwin M, Loi R, Cruz FF, Dunaway CW, Steele C, Weiss DJ (2014) Mesenchymal stromal cells mediate aspergillus hyphal extract-induced allergic airways inflammation by inhibition of the Th17 signaling pathway. Stem Cells Transl Med 3(2):194–205

Lau AN, Goodwin M, Kim CF, Weiss DJ (2012) Stem cells and regenerative medicine in lung biology and diseases. Mol Ther 20:1116–1130

Le Blanc K, Mougiakakos D (2012) Multipotent mesenchymal stromal cells and the innate immune system. Nat Rev Immunol 12:383–396

Lee RH, Pulin AA, Seo MJ et al (2009a) Intravenous hMSCs improve myocardial infarction in mice because cells embolized in lung are activated to secrete the anti-inflammatory protein TSG-6. Cell Stem Cell 5(1):54–63

Lee JW, Fang X, Gupta N et al (2009b) Allogeneic human mesenchymal stem cells for treatment of E. coli endotoxin-induced acute lung injury in the ex vivo perfused human lung. Proc Natl Acad Sci U S A 106:16357–16362

Lee S, Jang A, Kim Y, Cha J, Kim T, Jung S, Park S, Lee Y, Won J, Kim Y, Park C (2010) Modulation of cytokine and nitric oxide by mesenchymal stem cell transfer in lung injury/fibrosis. Respir Res 11:16

Lee SH, Jang AS, Kwon JH, Park SK, Won JH, Park CS (2011) Mesenchymal stem cell transfer suppresses airway remodeling in a toluene diisocyanate-induced murine asthma model. Allergy Asthma Immunol Res 3(3):205–211

Lee C, Mitsialis SA, Aslam M, Vitali SH, Vergadi E, Konstantinou G, Sdrimas K, Fernandez-Gonzalez A, Kourembanas S (2012) Exosomes mediate the cytoprotective action of mesenchymal stromal cells on hypoxia-induced pulmonary hypertension. Circulation 126: 2601–2611

Lee JW, Krasnodembskaya A, McKenna DH, Song Y, Abbott J, Matthay MA (2013) Therapeutic effects of human mesenchymal stem cells in ex vivo human lungs injured with live bacteria. Am J Respir Crit Care Med 187(7):751–760

Li M, Ikehara S (2013) Bone-marrow-derived mesenchymal stem cells for organ repair. Stem Cells Int 2013:132642

Li H, Xu Y, Fu Q, Li C (2012) Effects of multiple agents on epithelial differentiation of rabbit adipose-derived stem cells in 3D culture. Tissue Eng Part A 18(17–18):1760–1770

Liang OD, Mitsialis SA, Chang MS, Vergadi E, Lee C, Aslam M, Fernandez-Gonzalez A, Liu X, Baveja R, Kourembanas S (2011) Mesenchymal stromal cells expressing heme oxygenase-1 reverse pulmonary hypertension. Stem Cells 29:99–107

Loebinger MR, Eddaoudi A, Davies D, Janes SM (2009a) Mesenchymal stem cell delivery of TRAIL can eliminate metastatic cancer. Cancer Res 69:4134–4142

Loebinger MR, Kyrtatos PG, Turmaine M, Price AN, Pankhurst Q, Lythgoe MF, Janes SM (2009b) Magnetic resonance imaging of mesenchymal stem cells homing to pulmonary metastases using biocompatible magnetic nanoparticles. Cancer Res 69:8862–8867

Loi R, Beckett T, Goncz KK, Suratt BT, Weiss DJ (2006) Limited restoration of cystic fibrosis lung epithelium in vivo with adult bone marrow-derived cells. Am J Respir Crit Care Med 173(2):171–179

Loivos LP, Fonseca LMB, Lima MA, Rocco PRM, Silva JRL, Morales MM (2010) Phase-1 study of autologous bone marrow cells intrabronchial instillation for patients silicosis. ClinicalTrials.gov Identifier: NCT01239862

Lopes-Pacheco M, Xisto DG, Ornellas FM, Antunes MA, Abreu SC, Rocco PR, Takiya CM, Morales MM (2013) Repeated administration of bone marrow-derived cells prevents disease progression in experimental silicosis. Cell Physiol Biochem 32(6):1681–1694

Lotfinegad P, Shamsasenjan K, Movassaghpour A, Majidi J, Baradaran B (2014) Immunomodulatory nature and site specific affinity of mesenchymal stem cells: a hope in cell therapy. Adv Pharm Bull 4(1):5–13

Ma N, Gai H, Mei J, Ding FB, Bao CR, Nguyen DM, Zhong H (2011) Bone marrow mesenchymal stem cells can differentiate into type II alveolar epithelial cells in vitro. Cell Biol Int 35(12):1261–1266

Mafi R, Hindocha S, Mafi P, Griffin M, Khan WS (2011) Sources of adult mesenchymal stem cells applicable for musculoskeletal applications—a systematic review of the literature. Open Orthop J 5(Suppl 2):242–248

Manning E, Pham S, Li S, Vazquez-Padron RI, Mathew J, Ruiz P, Salgar SK (2010) Interleukin-10 delivery via mesenchymal stem cells: a novel gene therapy approach to prevent lung ischemia-reperfusion injury. Hum Gene Ther 21:713–727

Maria OM, Tran S (2011) Human mesenchymal stem cells cultured with salivary gland biopsies adopt an epithelial phenotype. Stem Cells Dev 20(6):959–967

Matsuzuka T, Rachakatla RS, Doi C, Maurya DK, Ohta N, Kawabata A, Pyle MM, Pickel L, Reischman J, Marini F, Troyer D, Tamura M (2010) Human umbilical cord matrix-derived stem cells expressing interferon-beta gene significantly attenuate bronchioloalveolar carcinoma xenografts in SCID mice. Lung Cancer 70:28–36

Mei SH, McCarter SD, Deng Y, Parker CH, Liles WC, Stewart DJ (2007) Prevention of LPS-induced acute lung injury in mice by mesenchymal stem cells overexpressing angiopoietin 1. PLoS One Med 4:e269

Mei SHJ, Haitsma JJ, Dos Santos CC, Deng Y, Lai PFH, Slutsky AS, Liles WC, Stewart DJ (2010) Mesenchymal stem cells reduce inflammation while enhancing bacterial clearance and improving survival in sepsis. Am J Respir Crit Care Med 182:1047–1057

Moodley Y, Atienza D, Manuelpillai U, Samuel CS, Tchongue J, Ilancheran S, Boyd R, Trounson A (2009) Human umbilical cord mesenchymal stem cells reduce fibrosis of bleomycin-induced lung injury. Am J Pathol 175:303–313

Najimi M, Khuu DN, Lysy PA, Jazouli N, Abarca J, Sempoux C et al (2007) Adult-derived human liver mesenchymal-like cells as a potential progenitor reservoir of hepatocytes? Cell Transplant 16(7):717–728

Nakahara H, Bruder SP, Haynesworth SE, Holecek JJ, Baber MA, Goldberg VM et al (1990) Bone and cartilage formation in diffusion chambers by subcultured cells derived from the periosteum. Bone 11(3):181–188

Nemeth K, Leelahavanichkul A, Yuen PS, Mayer B, Parmelee A, Doi K, Robey PG, Leelahavanichkul K, Koller BH, Brown JM, Hu X, Jelinek I, Star RA, Mezey E (2009) Bone marrow stromal cells attenuate sepsis via prostaglandin E(2)-dependent reprogramming of host macrophages to increase their interleukin-10 production. Nat Med 15:42–49

Nemeth K, Keane-Myers A, Brown JM et al (2010) Bone marrow stromal cells use TGF-beta to suppress allergic responses in a mouse model of ragweed-induced asthma. Proc Natl Acad Sci U S A 107:5652–5657

Nystedt J, Anderson H, Tikkanen J et al (2013) Cell surface structures influence lung clearance rate of systemically infused mesenchymal stromal cells. Stem Cells 31(2):317–326. doi:10.1002/stem.1271

Ortiz LA, Gambelli F, McBride C, Gaupp D, Baddoo M, Kaminski N, Phinney DG (2003) Mesenchymal stem cell engraftment in lung is enhanced in response to bleomycin exposure and ameliorates its fibrotic effects. Proc Natl Acad Sci U S A 100:8407–8411

Ortiz LA, Dutreil M, Fattman C, Pandey AC, Torres G, Go K, Phinney DG (2007) Interleukin 1 receptor antagonist mediates the antiinflammatory and antifibrotic effect of mesenchymal stem cells during lung injury. Proc Natl Acad Sci U S A 104:11002–11007

Oswald J, Boxberger S, Jørgensen B, Feldmann S, Ehninger G, Bornhäuser M et al (2004) Mesenchymal stem cells can be differentiated into endothelial cells in vitro. Stem Cells 22(3):377–384

Ou-Yang H-F, Huang Y, Hu X-B, Wu C-G (2011) Suppression of allergic airway inflammation in a mouse model of asthma by exogenous mesenchymal stem cells. Exp Biol Med 236(12):1461–1467

Park HK, Cho KS, Park HY et al (2010) Adipose-derived stromal cells inhibit allergic airway inflammation in mice. Stem Cells Dev 19:1811–1818

Pati S, Gerber M, Menge TD et al (2011) Bone marrow derived mesenchymal stem cells inhibit inflammation and preserve vascular endothelial integrity in the lungs after hemorrhagic shock. PLoS One 6:e25171

Paunescu V, Deak E, Herman D, Siska IR, Tanasie G, Bunu C et al (2007) In vitro differentiation of human mesenchymal stem cells to epithelial lineage. J Cell Mol Med 11(3):502–508

Pawitan JA (2009) Prospect of adipose tissue derived mesenchymal stem cells in regenerative medicine. Cell Tissue Transplant Ther 2:7–9

Pierro M, Ionescu L, Montemurro T, Vadivel A, Weissmann G, Oudit G, Emery D, Bodiga S, Eaton F, Peault B, Mosca F, Lazzari L, Thebaud B (2013) Short-term, long-term and paracrine effect of human umbilical cord-derived stem cells in lung injury prevention and repair in experimental bronchopulmonary dysplasia. Thorax 68(5):475–484

Pittenger MF, Mackay AM, Beck SC et al (1999) Multilineage potential of adult human mesenchymal stem cells. Science 284:143–147

Prockop DJ, Oh JY (2012a) Medical therapies with adult stem/progenitor cells (MSCs): a backward journey from dramatic results in vivo to the cellular and molecular explanations. J Cell Biochem 113(5):1460–1469

Prockop DJ, Oh JY (2012b) Mesenchymal stem/stromal cells (MSCs): role as guardians of inflammation. Mol Ther 20:14–20

Rachakatla RS, Marini F, Weiss ML, Tamura M, Troyer D (2007) Development of human umbilical cord matrix stem cell-based gene therapy for experimental lung tumors. Cancer Gene Ther 14:828–835

Ribeiro-Paes JT, Bilaqui A, Greco OT, Ruiz MA, Marcelino MY, Stessuk T, de Faria CA, Lago MR (2011) Unicentric study of cell therapy in chronic obstructive pulmonary disease/pulmonary emphysema. Int J Chron Obstruct Pulmon Dis 6:63–71

Rojas M, Xu J, Woods CR, Mora AL, Spears W, Roman J, Brigham KL (2005) Bone marrow-derived mesenchymal stem cells in repair of the injured lung. Am J Respir Cell Mol Biol 33:145–152

Romanov YA, Svintsitskaya VA, Smirnov VN (2003) Searching for alternative sources of postnatal human mesenchymal stem cells: candidate MSC-like cells from umbilical cord. Stem Cells 21(1):105–110

Romieu-Mourez R, Coutu DL, Galipeau J (2012) The immune plasticity of mesenchymal stromal cells from mice and men: concordances and discrepancies. Front Biosci 4:824–837

Sabapathy V, Ravi S, Srivastava V, Srivastava A, Kumar S (2012) Long-term cultured human term placenta-derived mesenchymal stem cells of maternal origin displays plasticity. Stem Cells Int 2012:174328

Sabatini F, Petecchia L, Tavian M, Jodon De Villeroche V, Rossi GA, Brouty-Boye D (2005) Human bronchial fibroblasts exhibit a mesenchymal stem cell phenotype and multilineage differentiating potentialities. Lab Invest 85(8):962–971

Saito S, Nakayama T, Hashimoto N, Miyata Y, Egashira K, Nakao N, Nishiwaki S, Hasegawa M, Hasegawa Y, Naoe T (2011) Mesenchymal stem cells stably transduced with a dominant-negative inhibitor of CCL2 greatly attenuate bleomycin-induced lung damage. Am J Pathol 179:1088–1094

Schweitzer K, Johnstone BH, Garrison J, Rush N, Cooper S, Traktuev DO, Feng D, Adamowicz JJ, Van Demark M, Fisher AJ, Kamocki K, Brown MB, Presson RG Jr, Broxmeyer HE, March KL, Petrache I (2011) Adipose stem cell treatment in mice attenuates lung and systemic injury induced by cigarette smoking. Am J Respir Crit Care Med 183:215–225

Shi D, Wang D, Li X, Zhang H, Che N, Lu Z, Sun L (2012) Allogeneic transplantation of umbilical cord-derived mesenchymal stem cells for diffuse alveolar hemorrhage in systemic lupus erythematosus. Clin Rheumatol 31:841–846

Shigemura N, Okumura M, Mizuno S, Imanishi Y, Nakamura T, Sawa Y (2006) Autologous transplantation of adipose tissue-derived stromal cells ameliorates pulmonary emphysema. Am J Transplant 6:2592–2600

Souza S, Loivos L, Lima M, Szklo A, Goldenberg R, Rocco PRM, Silva JRL, Morales MM, Fonseca L, Gutffilen B (2012) Intrabronchial instillation of bone marrow derived mononuclear cells in silicotic patients: nuclear medicine analysis and follow-up. J Nucl Med 53(Suppl 1):606

Spees JL, Olson SD, Whitney MJ, Prockop DJ (2006) Mitochondrial transfer between cells can rescue aerobic respiration. Proc Natl Acad Sci U S A 103:1283–1288

Stoff-Khalili MA, Rivera AA, Mathis JM, Banerjee NS, Moon AS, Hess A, Rocconi RP, Numnum TM, Everts M, Chow LT, Douglas JT, Siegal GP, Zhu ZB, Bender HG, Dall P, Stoff A, Pereboeva L, Curiel DT (2007) Mesenchymal stem cells as a vehicle for targeted delivery of CRAds to lung metastases of breast carcinoma. Breast Cancer Res Treat 105:157–167

Sueblinvong V, Loi R, Eisenhauer PL, Bernstein IM, Suratt BT, Spees JL, Weiss DJ (2008) Derivation of lung epithelium from human cord blood-derived mesenchymal stem cells. Am J Respir Crit Care Med 177(7):701–711

Sun CK, Yen CH, Lin YC, Tsai TH, Chang LT, Kao YH, Chua S, Fu M, Ko SF, Leu S, Yip HK (2011a) Autologous transplantation of adipose-derived mesenchymal stem cells markedly reduced acute ischemia-reperfusion lung injury in a rodent model. J Transl Med 9:118

Sun J, Han ZB, Liao W, Yang SG, Yang Z, Yu J, Meng L, Wu R, Han ZC (2011b) Intrapulmonary delivery of human umbilical cord mesenchymal stem cells attenuates acute lung injury by expanding CD4+CD25+ Forkhead Boxp3 (FOXP3)+ regulatory T cells and balancing anti- and pro-inflammatory factors. Cell Physiol Biochem 27(5):587–596

Sutsko RP, Young KC, Ribeiro A, Torres E, Rodriguez M, Hehre D, Devia C, McNeice I, Suguihira C (2013) Long-term reparative effects of mesenchymal stem cell therapy following neonatal hyperoxia-induced lung injury. Pediatr Res 73(1):46–53

Thebaud B, Stewart DJ (2012) Exosomes: cell garbage can, therapeutic carrier, or trojan horse? Circulation 126(22):2553–2555

Tran TC, Kimura K, Nagano M, Yamashita T, Ohneda K, Sugimori H et al (2011) Identification of human placenta-derived mesenchymal stem cells involved in re-endothelialization. J Cell Physiol 226(1):224–235

Tropea KA, Leder E, Aslam M, Lau AN, Raiser DM, Lee JH, Balasubramaniam V, Fredenburgh LE, Mitsialis A, Kourembanas S, Kim CF (2012) Bronchioalveolar stem cells increase after mesenchymal stromal cell treatment in a mouse model of bronchopulmonary dysplasia. Am J Physiol Lung Cell Mol Physiol 302:L829–L837

Tzouvelekis A, Paspaliaris V, Koliakos G, Ntolios P, Bouros E, Oikonomou A, Zissimopoulos A, Boussios N, Dardzinski B, Gritzalis D, Antoniadis A, Froudarakis M, Kolios G, Bouros D (2013) A prospective, non-randomized, no placebo-controlled, phase Ib clinical trial to study the safety of the adipose derived stromal cells-stromal vascular fraction in idiopathic pulmonary fibrosis. J Transl Med 11:171

Umar S, de Visser YP, Steendijk P et al (2009) Allogenic stem cell therapy improves right ventricular function by improving lung pathology in rats with pulmonary hypertension. Am J Phys Heart Circ Phys 297:H1606–H1616

Viswanathan S, Keating A, Deans R, Hematti P, Prockop DJ, Stroncek D, Stacey G, Weiss DJ, Mason C, Rao M (2014) Soliciting strategies for developing cell-based reference materials to advance MSC research and clinical translation. Stem Cells Dev 23(11):1157–1167

Wang HS, Hung SC, Peng ST, Huang CC, Wei HM, Guo YJ et al (2004) Mesenchymal stem cells in the Wharton's jelly of the human umbilical cord. Stem Cells 22(7):1330–1337

Wang L, Tu XH, Zhao P, Song JX, Zou ZD (2012) Protective effect of transplanted bone marrow-derived mesenchymal stem cells on pancreatitis-associated lung injury in rats. Mol Med Rep 6:287–292

Weiss DJ (2013) Stem cells, cell therapies and bioengineering in lung biology and diseases: comprehensive review of the literature. Ann Am Thorac Soc 10(5):S45–S97

Weiss DJ (2014) Concise review: current status of stem cells and regenerative medicine in lung biology and diseases. Stem Cells 32(1):16–25. doi:10.1002/stem.1506

Weiss DJ, Ortiz LA (2013) Invited editorial: cell therapy trials for lung diseases: progress and cautions. Am J Respir Crit Care Med 188(2):123–125

Weiss DJ, Bates JHT, Gilbert T, Liles WC, Lutzko C, Rajagopal J, Prockop DJ (2013a) Vermont stem cell conference report: stem cells and cell therapies in lung biology and diseases. Ann Am Thorac Soc 10(5):S25–S44

Weiss DJ, Casaburi R, Flannery R, LeRoux-Williams M, Tashkin DP (2013b) A placebo-controlled, randomized trial of mesenchymal stem cells in COPD. Chest 143(6):1590–1598

Xin H, Kanehira M, Mizuguchi H, Hayakawa T, Kikuchi T, Nukiwa T, Saijo Y (2007) Targeted delivery of CX3CL1 to multiple lung tumors by mesenchymal stem cells. Stem Cells 25:1618–1626

Xu J, Qu J, Cao L, Sai Y, Chen C, He L, Yu L (2008) Mesenchymal stem cell-based angiopoietin-1 gene therapy for acute lung injury induced by lipopolysaccharide in mice. J Pathol 214(4): 472–481

Xu YL, Liu YL, Wang Q, Li G, Lu XD, Kong B (2012) Intravenous transplantation of mesenchymal stem cells attenuates oleic acid induced acute lung injury in rats. Chin Med J 125(11): 2012–2018

Yagi H, Soto-Gutierrez A, Kitagawa Y, Tilles AW, Tompkins RG, Yarmush ML (2010a) Bone marrow mesenchymal stromal cells attenuate organ injury induced by LPS and burn. Cell Transplant 19:823–830

Yagi H, Soto-Gutierrez A, Navarro-Alvarez N, Nahmias Y, Goldwasser Y, Kitagawa Y, Tilles AW, Tompkins RG, Parekkadan B, Yarmush ML (2010b) Reactive bone marrow stromal cells attenuate systemic inflammation via sTNFR1. Mol Ther 18:1857–1864

Yan X, Liu Y, Han Q, Jia M, Liao L, Qi M, Zhao RC (2007) Injured microenvironment directly guides the differentiation of engrafted Flk-1(+) mesenchymal stem cell in lung. Exp Hematol 35(9):1466–1475

Yan C, Qu P, Du H (2012) Myeloid-specific expression of Stat3C results in conversion of bone marrow mesenchymal stem cells into alveolar type II epithelial cells in the lung. Sci China Life Sci 55(7):576–590

Yang Z, Sharma AK, Marshall M, Kron IL, Laubach V (2009) NADPH oxidase in bone marrow-derived cells mediates pulmonary ischemia-reperfusion injury. Am J Respir Cell Mol Biol 40:375–381

You Q, Cai L, Zheng J, Tong X, Zhang D, Zhang Y (2008) Isolation of human mesenchymal stem cells from third-trimester amniotic fluid. Int J Gyn Obstet 103(2):149–152

Zannettino AC, Paton S, Arthur A, Khor F, Itescu S, Gimble JM et al (2008) Multipotential human adipose-derived stromal stem cells exhibit a perivascular phenotype in vitro and in vivo. J Cell Physiol 214(2):413–421

Zhang X, Zhao P, Kennedy C, Chen K, Wiegand J, Washington G, Marrero L, Cui Y (2008) Treatment of pulmonary metastatic tumors in mice using lentiviral vector-engineered stem cells. Cancer Gene Ther 15:73–84

Zhang H, Fang J, Su H, Yang M, Lai W, Mai Y, Wu Y (2010) Bone marrow mesenchymal stem cells attenuate lung inflammation of hyperoxic newborn rats. Pediatr Transplant 16(6): 589–598

Zhang H, Liu X, Huang S, Bi X, Wang H, Xie L, Wang Y, Cao X, Xiao F, Yang Y, Guo Z (2012a) Microvesicles derived from human umbilical cord mesenchymal stem cells stimulated by hypoxia promote angiogenesis both in vitro and in vivo. Stem Cells Dev 21:3289–3297

Zhang X, Wang H, Shi Y, Peng W, Zhang S, Zhang W, Xu J, Mei Y, Feng Z (2012b) Role of bone marrow-derived mesenchymal stem cells in the prevention of hyperoxia-induced lung injury in newborn mice. Cell Biol Int 36(6):589–594

Zhang S et al (2013) Comparison of the therapeutic effects of human and mouse adipose-derived stem cells in a murine model of lipopolysaccharide-induced acute lung injury. Stem Cell Res Ther 4(1):13. doi:10.1186/scrt161

Zhao F, Zhang YF, Liu YG, Zhou JJ, Li ZK, Wu CG, Qi H (2008) Therapeutic effects of bone marrow-derived mesenchymal stem cells engraftment on bleomycin-induced lung injury in rats. Transplant Proc 40:1700–1705

Zhen G, Xue Z, Zhao J, Gu N, Tang Z, Xu Y, Zhang Z (2010) Mesenchymal stem cell transplantation increases expression of vascular endothelial growth factor in papain-induced emphysematous lungs and inhibits apoptosis of lung cells. Cytotherapy 12:605–614

Zuk PA, Zhu M, Ashjian P, De Ugarte DA, Huang JI, Mizuno H et al (2002) Human adipose tissue is a source of multipotent stem cells. Mol Biol Cell 13(12):4279–4295

Chapter 7
The Hope for iPSC in Lung Stem Cell Therapy and Disease Modeling

Tushar Menon and Amy L. Firth

Abbreviations

3D	Three dimensional
AAV	Adeno-associated virus
AD	Alzheimer's disease
AFE	Anterior foregut endoderm
ALI	Air–liquid interface
ATI/II	Alveolar type I /II cells
BMP4	Bone morphogenic protein 4
Bromo cAMP	Bromo cyclin adenosine monophosphate
CC10	Club cell 10 kDa protein
CD54	Cluster of differentiation factor 54 (aka: ICAM-1 (intercellular adhesion molecule 1))
CDX2	Caudal type homeobox 2
CFTR	Cystic fibrosis transmembrane regulator
ChIP	Chromatin immuno precipitation
CK5	Cytokeratin 5
COPD	Chronic obstructive pulmonary disease
CRISPR	Clustered regularly interspersed short palindromic repeats
DE	Definitive endoderm
DSB	Double-stranded break
EGF	Epidermal growth factor
ESCs	Embryonic stem cells
FD	Familial dysautonomia

T. Menon • A.L. Firth, Ph.D. (✉)
Laboratory of Genetics, The Salk Institute for Biological Studies,
10010 North Torrey Pines Road, La Jolla, CA 92037, USA
e-mail: afirth@salk.edu

A. Firth, J.X.-J. Yuan (eds.), *Lung Stem Cells in the Epithelium and Vasculature*,
Stem Cell Biology and Regenerative Medicine, DOI 10.1007/978-3-319-16232-4_7

FGF10	Fibroblast growth factor 10
FOXA2	Forkhead box A2
FOXJ1	Forkhead box J1
gRNA	Guide ribonucleic acid
HR	Homologous recombination
iPSC	Induced pluripotent stem cells
KGF	Keratinocyte growth factor (aka: FGF7)
KLF4	Kruppel-like factor 4
MAPK/ERK	Mitogen-activated protein kinase/extracellular signal-regulated kinase
MCC	Multiciliated cell
MUC5AC	Mucin 5 AC
NHEJ	Non-homologous end joining
Nkx2.1	NK2 homeobox 1 (aka: TTF1 thyroid transcription factor 1)
OCT4	Octamer-binding transcription factor 4 (aka: POU5F1 (POU domain class 5, transcription factor 1))
PCD	Primary ciliary dyskinesia
RA	Retinoic acid
SCZD	Schizophrenia
SHH	Sonic hedgehog
SMA	Spinal muscular atrophy
SOX2	Sex-determining region Y box2
T-1α	aka: Podoplanin
TALEN	Transcription activator-like effector nuclease
TGFβ	Transforming growth factor beta
Wnt3a	Wingless-type MMTV integration site family member 3A
ZFNs	Zinc finger nucleases

7.1 Introduction

Chronic lung disease affects over 35 million people in the USA and kills nearly 400,000 Americans each year, accounting for one in every six deaths according to the American Lung Association. This makes it the third highest cause of death in the USA, following only cardiovascular disease and cancer. However, death rates from pulmonary disease continue to increase, while those from cardiovascular and cancerous diseases are on the decline. In addition to this high prevalence and mortality, pulmonary disease imposes a huge financial burden, costing $95 million in direct health care costs and an additional $59 million in indirect costs, amounting to a total deficit of $154 million to the US economy. The situation is exacerbated by a fast deteriorating environment and a rapidly aging population, which will only serve to increase these numbers, as the most prevalent manifestations of lung disease, such as chronic obstructive pulmonary disease (COPD), are conditions affecting older patients. There is thus an immense and immediate need for better treatment and study of pulmonary disease.

Modeling human lung disease currently relies on the isolation of primary bronchial epithelial cells from the lungs of deceased patients, which is of both limited quantity and unreliable quality. These lungs are often at the end stages of the disease and have been exposed to a variety of therapeutic agents and external environmental factors and are often poorly characterized and phenotypically variable. Other cellular models, such as transformed cell lines, also fall short as they lack most of the key functional characteristics of the pulmonary system and so do not adequately represent the relevant biology of the lung or the diversity of its human diseases. Animal models have contributed greatly to the better understanding of lung biology, but have their limitations when it comes to modeling human disease, as in the case of cystic fibrosis, where CFTR knockout mouse models do not faithfully recapitulate the pathogenesis or symptoms of the human disease (Ratjen and Doring 2003).

Chronic lung disease is characterized by pathological fibrosis and the consequent loss of lung tissue due to impaired epithelial and endothelial regeneration (Moodley et al. 2013). To date, no therapeutic approaches have been developed to effectively repair and regenerate damaged lung tissue. While whole-organ transplantation is a valid option for certain terminal conditions, it remains a challenge due to paucity of donor organs and the clinical complications associated with such surgery (McCurry et al. 2009). Cell-based transplant approaches have no clinical utility due to the inability to engraft any type of lung stem cell or progenitor in animal models of lung injury, owing to the complexity of the lung architecture (Green et al. 2013). There is thus a major unmet clinical need for a cell-based system to enable a more comprehensive study of the pathogenesis of pulmonary disease and for the development of novel therapeutic approaches.

A reproducible model of human lung disease from a self-renewing population of cells would create the opportunity to study human lung disease more extensively. Utilizing new gene editing technology, the generation of gene-corrected respiratory epithelial progenitor cell from a patient with genetic disease could also proffer a potential therapeutic approach.

7.2 Induced Pluripotency: Heralding a New Era for Disease Modeling and Therapy

Mammalian development begins with the concurrent loss of pluripotency and commitment to specific lineages of the pluripotent cells within the inner cell mass (ICM) of the blastocyst; and from then on proceeds along a path of progressive restriction of cell fate as the cell specializes and differentiates to give rise to the cell types, tissues, and organs that comprise the adult body (Sommer and Mostoslavsky 2013). The reproduction and propagation of this transient developmental pluripotent state and subsequent differentiation to a variety of cellular lineages in vitro represents the ultimate experimental challenge in cellular and developmental biology. The derivation of the first embryonic stem cell (ESC) lines from mice (Martin 1981; Evans and

Kaufman 1981) and later humans (Thomson et al. 1998) represented major breakthroughs in this direction. The demonstration that these pluripotent stem cells could be cultured in a dish with unlimited self-renewal capacity and could be differentiated into a variety of cell types (Bradley et al. 1984) displayed their great potential for regenerative medicine and for modeling development and disease in a dish (Murry and Keller 2008). However, the use of ESCs derived from human embryos presented a major ethical dilemma and consequently a practical bottleneck in terms of their development for clinical use. It therefore became necessary to investigate alternate methods to derive pluripotent stem cells.

The experimental reversal of terminally differentiated somatic cells to a pluripotent state had previously been demonstrated by pioneering nuclear transfer experiments in frogs (Gurdon 1975) and mammals (Wilmut et al. 1997). But processes like somatic cell nuclear transfer and cell fusion were inherently too technically challenging and inefficient for clinical development. However, the fundamental hypothesis that this sort of nuclear reprogramming is driven by soluble factors present in pluripotent cells, like those of the oocytes used in Gurdon's experiments, motivated the seminal work of Dr. Shinya Yamanaka resulting ultimately in the discovery of artificially induced pluripotency in 2006. In initial experiments, 24 transcription factors were overexpressed in mouse fibroblasts, resulting in the formation of rare ESC-like colonies. This list was then narrowed down by a process of elimination to just four transcription factors essential for reprogramming somatic cells to induced pluripotency: Oct4, Sox2, Klf4, and c-Myc, now known as the Yamanaka factors for iPSC reprogramming (Takahashi and Yamanaka 2006; Takahashi et al. 2006). For their pioneering work on inducing pluripotency by nuclear reprogramming, Yamanaka and Gurdon jointly won the Nobel Prize for Physiology or Medicine in 2012.

Initial reprogramming experiments were done using gammaretroviruses to introduce the Yamanaka factors, and these were effective in generating iPSCs that met the widely accepted benchmark for true pluripotency; of being capable of generating teratomas upon transplantation into nude mice and producing viable chimeric mice upon injection into developing embryos (Maherali et al. 2007; Okita et al. 2007; Wernig et al. 2007). The retroviral system was also used to generate the first human iPSC lines (Takahashi et al. 2007; Park et al. 2008a, b). Since then there has been a plethora of alternative approaches for generating iPSC ranging from integrating lentivirus to alternative non-integrating approaches. These are summarized in Table 7.1.

While generating iPSC has become rote procedure in most laboratories, there is still much variability and need for optimization of reprogramming protocols based on the source tissue/cell type of origin, sample heterogeneity based on patient/disease-specific variation, and reprogramming methodology. iPSC lines generated from even the same cellular origin often display differences in pluripotent characteristics and differentiation potential to various tissues. But still, an iPSC-based approach is proving to be the most viable way to study and treat certain diseases in a patient-specific manner.

Table 7.1 A table summarizing the advantages and disadvantages of the various methodologies that have been used to reprogram somatic cells to iPSC. References for each method are provided in the right hand column

	Advantages	Disadvantages	References
Maloney-based retrovirus	Self-silencing in pluripotent cells, efficient	Genomic integration, transduces only dividing cells	Takahashi and Yamanaka (2006)
Polycistronic retrovirus	Only single integration event needed	Genomic integration, lower titers/efficiency	Rodriguez-Piza et al. (2010)
HIV-based lentivirus	Transduces dividing and non-dividing cells, efficient	Genomic integration, no silencing in pluripotent state	Yu et al. (2007)
Inducible lentivirus	Temporal control over reprogramming factors	Genomic integration, leaky expression of factors	Brambrink et al. (2008)
Polycistronic lentivirus	Single integration, all factors in one vector	Genomic integration, uneven expression of factors	Chin et al. (2010)
Inducible poly-cistronic lentivirus	Single integration, temporal control of reprogramming	Genomic integration, leaky/uneven expression of factors	Carey et al. (2009)
Integrase-deficient lentivirus	Lower frequency of genomic integration	Lower expression of reprogramming factors	Nightingale et al. (2006)
Excisable lentivirus	Excision of reprogramming factors with minimal genomic footprint	Screening required for multiple excision events, single loxP site left behind	Soldner et al. (2009)
Excisable poly-cistronic lentivirus	Single excision event with minimal genomic footprint	Screening required, uneven expression of factors	Chang et al. (2009)
Adenovirus	No genomic integration	Multiple infections needed, delayed reprogramming	Stadtfeld et al. (2008)
PiggyBac transposon	Complete excision from genome	Still needs secondary screening for excised clones	Kaji et al. (2009)
Inducible PiggyBac transposon	Complete excision, temporal control	Leaky expression, secondary screening	Woltjen et al. (2009)
Transient transfection of DNA plasmids	No genomic integration, no viral vectors	Multiple rounds of transfections needed, low expression of factors, delayed reprogramming	Okita et al. (2008), Yu et al. (2009)
Minicircles	Higher expression than DNA plasmids, integration-free, no viral vectors	Inefficient compared to viral methods, repeated transfections needed	Jia et al. (2010)
Sendaivirus	Non-integrating viral approach; efficient, fast reprogramming	Requires clearance of virus by multiple passaging of iPSC	Fusaki et al. (2009), Seki et al. (2010)

(continued)

Table 7.1 (continued)

	Advantages	Disadvantages	References
Small molecules	Transient, dosage controllable, simple methodology	Nonspecific effects/ toxicity, reprogramming not yet possible without factors	Huangfu et al. (2008)
RNA transfection	Non-integrating, transient expression of factors, DNA-free approach	Requires multiple transfections, inefficient	Warren et al. (2010)
Protein transfection	Transient, direct delivery of factors, nuclease-free	Multiple transfections needed, inefficient, difficult to reproduce	Zhou et al. (2009), Kim et al. (2009)

iPSCs provide an unlimited source of patient-derived and disease-specific cells, which can then be differentiated into a variety of different cell types. This makes it an ideal model system to study disease pathologies and also to develop new preclinical approaches towards therapies. When paired with new, powerful next-generation sequencing and gene-editing technologies, the possibilities are endless for developing patient-specific, personalized medicine-based treatments for genetic and other types of diseases.

Due to the early development of robust protocols for differentiating iPSC to neuronal lineages, their use in modeling neurological disorders has been at the forefront of the iPSC disease-modeling revolution. The earliest iPSC-based disease models were elucidated for inherited neurological disorders such as spinal muscular atrophy (SMA) (Ebert et al. 2009), familial dysautonomia (FD) (Lee et al. 2009), and Rett's syndrome (Marchetto et al. 2010). In all these cases, the pathology resulting from the disease-causing genetic mutations was elucidated at the cellular and molecular level using iPSC and then suitable drug treatments were discovered based on tests done in these disease-specific iPSC. These tests would otherwise not be possible due to such genetic diseases being rare in occurrence and there being no truly representative model for the disease before the advent of iPSC. The utility of iPSC-based models has been extended beyond inherited genetic diseases to adult-onset or complex neurodegenerative diseases as well. Schizophrenia (SCZD) (Brennand et al. 2011), Parkinson's disease (PD) (Nguyen et al. 2011), and Alzheimer's disease (AD) (Yagi et al. 2011) have all been accurately modeled using iPSC, down to the cellular pathology and molecular signature of the disease, and been used to find effective drug treatments against these diseases.

Another organ system that was otherwise very difficult to model was the cardiovascular system, due to lack of access to the human tissue and poor in vitro culturing capability of cells of cardiac origin. iPSC were generated from patients suffering from cardiovascular conditions like long QT syndrome (Moretti et al. 2010; Itzhaki et al. 2011; Yazawa et al. 2011) and LEOPARD syndrome (Carvajal-Vergara et al. 2010) and then converted into cardiomyocytes that could be used for screening for pharmacological agents capable of rescuing the observed electrophysiological defects these cells displayed. Metabolic diseases have also been modeled, for example, iPSC-derived hepatocytes from patients with hypercholesterolemia (Rashid et al. 2010), and insulin-producing cells from Type I diabetic patients (Maehr et al. 2009). Since these initial efforts, a large number of diseases have been modeled successfully using iPSC (Table 7.2).

Table 7.2 A summary of the types of genetic diseases that have been modeled using iPSC. The nature of the disease-specific genetic defect, the terminal cell type derived from the patient iPSC and the associated references are also listed

Disease type	Disease modeled	Genetic defect	Cell type derived	References
Neurological	Amyotrophic lateral sclerosis (ALS)	Heterozygous L144F mutation in *SOD1*	Motor neurons and glial cells	Dimos et al. (2008)
Neurological	Spinal muscular atrophy (SMA)	Mutations in *SMN1*	Neurons and astrocytes	Ebert et al. (2009)
Neurological	Parkinson's disease	Mutations in *SNCA* and *LRRK2*	Dopaminergic neurons	Park et al. (2008a, b), Soldner et al. (2009)
Neurological	Down's syndrome	Chromosome 21 trisomy	Teratomas	Park et al. (2008a, b)
Neurological	Familial dysautonomia	Mutation in *IKBKAP*	CNS, neurons, hematopoietic, endothelial and endodermal cells	Lee and Studer (2011)
Neurological	Rett's syndrome	Heterozygous mutation in *MECP2*	Neural progenitor cells	Marchetto et al. (2010)
Neurological	Schizophrenia	Complex trait	Neurons	Brennand et al. (2011)
Hematological	Fanconi's anemia	*FAA* and *FAD2*	Hematopoietic cells	Raya et al. (2009)
Hematological	β-Thalassemia	Homozygous deletion in β-globin gene	Hematopoietic cells	Ye et al. (2009a)
Hematological	*ADA* SCID	Mutation in *ADA*	None	Park et al. (2008a, b)
Hematological	Sickle-cell anemia	Homozygous HbS mutation	None	Ye et al. (2009a), Somers et al. (2010)
Neurological	Adrenoleukodystrophy (ALD)	Mutation in *ABCD1*	Oligodendrocytes	Jang et al. (2011)
Neurological	Huntington's disease	CAG repeats in *huntingtin* gene	None	Park et al. (2008a, b)
Neurological	Fragile X syndrome	Trisomy 21	None	Urbach et al. (2010)

(continued)

Table 7.2 (continued)

Disease type	Disease modeled	Genetic defect	Cell type derived	References
Hematological	Polycythaemia vera	Heterozygous V617F mutation in *JAK2*	Hematopoietic progenitors	Ye et al. (2009b)
Hematological	Primary myelofibrosis	Heterozygous mutation in *JAK2*	None	Ye et al. (2009b)
Metabolic	Type 1 diabetes	Multifactorial	β-Cell like cells	Maehr et al. (2009)
Metabolic	Gaucher's disease	Mutation in *GBA*	None	Park et al. (2008a, b)
Metabolic	α1-Antitrypsin deficiency	Homozygous mutation in α1-Antitrypsin	Hepatocyte-like cells	Rashid et al. (2010)
Metabolic	Glycogen storage disease	Mutation in glucose-6-phosphate gene	Hepatocyte-like cells	Rashid et al. (2010), Ghodsizadeh et al. (2010)
Metabolic	Familial hypercholesterolemia	Autosomal dominant mutation in *LDLR*	Hepatocyte-like cells	Rashid et al. (2010)
Metabolic	Pompe disease	Knockout of *Gaa*	Skeletal muscle cells	Kawagoe et al. (2011)
Metabolic	Hurler syndrome	Genetic defect in *IDUA*	Hematopoietic cells	Tolar et al. (2011a)
Cardiovascular	LEOPARD syndrome	Heterozygous mutation in *PTPN11*	Cardiomyocytes	Carvajal-Vergara et al. (2010)
Cardiovascular	Type 1 long QT syndrome	Dominant mutation in *KCNQ1*	Cardiomyocytes	Moretti et al. (2010)
Cardiovascular	Type 2 long QT syndrome	Missense mutation in *KCNH2*	Cardiomyocytes	Itzhaki et al. (2011)
Cardiovascular	Dilated cardiomyopathy (DCM)	Mutation in *TNNT2*	Cardiomyocytes	Sun et al. (2012)???
Cardiovascular	Arrhythmogenic right ventricular cardiomyopathy (ARVC)	Mutation in *PKP2*	Cardiomyocytes	Ma et al. (2013)
Primary immuno-deficiency	Severe combined immunodeficiency (SCID)	Mutation in *RAG1*	None	Pessach et al. (2011)

(continued)

Table 7.2 (continued)

Disease type	Disease modeled	Genetic defect	Cell type derived	References
Primary immuno-deficiency	Cartilage-hair hypoplasia (CHH)	Mutation in *RMRP*	None	Pessach et al. (2011)
Primary immuno-deficiency	Herpes simplex encephalitis (HSE)	Mutation in *STAT1* or *TLR3*	Mature CNS cell types	Pessach et al. (2011)
Muscular	Duchenne's muscular dystrophy	Deletion in *dystrophin* gene	None	Tchieu et al. (2010)
Integumentary	Dyskeratosis congenital (DC)	Deletion in *DKC1*	None	Agarwal et al. (2010)
Pulmonary	Cystic fibrosis	Deletion in CFTR	None	Warren et al. (2010)
CNS	Friedreich's ataxia (FRDA)	Trinucleotide GAA repeat expansion in *FXN*	Neurons and cardiomyocytes	Liu et al. (2011)
Retinal	Retinitis pigmentosa	Mutations in *RP9*, *RP1*, *PRPH2* or *RHO*	Retinal and photoreceptors progenitors, RPE cells and rod photoreceptors	Jin et al. (2011)
Skin	Recessive dystrophic epidermolysis bullosa	Mutation in *COL7A1*	Hematopoietic cells, keratinocyte	Tolar et al. (2011b)
Skin	Scleroderma	Unknown	None	Somers et al. (2010)
Osteopathic	Osteogenesis imperfect	Mutation in *COL1A2*	None	Khan et al. (2010)
Pulmonary	Pulmonary hypertension (IPAH)	Wnt signaling	Endothelial cells	West et al. (2014)
Pulmonary	Pulmonary alveolar proteinosis	Mutation in *CSF2RA*	Monocytes, macrophages	Suzuki et al. (2014), Lachmann et al. (2014)

7.2.1 The Genome Editing Revolution

Recent advances in genome editing technologies have enabled a quantum leap forward in terms of allowing for the unprecedented direct manipulation of the DNA genome of model organisms. Instead of introducing exogenous DNA elements or indirectly targeting gene products by RNA interference to deduce their physiological functions, we can now directly modify genes in situ and knockout genes by targeting their DNA sequence within the nuclear genome. This has been facilitated by a sea-change in the underlying technology, from traditional gene targeting by homologous recombination (HR) (Capecchi 1989), which was inherently very inefficient (targeting only 1 in 10^6 to 10^9 cells), to the era of highly precise and efficient genome editing by engineered designer nucleases, which hold immense potential to transform basic science, biotechnology, and medicine.

It had been known for a while that the generation of double-strand breaks (DSB) in the vicinity of the targeted DNA sequence greatly stimulates the efficiency of HR-mediated recombination events (Rudin et al. 1989; Plessis et al. 1992; Rouet et al. 1994). Chandrasegaran and colleagues first targeted specific DNA sequences using zinc finger DNA-binding domains, common to many eukaryotic transcription factors, and linked them to the endonuclease domain of the FokI Type IIS enzyme to create custom-engineered nucleases, called zinc finger nuclease (ZFNs) (Kim et al. 1996). Carroll then co-opted these ZFNs to target genomic sequences and achieve efficient, targeted HR at these sites (Bibikova et al. 2001, 2003). They also showed that in the absence of HR, error-prone non-homologous end joining (NHEJ) at the site of targeted DSBs, such as those generated by ZFNs, could result in deleterious indels that could be used to knockout genes by targeting their genomic sequences (Bibikova et al. 2002). Thus using custom-engineered nucleases, one could achieve highly efficient gene knockout, knock-in, or modification for the very first time.

Efforts had also been undertaken to protein engineer mega-nucleases derived from mobile genetic elements to target custom DNA sequences (Smith et al. 2006); however, this task proved arduous due to the lack of correspondence between their protein residues and their target DNA sequence specificity. ZFNs, on the other hand, could be manipulated to target specific DNA sequences by manipulating the modular assembly of the ZF domains comprising them (Urnov et al. 2005; Miller et al. 2007). However, due to the fact that each ZF domain recognizes a triplet of base pairs, one could not target many DNA sequences which lacked suitable ZFN-targeting specificity, as ZFs targeting all the combination of triplets have still not been engineered. In addition, assembling certain ZF domains adjacently altered their DNA-binding properties in unpredictable ways.

This was largely overcome by the discovery of TALENs, which linked the modular TAL DNA-binding domains found in the plant pathogen Xanthomonas to the FokI endonuclease domain, much like ZFNs (Moscou and Bogdanove 2009; Boch et al. 2009; Christian et al. 2010). Each TAL domain recognizes a single, specific base and occurs naturally as modularly assembled arrays, so TALENs have greater specificity, versatility, and utility for targeting the genome. However, both ZFN and TALEN technologies hit a roadblock *en route* to more popular and widespread adoption due to the laborious and costly process that assembling such arrays of protein domains entails, as both these technologies rely on protein-based recognition of target DNA sequences.

Given the difficulties of engineering arrays of modular DNA-binding domains, a different mode of DNA recognition would simplify the development of custom nucleases (Hsu et al. 2014). This is where CRISPR (clustered regularly interspersed short palindromic repeats) has revolutionized the field of genome editing by custom-engineered nucleases. A naturally occurring mechanism against viral infection found in a large variety of prokaryotes, the CRISPR system is comprised of a DNA endonuclease, Cas9, whose specificity to target sequences is defined by an RNA component, which binds to target DNA by Watson–Crick base pairing. This unique modality of DNA recognition and binding via a guide RNA (gRNA) of predefined sequence makes the engineering of custom CRISPR nucleases as easy as designing an oligonucleotide complementary to the desired target DNA sequence and expressing it as an RNA transcript.

The origins of CRISPR date back to 1987 when Ishino et al (1987) reported the discovery of clustered CRISPR repeats downstream of the *iap* gene in *E. coli* (Ishino et al. 1987). As more microbial genomes were sequenced, the presence of similar repeat elements was reported in a large variety of bacteria and archaea and came to be recognized as a unique family of clustered repeat elements (Mojica et al. 2000). These were subsequently named CRISPR and were subdivided into three types of systems (I–III) based on the identity of the cluster of genes found adjacent to these CRISPR loci, which encoded for the protein endonuclease components of the CRISPR system, called the *cas* proteins (Jansen et al. 2002; Haft et al. 2005). Type I and III CRISPR systems have multiple Cas proteins, but the Type II CRISPR system was comprised of significantly less discrete components and so was the most attractive prospect for development of customized molecular engineering and for adaptation towards targeted genome editing.

Systematic analyses of the CRISPR sequences identified the source of these "spacer" elements to be extrachromosomal in nature, derived from bacteriophage-associated origins (Mojica et al. 2005; Pourcel et al. 2005). This gave rise to the theory that CRISPR systems functioned as an innate immunity-based antiviral defense mechanism. The first experimental proof of CRISPR functioning as an adaptive immune system came in 2007, when the food ingredient company Danisco was studying the dairy production bacterium *Streptococcus thermophiles*, ostensibly for its use in industrial production of yogurt. They found that the Type II CRISPR system provides antiviral defense by using the spacer elements to dictate target sequence specificity and the Cas proteins to degrade the invasive phage DNA (Barrangou et al. 2007). Subsequently, several salient features of the CRISPR system came to be elucidated; such as the transcription of CRISPR arrays into short crRNA hairpins that guided Cas9 DNA endonuclease activity (Brouns et al. 2008) and the identification of the PAM motif, characteristic of each CRISPR system, that was found to be essential for recognition and cleavage of the target DNA (Bolotin et al. 2005; Deveau et al. 2008). The discovery that the Type II CRISPR system depended on only a single enzymatic component, Cas9, (Garneau et al. 2010) and the fuller understanding of the processing of the crRNA/tracrRNA dual RNA element that guides its endonuclease activity reduced the CRISPR system to its minimal functional components and thus illuminated its potential as a powerful genome editing tool.

The breakthrough studies by Charpentier and Doudna in 2012 first demonstrated the adaptation of the bacterial CRISPR system to target and cleave DNA in vitro, using a single, short gRNA hairpin by incorporating the tracrRNA and crRNA into a single oligonucleotide element (Jinek et al. 2012). This transformed the field of gene targeting, as for the first time a simple and effective methodology was developed to target defined DNA sequences using an RNA-guided, customizable endonuclease. The immense potential of this technology was demonstrated as early as 2013, when the CRISPR system was successfully utilized to achieve efficient and precise genome editing in mammalian cells (Cong et al. 2013; Mali et al. 2013). Since then the CRISPR field has veritably exploded, with new applications and developments being reported in a plethora of model systems.

CRISPR can be used to knockout genes by inducing the error-prone NHEJ repair mechanism to produce deleterious indel mutations within the coding sequences of target genes. By introducing a homologous DNA sequence, DSBs created by CRISPR can be used to modify or knock-in specific DNA sequences at targeted genomic loci by invoking HR-mediated repair. By using a catalytically inactive version of Cas9 (dCas9), which binds to but does not cleave target sequences, one can silence or activate transcription by linking transcriptional repressor or activator domains to the dCas9 protein, or even by interference of transcription by steric hindrance due to the binding of dCas9 (Qi et al. 2013; Maeder et al. 2013; Ran et al. 2013). The use of effector fusions to dCas9 can be utilized for dynamic visualization of genomic loci (Chen et al. 2013) to edit histone modifications in a locus-specific manner (Mendenhall et al. 2013) or to even alter the 3D organization of the genome and its associated chromatin. Using chromatin immunoprecipitation (ChIP), dCas9 could serve as a unique protein tag by which to isolate sequence-specific genomic DNA elements, in the context of its associated chromatin, to study transcription factors, histone modifications, or other chromatin characteristics associated with these genomic loci.

The ability to use CRISPR in multiplexed genome editing by simultaneously introducing gRNAs targeting a variety of genomic sequences is another unique feature of this technology that sets it apart (Cong et al. 2013; Mali et al. 2013). By using tandem gRNAs, one can produce deletions in the genome ranging from a few hundred bases to a few hundreds of thousands of bases (Xiao et al. 2013; Zhou et al. 2014). This multiplexing capability is also taken advantage of in creating a more high-fidelity version of Cas9, by perturbing one of its two DNA endonuclease domains, rendering it into a nickase of double-stranded DNA (Ran et al. 2013). Using a tandem pair of gRNAs offset by an appropriate length of intervening DNA sequence, one can essentially create staggered DSBs in the dsDNA using a pair of nickase molecules, much like ZFNs or TALENs. This improves the specificity of Cas9 by reducing off-target effects, due to the requirement for binding and cleavage by two discrete gRNA recognition elements instead of one.

The ability to easily generate arrays of gRNAs by simply designing unique oligonucleotide sequences against a multitude of genomic targets has been utilized in the generation of genome-wide CRISPR libraries using lentiviral delivery vectors (Wang et al. 2014; Shalem et al. 2014). This allows for genome-wide loss-of-function screens as have been described using RNAi, except now with the ability to actually knock out coding sequences at their genomic loci, and also target other non-coding RNA and regulatory DNA elements. One can link various effector functions to the dCas9 protein to carry out unprecedented transcriptional activation or epigenetic regulation screens at a genome-wide level.

Since its inception, CRISPR genome editing technology has been applied to an exponentially growing repertoire of cellular and animal models (Sander and Joung 2014). It has been used to generate a variety of transgenic animal models from mice to monkeys, where a specific genetic variation can be precisely phenocopied in vivo (Wang et al. 2013; Niu et al. 2014). Even more impressively, transgenic animal models can be generated by directly injecting the Cas9 protein and transcribed gRNA into fertilized zygotes, bypassing the typical ES cells stage for targeting one or multiple alleles, in a much shorter generation time than usual. For mice, novel transgenic models can be derived in a matter of weeks, instead of the traditional time frame of over a year. The ability to use this technology for a variety of organisms is transforming the landscape of experimental biology on almost a day-to-day basis.

CRISPR has already been used to introduce or correct specific mutations in isogenically engineered ESC and iPSC lines (Schwank et al. 2013; Wu et al. 2013). The unique advantage of this type of genome editing in iPSC lies in the fact that one can isolate and culture these cells clonally, and thus derived 100 % pure, isogenic cell lines containing the desired genomic manipulations. These engineered iPSC lines can be differentiated into suitable cell types to observe and study the resultant phenotype. For certain cell types that are otherwise hard to target, one can achieve efficient genome editing at the iPSC stage and then subsequently obtain the desired end cell type by differentiation of these iPSC. One can use CRISPR-mediated genome editing to create true genomic knock-in reporter iPSC lines, where fluorescence, epitope tags, or other detectable markers can be integrated downstream of key genes to monitor and detect their expression. Using such reporter lines, existing differentiation protocols can be vastly improved by detecting and isolating populations of cells expressing the appropriate markers characteristic of the cell types desired at each stage of differentiation. For example, existing lung differentiation protocols would benefit greatly from having reporter lines of genes like Sox2, Nkx2.1, FoxJ1, or FoxA2 to identify various lung progenitor and mature cell types and monitor and isolate these cell populations as they are generated during differentiation (Fig. 7.1).

One can already deliver CRISPR as DNA constructs, transcribed RNA, or packaged into viral vectors, like lentivirus or adeno-associated virus (AAV). In addition, inducible control of Cas9 expression will allow better and more precise genome editing. For iPSC, the method of choice is still nucleofection, which allows moderately efficient, but importantly, transient expression of CRISPR for genome editing. Most of the viral approaches, while more efficient, bear the risk of stably integrating a constitutively expressing Cas9 into the cellular genome. This is obviously undesirable due to the possibility of an accumulation of off-target cleavages by a constantly active Cas9, even though given the drastic conformational change in the structure of the Cas9 protein when it binds to gRNA, it seems unlikely it will have any residual activity when not bound to its RNA component (Jinek et al. 2014). Whether Cas9 has the ability to bind naturally occurring RNA transcripts within mammalian cells to catalyze DNA cleavage in a non-specific manner remains to be seen.

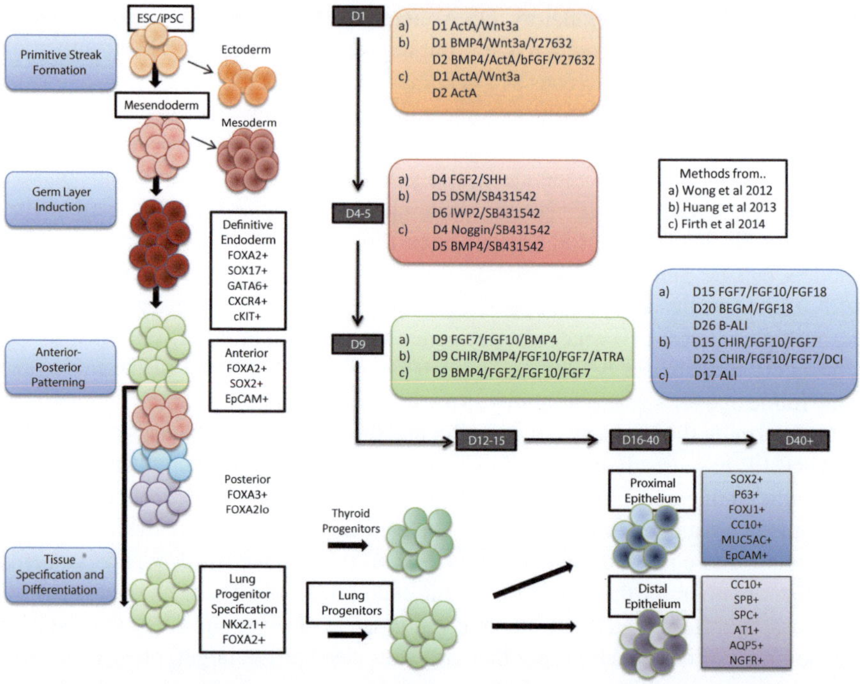

Fig. 7.1 Summary of differentiation from iPSC and hESC to airway epithelium. The diagram highlights the key phases in the development of the lung epithelium from pluripotent stem cells (iPSC/ESC) through definitive endoderm (DE), anterior foregut endoderm (AFE) to lung progenitors ($NKx2.1^+$ and $FOXA2^+$). From here, cells can be matured to generate the proximal conducting epithelium with Club, Ciliated, Goblet and Basal cells and the distal epithelium consisting of variant Club cells, Alveolar type I and II cells. The protocols used by (a) (Wong et al. 2012), (b) (Huang et al. 2013), and (c) (Firth et al. 2014) are included and readers are referred to the original papers and (Gomperts 2014) for more protocol detail

Recently, an inducible CRISPR system (iCRISPR) has been described, where TALENs were used to stably knock-in the Cas9 gene into the "safe harbor" AAVS1 genomic locus under a Dox-inducible promoter. This was used to derive human ESC and iPSC lines that can be stimulated to carry out CRISPR-mediated genome editing by Dox-induction of Cas9 expression and transient transfection of target-specific gRNA (Gonzalez et al. 2014). Such inducible expression systems will be invaluable for achieving efficient and precise genome editing, while providing optimal control over the off-target effects and temporal expression of CRISPR. Early reports are emerging of a stably integrated Cas9 mouse model, where one should be able to achieve organ and cell-specific genome editing in vivo by just introducing the gRNA into the appropriate cells, as the Cas9 is ubiquitously and constitutively expressed. As these delivery vectors and expression systems continue to improve, CRISPR technology will break through more biological barriers and technical roadblocks on its path to the conquest of the genome.

7.2.2 *Using iPSC to Model the Lung*

The seminal work of Yamanaka et al. in 2007 in generating pluripotent stem cells from somatic cells (induced pluripotent stem cells or iPSC) has opened the door for generating patient- or disease-specific pluripotent cells. These have the potential to differentiate to disease-relevant cell types to model disease and screen for novel therapeutic approaches (Takahashi et al. 2007). While iPSC have been widely adopted for differentiating into neural, cardiac, and other cell types discussed above, its implementation in lung biology has been limited. This is largely due to the complexity of lung structure and cell types, which hinders effective functional modeling of the lung in vitro. Therefore, it is necessary to develop cellular models that functionally represent all the different cell types that characterize the different regions of the lung, like the proximal airways, the conducting airways, and the distal alveoli, each of which is represented by a unique combination of cell types and tissue organization.

To be able to utilize iPSC to model lung disease, an effective differentiation protocol is necessary to be able to generate a pseudostratified polarized respiratory epithelium in a dish. In particular, this requires the generation of a specialized post-mitotic multiciliated cell. The differentiation of ESCs and iPSC to airway epithelial cells has recently received increased attention (Wong et al. 2012; Kadzik and Morrisey 2012), and was recently advanced further by Firth et al., 2014, in a study describing the differentiation of human iPSC to a functional respiratory epithelium, demonstrating the generation of multiciliated cells, Clara goblet, and basal cells in a polarized epithelial layer.

Directed differentiation towards endodermal cell lineages like the lung, as well as hepatocytes and pancreatic cells, has been previously attempted by recapitulating the paradigm of endodermal development in the embryo (D'Amour et al. 2006; Gouon-Evans et al. 2006; Green et al. 2011). Activin A can be used to mimic nodal signaling during gastrulation, and thus induce enriched definitive endoderm (DE) from ESC, marked by expression of transcription factors Sox17 and FoxA2 (Gadue et al. 2006; Yasunaga et al. 2005). Addition of Wnt3a, which is required during gastrulation for primitive streak formation (Tam and Loebel 2007), also enhances this process (Nostro and Keller 2012). In order to follow the lung developmental process along the pathway of endodermal maturation, one then has to induce expression of transcription factors Sox2 and FoxA2, characteristic of the anterior foregut endoderm (AFE), while suppressing posterior marker CDX2. Green et al. first described such an endodermal development-mimetic process, by dual inhibition of TGF-β and BMP in ESC-derived DE, by using a pharmacological inhibitor of ActivinA/nodal and TGF-β signaling, along with a physiological inhibitor of BMP, Noggin. This led to increased expression of the foregut marker Sox2 and maintenance of the endodermal marker FoxA2, along with the concomitant suppression of posterior marker CDX2, all indicative of AFE specification (Green et al. 2011).

The AFE gives rise to the pharyngeal endoderm as it develops along the anterio-posterior axis, and consequent dorso-ventral organ patterning gives rise to the lung field, marked by expression of Nkx2.1. Treatment of hPSC-derived AFE with a combination of Wnt3a, KGF, Fgf10, Bmp4, and EGF results in an increase in expression of Nkx2.1, along with a decrease in Sox2 expression, indicating ventralization of the AFE (Green et al. 2011). Following the observation of retinoic acid (RA) signaling in the lung field, but not in the simultaneously developed pharyngeal pouch, Green et al. used RA to increase further the expression of lung markers like Nkx2.1, Nkx2.5, and Pax1, while downregulating pharyngeal cell markers. However, the frequency of Nkx2.1$^+$ FoxA2$^+$ cells was low and specific markers of mature airway epithelium were missing from this protocol.

This protocol was adapted in an Nkx2.1-GFP mouse ESC reporter line by Longmire et al. to purify Nkx2.1$^+$ cell populations from in vitro-derived AFE. The mouse experiments recapitulated the human differentiation protocol (Green et al. 2011) for the most part, except FGF2 was necessary in the ventralization cocktail. They found that the Nkx2.1$^+$ cells sorted from the AFE stage were not entirely mature and also displayed expression of thyroid cell markers. These observations could be due to intrinsic differences between the temporal, dynamic, and physiological characteristics of mouse and human lung development. Addition of a cocktail of 8-bromo-cAMP, dexamethasone, isobutyl-methylxanthine (DCI), KGF, transferrin, and sodium selenite improved distal specification, giving rise to cells expressing mature lung cell markers like surfactant proteins and CC10. Transplantation of sorted Nkx2.1$^+$ AFE cells into decellularized mouse lungs gave sporadic expression of markers found in type I alveolar epithelial cells. Gene expression analysis of these sorted cells had significant overlap with similar cells from the developing mouse lung (Longmire et al. 2012; Gomperts 2014).

Mou et al. also showed that dual TGF-β and BMP inhibition promotes the generation of AFE from DE and drives lung differentiation in favor of neural differentiation. Also, while previous studies generated DE through embryoid body formation, this study differentiated mESC in monolayers using inhibition of TGF-β only (Mou et al. 2012). However, their differentiation only yielded Nkx2.1$^+$ lung progenitors at low efficiencies (10–30 %) and these could only be matured into Nkx2.1$^+$p63$^+$ proximal airway epithelial cells in vivo upon adding KGF and BMP7, inhibiting Wnt and MAPK/ERK signaling and subcutaneous transplantation into mice.

Wong et al. were the first to differentiate human iPSC into mature airway epithelium by using an air–liquid interface (ALI) to mimic the post-natal airway epithelial niche. They were able to obtain higher levels of AFE induction by treatment of hPSC-derived DE with sonic hedgehog (SHH) and FGF2, reporting that 78 % of the resultant cells expressed Nkx2.1 (Wong et al. 2012). However, it should be noted that this high efficiency of Nkx2.1-expressing cells was not reproducible by the efforts of Huang et al. The treatment of these Nk2.1+ cells with FGF10, KGF, FGF18, and low concentration of BMP4, followed by ALI culture, induced proximal lung differentiation, with cells expressing markers of basal, ciliated, and mucous cells, but not club cells, nor any distal lung cells. Functional expression of CFTR was also observed on the apical surface of these iPSC-derived airway epithelia, and this was used as an indicator to compare differentiated lung cells from CF patient-derived iPSC against normal iPSC. The aberrant CFTR expression and function in the patient-derived lung epithelium was rescued using CF corrector compounds. This study elucidated a disease-specific airway epithelial model system, using patient-derived iPSC.

Huang et al. reported a further improvement in the generation of AFE from hPSC, by sequential inhibition of TGF-β/ BMP and then Wnt, to recapitulate developmental anteriorization and ventralization of the DE. This generated Nkx2.1 + FoxA2+ lung progenitor cells with 86 % efficiency, and by following a defined differentiation protocol involving 2 weeks of DCI treatment, these cells yielded over 50 % SP-B$^+$ lung epithelial cells with <5 % SP-C expression (Gomperts 2014). This protocol circumvented the use of ALI culture and appeared to yield a variety of cell types, more biased towards the distal respiratory epithelium, unlike the protocol described by Wong et al., where ALI culture yielded cells more characteristic of proximal airway epithelium (Huang et al. 2014). It has been suggested, due to similarities in their expression profiles, that this protocol by Huang et al. may generate lung epithelial cells that resemble the human fetal lung rather than the adult lung.

More recently, Ghaedi et al. described a protocol that generated alveolar epithelial cells at much higher efficiencies. They reported up to 97 % expression of SP-C, 95 % of mucin, 93 % of SP-B, and 89 % of CD54 in relatively homogenous populations of alveolar epithelium type-II (AEC-II) cells. They also reported that exposure of iPSC-derived AEC-II cells to the Wnt/β-catenin inhibitor, IWR-1, switched the phenotype from that of AEC-II cells to AEC-I cells, with over 90 % expression of type-I markers T-1α and calveolin-1. Under the appropriate culture conditions, iPSC-derived lung progenitors adhered to and repopulated decellularized lung extracellular matrix (Ghaedi et al. 2013). The same group also described the use of ALI culture in a rotating bioreactor for the large-scale production of alveolar epithelial cells for tissue engineering and drug discovery (Ghaedi et al. 2014).

Firth et al. reported the first robust differentiation of iPSC into mature multiciliated respiratory epithelium. iPSC were differentiated via definitive endoderm to AFE, pulmonary endoderm, and then matured in ALI culture to derive a functional pseudostratified polarized epithelium (Firth et al. 2014). Exposure of the basal surface of the pulmonary epithelium to liquid media and the apical surface to air in this ALI culture gave rise to a polarized epithelium, showing the presence of Club cells with CC10 positive vesicles, MUC5A/C positive Goblet cells, and CK5, p63, and PDPN positive basal cells. The basal layer expressing mesenchymal markers was found to be essential for the differentiation and maintenance of the polarized epithelial layer. Primary cilia were evident at the apical surface of the ALI-based epithelium, and when Notch signaling was inhibited, mature multiciliated cells were generated. These were characterized by robust pericentrin staining, indicating the assembly of multiple centrioles at the apical surface, expression of transcription factor FOXJ1, and multiple acetylated tubulin-labeled cilia projections found in individual cells. This iPSC-derived epithelium showed the presence of forskolin-induced chloride currents sensitive to CFTRinh172 in isolated epithelial cells by whole cell patch clamp technique, demonstrating functionality of the epithelial layer (Firth et al. 2014).

Motile multiciliated cells (MCCs) are a population of specialized cells which have exited cell cycle, assembled basal bodies, and project hundreds of motile cilia as they differentiate. Centrioles form the core of the centrosome and are a microtubule-based structure that anchors the cilium (Marshall 2008). The generation of MCC is critical to the function of a respiratory epithelium; their coordinated beating is essential for the movement of mucous and protection of the lung. Inhibition of notch has been shown to be essential for the development of MCC in xenopous, mouse, and human airways (Stubbs et al. 2006, 2012; Rock et al. 2011; Tsao et al. 2011; Marcet et al. 2011; Jurisch-Yaksi et al. 2013). Generation of such cells from pluripotent stem cells, in a Notch-dependent manner, is a major finding by Firth et al., as it provides the opportunity for in-depth study of the development of these cells in the human system and may lead to the discovery of new mechanisms and therapeutic approaches for diseases such as primary ciliary dyskinesia (PCD), which have been difficult to model and understand with the research tools currently available (Noone et al. 2004).

Directed differentiation of iPSC thus provides a renewable source of human airway epithelial cells including multiciliated cells, which can be utilized to study human respiratory diseases that have previously been difficult to study and model in vitro. iPSC provide an unlimited source of cells and also proffer the opportunity for gene editing and clonal expansion of cells for disease modeling. The ability to isolate and culture iPSC clonally is particularly useful for genome editing, as it allows for homogenous gene editing to give rise to isogenic clonal populations of cells bearing the desired engineered genotype. There are several lung diseases with a known genetic origin, such as cystic fibrosis and PCD, which could be corrected by replacement of the defective gene by the correct gene by gene editing technology (Wood et al. 2011). In addition, one can introduce disease-causing mutations to see if they are causative of the pathological phenotype and thus establish de novo models of genetic diseases that are relatively rare in the human population and therefore hard to obtain patient-derived samples to generate iPSC from. It is hoped that patient-specific iPSC cells can be utilized to model the lung, not only to provide a platform for understanding the cellular and molecular mechanisms of respiratory diseases like CF, asthma, and bronchitis among others, but also to generate gene-corrected transplantable cell types capable of engraftment into the lung for direct clinical intervention.

7.2.3 The Way Forward with iPSC

iPSC thus represent a major hope for modeling a variety of monogenic as well as complex, multifactorial diseases. Animal models, while useful, do not always recapitulate the human disease faithfully. Drugs found to be effective in animal models often do not work to treat the human condition (Inoue and Yamanaka 2011). Conversely, there are also examples of drugs found to be efficacious in humans that were not so in animal models (Tobert 2003). Thus, it is important to use human cells for testing drugs in pre-clinical trials. Other sources of primary human cells are often difficult to obtain and/or culture long-term in the laboratory. iPSCs can be generated from almost any source tissue and can be cultured and propagated indefinitely in vitro. Also, iPSC models allow one to have a patient-specific model of the disease, while also accounting for the genetic and phenotypic variation in the human population, making it an ideal model system to study human disease.

One of the main limitations of using iPSC to model disease is the inherent variability of existing differentiation protocols, giving rise to heterogeneous cell populations. This could be due to incomplete reprogramming (Soldner and Jaenisch 2012), epigenetic memory (Kim et al. 2010), or defective X-chromosome inactivation (Mekhoubad et al. 2012). In the case of the lung, the variety of different cell types and their varied distribution in the lung tissue, and the need to accurately reproduce this cellular heterogeneity for functional modeling, further confounds the situation. There is thus a need for more robust and reproducible differentiation protocols to differentiate iPSC to various cell types, or even direct transdifferentiation of one somatic cell type to another. However, a recent study analyzing various differentiation and trans-differentiation protocols using an in silico approach to analyze gene expression data from 56 different published reports found that directly converted cells fail to completely silence expression programs of the original cell type and are thus more incomplete than iPSC/ESC differentiation protocols for generating mature cells of specific lineages (Cahan et al. 2014).

While early lung differentiation protocols from human and mouse ESC depended on just a couple of phenotypic markers of mature lung epithelium, more recent reports have incorporated a more sophisticated understanding of embryonic lung development along with tools like reporter lines for lineage tracing to achieve much more robust and efficient generation of lung epithelial cells of various lineages including proximal and distal airways, alveolar cells type I and II, etc. These derived cells have been shown to repopulate decellularized whole lung scaffolds (Longmire et al. 2012) and can be generated from a variety of patients with both genetic and acquired lung diseases like CF, alpha-1-antitrypsin deficiency, sickle cell anemia, and scleroderma, for better modeling and study of these diseases (Somers et al. 2010; Pickering et al. 2005). These approaches continue to gain traction and are being improved upon, even though there is still no approved clinical use of ESC or iPSC for treatment of lung disease. However, the variability in efficiency and homogeneity of the generated cells in all these current protocols continues to be a limitation for more widespread and uniform disease modeling using iPSC.

To improve these protocols further, it is important to reproduce the biological niche within which these cell types exist, so that phenotypes characteristic of the holistic human conditions are more accurately reproduced. Tissue- and organ-engineering approaches will go a long way in enabling this goal. There has been significant progress lately using both synthetic scaffolds as well as decellularized cadaveric or donor tissues for mimicking the structure of the trachea and diaphragm ex vivo, resulting in increased clinical use of such engineered tissues (Fishman et al. 2012; Badylak et al. 2012). Three-dimensional (3D) scaffolds using synthetic or biomimetic materials have been used to develop ex vivo lung parenchymal and vascular systems, and impregnation of these scaffolds with stem and other lung cell types and/or implantation in vivo has been used to generate functional lung tissue in animal models (Mondrinos et al. 2006; Nichols and Cortiella 2008). Such bioengineering of the lung is discussed in-depth in the subsequent chapter by Daniel Weiss on Ex Vivo Lung Bioengineering.

In spite of these challenges, generating functional 3D lung tissues ex vivo remains a major avenue of hope for the better utilization of iPSC-derived lung models. Other approaches, such as coating iPSC-derived lung epithelial cells onto porous polydimethylsiloxane chips to mimic alveolar architecture, may be useful for high-throughput drug screening and modeling alveolar physiology. But to model the entire lung ex vivo, significant improvement of our understanding and ability to recapitulate the dynamic interactions of the various cell types involved in the complex 3D cellular architecture of the lung is still required.

It is also necessary to carry out iPSC-based disease modeling experiments with the adequate isogenic controls. Since many diseases being modeled with iPSC are genetic conditions, caused by specific mutations, it is important to use control iPSC that have the wild-type sequence, but are otherwise perfectly isogenic. This is where gene editing technologies like CRISPR and TALENs will play a huge role in precise manipulation of the genome, to eliminate genetic variability accounting for observed phenotypes. Genome editing can also be used to introduce specific disease-causing mutations, when patient source material is unavailable, and to precisely correct such mutations in isogenic iPSC lines, to try and reverse the pathogenic phenotype towards gene/cell therapy.

Along these lines, iPSC have a completely separate, unique set of advantages towards cell transplantation therapy. They provide a limitless source of autologous cells that can be precisely gene-corrected in a customized manner, differentiated into desired cell types, and transplanted into patients with minimal risk of immune rejection. Proof-of-concept studies for this approach would involve generating disease-specific iPSC, correcting the causative genetic mutation by homologous recombination, differentiating cells into a transplantable cell type, and carrying out rescue transplantation in suitable animal models of the disease. Advanced bioengineering approaches like 3D-bioprinting of cells, decellularized organ and tissue engineering, development of novel biomaterials for scaffolding cell culture, and the use of animal models to generate and grow human organs are revolutionizing the possibilities for iPSC in cell transplantation therapy.

Another new application of iPSC in medicine is using them as renewable source of cells for toxicity studies of drugs in development before going to clinical trials (Inoue et al. 2014). Proof-of-concept toxicity studies conducted in iPSC-derived cell types (Guo et al. 2011; Medine et al. 2013) support the idea of larger scale drug toxicity screens in human cells. iPSC-derived hepatocytes, for example, can be used to test hepatotoxicity (Scott et al. 2013). Similarly, iPSC-derived cardiomyocytes can be screened by electrophysiological methods for cardiotoxic side effects of drugs, in addition to screening for the effectiveness of arrhythmogenic drugs in their target cell type (Guo et al. 2011; Lahti et al. 2012). iPSC can be used as a representative cohort of the genetic and phenotypic variation of the human population, as well as account for factors like environment and age that might affect outcomes of drug trials. For this, it will be necessary to build large “iPSC banks” where an appropriate subset of the test population, both diseased and wild-type controls, can be represented for drug responsiveness and safety testing. This way, a variety of iPSC-derived cells, from varied genetic and physical backgrounds, can be tested at a stage between the drug discovery and the development phase, before investing the large amount of financial and physical resources that a clinical trial entails. Such “iPSC clinical trials” (Inoue et al. 2014) would enable identification of patient groups that are more responsive to a drug, and thus inform the process of selection and organization of a Phase II clinical trial better, by representing the disease-relevant SNPs and other genetic variation found in the human population. As a surrogate approach to building and using an extensive iPSC bank to represent all this genomic variation for such trials, which will undoubtedly be an expensive and labor-intensive affair, newly described genome-wide CRISPR libraries (Shalem et al. 2014; Wang et al. 2014) can also be used in iPSC to introduce mutations across the genome and observe their effects on drug response and other treatments. This will also enable discovery of novel genes and genomic elements involved in certain pathologies and their response to drugs.

7.3 Summary

The advent of iPSC has revolutionized our access to human biology by providing the unique potential to study any cell type in the human body. All the current methodologies for studying lung disease, while having provided invaluable information, do have limitations. While iPSC may have their own limitations, the ability to edit the genome provides an opportunity to study cellular and molecular biology in an isogenic human system where the experimental conditions can have direct controls. There are still limited studies where iPSC have been differentiated into the lung epithelium, but as the field develops, more robust and reproducible protocols will become available, hopefully leading to a novel era of research in lung biology.

References

Agarwal S, Loh YH, McLoughlin EM, Huang J, Park IH, Miller JD, Huo H, Okuka M, Dos Reis RM, Loewer S, Ng HH, Keefe DL, Goldman FD, Klingelhutz AJ, Liu L, Daley GQ (2010) Telomere elongation in induced pluripotent stem cells from dyskeratosis congenita patients. Nature 464(7286):292–296. doi:10.1038/nature08792

Badylak SF, Weiss DJ, Caplan A, Macchiarini P (2012) Engineered whole organs and complex tissues. Lancet 379(9819):943–952. doi:10.1016/S0140-6736(12)60073-7

Barrangou R, Fremaux C, Deveau H, Richards M, Boyaval P, Moineau S, Romero DA, Horvath P (2007) CRISPR provides acquired resistance against viruses in prokaryotes. Science 315(5819):1709–1712. doi:10.1126/science.1138140

Bibikova M, Carroll D, Segal DJ, Trautman JK, Smith J, Kim YG, Chandrasegaran S (2001) Stimulation of homologous recombination through targeted cleavage by chimeric nucleases. Mol Cell Biol 21(1):289–297. doi:10.1128/MCB. 21.1.289-297.2001

Bibikova M, Golic M, Golic KG, Carroll D (2002) Targeted chromosomal cleavage and mutagenesis in Drosophila using zinc-finger nucleases. Genetics 161(3):1169–1175

Bibikova M, Beumer K, Trautman JK, Carroll D (2003) Enhancing gene targeting with designed zinc finger nucleases. Science 300(5620):764. doi:10.1126/science.1079512

Boch J, Scholze H, Schornack S, Landgraf A, Hahn S, Kay S, Lahaye T, Nickstadt A, Bonas U (2009) Breaking the code of DNA binding specificity of TAL-type III effectors. Science 326(5959):1509–1512. doi:10.1126/science.1178811

Bolotin A, Quinquis B, Sorokin A, Ehrlich SD (2005) Clustered regularly interspaced short palindrome repeats (CRISPRs) have spacers of extrachromosomal origin. Microbiology 151(Pt 8):2551–2561. doi:10.1099/mic. 0.28048-0

Bradley A, Evans M, Kaufman MH, Robertson E (1984) Formation of germ-line chimaeras from embryo-derived teratocarcinoma cell lines. Nature 309(5965):255–256

Brambrink T, Foreman R, Welstead GG, Lengner CJ, Wernig M, Suh H, Jaenisch R (2008) Sequential expression of pluripotency markers during direct reprogramming of mouse somatic cells. Cell Stem Cell 2(2):151–159. doi:10.1016/j.stem.2008.01.004

Brennand KJ, Simone A, Jou J, Gelboin-Burkhart C, Tran N, Sangar S, Li Y, Mu Y, Chen G, Yu D, McCarthy S, Sebat J, Gage FH (2011) Modelling schizophrenia using human induced pluripotent stem cells. Nature 473(7346):221–225. doi:10.1038/nature09915

Brouns SJ, Jore MM, Lundgren M, Westra ER, Slijkhuis RJ, Snijders AP, Dickman MJ, Makarova KS, Koonin EV, van der Oost J (2008) Small CRISPR RNAs guide antiviral defense in prokaryotes. Science 321(5891):960–964. doi:10.1126/science.1159689

Cahan P, Li H, Morris SA, Lummertz da Rocha E, Daley GQ, Collins JJ (2014) Cell net: network biology applied to stem cell engineering. Cell 158(4):903–915. doi:10.1016/j.cell.2014.07.020

Capecchi MR (1989) Altering the genome by homologous recombination. Science 244(4910): 1288–1292

Carey BW, Markoulaki S, Hanna J, Saha K, Gao Q, Mitalipova M, Jaenisch R (2009) Reprogramming of murine and human somatic cells using a single polycistronic vector. Proc Natl Acad Sci U S A 106(1):157–162. doi:10.1073/pnas.0811426106

Carvajal-Vergara X, Sevilla A, D'Souza SL, Ang YS, Schaniel C, Lee DF, Yang L, Kaplan AD, Adler ED, Rozov R, Ge Y, Cohen N, Edelmann LJ, Chang B, Waghray A, Su J, Pardo S, Lichtenbelt KD, Tartaglia M, Gelb BD, Lemischka IR (2010) Patient-specific induced pluripotent stem-cell-derived models of LEOPARD syndrome. Nature 465(7299):808–812. doi:10.1038/nature09005

Chang CW, Lai YS, Pawlik KM, Liu K, Sun CW, Li C, Schoeb TR, Townes TM (2009) Polycistronic lentiviral vector for "hit and run" reprogramming of adult skin fibroblasts to induced pluripotent stem cells. Stem Cells 27(5):1042–1049. doi:10.1002/stem.39

Chen B, Gilbert LA, Cimini BA, Schnitzbauer J, Zhang W, Li GW, Park J, Blackburn EH, Weissman JS, Qi LS, Huang B (2013) Dynamic imaging of genomic loci in living human cells by an optimized CRISPR/Cas system. Cell 155(7):1479–1491. doi:10.1016/j.cell.2013.12.001

Chin MH, Pellegrini M, Plath K, Lowry WE (2010) Molecular analyses of human induced pluripotent stem cells and embryonic stem cells. Cell Stem Cell 7(2):263–269. doi:10.1016/j.stem.2010.06.019

Christian M, Cermak T, Doyle EL, Schmidt C, Zhang F, Hummel A, Bogdanove AJ, Voytas DF (2010) Targeting DNA double-strand breaks with TAL effector nucleases. Genetics 186(2): 757–761. doi:10.1534/genetics.110.120717

Cong L, Ran FA, Cox D, Lin S, Barretto R, Habib N, Hsu PD, Wu X, Jiang W, Marraffini LA, Zhang F (2013) Multiplex genome engineering using CRISPR/Cas systems. Science 339(6121):819–823. doi:10.1126/science.1231143

D'Amour KA, Bang AG, Eliazer S, Kelly OG, Agulnick AD, Smart NG, Moorman MA, Kroon E, Carpenter MK, Baetge EE (2006) Production of pancreatic hormone-expressing endocrine cells from human embryonic stem cells. Nat Biotechnol 24(11):1392–1401. doi:10.1038/nbt1259

Deveau H, Barrangou R, Garneau JE, Labonte J, Fremaux C, Boyaval P, Romero DA, Horvath P, Moineau S (2008) Phage response to CRISPR-encoded resistance in Streptococcus thermophilus. J Bacteriol 190(4):1390–1400. doi:10.1128/JB.01412-07

Dimos JT, Rodolfa KT, Niakan KK, Weisenthal LM, Mitsumoto H, Chung W, Croft GF, Saphier G, Leibel R, Goland R, Wichterle H, Henderson CE, Eggan K (2008) Induced pluripotent stem cells generated from patients with ALS can be differentiated into motor neurons. Science 321(5893):1218–1221. doi:10.1126/science.1158799

Ebert AD, Yu J, Rose FF Jr, Mattis VB, Lorson CL, Thomson JA, Svendsen CN (2009) Induced pluripotent stem cells from a spinal muscular atrophy patient. Nature 457(7227):277–280. doi:10.1038/nature07677

Evans MJ, Kaufman MH (1981) Establishment in culture of pluripotential cells from mouse embryos. Nature 292(5819):154–156

Firth AL, Dargitz CT, Qualls SJ, Menon T, Wright R, Singer O, Gage FH, Khanna A, Verma IM (2014) Generation of multiciliated cells in functional airway epithelia from human induced pluripotent stem cells. Proc Natl Acad Sci U S A 111(17):E1723–E1730. doi:10.1073/pnas.1403470111

Fishman JM, Ansari T, Sibbons P, De Coppi P, Birchall MA (2012) Decellularized rabbit cricoarytenoid dorsalis muscle for laryngeal regeneration. Ann Otol Rhinol Laryngol 121(2):129–138

Fusaki N, Ban H, Nishiyama A, Saeki K, Hasegawa M (2009) Efficient induction of transgene-free human pluripotent stem cells using a vector based on Sendai virus, an RNA virus that does not integrate into the host genome. Proc Jpn Acad Ser B Phys Biol Sci 85(8):348–362

Gadue P, Huber TL, Paddison PJ, Keller GM (2006) Wnt and TGF-beta signaling are required for the induction of an in vitro model of primitive streak formation using embryonic stem cells. Proc Natl Acad Sci U S A 103(45):16806–16811. doi:10.1073/pnas.0603916103

Garneau JE, Dupuis ME, Villion M, Romero DA, Barrangou R, Boyaval P, Fremaux C, Horvath P, Magadan AH, Moineau S (2010) The CRISPR/Cas bacterial immune system cleaves bacteriophage and plasmid DNA. Nature 468(7320):67–71. doi:10.1038/nature09523

Ghaedi M, Calle EA, Mendez JJ, Gard AL, Balestrini J, Booth A, Bove PF, Gui L, White ES, Niklason LE (2013) Human iPS cell-derived alveolar epithelium repopulates lung extracellular matrix. J Clin Invest 123(11):4950–4962. doi:10.1172/JCI68793

Ghaedi M, Mendez JJ, Bove PF, Sivarapatna A, Raredon MS, Niklason LE (2014) Alveolar epithelial differentiation of human induced pluripotent stem cells in a rotating bioreactor. Biomaterials 35(2):699–710. doi:10.1016/j.biomaterials.2013.10.018

Ghodsizadeh A, Taei A, Totonchi M, Seifinejad A, Gourabi H, Pournasr B, Aghdami N, Malekzadeh R, Almadani N, Salekdeh GH, Baharvand H (2010) Generation of liver disease-specific induced pluripotent stem cells along with efficient differentiation to functional hepatocyte-like cells. Stem Cell Rev 6(4):622–632. doi:10.1007/s12015-010-9189-3

Gomperts BN (2014) Induction of multiciliated cells from induced pluripotent stem cells. Proc Natl Acad Sci U S A 111(17):6120–6121. doi:10.1073/pnas.1404414111

Gonzalez F, Zhu Z, Shi ZD, Lelli K, Verma N, Li QV, Huangfu D (2014) An iCRISPR platform for rapid, multiplexable, and inducible genome editing in human pluripotent stem cells. Cell Stem Cell 15(2):215–226. doi:10.1016/j.stem.2014.05.018

Gouon-Evans V, Boussemart L, Gadue P, Nierhoff D, Koehler CI, Kubo A, Shafritz DA, Keller G (2006) BMP-4 is required for hepatic specification of mouse embryonic stem cell-derived definitive endoderm. Nat Biotechnol 24(11):1402–1411. doi:10.1038/nbt1258

Green MD, Chen A, Nostro MC, d'Souza SL, Schaniel C, Lemischka IR, Gouon-Evans V, Keller G, Snoeck HW (2011) Generation of anterior foregut endoderm from human embryonic and induced pluripotent stem cells. Nat Biotechnol 29(3):267–272. doi:10.1038/nbt.1788

Green MD, Huang SX, Snoeck HW (2013) Stem cells of the respiratory system: from identification to differentiation into functional epithelium. Bioessays 35(3):261–270. doi:10.1002/bies.201200090

Guo L, Abrams RM, Babiarz JE, Cohen JD, Kameoka S, Sanders MJ, Chiao E, Kolaja KL (2011) Estimating the risk of drug-induced proarrhythmia using human induced pluripotent stem cell-derived cardiomyocytes. Toxicol Sci 123(1):281–289. doi:10.1093/toxsci/kfr158

Gurdon JB (1975) Nuclear transplantation and the analysis of gene activity in early amphibian development. Adv Exp Med Biol 62:35–44

Haft DH, Selengut J, Mongodin EF, Nelson KE (2005) A guild of 45 CRISPR-associated (Cas) protein families and multiple CRISPR/Cas subtypes exist in prokaryotic genomes. PLoS Comput Biol 1(6):e60. doi:10.1371/journal.pcbi.0010060

Hsu PD, Lander ES, Zhang F (2014) Development and applications of CRISPR-Cas9 for genome engineering. Cell 157(6):1262–1278. doi:10.1016/j.cell.2014.05.010

Huang SX, Islam MN, O'Neill J, Hu Z, Yang YG, Chen YW, Mumau M, Green MD, Vunjak-Novakovic G, Bhattacharya J, Snoeck HW (2013) Efficient generation of lung and airway epithelial cells from human pluripotent stem cells. Nat Biotechnol. doi:10.1038/nbt.2754

Huang SX, Islam MN, O'Neill J, Hu Z, Yang YG, Chen YW, Mumau M, Green MD, Vunjak-Novakovic G, Bhattacharya J, Snoeck HW (2014) Efficient generation of lung and airway epithelial cells from human pluripotent stem cells. Nat Biotechnol 32(1):84–91. doi:10.1038/nbt.2754

Huangfu D, Osafune K, Maehr R, Guo W, Eijkelenboom A, Chen S, Muhlestein W, Melton DA (2008) Induction of pluripotent stem cells from primary human fibroblasts with only Oct4 and Sox2. Nat Biotechnol 26(11):1269–1275. doi:10.1038/nbt.1502

Inoue H, Yamanaka S (2011) The use of induced pluripotent stem cells in drug development. Clin Pharmacol Ther 89(5):655–661. doi:10.1038/clpt.2011.38

Inoue H, Nagata N, Kurokawa H, Yamanaka S (2014) iPS cells: a game changer for future medicine. EMBO J 33(5):409–417. doi:10.1002/embj.201387098

Ishino Y, Shinagawa H, Makino K, Amemura M, Nakata A (1987) Nucleotide sequence of the iap gene, responsible for alkaline phosphatase isozyme conversion in Escherichia coli, and identification of the gene product. J Bacteriol 169(12):5429–5433

Itzhaki I, Maizels L, Huber I, Zwi-Dantsis L, Caspi O, Winterstern A, Feldman O, Gepstein A, Arbel G, Hammerman H, Boulos M, Gepstein L (2011) Modelling the long QT syndrome with induced pluripotent stem cells. Nature 471(7337):225–229. doi:10.1038/nature09747

Jang J, Kang HC, Kim HS, Kim JY, Huh YJ, Kim DS, Yoo JE, Lee JA, Lim B, Lee J, Yoon TM, Park IH, Hwang DY, Daley GQ, Kim DW (2011) Induced pluripotent stem cell models from X-linked adrenoleukodystrophy patients. Ann Neurol 70(3):402–409. doi:10.1002/ana.22486

Jansen R, Embden JD, Gaastra W, Schouls LM (2002) Identification of genes that are associated with DNA repeats in prokaryotes. Mol Microbiol 43(6):1565–1575

Jia F, Wilson KD, Sun N, Gupta DM, Huang M, Li Z, Panetta NJ, Chen ZY, Robbins RC, Kay MA, Longaker MT, Wu JC (2010) A nonviral minicircle vector for deriving human iPS cells. Nat Methods 7(3):197–199. doi:10.1038/nmeth.1426

Jin ZB, Okamoto S, Osakada F, Homma K, Assawachananont J, Hirami Y, Iwata T, Takahashi M (2011) Modeling retinal degeneration using patient-specific induced pluripotent stem cells. PLoS One 6(2):e17084. doi:10.1371/journal.pone.0017084

Jinek M, Chylinski K, Fonfara I, Hauer M, Doudna JA, Charpentier E (2012) A programmable dual-RNA-guided DNA endonuclease in adaptive bacterial immunity. Science 337(6096):816–821. doi:10.1126/science.1225829

Jinek M, Jiang F, Taylor DW, Sternberg SH, Kaya E, Ma E, Anders C, Hauer M, Zhou K, Lin S, Kaplan M, Iavarone AT, Charpentier E, Nogales E, Doudna JA (2014) Structures of Cas9 endonucleases reveal RNA-mediated conformational activation. Science 343(6176):1247997. doi:10.1126/science.1247997

Jurisch-Yaksi N, Rose AJ, Lu H, Raemaekers T, Munck S, Baatsen P, Baert V, Vermeire W, Scales SJ, Verleyen D, Vandepoel R, Tylzanowski P, Yaksi E, de Ravel T, Yost HJ, Froyen G, Arrington CB, Annaert W (2013) Rer1p maintains ciliary length and signaling by regulating gamma-secretase activity and Foxj1a levels. J Cell Biol 200(6):709–720. doi:10.1083/jcb.201208175

Kadzik RS, Morrisey EE (2012) Directing lung endoderm differentiation in pluripotent stem cells. Cell Stem Cell 10(4):355–361. doi:10.1016/j.stem.2012.03.013

Kaji K, Norrby K, Paca A, Mileikovsky M, Mohseni P, Woltjen K (2009) Virus-free induction of pluripotency and subsequent excision of reprogramming factors. Nature 458(7239):771–775. doi:10.1038/nature07864

Kawagoe S, Higuchi T, Meng XL, Shimada Y, Shimizu H, Hirayama R, Fukuda T, Chang H, Nakahata T, Fukada S, Ida H, Kobayashi H, Ohashi T, Eto Y (2011) Generation of induced pluripotent stem (iPS) cells derived from a murine model of Pompe disease and differentiation of Pompe-iPS cells into skeletal muscle cells. Mol Genet Metab 104(1–2):123–128. doi:10.1016/j.ymgme.2011.05.020

Khan IF, Hirata RK, Wang PR, Li Y, Kho J, Nelson A, Huo Y, Zavaljevski M, Ware C, Russell DW (2010) Engineering of human pluripotent stem cells by AAV-mediated gene targeting. Mol Ther 18(6):1192–1199. doi:10.1038/mt.2010.55

Kim YG, Cha J, Chandrasegaran S (1996) Hybrid restriction enzymes: zinc finger fusions to Fok I cleavage domain. Proc Natl Acad Sci U S A 93(3):1156–1160

Kim D, Kim CH, Moon JI, Chung YG, Chang MY, Han BS, Ko S, Yang E, Cha KY, Lanza R, Kim KS (2009) Generation of human induced pluripotent stem cells by direct delivery of reprogramming proteins. Cell Stem Cell 4(6):472–476. doi:10.1016/j.stem.2009.05.005

Kim K, Doi A, Wen B, Ng K, Zhao R, Cahan P, Kim J, Aryee MJ, Ji H, Ehrlich LI, Yabuuchi A, Takeuchi A, Cunniff KC, Hongguang H, McKinney-Freeman S, Naveiras O, Yoon TJ, Irizarry RA, Jung N, Seita J, Hanna J, Murakami P, Jaenisch R, Weissleder R, Orkin SH, Weissman IL, Feinberg AP, Daley GQ (2010) Epigenetic memory in induced pluripotent stem cells. Nature 467(7313):285–290. doi:10.1038/nature09342

Lachmann N, Happle C, Ackermann M, Luttge D, Wetzke M, Merkert S, Hetzel M, Kensah G, Jara-Avaca M, Mucci A, Skuljec J, Dittrich AM, Pfaff N, Brennig S, Schambach A, Steinemann D, Gohring G, Cantz T, Martin U, Schwerk N, Hansen G, Moritz T (2014) Gene correction of human induced pluripotent stem cells repairs the cellular phenotype in pulmonary alveolar proteinosis. Am J Respir Crit Care Med 189(2):167–182. doi:10.1164/rccm.201306-1012OC

Lahti AL, Kujala VJ, Chapman H, Koivisto AP, Pekkanen-Mattila M, Kerkela E, Hyttinen J, Kontula K, Swan H, Conklin BR, Yamanaka S, Silvennoinen O, Aalto-Setala K (2012) Model for long QT syndrome type 2 using human iPS cells demonstrates arrhythmogenic characteristics in cell culture. Dis Model Mech 5(2):220–230. doi:10.1242/dmm.008409

Lee G, Studer L (2011) Modelling familial dysautonomia in human induced pluripotent stem cells. Philos Trans R Soc Lond B Biol Sci 366(1575):2286–2296. doi:10.1098/rstb.2011.0026

Lee G, Papapetrou EP, Kim H, Chambers SM, Tomishima MJ, Fasano CA, Ganat YM, Menon J, Shimizu F, Viale A, Tabar V, Sadelain M, Studer L (2009) Modelling pathogenesis and treatment of familial dysautonomia using patient-specific iPSCs. Nature 461(7262):402–406. doi:10.1038/nature08320

Liu J, Verma PJ, Evans-Galea MV, Delatycki MB, Michalska A, Leung J, Crombie D, Sarsero JP, Williamson R, Dottori M, Pebay A (2011) Generation of induced pluripotent stem cell lines from Friedreich ataxia patients. Stem Cell Rev 7(3):703–713. doi:10.1007/s12015-010-9210-x

Longmire TA, Ikonomou L, Hawkins F, Christodoulou C, Cao Y, Jean JC, Kwok LW, Mou H, Rajagopal J, Shen SS, Dowton AA, Serra M, Weiss DJ, Green MD, Snoeck HW, Ramirez MI, Kotton DN (2012) Efficient derivation of purified lung and thyroid progenitors from embryonic stem cells. Cell Stem Cell 10(4):398–411. doi:10.1016/j.stem.2012.01.019

Ma D, Wei H, Lu J, Ho S, Zhang G, Sun X, Oh Y, Tan SH, Ng ML, Shim W, Wong P, Liew R (2013) Generation of patient-specific induced pluripotent stem cell-derived cardiomyocytes as a cellular model of arrhythmogenic right ventricular cardiomyopathy. Eur Heart J 34(15):1122–1133. doi:10.1093/eurheartj/ehs226

Maeder ML, Linder SJ, Cascio VM, Fu Y, Ho QH, Joung JK (2013) CRISPR RNA-guided activation of endogenous human genes. Nat Methods 10(10):977–979. doi:10.1038/nmeth.2598

Maehr R, Chen S, Snitow M, Ludwig T, Yagasaki L, Goland R, Leibel RL, Melton DA (2009) Generation of pluripotent stem cells from patients with type 1 diabetes. Proc Natl Acad Sci U S A 106(37):15768–15773. doi:10.1073/pnas.0906894106

Maherali N, Sridharan R, Xie W, Utikal J, Eminli S, Arnold K, Stadtfeld M, Yachechko R, Tchieu J, Jaenisch R, Plath K, Hochedlinger K (2007) Directly reprogrammed fibroblasts show global epigenetic remodeling and widespread tissue contribution. Cell Stem Cell 1(1):55–70. doi:10.1016/j.stem.2007.05.014

Mali P, Esvelt KM, Church GM (2013) Cas9 as a versatile tool for engineering biology. Nat Methods 10(10):957–963. doi:10.1038/nmeth.2649

Marcet B, Chevalier B, Coraux C, Kodjabachian L, Barbry P (2011) MicroRNA-based silencing of Delta/Notch signaling promotes multiple cilia formation. Cell Cycle 10(17):2858–2864

Marchetto MC, Carromeu C, Acab A, Yu D, Yeo GW, Mu Y, Chen G, Gage FH, Muotri AR (2010) A model for neural development and treatment of Rett syndrome using human induced pluripotent stem cells. Cell 143(4):527–539. doi:10.1016/j.cell.2010.10.016

Marshall WF (2008) Basal bodies platforms for building cilia. Curr Top Dev Biol 85:1–22. doi:10.1016/S0070-2153(08)00801-6

Martin GR (1981) Isolation of a pluripotent cell line from early mouse embryos cultured in medium conditioned by teratocarcinoma stem cells. Proc Natl Acad Sci U S A 78(12): 7634–7638

McCurry KR, Shearon TH, Edwards LB, Chan KM, Sweet SC, Valapour M, Yusen R, Murray S (2009) Lung transplantation in the United States, 1998–2007. Am J Transplant 9(4 Pt 2): 942–958. doi:10.1111/j.1600-6143.2009.02569.x

Medine CN, Lucendo-Villarin B, Storck C, Wang F, Szkolnicka D, Khan F, Pernagallo S, Black JR, Marriage HM, Ross JA, Bradley M, Iredale JP, Flint O, Hay DC (2013) Developing high-fidelity hepatotoxicity models from pluripotent stem cells. Stem Cells Transl Med 2(7):505–509. doi:10.5966/sctm.2012-0138

Mekhoubad S, Bock C, de Boer AS, Kiskinis E, Meissner A, Eggan K (2012) Erosion of dosage compensation impacts human iPSC disease modeling. Cell Stem Cell 10(5):595–609. doi:10.1016/j.stem.2012.02.014

Mendenhall EM, Williamson KE, Reyon D, Zou JY, Ram O, Joung JK, Bernstein BE (2013) Locus-specific editing of histone modifications at endogenous enhancers. Nat Biotechnol 31(12):1133–1136. doi:10.1038/nbt.2701

Miller JC, Holmes MC, Wang J, Guschin DY, Lee YL, Rupniewski I, Beausejour CM, Waite AJ, Wang NS, Kim KA, Gregory PD, Pabo CO, Rebar EJ (2007) An improved zinc-finger nuclease architecture for highly specific genome editing. Nat Biotechnol 25(7):778–785. doi:10.1038/nbt1319

Mojica FJ, Diez-Villasenor C, Soria E, Juez G (2000) Biological significance of a family of regularly spaced repeats in the genomes of Archaea, Bacteria and mitochondria. Mol Microbiol 36(1):244–246

Mojica FJ, Diez-Villasenor C, Garcia-Martinez J, Soria E (2005) Intervening sequences of regularly spaced prokaryotic repeats derive from foreign genetic elements. J Mol Evol 60(2): 174–182. doi:10.1007/s00239-004-0046-3

Mondrinos MJ, Koutzaki S, Jiwanmall E, Li M, Dechadarevian JP, Lelkes PI, Finck CM (2006) Engineering three-dimensional pulmonary tissue constructs. Tissue Eng 12(4):717–728. doi:10.1089/ten.2006.12.717

Moodley Y, Thompson P, Warburton D (2013) Stem cells: a recapitulation of development. Respirology 18(8):1167–1176. doi:10.1111/resp.12186

Moretti A, Bellin M, Welling A, Jung CB, Lam JT, Bott-Flugel L, Dorn T, Goedel A, Hohnke C, Hofmann F, Seyfarth M, Sinnecker D, Schomig A, Laugwitz KL (2010) Patient-specific induced pluripotent stem-cell models for long-QT syndrome. N Engl J Med 363(15):1397–1409. doi:10.1056/NEJMoa0908679

Moscou MJ, Bogdanove AJ (2009) A simple cipher governs DNA recognition by TAL effectors. Science 326(5959):1501. doi:10.1126/science.1178817

Mou H, Zhao R, Sherwood R, Ahfeldt T, Lapey A, Wain J, Sicilian L, Izvolsky K, Musunuru K, Cowan C, Rajagopal J (2012) Generation of multipotent lung and airway progenitors from mouse ESCs and patient-specific cystic fibrosis iPSCs. Cell Stem Cell 10(4):385–397. doi:10.1016/j.stem.2012.01.018

Murry CE, Keller G (2008) Differentiation of embryonic stem cells to clinically relevant populations: lessons from embryonic development. Cell 132(4):661–680. doi:10.1016/j.cell.2008.02.008

Nguyen HN, Byers B, Cord B, Shcheglovitov A, Byrne J, Gujar P, Kee K, Schule B, Dolmetsch RE, Langston W, Palmer TD, Pera RR (2011) LRRK2 mutant iPSC-derived DA neurons demonstrate increased susceptibility to oxidative stress. Cell Stem Cell 8(3):267–280. doi:10.1016/j.stem.2011.01.013

Nichols JE, Cortiella J (2008) Engineering of a complex organ: progress toward development of a tissue-engineered lung. Proc Am Thorac Soc 5(6):723–730. doi:10.1513/pats.200802-022AW

Nightingale SJ, Hollis RP, Pepper KA, Petersen D, Yu XJ, Yang C, Bahner I, Kohn DB (2006) Transient gene expression by nonintegrating lentiviral vectors. Mol Ther 13(6):1121–1132. doi:10.1016/j.ymthe.2006.01.008

Niu Y, Shen B, Cui Y, Chen Y, Wang J, Wang L, Kang Y, Zhao X, Si W, Li W, Xiang AP, Zhou J, Guo X, Bi Y, Si C, Hu B, Dong G, Wang H, Zhou Z, Li T, Tan T, Pu X, Wang F, Ji S, Zhou Q, Huang X, Ji W, Sha J (2014) Generation of gene-modified cynomolgus monkey via Cas9/RNA-mediated gene targeting in one-cell embryos. Cell 156(4):836–843. doi:10.1016/j.cell.2014.01.027

Noone PG, Leigh MW, Sannuti A, Minnix SL, Carson JL, Hazucha M, Zariwala MA, Knowles MR (2004) Primary ciliary dyskinesia: diagnostic and phenotypic features. Am J Respir Crit Care Med 169(4):459–467. doi:10.1164/rccm.200303-365OC

Nostro MC, Keller G (2012) Generation of beta cells from human pluripotent stem cells: Potential for regenerative medicine. Semin Cell Dev Biol 23(6):701–710. doi:10.1016/j.semcdb.2012.06.010

Okita K, Ichisaka T, Yamanaka S (2007) Generation of germline-competent induced pluripotent stem cells. Nature 448(7151):313–317. doi:10.1038/nature05934

Okita K, Nakagawa M, Hyenjong H, Ichisaka T, Yamanaka S (2008) Generation of mouse induced pluripotent stem cells without viral vectors. Science 322(5903):949–953. doi:10.1126/science.1164270

Park IH, Lerou PH, Zhao R, Huo H, Daley GQ (2008a) Generation of human-induced pluripotent stem cells. Nat Protoc 3(7):1180–1186. doi:10.1038/nprot.2008.92

Park IH, Arora N, Huo H, Maherali N, Ahfeldt T, Shimamura A, Lensch MW, Cowan C, Hochedlinger K, Daley GQ (2008b) Disease-specific induced pluripotent stem cells. Cell 134(5):877–886. doi:10.1016/j.cell.2008.07.041

Pessach IM, Ordovas-Montanes J, Zhang SY, Casanova JL, Giliani S, Gennery AR, Al-Herz W, Manos PD, Schlaeger TM, Park IH, Rucci F, Agarwal S, Mostoslavsky G, Daley GQ, Notarangelo LD (2011) Induced pluripotent stem cells: a novel frontier in the study of human primary immunodeficiencies. J Allergy Clin Immunol 127(6):1400–1407. doi:10.1016/j.jaci.2010.11.008, e1404

Pickering SJ, Minger SL, Patel M, Taylor H, Black C, Burns CJ, Ekonomou A, Braude PR (2005) Generation of a human embryonic stem cell line encoding the cystic fibrosis mutation deltaF508, using preimplantation genetic diagnosis. Reprod Biomed Online 10(3):390–397

Plessis A, Perrin A, Haber JE, Dujon B (1992) Site-specific recombination determined by I-SceI, a mitochondrial group I intron-encoded endonuclease expressed in the yeast nucleus. Genetics 130(3):451–460

Pourcel C, Salvignol G, Vergnaud G (2005) CRISPR elements in Yersinia pestis acquire new repeats by preferential uptake of bacteriophage DNA, and provide additional tools for evolutionary studies. Microbiology 151(Pt 3):653–663. doi:10.1099/mic.0.27437-0

Qi LS, Larson MH, Gilbert LA, Doudna JA, Weissman JS, Arkin AP, Lim WA (2013) Repurposing CRISPR as an RNA-guided platform for sequence-specific control of gene expression. Cell 152(5):1173–1183. doi:10.1016/j.cell.2013.02.022

Ran FA, Hsu PD, Lin CY, Gootenberg JS, Konermann S, Trevino AE, Scott DA, Inoue A, Matoba S, Zhang Y, Zhang F (2013) Double nicking by RNA-guided CRISPR Cas9 for enhanced genome editing specificity. Cell 154(6):1380–1389. doi:10.1016/j.cell.2013.08.021

Rashid ST, Corbineau S, Hannan N, Marciniak SJ, Miranda E, Alexander G, Huang-Doran I, Griffin J, Ahrlund-Richter L, Skepper J, Semple R, Weber A, Lomas DA, Vallier L (2010) Modeling inherited metabolic disorders of the liver using human induced pluripotent stem cells. J Clin Invest 120(9):3127–3136. doi:10.1172/JCI43122

Ratjen F, Doring G (2003) Cystic fibrosis. Lancet 361(9358):681–689. doi:10.1016/S0140-6736(03)12567-6

Raya A, Rodriguez-Piza I, Guenechea G, Vassena R, Navarro S, Barrero MJ, Consiglio A, Castella M, Rio P, Sleep E, Gonzalez F, Tiscornia G, Garreta E, Aasen T, Veiga A, Verma IM, Surralles J, Bueren J, Izpisua Belmonte JC (2009) Disease-corrected haematopoietic progenitors from Fanconi anaemia induced pluripotent stem cells. Nature 460(7251):53–59. doi:10.1038/nature08129

Rock JR, Gao X, Xue Y, Randell SH, Kong YY, Hogan BL (2011) Notch-dependent differentiation of adult airway basal stem cells. Cell Stem Cell 8(6):639–648. doi:10.1016/j.stem.2011.04.003

Rodriguez-Piza I, Richaud-Patin Y, Vassena R, Gonzalez F, Barrero MJ, Veiga A, Raya A, Izpisua Belmonte JC (2010) Reprogramming of human fibroblasts to induced pluripotent stem cells under xeno-free conditions. Stem Cells 28(1):36–44. doi:10.1002/stem.248

Rouet P, Smih F, Jasin M (1994) Introduction of double-strand breaks into the genome of mouse cells by expression of a rare-cutting endonuclease. Mol Cell Biol 14(12):8096–8106

Rudin N, Sugarman E, Haber JE (1989) Genetic and physical analysis of double-strand break repair and recombination in Saccharomyces cerevisiae. Genetics 122(3):519–534

Sander JD, Joung JK (2014) CRISPR-Cas systems for editing, regulating and targeting genomes. Nat Biotechnol 32(4):347–355. doi:10.1038/nbt.2842

Schwank G, Koo BK, Sasselli V, Dekkers JF, Heo I, Demircan T, Sasaki N, Boymans S, Cuppen E, van der Ent CK, Nieuwenhuis EE, Beekman JM, Clevers H (2013) Functional repair of CFTR by CRISPR/Cas9 in intestinal stem cell organoids of cystic fibrosis patients. Cell Stem Cell 13(6):653–658. doi:10.1016/j.stem.2013.11.002

Scott E, Loya K, Mountford J, Milligan G, Baker AH (2013) MicroRNA regulation of endothelial homeostasis and commitment-implications for vascular regeneration strategies using stem cell therapies. Free Radic Biol Med 64:52–60. doi:10.1016/j.freeradbiomed.2013.04.037

Seki T, Yuasa S, Oda M, Egashira T, Yae K, Kusumoto D, Nakata H, Tohyama S, Hashimoto H, Kodaira M, Okada Y, Seimiya H, Fusaki N, Hasegawa M, Fukuda K (2010) Generation of induced pluripotent stem cells from human terminally differentiated circulating T cells. Cell Stem Cell 7(1):11–14. doi:10.1016/j.stem.2010.06.003

Shalem O, Sanjana NE, Hartenian E, Shi X, Scott DA, Mikkelsen TS, Heckl D, Ebert BL, Root DE, Doench JG, Zhang F (2014) Genome-scale CRISPR-Cas9 knockout screening in human cells. Science 343(6166):84–87. doi:10.1126/science.1247005

Smith AD, Sumazin P, Xuan Z, Zhang MQ (2006) DNA motifs in human and mouse proximal promoters predict tissue-specific expression. Proc Natl Acad Sci U S A 103(16):6275–6280. doi:10.1073/pnas.0508169103

Soldner F, Jaenisch R (2012) Medicine. iPSC disease modeling. Science 338(6111):1155–1156. doi:10.1126/science.1227682

Soldner F, Hockemeyer D, Beard C, Gao Q, Bell GW, Cook EG, Hargus G, Blak A, Cooper O, Mitalipova M, Isacson O, Jaenisch R (2009) Parkinson's disease patient-derived induced pluripotent stem cells free of viral reprogramming factors. Cell 136(5):964–977. doi:10.1016/j.cell.2009.02.013

Somers A, Jean JC, Sommer CA, Omari A, Ford CC, Mills JA, Ying L, Sommer AG, Jean JM, Smith BW, Lafyatis R, Demierre MF, Weiss DJ, French DL, Gadue P, Murphy GJ, Mostoslavsky G, Kotton DN (2010) Generation of transgene-free lung disease-specific human induced pluripotent stem cells using a single excisable lentiviral stem cell cassette. Stem Cells 28(10):1728–1740. doi:10.1002/stem.495

Sommer CA, Mostoslavsky G (2013) The evolving field of induced pluripotency: recent progress and future challenges. J Cell Physiol 228(2):267–275. doi:10.1002/jcp.24155

Stadtfeld M, Nagaya M, Utikal J, Weir G, Hochedlinger K (2008) Induced pluripotent stem cells generated without viral integration. Science 322(5903):945–949. doi:10.1126/science.1162494

Stubbs JL, Davidson L, Keller R, Kintner C (2006) Radial intercalation of ciliated cells during Xenopus skin development. Development 133(13):2507–2515. doi:10.1242/dev.02417

Stubbs JL, Vladar EK, Axelrod JD, Kintner C (2012) Multicilin promotes centriole assembly and ciliogenesis during multiciliate cell differentiation. Nat Cell Biol 14(2):140–147. doi:10.1038/ncb2406

Sun N, Yazawa M, Liu J, Han L, Sanchez-Freire V, Abilez OJ, Navarrete EG, Hu S, Wang L, Lee A, Pavlovic A, Lin S, Chen R, Hajjar RJ, Snyder MP, Dolmetsch RE, Butte MJ, Ashley EA, Longaker MT, Robbins RC, Wu JC (2012) Patient-specific induced pluripotent stem cells as a model for familial dilated cardiomyopathy. Sci Transl Med 4(130):130ra47. doi:10.1126/scitranslmed.3003552

Suzuki T, Mayhew C, Sallese A, Chalk C, Carey BC, Malik P, Wood RE, Trapnell BC (2014) Use of induced pluripotent stem cells to recapitulate pulmonary alveolar proteinosis pathogenesis. Am J Respir Crit Care Med 189(2):183–193. doi:10.1164/rccm.201306-1039OC

Takahashi K, Yamanaka S (2006) Induction of pluripotent stem cells from mouse embryonic and adult fibroblast cultures by defined factors. Cell 126(4):663–676. doi:10.1016/j.cell.2006.07.024

Takahashi K, Ichisaka T, Yamanaka S (2006) Identification of genes involved in tumor-like properties of embryonic stem cells. Methods Mol Biol 329:449–458. doi:10.1385/1-59745-037-5:449

Takahashi K, Okita K, Nakagawa M, Yamanaka S (2007) Induction of pluripotent stem cells from fibroblast cultures. Nat Protoc 2(12):3081–3089. doi:10.1038/nprot.2007.418

Tam PP, Loebel DA (2007) Gene function in mouse embryogenesis: get set for gastrulation. Nat Rev Genet 8(5):368–381. doi:10.1038/nrg2084

Tchieu J, Kuoy E, Chin MH, Trinh H, Patterson M, Sherman SP, Aimiuwu O, Lindgren A, Hakimian S, Zack JA, Clark AT, Pyle AD, Lowry WE, Plath K (2010) Female human iPSCs retain an inactive X chromosome. Cell Stem Cell 7(3):329–342. doi:10.1016/j.stem.2010.06.024

Thomson JA, Itskovitz-Eldor J, Shapiro SS, Waknitz MA, Swiergiel JJ, Marshall VS, Jones JM (1998) Embryonic stem cell lines derived from human blastocysts. Science 282(5391):1145–1147

Tobert JA (2003) Lovastatin and beyond: the history of the HMG-CoA reductase inhibitors. Nat Rev Drug Discov 2(7):517–526. doi:10.1038/nrd1112

Tolar J, Park IH, Xia L, Lees CJ, Peacock B, Webber B, McElmurry RT, Eide CR, Orchard PJ, Kyba M, Osborn MJ, Lund TC, Wagner JE, Daley GQ, Blazar BR (2011a) Hematopoietic differentiation of induced pluripotent stem cells from patients with mucopolysaccharidosis type I (Hurler syndrome). Blood 117(3):839–847. doi:10.1182/blood-2010-05-287607

Tolar J, Xia L, Riddle MJ, Lees CJ, Eide CR, McElmurry RT, Titeux M, Osborn MJ, Lund TC, Hovnanian A, Wagner JE, Blazar BR (2011b) Induced pluripotent stem cells from individuals with recessive dystrophic epidermolysis bullosa. J Invest Dermatol 131(4):848–856. doi:10.1038/jid.2010.346

Tsao PN, Wei SC, Wu MF, Huang MT, Lin HY, Lee MC, Lin KM, Wang IJ, Kaartinen V, Yang LT, Cardoso WV (2011) Notch signaling prevents mucous metaplasia in mouse conducting airways during postnatal development. Development 138(16):3533–3543. doi:10.1242/dev.063727

Urbach A, Bar-Nur O, Daley GQ, Benvenisty N (2010) Differential modeling of fragile X syndrome by human embryonic stem cells and induced pluripotent stem cells. Cell Stem Cell 6(5):407–411. doi:10.1016/j.stem.2010.04.005

Urnov FD, Miller JC, Lee YL, Beausejour CM, Rock JM, Augustus S, Jamieson AC, Porteus MH, Gregory PD, Holmes MC (2005) Highly efficient endogenous human gene correction using designed zinc-finger nucleases. Nature 435(7042):646–651. doi:10.1038/nature03556

Wang H, Yang H, Shivalila CS, Dawlaty MM, Cheng AW, Zhang F, Jaenisch R (2013) One-step generation of mice carrying mutations in multiple genes by CRISPR/Cas-mediated genome engineering. Cell 153(4):910–918. doi:10.1016/j.cell.2013.04.025

Wang T, Wei JJ, Sabatini DM, Lander ES (2014) Genetic screens in human cells using the CRISPR-Cas9 system. Science 343(6166):80–84. doi:10.1126/science.1246981

Warren L, Manos PD, Ahfeldt T, Loh YH, Li H, Lau F, Ebina W, Mandal PK, Smith ZD, Meissner A, Daley GQ, Brack AS, Collins JJ, Cowan C, Schlaeger TM, Rossi DJ (2010) Highly efficient reprogramming to pluripotency and directed differentiation of human cells with synthetic modified mRNA. Cell Stem Cell 7(5):618–630. doi:10.1016/j.stem.2010.08.012

Wernig M, Meissner A, Foreman R, Brambrink T, Ku M, Hochedlinger K, Bernstein BE, Jaenisch R (2007) In vitro reprogramming of fibroblasts into a pluripotent ES-cell-like state. Nature 448(7151):318–324. doi:10.1038/nature05944

West JD, Austin ED, Gaskill C, Marriott S, Baskir R, Bilousova G, Jean JC, Hemnes AR, Menon S, Bloodworth NC, Fessel JP, Kropski JA, Irwin D, Ware LB, Wheeler L, Hong CC, Meyrick B, Loyd JE, Bowman AB, Ess KC, Klemm DJ, Young PP, Merryman WD, Kotton D, Majka SM (2014) Identification of a common Wnt-associated genetic signature across multiple cell types in pulmonary arterial hypertension. Am J Physiol Cell Physiol 307(5):C415–C430. doi:10.1152/ajpcell.00057.2014

Wilmut I, Schnieke AE, McWhir J, Kind AJ, Campbell KH (1997) Viable offspring derived from fetal and adult mammalian cells. Nature 385(6619):810–813. doi:10.1038/385810a0

Woltjen K, Michael IP, Mohseni P, Desai R, Mileikovsky M, Hamalainen R, Cowling R, Wang W, Liu P, Gertsenstein M, Kaji K, Sung HK, Nagy A (2009) piggyBac transposition reprograms fibroblasts to induced pluripotent stem cells. Nature 458(7239):766–770. doi:10.1038/nature07863

Wong AP, Bear CE, Chin S, Pasceri P, Thompson TO, Huan LJ, Ratjen F, Ellis J, Rossant J (2012) Directed differentiation of human pluripotent stem cells into mature airway epithelia expressing functional CFTR protein. Nat Biotechnol 30(9):876–882. doi:10.1038/nbt.2328

Wood AJ, Lo TW, Zeitler B, Pickle CS, Ralston EJ, Lee AH, Amora R, Miller JC, Leung E, Meng X, Zhang L, Rebar EJ, Gregory PD, Urnov FD, Meyer BJ (2011) Targeted genome editing across species using ZFNs and TALENs. Science 333(6040):307. doi:10.1126/science.1207773

Wu Y, Liang D, Wang Y, Bai M, Tang W, Bao S, Yan Z, Li D, Li J (2013) Correction of a genetic disease in mouse via use of CRISPR-Cas9. Cell Stem Cell 13(6):659–662. doi:10.1016/j.stem.2013.10.016

Xiao A, Wang Z, Hu Y, Wu Y, Luo Z, Yang Z, Zu Y, Li W, Huang P, Tong X, Zhu Z, Lin S, Zhang B (2013) Chromosomal deletions and inversions mediated by TALENs and CRISPR/Cas in zebrafish. Nucleic Acids Res 41(14):e141. doi:10.1093/nar/gkt464

Yagi T, Ito D, Okada Y, Akamatsu W, Nihei Y, Yoshizaki T, Yamanaka S, Okano H, Suzuki N (2011) Modeling familial Alzheimer's disease with induced pluripotent stem cells. Hum Mol Genet 20(23):4530–4539. doi:10.1093/hmg/ddr394

Yasunaga M, Tada S, Torikai-Nishikawa S, Nakano Y, Okada M, Jakt LM, Nishikawa S, Chiba T, Era T (2005) Induction and monitoring of definitive and visceral endoderm differentiation of mouse ES cells. Nat Biotechnol 23(12):1542–1550. doi:10.1038/nbt1167

Yazawa M, Hsueh B, Jia X, Pasca AM, Bernstein JA, Hallmayer J, Dolmetsch RE (2011) Using induced pluripotent stem cells to investigate cardiac phenotypes in Timothy syndrome. Nature 471(7337):230–234. doi:10.1038/nature09855

Ye L, Chang JC, Lin C, Sun X, Yu J, Kan YW (2009a) Induced pluripotent stem cells offer new approach to therapy in thalassemia and sickle cell anemia and option in prenatal diagnosis in genetic diseases. Proc Natl Acad Sci U S A 106(24):9826–9830. doi:10.1073/pnas.0904689106

Ye Z, Zhan H, Mali P, Dowey S, Williams DM, Jang YY, Dang CV, Spivak JL, Moliterno AR, Cheng L (2009b) Human-induced pluripotent stem cells from blood cells of healthy donors and patients with acquired blood disorders. Blood 114(27):5473–5480. doi:10.1182/blood-2009-04-217406

Yu J, Vodyanik MA, Smuga-Otto K, Antosiewicz-Bourget J, Frane JL, Tian S, Nie J, Jonsdottir GA, Ruotti V, Stewart R, Slukvin II, Thomson JA (2007) Induced pluripotent stem cell lines derived from human somatic cells. Science 318(5858):1917–1920. doi:10.1126/science.1151526

Yu J, Hu K, Smuga-Otto K, Tian S, Stewart R, Slukvin II, Thomson JA (2009) Human induced pluripotent stem cells free of vector and transgene sequences. Science 324(5928):797–801. doi:10.1126/science.1172482

Zhou H, Wu S, Joo JY, Zhu S, Han DW, Lin T, Trauger S, Bien G, Yao S, Zhu Y, Siuzdak G, Scholer HR, Duan L, Ding S (2009) Generation of induced pluripotent stem cells using recombinant proteins. Cell Stem Cell 4(5):381–384. doi:10.1016/j.stem.2009.04.005

Zhou J, Wang J, Shen B, Chen L, Su Y, Yang J, Zhang W, Tian X, Huang X (2014) Dual sgRNAs facilitate CRISPR/Cas9-mediated mouse genome targeting. FEBS J 281(7): 1717–1725. doi:10.1111/febs.12735

Chapter 8
Ex Vivo Lung Bioengineering

Darcy E. Wagner, Franziska E. Uhl, Melanie Königshoff, and Daniel J. Weiss

Abbreviations

2D	Two-dimensional
3D	Three-dimensional
α-SMA	Alpha smooth muscle actin
A549	Adenocarcinomic human alveolar basal epithelial cells
A9	Transformed subcutaneous murine fibroblasts
AFM	Atomic force microscopy
ALI	Air-liquid-interface
ATI	Type 1 alveolar epithelial cells
ATII	Type 2 alveolar epithelial cells
Aqp5	Aquaporin 5
BMMSCs	Porcine bone marrow-derived mesenchymal stem cells
C10	Immortalized murine alveolar epithelial cells
$CaCl_2$	Calcium chloride
CBF12	Human endothelial progenitor cells
CC10	Club Cell 10 kDa Protein (=CCSP)
CCSP	Club cell secretory protein (=CC10)
CD	Cluster of differentiation
CD206	Mannose receptor C 1 (cluster of differentiation factor 206)
CHAPS	3-[(3-cholamidopropyl)dimethylammonio]-1-propanesulfonate

D.E. Wagner • F.E. Uhl • M. Königshoff
Comprehensive Pneumology Center, Munich, Germany

D.J. Weiss (✉)
Department of Medicine, Pulmonary and Critical Care, University of Vermont, 149 Beaumont Avenue, Room 226 Health Sciences Research Facility, Burlington, VT 05405, USA
e-mail: daniel.weiss@med.uvm.edu

A. Firth, J.X.-J. Yuan (eds.), *Lung Stem Cells in the Epithelium and Vasculature*, Stem Cell Biology and Regenerative Medicine, DOI 10.1007/978-3-319-16232-4_8

Ck	Cytokeratin
cmH_2O	Centimeters of water (pressure)
Col1α1	Collagen 1
COPD	Chronic obstructive pulmonary disease
CT	Computed tomography
Ctnnb1	Beta catenin
DAPI	4′,6-diamidino-2-phenylindole
DMEM	Dulbecco's modified Eagle's medium
DNA	Deoxyribonucleic acid
DNase	Deoxyribonuclease
E17	Embryonic day 17
ECM	Extracellular matrix
EDTA	Ethylenediaminetetracetic acid
ESC	Embryonic stem cells
EVLP	Ex vivo lung perfusion
FBS	Fetal bovine serum
FDA	Food and Drug Administration
GAGS	Glycosaminoglycans
GFP	Green fluorescent protein
H&E	Hematoxylin–eosin
H_2O	Water
hAEC	Human alveolar epithelial cells
hAT-MSC	Human adipose-derived mesenchymal stem cells
hBE	Human bronchial epithelial cells
hBM-MSC	Human bone marrow-derived mesenchymal stem cells
hFLC	Human fetal lung cells
hiPS	Human induced pluripotent cells
hLF	Human lung fibroblasts
hMSC	Human bone marrow-derived mesenchymal stem cells
HUVEC	Human umbilical vein endothelial cells
IPF	Idiopathic pulmonary fibrosis
iPS	Induced pluripotent stem cells
mESCs	Murine embryonic stem cells
$MgSO_4$	Magnesium sulfate
mM	Millimolar
mmHg	Millimeters of mercury
MRC5	Human fetal lung fibroblast cell line
MSC	Mesenchymal stem cells
NaCl	Sodium chloride
$NaHCO_3$	Sodium bicarbonate
Nkx2-1	Nkx2 homeobox 1 protein (aka TTF1—thyroid transcription factor 1)
PAEC	Pulmonary alveolar epithelial cells
PBS	Phosphate-buffered saline
PCR	Polymerase chain reaction
PDGFRα	Platelet derived growth factor receptor alpha

RGD	Arg-Gly-Asp
rhCHI3L1	Recombinant human chitinase 3-like 1 protein
RNase	Ribonuclease
SAEC	Small airway epithelial cells
SDC	Sodium deoxycholate
SDS	Sodium dodecyl sulfate
SEM	Scanning electron microscopy
SPA	Surfactant protein A
SPC	Surfactant protein C
TEM	Transmission electron microscopy
T1α	T1alpha, podoplanin
TGF-β	Transforming growth factor beta
Ttf-1	Thyroid transcription factor 1 (Nkx2.1)
VEGF	Vascular endothelial growth factor

8.1 Introduction

Chronic lung diseases such as chronic obstructive pulmonary disease (COPD) and idiopathic pulmonary fibrosis (IPF) are predicted to significantly increase in prevalence and are an increasing major worldwide healthcare burden (Lopez et al. 2006; Eisner et al. 2010). These diseases have no cure and cause significant morbidity with high mortality rates. Lung transplantation remains the only treatment option, but there are not enough available lungs to meet transplantation demands. Further, transplantation recipients require lifelong immunosuppression and the 5-year survival after lung transplantation remains around 50 % (Lopez et al. 2006; Eisner et al. 2010). Alternative options are desperately needed for this patient population.

Recent advances in ex vivo lung bioengineering of functional lung tissue that could be transplanted into patients with end-stage lung disease show promise. The basic principle behind ex vivo lung bioengineering involves coupling a biologic or synthetic scaffold (or a combination thereof) with an appropriate cell source(s) and appropriate ex vivo culture strategy (Fig. 8.1). The use of autologous stem, progenitor, or other cells obtained from the eventual transplant recipient would eliminate the need for immunosuppressive drugs. While this approach is not yet feasible in lungs, comparable approaches have been successfully utilized clinically to regenerate tissues including trachea, skin, and bone (Macchiarini et al. 2008; Keane and Badylak 2014). However, lung tissue is a considerably more complex organ and is comprised of a variety of cell types which must be recapitulated in vivo for functional tissue regeneration. Therefore, progress has significantly lagged behind the advances made in other organs.

Synthetic scaffolds are one potential option and a number of different materials and manufacturing technologies have been evaluated for lung (Mondrinos et al. 2006, 2007; Lin et al. 2006; Andrade et al. 2007; Cortiella et al. 2006; Ingenito et al. 2010; Miller et al. 2010; Nichols and Cortiella 2008; Roomans 2010; Tsunooka

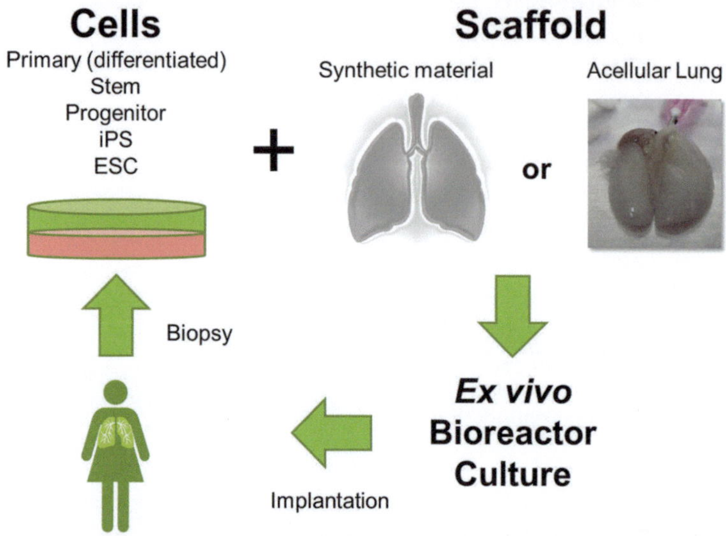

Fig. 8.1 Schematic of ex vivo organ engineering. Autologous cells are obtained by a biopsy from the eventual transplant recipient and expanded in ex vivo culture. A scaffold, either synthetic or an acellular lung, is manufactured and repopulated ex vivo by the usage of a bioreactor to create a functional tissue suitable for re-implantation

et al. 2011). The porous structure of lung, like many other tissues, presents a unique manufacturing engineering challenge. Both additive (i.e., processes which generate final products layer-by-layer or unit-by-unit manufacturing) and subtractive processes (i.e., traditional manufacturing processes whereby material is removed to form the final product) are current areas of active research (Melchels et al. 2012). Additive processes, such as rapid prototyping three dimensional (3D) bioprinting techniques (Murphy and Atala 2014) are traditionally viewed as more advantageous for generating scaffolds with interconnected pores; however, subtractive processes such as porogen-forming techniques and sphere-templating have produced promising initial results (Bryant et al. 2007; Ling et al. 2014). Alternatively, hydrogels and electrospun scaffolds have also been proposed as potential scaffolds (Dunphy et al. 2014; Fischer et al. 2011). However, despite a variety of available state-of-the-art 3D printing and other technologies, these are so far unable to recapitulate the complex 3-dimensional architecture of the lung. Further, successful transplantation strategies and clinical use of synthetic lung scaffolds remain unknown.

An exciting new and active area of research involves the use of acellular lung scaffolds derived from cadaveric or failed transplant lungs. Acellular tissues are generated by removing cells from native organs while preserving the 3-dimensional macroarchitecture of the innate extracellular matrix (ECM) proteins (Badylak et al. 2012; Baiguera et al. 2012; Fishman et al. 2011; Haag et al. 2012; Haykal et al. 2012; Hinderer et al. 2012; Jungebluth et al. 2012a, b; Krawiec and Vorp 2012;

Orlando et al. 2011; Ott et al. 2008; Totonelli et al. 2012; Wertheim et al. 2012). In the instance of lung, vasculature and airspaces are retained. This technique was originally described for lung tissue by Lwebuga-Mukasa et al. in 1986 where they utilized a decellularized rat lung to study type II alveolar epithelial (ATII) cell behavior on a native basement membrane (Lwebuga-Mukasa et al. 1986). Whole organ decellularization as a platform for organ regeneration was first described in heart in 2008 (Ott et al. 2008), and beginning in 2010, several groups described similar techniques in lung (Ott et al. 2010; Petersen et al. 2010; Price et al. 2010; Cortiella et al. 2010; Daly et al. 2012a; Wallis et al. 2012). The use of acellular lungs has since expanded beyond their usage in regenerative medicine and has become an incredibly powerful in vitro tool for studying cell–ECM interactions in a more in vivo-like culture system (Sokocevic et al. 2013; Wagner et al. 2014a, b; Booth et al. 2012; Zhou et al. 2014; Parker et al. 2014).

In this chapter, we will address some of the critical factors involved in the theoretical and practical considerations for use of decellularized whole lungs (alternatively referred to as acellular) for ex vivo lung regeneration. These include decellularization and recellularization procedures as well as the challenges which remain in creating a translationally feasible strategy for the clinic (Fig. 8.1).

8.2 Decellularization

8.2.1 Methods of Decellularization

Acellular biologic scaffolds have been created from a variety of different tissues, including skin, esophagus, trachea, heart, and lung (Keane and Badylak 2014) Creation of biologic scaffolds involves removing the endogenous cell population while maintaining the gross structure of the ECM and its composition (Fig. 8.2). While a variety of methods have been utilized to generate acellular scaffolds, those that minimize destruction to ECM components and loss of mechanical properties are viewed as most desirable, as these are critical inputs to regulating cellular behavior. The most common techniques utilize a series or combination of chemical agents (e.g., detergents, solvents, acids/bases, and hypotonic/hypertonic solutions), biological agents (e.g., enzymes or chelating agents), and physical methods (e.g., freeze/thaw cycles, sonication) and vary in duration based on the species, organ, and method of decellularization [reviewed in (Crapo et al. 2011)] (see Table 8.1 for a list of commonly utilized reagents). In general, most protocols last from 1 to 7 days.

In the lung, maintenance of both large and small airways and vessels, as well as the more delicate alveolar, capillary, and lymphatic systems, is vital. Thus far, several techniques have been reported for decellularization of healthy mouse, rat, porcine, non-human primate, and human lungs (Tables 8.2 and 8.3). Perfusion decellularization has been most commonly utilized for whole organ decellularization, but there have also been reports of excising segments from native lung and decellularizing smaller segments (O'Neill et al. 2013; Nakayama et al. 2013; Zhou

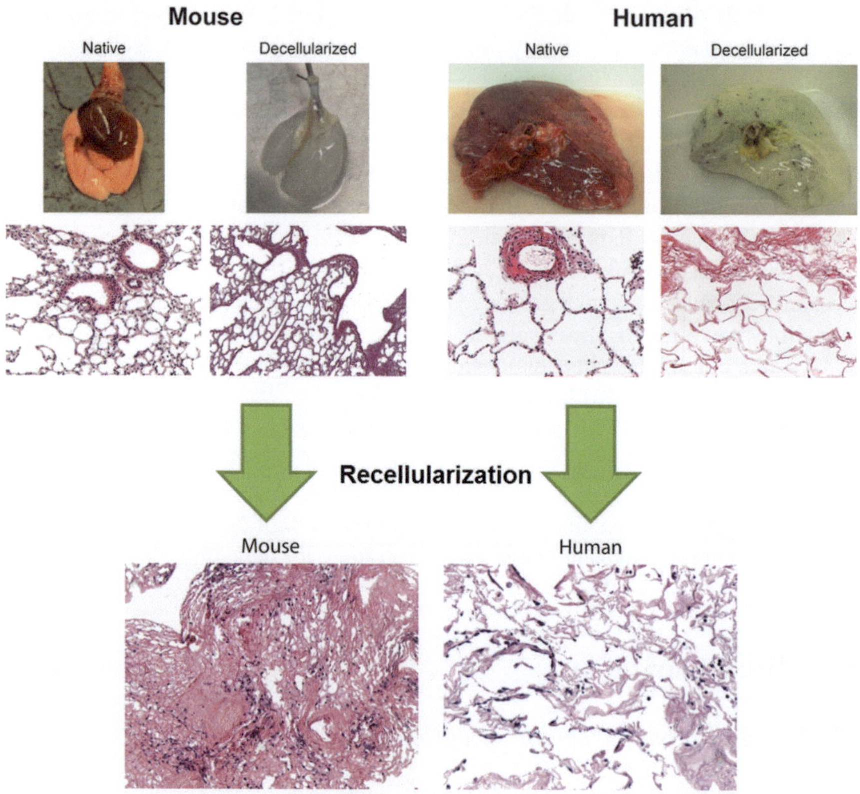

Fig. 8.2 Overview of the decellularization and recellularization process. Representative images of native and decellularized lungs from mice and humans (*upper panel*) demonstrating loss of pigmentation following decellularization, whereby the lungs become translucent *white* in color. H&E staining reveals complete cellular removal and gross maintenance of histological architecture. Histological analysis following recellularization with murine alveolar epithelial cells (C10) (*left*) and human bronchial epithelial cells (HBE) (*right*) into acellular mouse and human lung slices. Cells can be seen to have attached to the acellular lungs after 1 day of slice culture

et al. 2014; Parker et al. 2014). Detergents are most commonly utilized for perfusion-based lung decellularization and several works have compared differences in the proteomic composition and mechanical properties of the final acellular scaffold and potency for recellularization in the different detergent-based protocols (Wallis et al. 2012; Gilpin et al. 2014; O'Neill et al. 2013). The most commonly utilized detergents for lung are either the ionic detergents sodium deoxycholate (SDC) or sodium dodecyl sulfate (SDS) in combination with the nonionic detergent Triton X-100 (Bonenfant et al. 2013; Bonvillain et al. 2012; Booth et al. 2012; Daly et al. 2012a; Jensen et al. 2012; Longmire et al. 2012; Sokocevic et al. 2013; Wagner et al. 2014a, b; Wallis et al. 2012; Gilpin et al. 2014; Ott et al. 2010; Song et al. 2011; Parker et al. 2014). Alternatively, zwitterionic detergents such as 3-[(3-cholamidopropyl)

Table 8.1 Chemical agents commonly used in whole organ decellularization

Agent	Properties
Triton X-100	Nonionic detergent used to solubilize proteins; mild non-denaturing detergent
Sodium deoxycholate (SDC)	Water-soluble ionic detergent used for disrupting and dissociating protein interaction
Sodium dodecyl sulfate (SDS)	Anionic surfactant used for lysing cells and unraveling proteins
3-[(3-cholamidopropyl) dimethylammonio]-1-propanesulfonate (CHAPS)	Non-denaturing zwitterionic detergent used to solubilize proteins
Ethylenediaminetetracetic acid (EDTA)	Chelating agent that binds to calcium and prevents joining of cadherins between cells, preventing clumping of cells grown in liquid suspension, and detaching adherent cells. Can also be used to inhibit metalloproteinases
Peracetic acid	Removes residual nucleic acids, sterilization agent
Antibiotics	Typically Penicillin, Streptomycin, and an anti-mycotic Amphotericin
Other	DNase, RNase, and heparin

dimethylammonio]-1-propanesulfonate (CHAPS) have also been used (Petersen et al. 2010, 2011, 2012; O'Neill et al. 2013). Further, most protocols incorporate hypertonic lysis of cells with sodium chloride (NaCl) as well as a DNase and/or RNase step to clear residual DNA and RNA which is difficult to remove. There is, as yet, no consensus on the best route of administration and removal of decellularization agents (i.e., vascular versus airway perfusion or a combination). While both vascular-only perfusion and a combination of vascular and airway perfusion have produced acellular scaffolds capable of supporting recellularization, how differences in protocols and routes of administration for decellularization reagents might affect recellularization protocols or potential immunogenicity of implanted scaffolds is not yet known.

While standards do not yet exist for generating decellularized scaffolds, Crapo et al. proposed three minimal criteria: (1) <50 ng dsDNA per 1 mg ECM dry weight; (2) <200 bp DNA fragment length; (3) absence of visible nuclear content in histological sections by 4′,6-diamidino-2-phenylindole (DAPI) or hematoxylin–eosin (H&E) staining (Crapo et al. 2011). However, these criteria do not set forth additional characteristics which acellular scaffolds should have, such as preservation of specific ECM components, mechanical properties, or biological activity.

8.2.2 Scaling Up Decellularization Protocols for the Clinic

Scaling up decellularization protocols from rodent models to potential large animal xenogeneic sources (e.g., porcine) or human scaffolds presents additional challenges. In addition to anatomical differences, the larger size is a significant

Table 8.2 Compiled studies of ex vivo lung bioengineering using decellularized whole lung scaffolds

Reference	Scaffold	Study objective	Method of decellularization	Length of decellularization process	Endpoint assessments
Kuttan et al. (1981)	Alveolar basement membrane (calf, dog, rabbit, adult/newborn rat)	Basement membrane	Filtered distal lung homogenate, saline, 4 % Triton X-100 with protease inhibitors, $NaHCO_3$ rinse, distilled H_2O rinse	26–52 h depending on homogenate volume	Histology, immunofluorescence, electron microscopy, amino acid analysis, carbohydrate analysis
Lwebuga-Mukasa et al. (1986)	Acellular alveolar versus amniotic basement membranes	Differentiation on different basement membranes	Distilled H_2O, 0.1 % Triton X-100, 2 % SDC, NaCl, pancreatic DNase type 1S	>2 days	Cell attachment and morphology
Price et al. (2010)	Mouse (female C57/BL6) acellular lungs	Effect of matrix on spatial engraftment of E17 fetal lung homogenate	Airway and vascular perfusion: distilled H_2O, 0.1 % Triton X-100, SDC, NaCl, porcine pancreatic DNase	3 days (approximately 63 h)	Histology, quantification of ECM proteins, immunofluorescence, SEM, function with flexivent, bioreactor with fetal type II cells
Petersen et al. (2010)	Rat acellular lungs (male Fischer 344)	Development of bioartificial lung for orthotopic transplantation	Vascular perfusion only (1–5 mL/min with less than 20 mmHg arterial pressure) CHAPS, NaCl, EDTA, PBS	4 h	Histology, immunofluorescence, DNA quantification assay, collagen assay, GAG assay, western blots, SEM, TEM, micro-CT imaging
Cortiella et al. (2010)	Rat acellular lung (Sprague Dawley)	Comparison of matrices including decellularized rat lung in ability to support mESCs	Fast freeze/thaw cycles, 1 % SDS, DNase, RNase, PBS, Penicillin/Streptomycin, Amphotericin, DMEM	>6 weeks	Quantification of DNA, immunohistochemistry, confocal microscopy, flow cytometry, 2 photon microscopy, presence of SPA
Ott et al. (2010)	Rat acellular lung (Sprague Dawley)	Development of bioartificial lung for orthotopic transplantation	Vascular perfusion only: pulmonary artery pressure kept constant At 80 cmH_2O, heparinized PBS with 0.1 % SDS, deionized water, Triton X-100, and PBS with Penicillin, Streptomycin, Amphotericin B	3 days (approximately 75 h) including incubation with antibiotics	Histology, morphology, mechanical function, fluoroscopy, gas exchange, transplantation, protein analysis

Song et al. (2011)	Rat acellular lung (Sprague Dawley)	Orthotopic transplantation	Vascular perfusion only: pulmonary artery pressure kept constant at 80 cmH_2O, heparinized PBS with 0.1 % SDS, deionized water, Triton X-100, and PBS with Penicillin, Streptomycin, Amphotericin B	3 days (approximately 75 h) including incubation with antibiotics	Histology, immunohistochemistry, morphology, fluoroscopy, functional analysis, transplantation seeded lungs with fetal pulmonary cells and pulmonary artery and vein with endothelial cells
Shamis et al. (2011)	Rat acellular liver and lung (Lewis)	Cellular differentiation on 3D in vitro scaffold	Lung lobes cut into 300 μm thick, 0.5 % Triton X-100, 10 mm ammonia, mechanical disruption, PBS, distilled water	N/A	Histology, TEM, environmental scanning, PCR, immunohistochemistry, liquid chromatography with tandem mass spectrometry
Daly et al. (2012a)	Mouse acellular lung (C57BL/6; BALB/C)	Initial binding and recellularization of MSCs in acellular scaffold; directed seeding with integrin blocking	Airway and vascular perfusion: distilled H_2O, 0.1 % Triton X-100, 2 % SDC, NaCl, pancreatic DNase type 1S	3 days (approximately 72 h)	Histology, immunofluorescence, EM, perfusion to assess vascular continuity, mass spectrometry, western blot, lung mechanics with flexivent, innoculation of bone marrow-derived MSCs
Wallis et al. (2012)	Mouse acellular lung and lung slices (BALB/C)	Comparison of detergent-based decellularization protocols	Airway and vascular perfusion. Three different protocols tested: (1) H_2O, 0.1 % Triton X-100, 2 % SDC, NaCl, porcine pancreatic DNase; (2) PBS, 0.1 %SDS, 0.1 %Triton X-100; (3) PBS, CHAPS, NaCl, EDTA, DNase, FBS	3 days (approximately 72 h)	Immunohistochemistry, mass spectrometry, western, mechanical analysis, gelatinase, DNase, RNase, comparative recellularization with MSCs and C10s

(continued)

Table 8.2 (continued)

Reference	Scaffold	Study objective	Method of decellularization	Length of decellularization process	Endpoint assessments
Bonvillain et al. (2012)	Normal rhesus macaque acellular lung	Initial binding and recellularization of MSCs in acellular scaffold	Airway and vascular perfusion: PBS, EDTA, Penicillin/ Streptomycin at initial harvest: pulmonary artery: PBS + heparin + sodium nitroprusside with pressures 25–30 mmHg; then trachea and vasculature: deionized H_2O, 0.1 % Triton X-100, 2 %SDC, NaCl, bovine pancreatic DNase	2–3 days (approximately 48–72 h)	Histology, morphology, immunohistochemistry, western blot, genomic DNA, proteomics, seeding with bone marrow and adipose-derived rhesus MSCs
Longmire et al. (2012)	Mouse acellular lung and lung slices (C57/BL6)	Seeding with and differentiation of mESCs-derived endodermal lung precursors	Airway and vascular perfusion: distilled H_2O, 0.1 % Triton X-100, 2 % SDC, NaCl, pancreatic DNase type 1S	3 days (approximately 72 h)	Evaluation of ability to differentiate mESCs into lung precursor cells
Jensen et al. (2012)	Mouse acellular lung (C57BL/6)	Comparison of timing of decellularization, coating of decellularized matrices, and support of mESCs differentiated into alveolar epithelial cells	Airway and vascular perfusion: 0.1 % Triton X-100, 2 % SDC, NaCl, porcine pancreatic DNase, PBS	1 vs. 3 days (approximately 24 h vs. 50 h)	Histology, morphology, EM, western blot, gelatinase assay, immunofluorescence, mechanical properties with flexivent, support of differentiated mESCs within scaffold, subcutaneous implantation of scaffold
Petersen et al. (2012)	Rat acellular lung (Fisher 344)	Comparison of different detergent-based decellularization protocols	Two approaches: (1) vascular perfusion CHAPS, NaCl, EDTA, PBS; (2) NaCl, EDTA, SDS	4 h	Histology, collagen assay, elastin assay, GAG assay, DNA assay, mechanical testing with linear strips

Mishra et al. (2012)	Rat acellular lung	Creation of perfusable human lung cancer nodules	Vascular perfusion only: pulmonary artery pressure kept constant at 80 cmH_2O, heparinized PBS with 0.1 % SDS, deionized water, Triton X-100, and PBS with Penicillin, Streptomycin, Amphotericin B	3 days (approximately 72 h) including incubation with antibiotics	Recellularization with human A549, H460, or H1299 and cultured with perfused, oxygenated media for 7–14 days
Bonenfant et al. (2013)	Mouse acellular lung and lung slices (C57BL/6)	Effect of time to necropsy, length of storage, and two different methods of sterilization of construct	Airway and vascular perfusion: distilled H_2O, 0.1 % Triton X-100, 2 % SDC, NaCl, pancreatic deoxyribonuclease type 1S, $MgSO_4$, $CaCl_2$, Penicillin, Streptomycin	3 days (approximately 72 h)	Histology, immunohistochemistry, morphology, mass spectrometry, seeded lungs with MSCs and C10 epithelial cell line
Sokocevic et al. (2013)	Mouse acellular lung and lung slices (C57BL/6)	Effect of recipient age and elastase, or bleomycin injury on decellularization and recellularization	Airway and vascular perfusion: distilled H_2O, 0.1 % Triton X-100, 2 % SDC, NaCl, pancreatic deoxyribonuclease type 1S, $MgSO_4$, $CaCl_2$, Penicillin, Streptomycin	3 days (approximately 72 h)	Histology, immunohistochemistry, mass spectrometry, inoculation with MSCs and C10 cells
Sun et al. (2014)	Rat and mouse acellular lung slices	Engraftment and survival of fibroblasts through a β1-integrin and FAK-dependent pathway through ERK	CHAPS, NaCl, EDTA	1 day	Histology, SEM, immunohistochemistry, DNA assay, recellularization with mouse A9 cells

A9 transformed subcutaneous murine fibroblasts, *AFM* atomic force microscopy, *C10* immortalized murine alveolar epithelial cells, *$CaCl_2$* calcium chloride, *CHAPS* 3-[(3-cholamidopropyl)dimethylammonio]-1-propanesulfonate hydrate, *cmH_2O* centimeters of water (pressure), *CT* computed tomography, *DMEM* Dulbecco's modified Eagle's medium, *DNase* deoxyribonuclease, *DNA* deoxyribonucleic acid, *E17* embryonic day 17, *ECM* extracellular matrix, *EDTA* ethylenediaminetetraacetic acid, *FBS* fetal bovine serum, *GAG* glycosaminoglycan, *H_2O* water, *mESCs* murine embryonic stem cells, *$MgSO_4$* magnesium sulfate, *mM* millimolar, *mmHg* millimeters of mercury, *MSC* mesenchymal stem cells, *NaCl* sodium chloride, *$NaHCO_3$* sodium bicarbonate, *PBS* phosphate-buffered saline, *PCR* polymerase chain reaction, *RNase* ribonuclease, *SDC* sodium deoxycholate, *SDS* sodium dodecyl sulfate, *SEM* scanning electron microscopy, *SPA* surfactant protein A, *TEM* transmission electron microscopy

Table 8.3 Summary of decellularization methods for human and porcine lungs

Reference	Species	Decellularization agents	Perfusion parameters	Instillation route	Days
Petersen et al. (2010)	Human	CHAPS, NaCl, and EDTA	Constant pressure (25 mmHg)	Airway and vasculature	1
Booth et al. (2012)	Human/IPF	Triton X-100, SDC, NaCl, DNase	Unspecified	Airway and vasculature	3
Gilpin et al. (2014)	Human/porcine	SDS, Triton X-100	Constant pressure (30 cmH_2O)	Vascular	4–7
O'Neill et al. (2013)	Human/porcine	(a) SDS or (b) CHAPS or (c) Tween-20, SDC, peracetic acid	None—lung segments and agitation	N/A	1
Nichols et al. (2013)	Human/porcine	Freeze/thaw; graded SDS perfusion	Varying flow rates (100–500 mL/h)	Airway and vasculature	7
Price et al. (2015)	Porcine	Triton X-100, SDC, NaCl	12–25 mL/min (15 mmHg)	Airway and vasculature	1
Parker et al. (2014)	Human/IPF	SDS, Triton, NaCl	None—thin lung slices	N/A	2
Wagner et al. (2014b)	Human/porcine	Triton X-100, SDC, NaCl, DNase peracetic acid	Constant flow rates (1 L, 2 L, 3 L/min)	Airway and vasculature	3
Wagner et al. (2014a)	Human/COPD	Triton X-100, SDC, NaCl, DNase peracetic acid	Constant flow rate 2 L/min	Airway and vasculature	3

CHAPS 3-[(3-cholamidopropyl)dimethylammonio]-1-propanesulfonate, *COPD* chronic obstructive pulmonary disease, *IPF* idiopathic pulmonary fibrosis, *NaCl* sodium chloride, *SDC* sodium deoxycholate, *SDS* sodium dodecyl sulfate

consideration for the actual decellularization technique. While rodent and macaque lungs can be decellularized by hand, higher pressures and volumes must be utilized for sufficient inflation of perfusion pathways (e.g., vasculature, airways, etc.) in larger organs to ensure that perfused solutions and the ensuing cellular debris are cleared from the lungs. All published protocols to date for decellularizing whole large animal or human lungs utilize perfusion pumps to generate acellular scaffolds which can support recellularization (Wagner et al. 2014a, b; Nichols et al. 2013; Booth et al. 2012; Petersen et al. 2010; Price et al. 2015; Gilpin et al. 2014). While not clinically translational for transplantation, human and porcine lung segments have also been decellularized using small segments (Zhou et al. 2014; O'Neill et al. 2013; Nakayama et al. 2013; Parker et al. 2014) (Table 8.3).

A variety of post-decellularization techniques can be utilized to assess differences in potential decellularization techniques, including histologic, immunofluorescent staining, and DNA detection and quantification (Fig. 8.2). Further, proteomic analysis is a powerful tool which can be used to help delineate differences between protein loss and retention in protocols and can aid in the selection of optimal protocol parameters such as flow rates or pressures (Wagner et al. 2014b). For example, proteomic analysis can help delineate the impact of different steps and parameters in decellularization protocols (e.g., flow rate, pressure, rinse volumes, etc.) which preferentially retain certain ECM components or minimize retention of cellular-associated proteins (Wagner et al. 2014a, b; Booth et al. 2012).

8.2.3 Residual ECM and Other Proteins

The lung ECM itself has long been known to provide instructional cues during prenatal development as well as postnatally in maintaining tissue homeostasis and directing remodeling responses after injury (Fernandes et al. 2010; Lin et al. 2010; Mariani et al. 1997; Nguyen and Senior 2006; Rippon et al. 2006). While synthetic scaffolds offer the advantages of precise control and the ability to be more readily mass produced and stored, synthetic materials lack the biologically inductive capabilities observed in acellular scaffolds. Acellular scaffolds retain native integrin-binding sites in their correct spatial arrangement, in addition to preserving organ macro-architecture. A number of studies have shown their ability to induce phenotypic differentiation in the absence of the use of additional growth factors. While synthetic materials could be engineered to include specific integrin-binding sites to enhance cell adhesion [e.g., Arg-Gly-Asp (RGD) binding sites], it remains unknown what specific integrin-binding sites need to be included and in what spatial arrangement they need to be. Hybrid materials, consisting of synthetic and acellular matrix components, are also an attractive possibility and could be utilized to enhance cell adhesion and biological activity while taking advantage of the ability to more precisely manufacture scaffolds or scaffold components with synthetic materials (Ingenito et al. 2010, 2012; Cortiella et al. 2010; Nichols et al. 2013; Fischer et al. 2011). (Differences between acellular and synthetic scaffold approaches summarized in Table 8.4.)

Table 8.4 Comparison of biologic versus synthetic scaffold approach for ex vivo bioengineering

	Biologic (acellular) scaffold		Synthetic scaffold	
Differentiation and engraftment cues	+	Retains native integrin-binding sites	–	Lacks specific integrin-binding sites (must be engineered into scaffolds)
Immunogenicity	+	Antigen removal during decellularization	+/–	Unknown/variable depending on material
Manufacturability	+	Native architecture largely retained	–	Complex architecture
	–	Large variability between donor scaffolds	+	Precise control possible (i.e., repeatability)
Long-term storage	–	Degradation with long-term storage	+	Improved storage stability

Owing to the importance of ECM components, retention of key ECM components is a critical parameter to control and assess as an endpoint when evaluating potential decellularization protocols. The precise combination of ECM proteins that must be retained to preserve the ability of the acellular scaffold to give organotypic cues for cellular differentiation and functional tissue level assembly remains unknown. The major structural and functional molecules in the ECM include both proteins such as collagens, elastin, fibronectin, and laminins as well as a variety of glycoproteins including the glycosaminoglycans (GAGs). Collagens are important structural components of the lung and are responsible for overall mechanical strength, while elastin gives the lung its elastic properties of reversible distension and intrinsic recoil. GAGs help control macromolecular and cellular movement across the basal lamina and may play a role in the mechanical integrity of the lung. Matrix molecules are generally highly conserved proteins in eukaryotic organisms. This may theoretically explain the lack of an adverse immune response seen in xenotransplantation of other decellularized organs such as skin, trachea, and esophagus (Jungebluth et al. 2012a, b; Keane and Badylak 2014; Fishman et al. 2013).

There are a variety of techniques which have been used to evaluate ECM components, including histology, immunohistochemistry, Western Blotting, mass spectrometry-based proteomics, and component-specific assays such as Picrosirius, Fastin elastase, etc. (Table 8.2). The majority of decellularization techniques used for lung results in loss of elastin and sulfated GAGs in all species (Ott et al. 2010; Petersen et al. 2010; Price et al. 2010; Bonenfant et al. 2013; Bonvillain et al. 2012; Daly et al. 2012a; Jensen et al. 2012; Sokocevic et al. 2013; Wallis et al. 2012). In head to head comparison studies in perfused lungs, SDS and SDC have been found to retain more elastin as compared to CHAPS-based protocols (Petersen et al. 2012; Wallis et al. 2012). However, despite the differences in retention of ECM components, inoculated cells appear to behave similar in the repopulation assays currently used (including histological and immunofluorescence evaluation). Therefore, it remains unknown if there is an optimal decellularization protocol, and if so, which is best suited for translation to the clinic.

In contrast to traditional protein detection methods, mass spectrometry-based proteomics is gaining traction as an invaluable tool for assessing acellular scaffolds. In addition to detecting ECM composition and residual proteins in acellular scaffolds, it has been used for distinguishing differences between decellularization methods or lung origin, including disease states or donor age (Wallis et al. 2012; Bonvillain et al. 2012; Sokocevic et al. 2013; Wagner et al. 2014a, b; Booth et al. 2012; Nakayama et al. 2013; Gilpin et al. 2014). One particularly striking and consistent result among the various groups utilizing this analysis approach is the amount and breadth of non-ECM proteins detected in the scaffold following decellularization. In particular, cytoskeletal elements and histones appear to be retained in the scaffolds, while lesser secreted proteins are detected. This suggests that transmembrane proteins and their associated cytoskeletal elements may remain anchored to the ECM with the currently used decellularization protocols. The impact of these residual proteins on recellularization, including potential immunogenicity, remains unknown.

8.2.4 Mechanical Assessments of Decellularized Scaffolds

A variety of in vitro assessments have been utilized to begin to establish metrics for assessing the potential functionality of acellular scaffolds. Investigators have explored both micro- (Melo et al. 2014a, b; Nonaka et al. 2014; Booth et al. 2012) and macro-scale (Daly et al. 2012a; Wallis et al. 2012; Nichols et al. 2013) mechanical measurements of acellular lungs as well as force tension relationships in linear strips of decellularized lungs (Petersen et al. 2012; O'Neill et al. 2013). While techniques such as atomic force microscopy (AFM) are useful in obtaining topographical information and initially assessing mechanical properties of the scaffolds (Melo et al. 2014a, b; Nonaka et al. 2014; Booth et al. 2012), these results do not necessarily translate to recellularization and functional performance. Traditional lung mechanics testing of acellular scaffolds has shown that in the absence of cells and surfactants, acellular scaffolds are stiffer than their naïve counterparts (Daly et al. 2012a). Introduction of exogenous surfactant into the acellular scaffolds can partially restore lung compliance (Daly et al. 2012a). This is an important finding and indicates that during recellularization strategies, serial measurements of lung mechanics could be used as a non-invasive and non-destructive means to assess functionality of the regenerating scaffold. For example, decreases in elastance could be used as a measurement of de novo surfactant production.

8.3 Recellularization

8.3.1 Recellularization of Acellular Scaffolds for Bioengineering New Lung

The lung is comprised of many different cell subtypes and all uniquely contribute to some critical aspect of lung function (Morrisey and Hogan 2010). This heterogeneous cell population is replenished by resident stem or progenitor cells following injury. Regeneration of lungs suitable for transplantation will require some minimal restoration of these subtypes for subsequent long-term functionality. While a variety of cell sources are being investigated for recellularizing acellular scaffolds, obtaining sufficient cell numbers with any source remains a significant open question. The ideal solution is to use an autologously derived source of cells to minimize post-transplantation immune complications, which are a significant cause of morbidity in transplanted patients. One potential source is the use of fully differentiated primary adult cells, which ideally would come from an autologous source. However, using this approach, the multiple cell types would need to be isolated from the eventual transplant recipient, grown to sufficient numbers ex vivo and then further be able to be used in a recellularization approach to restore functionality. While it has been shown that a strategy such as integrin blocking can be used to direct initial cell engraftment of a single cell population (Daly et al. 2012a; Sun et al. 2014), scaling this clinically and further adding the complex challenge of uniquely directing the right cell population to a specific architectural location would be challenging. Alternatively, autologous endogenous lung progenitor cells from the various compartments could be utilized (e.g., distal and proximal epithelial progenitor cells, endothelial progenitor cells, etc.) along with stromal cells to recellularize acellular scaffolds. However, the same challenges of obtaining sufficient cell numbers for an initial seeding strategy and directing cells to their correct compartment remain. In both instances, however, it remains unknown if normal cells could be obtained from a patient with a preexisting lung disease or if isolated diseased cells could be gene-corrected prior to subsequent recellularization. However, recent work indicates that the scaffold may more significantly contribute to phenotype than cell origin (Parker et al. 2014). An allogeneic cell source could also be used, but this re-introduces the potential for immune complications following transplantation. Furthermore, the identification of bona fide lung progenitor cells in the adult human lung remains controversial.

A potentially more appealing autologous approach is the use of induced pluripotent stem (iPS) cells, which are derived from reprogramming somatic cells to a stem-cell-like state. While iPS cells avoid the ethical controversies surrounding the use of embryonic stem cells (ESCs)—stem cells derived from the inner blastocyst of in vitro fertilized embryos—they have been shown to retain epigenetic memory of their tissue origin and have been shown to form teratomas. iPS cells are typically derived from dermal fibroblasts, and thus, differentiating them into the various lung cell types has been challenging. However, despite this limitation, recent work has demonstrated that human iPS cells can be differentiated into cells expressing a distal pulmonary epithelial cell phenotype and seeded into acellular human lung scaffolds (Ghaedi et al. 2013; Huang et al. 2014). These results further encourage the use of this approach in moving towards the clinic.

Other potential approaches include the use of fetal homogenates or ESCs. As previously mentioned, ethical concerns remain for either of these approaches, as well as the potential for teratoma formation with ESCs. While initial studies have shown that ESCs can engraft in acellular murine lungs (Longmire et al. 2012; Cortiella et al. 2010), seeding into acellular lungs was not sufficient to induce differentiation and thus optimized in vitro differentiation protocols must be used in conjunction with seeding and repopulation strategies. ESC-derived murine Nkx2-1GFP+ progenitor cells were able to recellularize acellular murine lungs and form alveolar structures, while in contrast, seeding with undifferentiated ESCs resulted in nonspecific cell masses in distal regions of acellular lungs. Fetal homogenates have the distinct advantage of containing all necessary cell populations which have been shown to have some capacity for self-assembly. Further, these cells have been successfully used in the current rodent models of ex vivo regeneration and transplantation. However, in both instances, ethical concerns remain in obtaining these cells and the use of immunosuppressive drugs post-transplantation remains unknown. Tables 8.5 and 8.6 are a summary of recellularization approaches in animal and human models and the phenotypic adoption of seeded cells.

Table 8.5 Distribution and phenotype of cells seeded onto animal models of acellular scaffolds

Reference	Cells used for seeding	Scaffold	Route	Duration (days)	Distribution	Final phenotype
Lwebuga-Mukasa et al. (1986)	AECII	Acellular alveolar vs. amniotic basement membranes	Direct seeding	8	N/A	Alveolar matrices: ATI; amnionic membranes ATII
Cortiella et al. (2010)	mESC	Rat (Sprague Dawley) acellular lung	Trachea	21	Proximal-distal regional specific CC10, proSP C expression)	Tracheo-bronchial: CC10, Ck18; distal lung: proSPC, CD31, Pdgfrα
Ott et al. (2010)	HUVEC (DsRed)	Rat acellular lung	Pulmonary artery	9	All vessels	Endothelial cells
	A549	Rat acellular lung	Trachea	9	Airways/alveoli	Airway/alveolar epithelium
	HUVEC (DsRed)	Rat acellular lung	Pulmonary artery	9	Entire vasculature	Endothelial cells
	Rat fetal lung cells (GD19-20)	Rat acellular lung	Trachea	9	Airways/alveoli	proSp-a, proSFC, Ttf-1/Nkx2.1 (AECII); T1α (AECI); vimentin (fibroblast)
Petersen et al. (2010)	Neonatal (7d) lung epithelial cells (rat)	Rat acellular lungs (Fischer 344)	Trachea	8	Alveolar, small airways	CCSP (Club cell), proSPC (ATII), Aqp5 (ATI), Ck14 (basal cell)
	Lung vascular endothelium (rat)	Rat acellular lungs (Fischer 344)	Pulmonary artery	7	Microvascular	CD31
Price et al. (2010)	Fetal lung (E17)	Ms acellular lungs	Tracheal	7	Alveolar	CK18+/proSPC+ (ATII); CD11b, Aqp5, CCSP, CD31, and vimentin
Daly et al. (2012a)	mBM-MSCs	Ms acellular lung	Trachea	28	Parenchymal > airway (squamous)	MSCs: no evidence for transdifferentiation
	C10-hAECII (non-tumorigenic)	Ms acellular lung	Trachea	28	Parenchymal	N/A
Ott et al. (2010)	Rat fetal (GD17-20) pneumocytes	Rat acellular lung	Trachea	14	Alveolar/distal bronchioles > trachea/bronchi	CCSP (airways); Ttf-1, proSPC (alveolar)
	HUVEC	Athymic nude rat	Pulmonary artery	14	Proximal to distal vasculature	$CD31^{pos}$

(continued)

Table 8.5 (continued)

Reference	Cells used for seeding	Scaffold	Route	Duration (days)	Distribution	Final phenotype
Shamis et al. (2011)	Ms AECII (primary or P2)	Acellular lung microscaffold	Direct seeding	22	Alveolar	proSPC/SPC (ATII-like, from primary or cultured ATII); Aqp5, PDPN (ATI); CCSP (primary)
Wallis et al. (2012)	mBM-MSCs	Ms acellular lung	Trachea	14	Alveolar	MSCs
	C10-hAECII (non-tumorigenic)	Ms acellular lung	Trachea	14	Large and small airways	Squamous morphology
Bonvillain et al. (2012)	Rhesus BM-MSCs	Rhesus macaque	Secondary bronchus	7	Alveolar septae, terminal bronchioles, respiratory bronchioles	MSCs phenotype
	Rhesus AD-MSCs	Rhesus macaque	Secondary bronchus	7	Alveolar septae, terminal and respiratory bronchioles	MSCs phenotype
Jensen et al. (2012)	mESCs diff. to Ttf1^{pos}/proSP-C^{pos}	Ms acellular lung	Immersion	14	Alveolar	Ttf1/Nkx2.1, proSPC (alveolar); Pdgfrα (mesenchymal)
Longmire et al. (2012)	mESCs	Ms acellular lung	Trachea	10	Hypercellular sheets (alveolar)	Ciliated cells (airways); T1α^{neg} (alveoli)
	Nkx2.1^{GFP}	Ms acellular lung	Trachea	10	Alveolar	Nkx2.1/T1α (alveoli)
Bonenfant et al. (2013)	Mouse BM-MSCs	Ms acellular lung	Trachea	28	Alveolar	N/A
	C10-mAECII (non-tumorigenic)	Ms acellular lung	Trachea	28	Alveolar	N/A
Sokocevic et al. (2013)	mBM-MSCs	Ms acellular lung	Trachea	28	Alveolar	N/A
	C10-mAECII (non-tumorigenic)	Ms acellular lung	Trachea	28	Alveolar	N/A
Sun et al. (2014)	A9-murine immortalized fibroblasts	Rat and Ms acellular lung	Slices	14	Alveolar	Pro-Col1α1

Table 8.6 Recellularization studies in porcine and human acellular scaffolds

Reference	Species	Lung origin	Recellularization technique	Cell types	Summary
Petersen et al. (2010)	Human	Normal	Segment incubated with cells	A549; endothelial progenitor cells	Human scaffolds support recellularization
Booth et al. (2012)	Human	Normal/IPF	Thin slices	hLF	hLFs on IPF scaffolds increase α-SMA expression
Ghaedi et al. (2013)	Human	Normal	Thin slices	hiPS-ATII, ATII	hiPS-ATII cells adhered to acellular lung matrix and a subpopulation differentiated to ATI phenotype
Gilpin et al. (2014)	Human/porcine	Normal	Thin slices; single lobe recellularization SAEC and pressure-controlled perfusion	HUVEC, SAEC, PAEC	Cell survival for 3 days in an acellular perfused upper right lobe recellularized with PAECs
O'Neill et al. (2013)	Human/porcine	Normal	Thin slices	hMRC-5, hSAEC	Acellular and human porcine lungs supported cell viability
Ghaedi et al. (2014)	Human	Normal	Thin slices and rotating bioreactor	hiPS-ATII, ATII	Rotating bioreactor culture at ALI enhanced ATII to ATI differentiation
Mendez et al. (2014)	Human	Normal	Thin slices	hBM-MSCs, hAT-MSCs	Both MSC populations attached to the acellular lung matrix; culture on acellular lung matrices enhanced SPC, Aqp5, and caveolin-1
Nichols et al. (2013)	Human/porcine	Normal	Excised segments (~0.5 cm^3) injected with cells	mESC, hFLC, BMMSCs, hAEC	Porcine/human lungs supported cellular attachment; 1 % SDS protocol minimized T cell activation
Parker et al. (2014)	Human	Normal/IPF	Thin slices	Normal and IPF-derived hLF	ECM contributed more significantly to IPF phenotype rather than cell-origin
Wagner et al. (2014b)	Human/porcine	Normal	Physiologic instillation and thin slices	hMSC, hBE, CBF12, hLF	Normal acellular porcine and human lungs are non-toxic
Wagner et al. (2014a)	Human	Normal/ COPD	Physiologic instillation and thin slices	hMSC, hBE, CBF12, hLF	COPD lungs do not support viability comparable to normal lungs

BMMSCs porcine bone marrow-derived mesenchymal stem cells, *CBF12* human endothelial progenitor cells, *hAEC* human alveolar epithelial cells, *hAT-MSC* human adipose-derived mesenchymal stem cells, *hBE* human bronchial epithelial cells, *hBM-MSC* human bone marrow-derived mesenchymal stem cells, *hFLC* human fetal lung cells, *hiPS* human-induced pluripotent cells, *hLF* human lung fibroblasts, *hMSC* human bone marrow-derived mesenchymal stem cells, *HUVEC* human umbilical vein endothelial cells, *mESC* murine embryonic stem cells, *PAEC* pulmonary alveolar epithelial cells, *SAEC* human small airway epithelial cells

8.3.2 *Acellular Lungs as Ex Vivo Models of Disease*

In addition to their potential use as scaffolds for tissue engineering, there has been rapid growth in the use of acellular lungs as ex vivo models which more closely recapitulate diseased in vivo environments. These experiments provide a new opportunity for insight into cell–ECM interactions capable of driving disease phenotypes. Human fibroblasts from normal human lungs seeded onto acellular scaffolds derived from fibrotic lungs were found to increase their alpha-smooth muscle actin (α-SMA) expression (Booth et al. 2012), and the ECM was found to contribute more significantly to IPF-correlated gene expression changes in fibroblasts rather than cell-origin (i.e., from IPF or normal lungs) (Parker et al. 2014). However, many cell-associated proteins, characteristic of pulmonary fibrosis [e.g., transforming growth factor beta (TGF-β), beta-catenin (Ctnnb1), etc.], are retained in decellularized mouse lungs following bleomycin injury (Sokocevic et al. 2013). In addition, ECM-associated proteins and matrikines (ECM-derived peptides which are liberated by partial proteolysis of ECM macromolecules) are detectable by proteomic approaches following decellularization (Nakayama et al. 2013; Bonenfant et al. 2013; Booth et al. 2012; Daly et al. 2012a; Sokocevic et al. 2013; Wagner et al. 2014a, b; Wallis et al. 2012). These proteins, in addition to the detected ECM components, may significantly contribute to the phenotypic changes observed by several groups in recellularization assays. In particular, observation of acquisition of a more fibrotic phenotype by normal fibroblasts in acellular human IPF lungs may be attributed to these residual proteins, rather than the ECM components alone (Booth et al. 2012; Parker et al. 2014).

Similarly, in acellular lungs derived from murine models of emphysema and from human patients with COPD, cells were unable to remain comparably viable as the same cells seeded into healthy acellular scaffolds (Sokocevic et al. 2013; Wagner et al. 2014a). This suggests that either the matrix is impaired in COPD or that the residual protein composition is significantly altered as compared to normal acellular lungs. These studies generate exciting insight into the potential role of the matrix and matrix-associated proteins in driving disease phenotypes and provide proof of concept for use of acellular lungs as a novel platform for studying cell–matrix interactions.

A further novel use of acellular scaffolds in disease models has been utilized to study the role of macrophages in IPF using a Transwell culture setup of thin acellular lung slices recellularized in the Transwell insert, with macrophage co-culture (Zhou et al. 2014). Decellularized mouse lung slices seeded with murine fibroblasts were co-cultured with $CD206^+$ or $CD206^-$ macrophages from day 14 of murine lungs following bleomycin-induced lung injury (or in the absence of macrophages). $CD206^+$ macrophages were found to increase fibroblast proliferation and survival in the lung slices. However, there was no induction of α-SMA expression. Nonetheless, this study takes advantage of the ability to selectively study cells and cell combinations in isolation using acellular lungs. Similarly, the human fibroblast cell line MRC5 was seeded onto slices of normal human decellularized lung slices and stimulated with recombinant human chitinase 3-like 1 protein (rhCHI3L1), a prototypic-chitinase-like protein recently shown to be elevated in human IPF. The addition of rhCHI3L1 induced α-SMA expression in the MRC5 cells and they adopted a contractile phenotype, as assessed by histology (Zhou et al. 2014).

In addition to repopulation assays, it has also been suggested that recellularized acellular scaffolds could also be used for studying infectious diseases (Crabbé et al. 2014) and used as models for cancer development (Mishra et al. 2012). Thus, studies to date have likely only begun to demonstrate the utility of acellular tissue as ex vivo models of disease which more closely recapitulate in vivo microenvironments than traditional in vitro setups.

8.3.3 Implantation of Recellularized Scaffolds

Important proof of concept studies has shown that recellularized scaffolds can be implanted and participate in gas exchange for short time periods. Decellularized rat lungs re-endothelialized with human umbilical vein endothelial cells (HUVEC) and recellularized with fetal rat lung homogenates and A549 epithelial cells were transplanted into rats that had undergone previous pneumonectomy (Ott et al. 2010; Petersen et al. 2010). While the ex vivo regenerated lungs were shown to contribute to gas exchange following transplantation, the transplants developed significant pulmonary edema and/or hemorrhage resulting in respiratory failure after several hours. In a further study, survival for 14 days was achieved after implantation, but lung function progressively declined and the histologic appearance of the graft at necropsy demonstrated significant atelectasis and appeared fibrotic (Song et al. 2011). These studies provide proof of concept that acellular lungs can be recellularized, surgically implanted, and can provide minimal restoration of function; however, they also demonstrate the challenges that remain in translating towards the clinic. A recellularized acellular lung needs to meet a number of functional requirements in order to be clinically translatable: adequate gas exchange, intact alveolar and vascular compartments, unidirectional mucociliary clearance, and the ability to maintain physiologic airway pressures and volumes. Thus far, there has been a compartmentalized approach to the respiratory system, separating regeneration of the trachea, vasculature, proximal airways, and distal lung. An animal model which accomplishes restoration of all of these functions has not been achieved to date and it will likely be several years before this can be achieved.

8.3.4 Immunogenicity of Implanted Scaffolds

The a priori assumption for clinical use of decellularized lung scaffolds is that because the cellular material has been removed and thus the cell-associated immunogens will be removed, acellular scaffolds will be non-immunogenic. However, some ECM and other proteins identified in the remaining decellularized scaffolds are known to be immunogenic (Franz et al. 2011; Badylak and Gilbert 2008; Shaw and Filbert 2009; Daly et al. 2012b). It has also been observed that cells inoculated into decellularized scaffolds secrete ECM and other proteins (Daly et al. 2012a). Thus, inoculated cells may considerably remodel the scaffold and generate their own basement membrane, shielding the denuded basement membrane, which can be immunogenic (Iwata et al. 2008).

Nevertheless, some of these remaining proteins may actual be beneficial with regard to their ability to induce an immune response. A growing body of literature suggests that decellularized scaffolds can polarize macrophages to the anti-inflammatory M2 phenotype, which is viewed as a more permissive regenerative phenotype (Badylak and Gilbert 2008; Brown et al. 2009, 2012a, b; Keane and Badylak 2014; Fishman et al. 2013). Further, recent work in lung repair and regeneration has demonstrated the critical role that the immune system has in orchestrating normal repair and regeneration in adult lungs (Hogan et al. 2014). To date, with the exception of the use of fetal homogenates, no recellularization studies have included immune cells. Thus, it is unknown whether retention of these immunogenic components may actually be beneficial in a regeneration strategy.

8.3.5 Environmental Factors in Ex Vivo Lung Regeneration

The majority of published work focuses on decellularization methods, lung origin (i.e., disease state or age), and cell sources. There have been limited investigations into the addition of growth factors and especially a lack of studies examining the role of environmental cues, such as mechanical stimuli and oxygen tension in regeneration schemes. These are critical factors known to play roles in both embryonic development and post-natal repair and regeneration (Colom et al. 2014; Ingber 2006a, b). Traditional in vitro cell culture is performed at 20 % oxygen; however, physiologic oxygen levels in individual cells vary depending on the tissue type, tissue density, and proximity to blood vessels (Simon and Keith 2008). It has long been known that hypoxia can mediate angiogenesis and that vascular endothelial growth factor (VEGF) is upregulated in hypoxia (Shweiki et al. 1992). During embryonic development, the lung environment is hypoxic (1–5 % oxygen) (Simon and Keith 2008). Lowering oxygen tension to levels typically encountered by cells in the developing embryo has been shown to enhance in vitro differentiation of ESC and iPS cells to Nkx2-1+ lung/thyroid progenitor cells (Garreta et al. 2014). Studies of cellular differentiation in acellular scaffolds are needed to further clarify the role of oxygen tension in an ex vivo regeneration strategy.

There is also a large and growing body of literature that delineates the importance of mechanical stimuli on embryonic lung development as well as normal and diseased tissue repair and regeneration in vivo (Ingber 2006a, b). Mechanical stretch is known to induce upregulation of surfactant protein C (SPC) mRNA and protein expression in ATII cells, while shear stress on endothelial cells is critical for VEGF expression (Sanchez-Esteban et al. 2001). Several studies have examined the biological consequences of mechanotransduction on fetal or adult lung cells in vitro (Sanchez-Esteban et al. 1998; Boudreault and Tschumperlin 2010; Huang et al. 2012; Weibel 2013), but there is no available information on effects of stretch on development of lung epithelial tissue from embryonic or adult stem cells or from endogenous lung progenitor cells. We have observed upregulation of lung epithelial genes in murine bone marrow-derived mesenchymal stem cells seeded into acellular mouse lungs and ventilated (Wagner et al. unpublished data). In particular, we found that SPC mRNA was significantly upregulated at physiologic tidal volumes; a result we also observed in human ATII cells ventilated in small segments of acellular human lung (Wagner et al. unpublished data) using an artificial pleural coating on excised acellular segments, permitting ventilation (Wagner et al. 2014c). While perfusion parameters have not yet been studied in detail, cultivation of a recellularized human lobe was done under perfusion conditions (Gilpin et al. 2014) and a rotating bioreactor culture was found to have positive effects on differentiating iPS cells into distal lung epithelial cells (Ghaedi et al. 2014).

In addition to utilizing a scaffold from a suitable source and using an optimized decellularization protocol, precise control of the mechanical and gaseous environment with bioreactor technologies (e.g., stimuli mimicking stretch from breathing and shear stress induced by blood flow or breathing) will be necessary for a successful regeneration scheme.

8.4 Lessons Learned from Ex Vivo Organ and Tissue Culture

Despite rigorous research efforts, it remains challenging to keep normal, healthy tissue slices and organ explants viable for in vitro studies (i.e., lung slices) longer than a few days. Further, it is difficult to maintain adequate tissue viability for more than a few hours in candidate donor lungs for transplantation. This makes the challenge of generating functional lung tissue ex vivo even more daunting. While it is largely recognized that sophisticated bioreactor technologies will be needed for ex vivo lung tissue regeneration, few studies to date have looked into the various logistical aspects of bioreactor culture schemes such as media formulation and perfusion or ventilation parameters. However, several studies have already demonstrated the importance of incorporating these technologies with ex vivo schemes to maintain or enhance phenotypes. Culture of hATII cells and hiPS-ATII cells on acellular human scaffolds in a rotating bioreactor at air-liquid-interface (ALI) proved to be beneficial in maintaining phenotypic expression of distal epithelial lung cells (Ghaedi et al. 2014) and recellularization of an entire human lung lobe with human small airway epithelial cells (SAECs) and constant media perfusion showed that cell viability could be maintained for 3 days (Gilpin et al. 2014). However, both of these studies utilized short timepoints and longer ex vivo schemes will likely be critical for generating functional lung tissue (Murphy and Atala 2014). Fortunately, there is a rich history of ex vivo tissue culture studies from which the field of ex vivo bioengineering can greatly benefit from.

8.4.1 Precision Cut Tissue Slices

The concept of "precision cut tissue slices" was established in the mid-1980s, when Smith et al. first reported a method to slice liver into 250 μm thin sections with low variation in thickness (<5 %) for cultivation and ex vivo analysis (Smith et al. 1985). Previously, slices (predominately from liver) were produced by manually cutting with razorblades. This produced tissues with highly variable thickness (one to several millimeter) and further poses a risk of malnutrition and lack of oxygenation for cells inside the tissue slice (Sanderson 2011). The development of the Krumdieck tissue slicer helped overcome many of these problems by allowing for precise cutting of tissue <1 mm in thickness (in general, 25–300 μm thickness slices are generated) (Sanderson 2011; Liberati et al. 2010). In addition to the tissue slicers (e.g., Alabama Research and Development or Leica), vibratomes (e.g., Zeiss, Leica) can also be used. The main difference between these devices is the way the slices are produced. In the Krumdieck tissue slicers, a core is drilled from the tissue and is further sliced by a rotating knife perpendicular to the core axis, while the vibratome uses a vibrating knife to cut the tissue and has lower mechanical impact (Rice et al. 2013).

Slices have been produced from several organs including brain (Pilaz and Silver 2014; Wang et al. 2014), heart (Camelliti et al. 2011; Bussek et al. 2009; Parrish et al. 1995), liver (Hammad et al. 2014; Satoh et al. 2005; de Graaf et al. 2010; De Kanter et al. 1999; Parrish et al. 1995), kidney (Rice et al. 2013; De Kanter et al. 1999; Parrish et al. 1995), organ of Corti (Shim 2011), and lung (De Kanter et al. 1999; Morin et al. 2013; Martin et al. 1996; Liberati et al. 2010; Delmotte and Sanderson 2006; Davidovich et al. 2013; Parrish et al. 1995). Furthermore, this technique was applied to organs from various species including mice (Henjakovic et al. 2008a; Schleputz et al. 2012; Schnorbusch et al. 2012; Delmotte and Sanderson 2006), rats (Schleputz et al. 2012; Davidovich et al. 2012; de Graaf et al. 2010; Martin et al. 1996; Moreno et al. 2006; Nguyen et al. 2013), guinea pigs (Schleputz et al. 2012; Ressmeyer et al. 2006; Bussek et al. 2009), sheep (Schleputz et al. 2012), dogs (Nguyen et al. 2013), ferrets (Nguyen et al. 2013), monkeys (Schleputz et al. 2012; Nguyen et al. 2013), and humans (De Kanter et al. 1999; Schleputz et al. 2012; Wohlsen et al. 2003; Camelliti et al. 2011; de Graaf et al. 2010).

For all organs, except for the lung, the stiffness of the tissue itself is sufficient for slicing. However, in lung, to achieve the stiffness needed for cutting, the lung tissue needs to be filled with a supporting material. Low melting agarose in the range of 1–3 % (w/v) at 37 °C can be used to inflate the tissue and is followed by a cooling period on ice to allow gelling. Slices have been used in diverse studies, including pulmonary physiology (Morin et al. 2013; Dassow et al. 2010; Wright and Churg 2008; Schleputz et al. 2011; Wyatt et al. 2012; Schnorbusch et al. 2012; Davidovich et al. 2012, 2013), pharmacology (Moreno et al. 2006; Morin et al. 2013; Seehase et al. 2011; Wohlsen et al. 2003; Held et al. 1999; Bussek et al. 2009), pathogenesis (Zhou et al. 2012; Vaira et al. 2010), toxicity (Parrish et al. 1995; Henjakovic et al. 2008b), cellular effects of mechanical stretch (Dassow et al. 2010; Davidovich et al. 2012, 2013; Rausch et al. 2011), cytokine release (Henjakovic et al. 2008b), viral infection and gene transfer (McBride et al. 2000; Nguyen et al. 2013), and viral exacerbations (Bauer et al. 2010; Marquardt et al. 2011). Typical experimental durations have been reported in the range of 24–72 h (Liberati et al. 2010; Sanderson 2011). Thereafter, it has been reported that slice cultures decrease in viability and therefore most experiments cannot be conducted for long-time culture. There is a great interest in prolonging this cultivation to have the possibility to study disease development.

8.4.2 *Ex Vivo Maintenance of Explanted Organs for Transplantation*

Many patients die on the waiting list for a transplant due to a shortage of donor organs. One potential area of active research is to improve perfusion techniques ex vivo to improve the quality of organs previously deemed as suboptimal organs, and thus increase the potential donor pool based on existing organ availability (Bruinsma et al. 2014). In liver transplantation, it is known that for grafts with suboptimal quality the traditionally used static cold preservation fails to maintain cell viability and results in poor graft performance. To overcome this limitation, a novel pretreatment strategy was utilized to make more organs suitable for transplantation. Livers perfused using pulsatile flow via dual perfusion of both the hepatic artery and the portal vein in a closed circuit at 37 °C for 6 h were found to have improved graft function (op den Dries et al. 2013). Process control was used to limit the pressure during perfusion to a mean of 50 mmHg in the hepatic artery and 11 mmHg for the portal vein, while keeping the organ at 37 °C. In another study, porcine livers were normothermically perfused (37 °C) for 3 days (Butler et al. 2002). Explanted livers were normal for all tested functional parameters (i.e., normal physiological levels of pH and electrolytes, continued hepatic protein synthesis, hemodynamic parameters in normal physiological range, and histology with no overall architectural change were observed). Unfortunately, no re-implantation was conducted and in vivo organ performance was not assessed. In another study dog kidneys were perfused at 5 °C for 5 days using different perfusate solutions and successfully re-implanted depending on the choice of perfusate solution (McAnulty et al. 1989). A perfusate with 0.5 mM calcium preserved mitochondrial function and increased post-transplant survival the best.

In lung transplantation, approximately 80 % of all lungs cannot be used due to potential explant injury. To make more of these lungs available for transplantation, a technique called normothermic (37 °C) ex vivo lung perfusion (EVLP) has been recently investigated. After 4 h of ventilation, lungs can be fully inflated, cooled to 10 °C, and stored in 4 °C Perfadex solution until transplantation. Using this technique, high-risk donor lungs that were found to be physiologically stable during the 4 h EVLP were transplanted with no significant drawbacks compared to normal donor lungs (Cypel et al. 2011).

Recently, the "XVIVO Perfusion System with STEEN Solution" (XVIVO Perfusion Inc. Englewood, Colorado, USA) was approved by the US Food and Drug

Administration (FDA). This system is applied when more time for evaluation of the functional suitability of a donor organ for transplantation is needed. Donor lungs are kept at body temperature while flushing the vasculature up to 4 h with a sterile solution (STEEN Solution). Thereby, the lungs are ventilated, cells maintain more physiologic oxygen levels, and waste products are removed. This makes it possible to transplant lungs once viewed as non-ideal and these lungs have been shown in a preclinical trial to have similar rates of organ rejection and 12-month survival rates compared to optimal donor organs These studies may help elucidate critical parameters which need to be controlled and optimized for long-term cultivation of organs for basic research. Collectively, these studies indicate that there are parameters which are critical to control ex vivo organ culture, whether it is simply for longer ex vivo culture for basic science or for preservation/maintenance for candidate transplant organs or regeneration schemes.

8.5 Discussion and Outlook

The difficulties encountered in maintaining ex vivo viability of freshly explanted healthy organs highlight the challenges which the ex vivo regeneration field faces. Ex vivo whole organ cultures experience decreases in viability, selective survival of specific cell types, and loss of phenotypic expression occur over time with current techniques. Even in very thin tissue slice models of naïve tissue, where lack of nutrition and oxygenation is theoretically not of major concern, cells can only maintain their functionality, proliferative capacity, and viability for short periods of time (up to 72 h). This is exacerbated in cultivating whole organs, such as lung, where the need for proper control of medium oxygenation, osmolarity, pH, ventilation, and tissue perfusion in three dimensions is required to keep the tissue viable for long-term cultivation.

Currently, media formulations optimized for two-dimensional (2D) cultivation of homogenous cell populations are utilized in cultivation of tissue slices in both precision cut tissue slices and recellularized acellular tissue slices. However, the media composition needed for whole organ cultivation needs to be optimized for multiple cell types. Furthermore, stem and progenitor cells should sustain their capability to differentiate and replenish damaged or absent cell compartments. Therefore, the media formulation should also be optimized to sustain these niches. To achieve this, different cell types and stem cells initially seeded into acellular scaffolds in an undifferentiated state will require the timed sequential addition of different growth factors, nutrients, and amino acids to regulate signaling pathways involved in cellular proliferation and differentiation. Means of surveying and controlling the cultivation conditions and media formulation are needed. The knowledge from bioprocess engineering may help to fulfil the needs of whole organ cultivation. To date, no study has been conducted addressing the composition of organ-specific cultivation medium supporting long-term cultivation and cellular maintenance in recellularizing lung scaffolds.

Additionally, currently used cultivation conditions for either ex vivo naïve tissue or recellularized acellular scaffolds do not resemble the in vivo environment due to lack of proper mechanical (stretch) and environmental stimuli (contact to certain media/air). For example in the lung, it has been shown that isolated ATII cells in tissue culture lose SPC expression over time and trans-differentiate into alveolar epithelial type I (ATI) cells (Ezzie et al. 2012; Flozak et al. 2010). A similar decrease in SPC expression was observed when we cultured naïve murine and human lung tissue slices for 7 days in submerged culture (Uhl et al., unpublished data). As it is known that mechanical stimulation induces SPC expression in ATII cells (Majumdar et al. 2012; Dietl et al. 2010; Douville et al. 2011), this suggests that ventilation of whole organ cultures or stretching of lung slices may be necessary to retain ATII cells in their progenitor state. On the other hand, non-physiologic ventilation may cause alveolar epithelial cell damage. In a healthy organ the tolerance of cells regarding mechanical stimuli may be different than in disease. This shows the importance of maintaining precise control of the environmental parameters to the regenerating lung tissue.

Reseeding of decellularized matrixes is usually done with one or only few different cell types. Each additional cell type adds complexity, making interpretation of results utilizing homogenates or multiple cell types challenging. Usage of stem and progenitor cells (e.g., ESCs, mesenchymal stem cells, or iPS cells) is appealing for recellularization strategies as these cells can potentially differentiate into the multitude of cell types needed in a specific area of the scaffold. The potential for this approach was demonstrated with the use of ESC-derived murine Nkx2-1^{GFP+} in acellular lungs. These cells repopulated distal airspaces and a subpopulation differentiated into Nkx2-1^{GFP-} and acquired a morphology characteristic of ATI cells and expressed the phenotypic ATI marker podoplanin (T1α). While encouraging, the necessity of regenerating the multitude of cell types in the lung remains a challenge.

Recellularization of decellularized organs will require precise process control. First, the cells need to be seeded onto the matrix and need to engraft and recellularize the whole tissue, which may be accomplished by migration and/or proliferation. During this initial seeding phase, ventilation and perfusion may not be feasible, and in fact may even be detrimental. The initial properties and composition of the organ are not comparable to the in vivo situation. For example, in decellularized lungs, there is a lack of surfactant in the alveoli prior to recellularization, and this dramatically affects mechanical properties (Daly et al. 2012a). It has also been shown that there is a loss of ECM components, such as elastin, following perfusion decellularization using most protocols (Wagner et al. 2013). The effect of the loss of these ECM components on initial engraftment and subsequent recellularization and regeneration remains unknown. The importance of preserving the native integrin-binding sites in recellularization schemes has already been demonstrated. Cells can be directed to certain ECM-binding sites through integrin blocking (Daly et al. 2012a) and fibroblasts seeded into acellular mouse lungs utilize a β-1-integrin-dependent pathway. Further, a collagen I and Matrigel solution has been used as a pre-treatment to coat the decellularized lungs via the trachea before cell seeding to enhance engraftment (Jensen et al. 2012). Alternatively, cells have also been injected in a hydrogel

(Pluronic-F127) (Cortiella et al. 2010; Nichols et al. 2013). Addressing the question of how the matrix should be prepared before inoculation might be an extremely important aspect not yet explored in detail.

For organ cultivation, restoration of graft function for transplantation, and ex vivo bioengineering, there are still major hurdles to overcome. Using state-of-the-art ex vivo preservation techniques, freshly explanted organs, such as kidney and liver, can only maintain viability and function for 5 days (Butler et al. 2002; McAnulty et al. 1989). Perfusion at physiologic flow rates is needed in conjunction with the appropriate perfusates tailored in their chemical composition for lungs. This will also be essential for ex vivo recellularization strategies using acellular or synthetic scaffolds. As cells first need to be distributed by migration and likely undergo differentiation inside the matrix, optimal media composition and environmental stimuli will be crucial for ex vivo bioengineering strategies. In order to control for and adapt these stimuli to the regenerating organ during the cultivation period, a range of ancillary technologies need to be integrated into existing bioreactor technologies (e.g., sensors, pumps, and analytic and process control systems). While the road to translating acellular scaffolds into the clinic is long, steady progress has been made in this relatively young field and is encouraging for its future prospect.

References

Andrade CF, Wong AP, Waddell TK, Keshavjee S, Liu M (2007) Cell-based tissue engineering for lung regeneration. Am J Physiol Lung Cell Mol Physiol 292(2):L510–L518

Badylak SF, Gilbert TW (2008) Immune response to biologic scaffold materials. Semin Immunol 20(2):109–116

Badylak SF, Weiss DJ, Caplan A, Macchiarini P (2012) Engineered whole organs and complex tissues. Lancet 379(9819):943–952

Baiguera S, Del Gaudio C, Jaus MO, Polizzi L, Gonfiotti A, Comin CE, Bianco A, Ribatti D, Taylor DA, Macchiarini P (2012) Long-term changes to in vitro preserved bioengineered human trachea and their implications for decellularized tissues. Biomaterials 33(14):3662–3672

Bauer CM, Zavitz CC, Botelho FM, Lambert KN, Brown EG, Mossman KL, Taylor JD, Stampfli MR (2010) Treating viral exacerbations of chronic obstructive pulmonary disease: insights from a mouse model of cigarette smoke and H1N1 influenza infection. PLoS One 5(10):e13251

Bonenfant NR, Sokocevic D, Wagner DE, Borg ZD, Lathrop MJ, Lam YW, Deng B, Desarno MJ, Ashikaga T, Loi R, Weiss DJ (2013) The effects of storage and sterilization on de-cellularized and re-cellularized whole lung. Biomaterials 34(13):3231–3245

Bonvillain RW, Danchuk S, Sullivan DE, Betancourt AM, Semon JA, Eagle ME, Mayeux JP, Gregory AN, Wang G, Townley IK, Borg ZD, Weiss DJ, Bunnell BA (2012) A nonhuman primate model of lung regeneration: detergent-mediated decellularization and initial in vitro recellularization with mesenchymal stem cells. Tissue Eng Part A 18(23–24):2437–2452

Booth AJ, Hadley R, Cornett AM, Dreffs AA, Matthes SA, Tsui JL, Weiss K, Horowitz JC, Fiore VF, Barker TH, Moore BB, Martinez FJ, Niklason LE, White ES (2012) Acellular normal and fibrotic human lung matrices as a culture system for in vitro investigation. Am J Respir Crit Care Med 186(9):866–876

Boudreault F, Tschumperlin DJ (2010) Stretch-induced mitogen-activated protein kinase activation in lung fibroblasts is independent of receptor tyrosine kinases. Am J Respir Cell Mol Biol 43(1):64–73

Brown BN, Valentin JE, Stewart-Akers AM, McCabe GP, Badylak SF (2009) Macrophage phenotype and remodeling outcomes in response to biologic scaffolds with and without a cellular component. Biomaterials 30(8):1482–1491

Brown BN, Londono R, Tottey S, Zhang L, Kukla KA, Wolf MT, Daly KA, Reing JE, Badylak SF (2012a) Macrophage phenotype as a predictor of constructive remodeling following the implantation of biologically derived surgical mesh materials. Acta Biomater 8(3):978–987

Brown BN, Ratner BD, Goodman SB, Amar S, Badylak SF (2012b) Macrophage polarization: an opportunity for improved outcomes in biomaterials and regenerative medicine. Biomaterials 33(15):3792–3802

Bruinsma BG, Yarmush ML, Uygun K (2014) Organomatics and organometrics: novel platforms for long-term whole-organ culture. Technology (Singap World Sci) 2(1):13

Bryant SJ, Cuy JL, Hauch KD, Ratner BD (2007) Photo-patterning of porous hydrogels for tissue engineering. Biomaterials 28(19):2978–2986

Bussek A, Wettwer E, Christ T, Lohmann H, Camelliti P, Ravens U (2009) Tissue slices from adult mammalian hearts as a model for pharmacological drug testing. Cell Physiol Biochem 24(5–6):527–536

Butler AJ, Rees MA, Wight DG, Casey ND, Alexander G, White DJ, Friend PJ (2002) Successful extracorporeal porcine liver perfusion for 72 hr. Transplantation 73(8):1212–1218

Camelliti P, Al-Saud SA, Smolenski RT, Al-Ayoubi S, Bussek A, Wettwer E, Banner NR, Bowles CT, Yacoub MH, Terracciano CM (2011) Adult human heart slices are a multicellular system suitable for electrophysiological and pharmacological studies. J Mol Cell Cardiol 51(3): 390–398

Colom A, Galgoczy R, Almendros I, Xaubet A, Farre R, Alcaraz J (2014) Oxygen diffusion and consumption in extracellular matrix gels: implications for designing three-dimensional cultures. J Biomed Mater Res A 102(8):2776–2784

Cortiella J, Nichols JE, Kojima K, Bonassar LJ, Dargon P, Roy AK, Vacant MP, Niles JA, Vacanti CA (2006) Tissue-engineered lung: an in vivo and in vitro comparison of polyglycolic acid and pluronic F-127 hydrogel/somatic lung progenitor cell constructs to support tissue growth. Tissue Eng 12(5):1213–1225

Cortiella J, Niles J, Cantu A, Brettler A, Pham A, Vargas G, Winston S, Wang J, Walls S, Nichols JE (2010) Influence of acellular natural lung matrix on murine embryonic stem cell differentiation and tissue formation. Tissue Eng Part A 16(8):2565–2580

Crabbé A, Ledesma MA, Nickerson CA (2014) Mimicking the host and its microenvironment in vitro for studying mucosal infections by Pseudomonas aeruginosa. Pathog Dis 71(1):1–19

Crapo PM, Gilbert TW, Badylak SF (2011) An overview of tissue and whole organ decellularization processes. Biomaterials 32(12):3233–3243

Cypel M, Yeung JC, Liu M, Anraku M, Chen F, Karolak W, Sato M, Laratta J, Azad S, Madonik M, Chow CW, Chaparro C, Hutcheon M, Singer LG, Slutsky AS, Yasufuku K, de Perrot M, Pierre AF, Waddell TK, Keshavjee S (2011) Normothermic ex vivo lung perfusion in clinical lung transplantation. N Engl J Med 364(15):1431–1440

Daly AB, Wallis JM, Borg ZD, Bonvillain RW, Deng B, Ballif BA, Jaworski DM, Allen GB, Weiss DJ (2012a) Initial binding and recellularization of decellularized mouse lung scaffolds with bone marrow-derived mesenchymal stromal cells. Tissue Eng Part A 18(1–2):1–16

Daly KA, Liu S, Agrawal V, Brown BN, Johnson SA, Medberry CJ, Badylak SF (2012b) Damage associated molecular patterns within xenogeneic biologic scaffolds and their effects on host remodeling. Biomaterials 33(1):91–101

Dassow C, Wiechert L, Martin C, Schumann S, Muller-Newen G, Pack O, Guttmann J, Wall WA, Uhlig S (2010) Biaxial distension of precision-cut lung slices. J Appl Physiol 108(3):713–721

Davidovich N, Huang J, Margulies SS (2012) Reproducible uniform equibiaxial stretch of precision-cut lung slices. Am J Physiol Lung Cell Mol Physiol 304(4):L210–L220

Davidovich N, Chhour P, Margulies SS (2013) Uses of remnant human lung tissue for mechanical stretch studies. Cell Mol Bioeng 6(2):175–182

de Graaf IA, Olinga P, de Jager MH, Merema MT, de Kanter R, van de Kerkhof EG, Groothuis GM (2010) Preparation and incubation of precision-cut liver and intestinal slices for application in drug metabolism and toxicity studies. Nat Protoc 5(9):1540–1551

De Kanter R, Olinga P, De Jager MH, Merema MT, Meijer DK, Groothius GM (1999) Organ slices as an in vitro test system for drug metabolism in human liver, lung and kidney. Toxicol In Vitro 13(4–5):737–744

Delmotte P, Sanderson MJ (2006) Ciliary beat frequency is maintained at a maximal rate in the small airways of mouse lung slices. Am J Respir Cell Mol Biol 35(1):110–117

Dietl P, Liss B, Felder E, Miklavc P, Wirtz H (2010) Lamellar body exocytosis by cell stretch or purinergic stimulation: possible physiological roles, messengers and mechanisms. Cell Physiol Biochem 25(1):1–12

Douville NJ, Zamankhan P, Tung YC, Li R, Vaughan BL, Tai CF, White J, Christensen PJ, Grotberg JB, Takayama S (2011) Combination of fluid and solid mechanical stresses contribute to cell death and detachment in a microfluidic alveolar model. Lab Chip 11(4):609–619

Dunphy SE, Bratt JAJ, Akram KM, Forsyth NR, El Haj AJ (2014) Hydrogels for lung tissue engineering: biomechanical properties of thin collagen–elastin constructs. J Mech Behav Biomed Mater 38:251–259

Eisner MD, Anthonisen N, Coultas D, Kuenzli N, Perez-Padilla R, Postma D, Romieu I, Silverman EK, Balmes JR, Committee on Nonsmoking COPD, Environmental and Occupational Health Assembly (2010) An official American Thoracic Society public policy statement: novel risk factors and the global burden of chronic obstructive pulmonary disease. Am J Respir Crit Care Med 182(5):693–718

Ezzie ME, Crawford M, Cho JH, Orellana R, Zhang S, Gelinas R, Batte K, Yu L, Nuovo G, Galas D, Diaz P, Wang K, Nana-Sinkam SP (2012) Gene expression networks in COPD: microRNA and mRNA regulation. Thorax 67(2):122–131

Fernandes H, Mentink A, Bank R, Stoop R, van Blitterswijk C, de Boer J (2010) Endogenous collagen influences differentiation of human multipotent mesenchymal stromal cells. Tissue Eng Part A 16(5):1693–1702

Fischer SN, Johnson JK, Baran CP, Newland CA, Marsh CB, Lannutti JJ (2011) Organ-derived coatings on electrospun nanofibers as ex vivo microenvironments. Biomaterials 32(2): 538–546

Fishman JM, De Coppi P, Elliott MJ, Atala A, Birchall MA, Macchiarini P (2011) Airway tissue engineering. Expert Opin Biol Ther 11(12):1623–1635

Fishman JM, Lowdell MW, Urbani L, Ansari T, Burns AJ, Turmaine M, North J, Sibbons P, Seifalian AM, Wood KJ, Birchall MA, De Coppi P (2013) Immunomodulatory effect of a decellularized skeletal muscle scaffold in a discordant xenotransplantation model. Proc Natl Acad Sci U S A 110(35):14360–14365

Flozak AS, Lam AP, Russell S, Jain M, Peled ON, Sheppard KA, Beri R, Mutlu GM, Budinger GR, Gottardi CJ (2010) Beta-catenin/T-cell factor signaling is activated during lung injury and promotes the survival and migration of alveolar epithelial cells. J Biol Chem 285(5): 3157–3167

Franz S, Rammelt S, Scharnweber D, Simon JC (2011) Immune responses to implants—a review of the implications for the design of immunomodulatory biomaterials. Biomaterials 32(28): 6692–6709

Garreta E, Melo E, Navajas D, Farré R (2014) Low oxygen tension enhances the generation of lung progenitor cells from mouse embryonic and induced pluripotent stem cells. Physiol Rep 2(7):e12075

Ghaedi M, Calle EA, Mendez JJ, Gard AL, Balestrini J, Booth A, Bove PF, Gui L, White ES, Niklason LE (2013) Human iPS cell-derived alveolar epithelium repopulates lung extracellular matrix. J Clin Invest 123(11):4950–4962

Ghaedi M, Mendez JJ, Bove PF, Sivarapatna A, Raredon MS, Niklason LE (2014) Alveolar epithelial differentiation of human induced pluripotent stem cells in a rotating bioreactor. Biomaterials 35(2):699–710

Gilpin SE, Guyette JP, Gonzalez G, Ren X, Asara JM, Mathisen DJ, Vacanti JP, Ott HC (2014) Perfusion decellularization of human and porcine lungs: bringing the matrix to clinical scale. J Heart Lung Transplant 33(3):298–308

Haag J, Baiguera S, Jungebluth P, Barale D, Del Gaudio C, Castiglione F, Bianco A, Comin CE, Ribatti D, Macchiarini P (2012) Biomechanical and angiogenic properties of tissue-engineered rat trachea using genipin cross-linked decellularized tissue. Biomaterials 33(3):780–789

Hammad S, Hoehme S, Friebel A, von Recklinghausen I, Othman A, Begher-Tibbe B, Reif R, Godoy P, Johann T, Vartak A, Golka K, Bucur PO, Vibert E, Marchan R, Christ B, Dooley S, Meyer C, Ilkavets I, Dahmen U, Dirsch O, Bottger J, Gebhardt R, Drasdo D, Hengstler JG (2014) Protocols for staining of bile canalicular and sinusoidal networks of human, mouse and pig livers, three-dimensional reconstruction and quantification of tissue microarchitecture by image processing and analysis. Arch Toxicol 88(5):1161–1183

Haykal S, Soleas JP, Salna M, Hofer SO, Waddell TK (2012) Evaluation of the structural integrity and extracellular matrix components of tracheal allografts following cyclical decellularization techniques: comparison of three protocols. Tissue Eng Part C Methods 18(8):614–623

Held HD, Martin C, Uhlig S (1999) Characterization of airway and vascular responses in murine lungs. Br J Pharmacol 126(5):1191–1199

Henjakovic M, Martin C, Hoymann HG, Sewald K, Ressmeyer AR, Dassow C, Pohlmann G, Krug N, Uhlig S, Braun A (2008a) Ex vivo lung function measurements in precision-cut lung slices (PCLS) from chemical allergen-sensitized mice represent a suitable alternative to in vivo studies. Toxicol Sci 106(2):444–453

Henjakovic M, Sewald K, Switalla S, Kaiser D, Muller M, Veres TZ, Martin C, Uhlig S, Krug N, Braun A (2008b) Ex vivo testing of immune responses in precision-cut lung slices. Toxicol Appl Pharmacol 231(1):68–76

Hinderer S, Schesny M, Bayrak A, Ibold B, Hampel M, Walles T, Stock UA, Seifert M, Schenke-Layland K (2012) Engineering of fibrillar decorin matrices for a tissue-engineered trachea. Biomaterials 33(21):5259–5266

Hogan Brigid LM, Barkauskas Christina E, Chapman Harold A, Epstein Jonathan A, Jain R, Hsia Connie CW, Niklason L, Calle E, Le A, Randell Scott H, Rock J, Snitow M, Krummel M, Stripp Barry R, Vu T, White Eric S, Whitsett Jeffrey A, Morrisey Edward E (2014) Repair and regeneration of the respiratory system: complexity, plasticity, and mechanisms of lung stem cell function. Cell Stem Cell 15(2):123–138

Huang Z, Wang Y, Nayak PS, Dammann CE, Sanchez-Esteban J (2012) Stretch-induced fetal type II cell differentiation is mediated via ErbB1-ErbB4 interactions. J Biol Chem 287(22): 18091–18102

Huang SX, Islam MN, O'Neill J, Hu Z, Yang YG, Chen YW, Mumau M, Green MD, Vunjak-Novakovic G, Bhattacharya J, Snoeck HW (2014) Efficient generation of lung and airway epithelial cells from human pluripotent stem cells. Nat Biotechnol 32(1):84–91

Ingber DE (2006a) Cellular mechanotransduction: putting all the pieces together again. FASEB J 20(7):811–827

Ingber DE (2006b) Mechanical control of tissue morphogenesis during embryological development. Int J Dev Biol 50(2–3):255–266

Ingenito EP, Sen E, Tsai LW, Murthy S, Hoffman A (2010) Design and testing of biological scaffolds for delivering reparative cells to target sites in the lung. J Tissue Eng Regen Med 4(4):259–272

Ingenito EP, Tsai L, Murthy S, Tyagi S, Mazan M, Hoffman A (2012) Autologous lung-derived mesenchymal stem cell transplantation in experimental emphysema. Cell Transplant 21(1):175–189

Iwata T, Philipovskiy A, Fisher AJ, Presson RG Jr, Chiyo M, Lee J, Mickler E, Smith GN, Petrache I, Brand DB, Burlingham WJ, Gopalakrishnan B, Greenspan DS, Christie JD, Wilkes DS (2008) Anti-type V collagen humoral immunity in lung transplant primary graft dysfunction. J Immunol 181(8):5738–5747

Jensen T, Roszell B, Zang F, Girard E, Matson A, Thrall R, Jaworski DM, Hatton C, Weiss DJ, Finck C (2012) A rapid lung de-cellularization protocol supports embryonic stem cell differentiation in vitro and following implantation. Tissue Eng Part C Methods 18(8):632–646

Jungebluth P, Bader A, Baiguera S, Moller S, Jaus M, Lim ML, Fried K, Kjartansdottir KR, Go T, Nave H, Harringer W, Lundin V, Teixeira AI, Macchiarini P (2012a) The concept of in vivo airway tissue engineering. Biomaterials 33(17):4319–4326

Jungebluth P, Moll G, Baiguera S, Macchiarini P (2012b) Tissue-engineered airway: a regenerative solution. Clin Pharmacol Ther 91(1):81–93

Keane TJ, Badylak SF (2014) Biomaterials for tissue engineering applications. Semin Pediatr Surg 23(3):112–118

Krawiec JT, Vorp DA (2012) Adult stem cell-based tissue engineered blood vessels: a review. Biomaterials 33(12):3388–3400

Kuttan R, Spall RD, Duhamel RC, Sipes IG, Meezan E, Brendel K (1981) Preparation and composition of alveolar extracellular matrix and incorporated basement membrane. Lung 159(6): 333–345

Liberati TA, Randle MR, Toth LA (2010) In vitro lung slices: a powerful approach for assessment of lung pathophysiology. Expert Rev Mol Diagn 10(4):501–508

Lin YM, Boccaccini AR, Polak JM, Bishop AE, Maquet V (2006) Biocompatibility of poly-DL-lactic acid (PDLLA) for lung tissue engineering. J Biomater Appl 21(2):109–118

Lin YM, Zhang A, Rippon HJ, Bismarck A, Bishop AE (2010) Tissue engineering of lung: the effect of extracellular matrix on the differentiation of embryonic stem cells to pneumocytes. Tissue Eng Part A 16(5):1515–1526

Ling T-Y, Liu Y-L, Huang Y-K, Gu S-Y, Chen H-K, Ho C-C, Tsao P-N, Tung Y-C, Chen H-W, Cheng C-H, Lin K-H, Lin F-H (2014) Differentiation of lung stem/progenitor cells into alveolar pneumocytes and induction of angiogenesis within a 3D gelatin—microbubble scaffold. Biomaterials 35(22):5660–5669

Longmire TA, Ikonomou L, Hawkins F, Christodoulou C, Cao Y, Jean JC, Kwok LW, Mou H, Rajagopal J, Shen SS, Dowton AA, Serra M, Weiss DJ, Green MD, Snoeck HW, Ramirez MI, Kotton DN (2012) Efficient derivation of purified lung and thyroid progenitors from embryonic stem cells. Cell Stem Cell 10(4):398–411

Lopez AD, Shibuya K, Rao C, Mathers CD, Hansell AL, Held LS, Schmid V, Buist S (2006) Chronic obstructive pulmonary disease: current burden and future projections. Eur Respir J 27(2):397–412

Lwebuga-Mukasa JS, Ingbar DH, Madri JA (1986) Repopulation of a human alveolar matrix by adult rat type II pneumocytes in vitro. A novel system for type II pneumocyte culture. Exp Cell Res 162(2):423–435

Macchiarini P, Jungebluth P, Go T, Asnaghi MA, Rees LE, Cogan TA, Dodson A, Martorell J, Bellini S, Parnigotto PP, Dickinson SC, Hollander AP, Mantero S, Conconi MT, Birchall MA (2008) Clinical transplantation of a tissue-engineered airway. Lancet 372(9655):2023–2030

Majumdar A, Arold SP, Bartolak-Suki E, Parameswaran H, Suki B (2012) Jamming dynamics of stretch-induced surfactant release by alveolar type II cells. J Appl Physiol 112(5):824–831

Mariani TJ, Sandefur S, Pierce RA (1997) Elastin in lung development. Exp Lung Res 23(2):131–145

Marquardt A, Halle S, Seckert CK, Lemmermann NA, Veres TZ, Braun A, Maus UA, Forster R, Reddehase MJ, Messerle M, Busche A (2011) Single cell detection of latent cytomegalovirus reactivation in host tissue. J Gen Virol 92(Pt 6):1279–1291

Martin C, Uhlig S, Ullrich V (1996) Videomicroscopy of methacholine-induced contraction of individual airways in precision-cut lung slices. Eur Respir J 9(12):2479–2487

McAnulty JF, Ploeg RJ, Southard JH, Belzer FO (1989) Successful five-day perfusion preservation of the canine kidney. Transplantation 47(1):37–41

McBride S, Rannie D, Harrison DJ (2000) Gene transfer to adult human lung tissue ex vivo. Gene Ther 7(8):675–678

Melchels FPW, Domingos MAN, Klein TJ, Malda J, Bartolo PJ, Hutmacher DW (2012) Additive manufacturing of tissues and organs. Prog Polym Sci 37(8):1079–1104

Melo E, Cardenes N, Garreta E, Luque T, Rojas M, Navajas D, Farre R (2014a) Inhomogeneity of local stiffness in the extracellular matrix scaffold of fibrotic mouse lungs. J Mech Behav Biomed Mater 37:186–195

Melo E, Garreta E, Luque T, Cortiella J, Nichols J, Navajas D, Farre R (2014b) Effects of the decellularization method on the local stiffness of acellular lungs. Tissue Eng Part C Methods 20(5):412–422

Mendez JJ, Ghaedi M, Steinbacher D, Niklason LE (2014) Epithelial cell differentiation of human mesenchymal stromal cells in decellularized lung scaffolds. Tissue Eng Part A 20:1735–1746

Miller C, George S, Niklason L (2010) Developing a tissue-engineered model of the human bronchiole. J Tissue Eng Regen Med 4(8):619–627

Mishra DK, Thrall MJ, Baird BN, Ott HC, Blackmon SH, Kurie JM, Kim MP (2012) Human lung cancer cells grown on acellular rat lung matrix create perfusable tumor nodules. Ann Thorac Surg 93(4):1075–1081

Mondrinos MJ, Koutzaki S, Jiwanmall E, Li M, Dechadarevian JP, Lelkes PI, Finck CM (2006) Engineering three-dimensional pulmonary tissue constructs. Tissue Eng 12(4):717–728

Mondrinos MJ, Koutzaki S, Lelkes PI, Finck CM (2007) A tissue-engineered model of fetal distal lung tissue. Am J Physiol Lung Cell Mol Physiol 293(3):L639–L650

Moreno L, Perez-Vizcaino F, Harrington L, Faro R, Sturton G, Barnes PJ, Mitchell JA (2006) Pharmacology of airways and vessels in lung slices in situ: role of endogenous dilator hormones. Respir Res 7:111

Morin JP, Baste JM, Gay A, Crochemore C, Corbiere C, Monteil C (2013) Precision cut lung slices as an efficient tool for in vitro lung physio-pharmacotoxicology studies. Xenobiotica 43(1):63–72

Morrisey EE, Hogan BLM (2010) Preparing for the first breath: genetic and cellular mechanisms in lung development. Dev Cell 18(1):8–23

Murphy SV, Atala A (2014) 3D bioprinting of tissues and organs. Nat Biotechnol 32(8):773–785

Nakayama KH, Lee CCI, Batchelder CA, Tarantal AF (2013) Tissue specificity of decellularized rhesus monkey kidney and lung scaffolds. PLoS One 8(5):e64134

Nguyen NM, Senior RM (2006) Laminin isoforms and lung development: all isoforms are not equal. Dev Biol 294(2):271–279

Nguyen DT, de Vries RD, Ludlow M, van den Hoogen BG, Lemon K, van Amerongen G, Osterhaus AD, de Swart RL, Duprex WP (2013) Paramyxovirus infections in ex vivo lung slice cultures of different host species. J Virol Methods 193(1):159–165

Nichols JE, Cortiella J (2008) Engineering of a complex organ: progress toward development of a tissue-engineered lung. Proc Am Thorac Soc 5(6):723–730

Nichols JE, Niles J, Riddle M, Vargas G, Schilagard T, Ma L, Edward K, La Francesca S, Sakamoto J, Vega S, Ogadegbe M, Mlcak R, Deyo D, Woodson L, McQuitty C, Lick S, Beckles D, Melo E, Cortiella J (2013) Production and assessment of decellularized pig and human lung scaffolds. Tissue Eng Part A 19(17–18):2045–2062

Nonaka PN, Uriarte JJ, Campillo N, Melo E, Navajas D, Farre R, Oliveira LV (2014) Mechanical properties of mouse lungs along organ decellularization by sodium dodecyl sulfate. Respir Physiol Neurobiol 200:1–5

O'Neill JD, Anfang R, Anandappa A, Costa J, Javidfar J, Wobma HM, Singh G, Freytes DO, Bacchetta MD, Sonett JR, Vunjak-Novakovic G (2013) Decellularization of human and porcine lung tissues for pulmonary tissue engineering. Ann Thorac Surg 96(3):1046–1055, discussion 1055–1046

op den Dries S, Karimian N, Sutton ME, Westerkamp AC, Nijsten MW, Gouw AS, Wiersema-Buist J, Lisman T, Leuvenink HG, Porte RJ (2013) Ex vivo normothermic machine perfusion and viability testing of discarded human donor livers. Am J Transplant 13(5):1327–1335

Orlando G, Baptista P, Birchall M, De Coppi P, Farney A, Guimaraes-Souza NK, Opara E, Rogers J, Seliktar D, Shapira-Schweitzer K, Stratta RJ, Atala A, Wood KJ, Soker S (2011) Regenerative medicine as applied to solid organ transplantation: current status and future challenges. Transpl Int 24(3):223–232

Ott HC, Matthiesen TS, Goh SK, Black LD, Kren SM, Netoff TI, Taylor DA (2008) Perfusion-decellularized matrix: using nature's platform to engineer a bioartificial heart. Nat Med 14(2):213–221

Ott HC, Clippinger B, Conrad C, Schuetz C, Pomerantseva I, Ikonomou L, Kotton D, Vacanti JP (2010) Regeneration and orthotopic transplantation of a bioartificial lung. Nat Med 16(8): 927–933

Parker MW, Rossi D, Peterson M, Smith K, Sikström K, White ES, Connett JE, Henke CA, Larsson O, Bitterman PB (2014) Fibrotic extracellular matrix activates a profibrotic positive feedback loop. J Clin Invest 124(4):1622–1635

Parrish AR, Gandolfi AJ, Brendel K (1995) Precision-cut tissue slices: applications in pharmacology and toxicology. Life Sci 57(21):1887–1901

Petersen TH, Calle EA, Zhao L, Lee EJ, Gui L, Raredon MB, Gavrilov K, Yi T, Zhuang ZW, Breuer C, Herzog E, Niklason LE (2010) Tissue-engineered lungs for in vivo implantation. Science 329(5991):538–541

Petersen TH, Calle EA, Colehour MB, Niklason LE (2011) Bioreactor for the long-term culture of lung tissue. Cell Transplant 20(7):1117–1126

Petersen TH, Calle EA, Colehour MB, Niklason LE (2012) Matrix composition and mechanics of decellularized lung scaffolds. Cells Tissues Organs 195(3):222–231

Pilaz LJ, Silver DL (2014) Live imaging of mitosis in the developing mouse embryonic cortex. J Vis Exp (88). doi:10.3791/51298

Price AP, England KA, Matson AM, Blazar BR, Panoskaltsis-Mortari A (2010) Development of a decellularized lung bioreactor system for bioengineering the lung: the matrix reloaded. Tissue Eng Part A 16(8):2581–2591

Price AP, Godin LM, Domek A, Cotter T, D'Cunha J, Taylor DA, Panoskaltsis-Mortari A (2015) Automated decellularization of intact, human-sized lungs for tissue engineering. Tissue Eng Part C Methods 21(1):94–103

Rausch SM, Haberthur D, Stampanoni M, Schittny JC, Wall WA (2011) Local strain distribution in real three-dimensional alveolar geometries. Ann Biomed Eng 39(11):2835–2843

Ressmeyer AR, Larsson AK, Vollmer E, Dahlen SE, Uhlig S, Martin C (2006) Characterisation of guinea pig precision-cut lung slices: comparison with human tissues. Eur Respir J 28(3):603–611

Rice WL, Van Hoek AN, Paunescu TG, Huynh C, Goetze B, Singh B, Scipioni L, Stern LA, Brown D (2013) High resolution helium ion scanning microscopy of the rat kidney. PLoS One 8(3):e57051

Rippon HJ, Polak JM, Qin M, Bishop AE (2006) Derivation of distal lung epithelial progenitors from murine embryonic stem cells using a novel three-step differentiation protocol. Stem Cells 24(5):1389–1398

Roomans GM (2010) Tissue engineering and the use of stem/progenitor cells for airway epithelium repair. Eur Cell Mater 19:284–299

Sanchez-Esteban J, Tsai SW, Sang J, Qin J, Torday JS, Rubin LP (1998) Effects of mechanical forces on lung-specific gene expression. Am J Med Sci 316(3):200–204

Sanchez-Esteban J, Cicchiello LA, Wang Y, Tsai SW, Williams LK, Torday JS, Rubin LP (2001) Mechanical stretch promotes alveolar epithelial type II cell differentiation. J Appl Physiol 91(2):589–595

Sanderson MJ (2011) Exploring lung physiology in health and disease with lung slices. Pulm Pharmacol Ther 24(5):452–465

Satoh K, Takahashi G, Miura T, Hayakari M, Hatayama I (2005) Enzymatic detection of precursor cell populations of preneoplastic foci positive for gamma-glutamyltranspeptidase in rat liver. Int J Cancer 115(5):711–716

Schleputz M, Uhlig S, Martin C (2011) Electric field stimulation of precision-cut lung slices. J Appl Physiol 110(2):545–554

Schleputz M, Rieg AD, Seehase S, Spillner J, Perez-Bouza A, Braunschweig T, Schroeder T, Bernau M, Lambermont V, Schlumbohm C, Sewald K, Autschbach R, Braun A, Kramer BW, Uhlig S, Martin C (2012) Neurally mediated airway constriction in human and other species: a comparative study using precision-cut lung slices (PCLS). PLoS One 7(10):e47344

Schnorbusch K, Lembrechts R, Brouns I, Pintelon I, Timmermans JP, Adriaensen D (2012) Precision-cut vibratome slices allow functional live cell imaging of the pulmonary neuroepithelial body microenvironment in fetal mice. Adv Exp Med Biol 758:157–166

Seehase S, Schleputz M, Switalla S, Matz-Rensing K, Kaup FJ, Zoller M, Schlumbohm C, Fuchs E, Lauenstein HD, Winkler C, Kuehl AR, Uhlig S, Braun A, Sewald K, Martin C (2011) Bronchoconstriction in non-human primates: a species comparison. J Appl Physiol 111:791–798

Shamis Y, Hasson E, Soroker A, Bassat E, Shimoni Y, Ziv T, Sionov RV, Mitrani E (2011) Organ-specific scaffolds for in vitro expansion, differentiation, and organization of primary lung cells. Tissue Eng Part C Methods 17(8):861–870

Shaw AS, Filbert EL (2009) Scaffold proteins and immune-cell signalling. Nat Rev Immunol 9(1):47–56

Shim K (2011) Vibratome sectioning for enhanced preservation of the cytoarchitecture of the mammalian organ of Corti. J Vis Exp (52). pii: 2793. doi:10.3791/2793

Shweiki D, Itin A, Soffer D, Keshet E (1992) Vascular endothelial growth factor induced by hypoxia may mediate hypoxia-initiated angiogenesis. Nature 359(6398):843–845

Simon MC, Keith B (2008) The role of oxygen availability in embryonic development and stem cell function. Nat Rev Mol Cell Biol 9(4):285–296

Smith PF, Gandolfi AJ, Krumdieck CL, Putnam CW, Zukoski CF III, Davis WM, Brendel K (1985) Dynamic organ culture of precision liver slices for in vitro toxicology. Life Sci 36(14):1367–1375

Sokocevic D, Bonenfant NR, Wagner DE, Borg ZD, Lathrop MJ, Lam YW, Deng B, Desarno MJ, Ashikaga T, Loi R, Hoffman AM, Weiss DJ (2013) The effect of age and emphysematous and fibrotic injury on the re-cellularization of de-cellularized lungs. Biomaterials 34(13): 3256–3269

Song JJ, Kim SS, Liu Z, Madsen JC, Mathisen DJ, Vacanti JP, Ott HC (2011) Enhanced in vivo function of bioartificial lungs in rats. Ann Thorac Surg 92(3):998–1005, discussion 1005–1006

Sun H, Calle E, Chen X, Mathur A, Zhu Y, Mendez J, Zhao L, Niklason L, Peng X, Peng H, Herzog EL (2014) Fibroblast engraftment in the decellularized mouse lung occurs via a beta1-integrin-dependent, FAK-dependent pathway that is mediated by ERK and opposed by AKT. Am J Physiol Lung Cell Mol Physiol 306(6):L463–L475

Totonelli G, Maghsoudlou P, Garriboli M, Riegler J, Orlando G, Burns AJ, Sebire NJ, Smith VV, Fishman JM, Ghionzoli M, Turmaine M, Birchall MA, Atala A, Soker S, Lythgoe MF, Seifalian A, Pierro A, Eaton S, De Coppi P (2012) A rat decellularized small bowel scaffold that preserves villus-crypt architecture for intestinal regeneration. Biomaterials 33(12):3401–3410

Tsunooka N, Hirayama S, Medin JA, Liles WC, Keshavjee S, Waddell TK (2011) A novel tissue-engineered approach to problems of the postpneumonectomy space. Ann Thorac Surg 91(3):880–886

Vaira V, Fedele G, Pyne S, Fasoli E, Zadra G, Bailey D, Snyder E, Faversani A, Coggi G, Flavin R, Bosari S, Loda M (2010) Preclinical model of organotypic culture for pharmacodynamic profiling of human tumors. Proc Natl Acad Sci U S A 107(18):8352–8356

Wagner DE, Bonvillain RW, Jensen T, Girard ED, Bunnell BA, Finck CM, Hoffman AM, Weiss DJ (2013) Can stem cells be used to generate new lungs? Ex vivo lung bioengineering with decellularized whole lung scaffolds. Respirology 18(6):895–911

Wagner DE, Bonenfant NR, Parsons CS, Sokocevic D, Brooks EM, Borg ZD, Lathrop MJ, Wallis JD, Daly AB, Lam YW, Deng B, DeSarno MJ, Ashikaga T, Loi R, Weiss DJ (2014a) Comparative decellularization and recellularization of normal versus emphysematous human lungs. Biomaterials 35(10):3281–3297

Wagner DE, Bonenfant NR, Sokocevic D, DeSarno MJ, Borg ZD, Parsons CS, Brooks EM, Platz JJ, Khalpey ZI, Hoganson DM, Deng B, Lam YW, Oldinski RA, Ashikaga T, Weiss DJ (2014b) Three-dimensional scaffolds of acellular human and porcine lungs for high throughput studies of lung disease and regeneration. Biomaterials 35(9):2664–2679

Wagner DE, Fenn S, Bonenfant N, Marks E, Borg Z, Saunders P, Oldinski R, Weiss D (2014c) Design and synthesis of an artificial pulmonary pleura for high throughput studies in acellular human lungs. Cell Mol Bioeng 7(2):184–195

Wallis JM, Borg ZD, Daly AB, Deng B, Ballif BA, Allen GB, Jaworski DM, Weiss DJ (2012) Comparative assessment of detergent-based protocols for mouse lung de-cellularization and re-cellularization. Tissue Eng Part C Methods 18(6):420–432

Wang H, Zhu J, Akkin T (2014) Serial optical coherence scanner for large-scale brain imaging at microscopic resolution. Neuroimage 84:1007–1017

Weibel ER (2013) It takes more than cells to make a good lung. Am J Respir Crit Care Med 187(4):342–346

Wertheim JA, Baptista PM, Soto-Gutierrez A (2012) Cellular therapy and bioartificial approaches to liver replacement. Curr Opin Organ Transplant 17(3):235–240

Wohlsen A, Martin C, Vollmer E, Branscheid D, Magnussen H, Becker WM, Lepp U, Uhlig S (2003) The early allergic response in small airways of human precision-cut lung slices. Eur Respir J 21(6):1024–1032

Wright JL, Churg A (2008) Short-term exposure to cigarette smoke induces endothelial dysfunction in small intrapulmonary arteries: analysis using guinea pig precision cut lung slices. J Appl Physiol 104(5):1462–1469

Wyatt TA, Sisson JH, Allen-Gipson DS, McCaskill ML, Boten JA, DeVasure JM, Bailey KL, Poole JA (2012) Co-exposure to cigarette smoke and alcohol decreases airway epithelial cell cilia beating in a protein kinase C epsilon-dependent manner. Am J Pathol 181(2):431–440

Zhou J, Alvarez-Elizondo MB, Botvinick E, George SC (2012) Local small airway epithelial injury induces global smooth muscle contraction and airway constriction. J Appl Physiol 112(4):627–637

Zhou Y, Peng H, Sun H, Peng X, Tang C, Gan Y, Chen X, Mathur A, Hu B, Slade MD, Montgomery RR, Shaw AC, Homer RJ, White ES, Lee C-M, Moore MW, Gulati M, Geun Lee C, Elias JA, Herzog EL (2014) Chitinase 3-like 1 suppresses injury and promotes fibroproliferative responses in mammalian lung fibrosis. Sci Transl Med 6(240):240ra276

Part II
Pulmonary Vasculature Stem Cells

Chapter 9
Endothelial Colony-Forming Cells in Pulmonary Arterial Hypertension

Kewal Asosingh, Jonathan Rose, and Serpil Erzurum

Abbreviations

CD	Cluster designation
CXCR-4	Chemokine (C-X-C motif) receptor 4
DiI-Ac-LDL	1,1′-dioctadecyl-3,3,3′,3′-tetramethyl-indocarbocyanine perchlorate acetylated-low density lipoprotein
ECFC	Endothelial colony-forming cells
EGM-2	Endothelial growth medium-2
eNOS	Endothelial nitric oxide synthase
Flt-1	FMS-like tyrosine kinas-1
HUVEC	Human umbilical vein endothelial cell
KDR	Kinase insert domain receptor
lacZ	Lactose operon-Z
NOD/SCID	Nonobese diabetic/severe combined immunodeficiency
SMC	Smooth muscle cells
VE-cadherin	Vascular endothelium-cadherin
vWF	von Willebrand factor
WKYMVm	Stands for peptide strand structure Trp-Lys-Tyr-Met-Val-D-Met-NH_2

K. Asosingh, Ph.D. (✉) • J. Rose
Department of Pathobiology, Lerner Research Institute, Cleveland Clinic
Lerner College of Medicine of Case Western Reserve University,
NC22, 9500 Euclid Ave, Cleveland, OH 44195, USA
e-mail: asosink@ccf.org

S. Erzurum, M.D. (✉)
Department of Pathobiology, Lerner Research Institute and Respiratory Institute, Cleveland Clinic Lerner College of Medicine of Case Western Reserve University,
NC22, 9500 Euclid Ave, Cleveland, OH 44195, USA
e-mail: erzurus@ccf.org

A. Firth, J.X.-J. Yuan (eds.), *Lung Stem Cells in the Epithelium and Vasculature*,
Stem Cell Biology and Regenerative Medicine, DOI 10.1007/978-3-319-16232-4_9

9.1 Definition and Origins of Endothelial Colony-Forming Cells

Endothelial colony forming-like cells were first described as the endothelial progenitor cells (EPCs) that were isolated from the peripheral blood mononuclear cell fraction 2–4 weeks after plating in specialized endothelial growth medium, i.e., late outgrowth EPC (Hur et al. 2004). These cells replicated rapidly to form a monolayer and sustained multiple population doublings without senescence. Late EPCs were resistant to apoptosis under conditions of serum starvation and exhibited strong expression of endothelial-specific genes such as vascular endothelium (VE)-cadherin, fms-related tyrosine kinase 1 (Flt-1), kinase insert domain receptor (KDR), endothelial nitric oxide synthase (eNOS), and von Willebrand factor (vWF). Ex vivo robust endothelial tube formation was observed on matrigel, and in vivo late EPCs were able to rescue mice from hindleg ischemic injury.

The terminology "endothelial colony forming cells or ECFCs" was introduced by Ingram et al. (2004). This meticulously detailed landmark study characterized circulating cells with the potential to give rise to endothelial cells. Highly proliferative cells capable of self-renewal and de novo vessel formation were detected in the mononuclear cell fraction of peripheral blood or umbilical cord blood after 5–22 days of culture in Endothelial Growth Medium-2 on rat tail collagen-I coated plates. ECFCs were described as well-circumscribed colonies of 50 or more cells with a cobblestone morphology, which expressed endothelial cell-surface antigens CD31, CD141, CD105, CD146, CD144, vWF, and flk-1, were negative for hematopoietic cell-specific surface antigens CD45 and CD14, incorporated acetylated low density lipoprotein, and could form capillary-like structures in Matrigel. Furthermore, ECFCs impregnated in a mixture of collagen and fibronectin gel, which were xenografted subcutaneously in immune-deficient mice, formed de novo blood vessels (Yoder et al. 2007). A complete hierarchy of ECFCs was revealed based on the proliferation capacity of the cells (Fig. 9.1). The most primitive high proliferative potential endothelial colony-forming cells (HPP-ECFCs) were defined by large colonies that formed secondary and tertiary HPP-ECFCs and low proliferative potential ECFCs (LPP-ECFCs) colonies on replating. These cells were enriched in umbilical cord blood and gave rise to secondary HPP-ECFCs. ECFCs with more than 50 cells, but unable to form secondary colonies on replating, were referred to as LPP-ECFCs. In a subsequent study (Ingram et al. 2005), ECFCs were similarly identified using proliferation hierarchy of single endothelial cells sorted from cultures of human umbilical vein or human aorta endothelium.

The ancestor cell of ECFCs is unknown. During development, endothelial cells originate from the hemangioblast, a common hematopoietic/endothelial stem cell (Samokhvalov et al. 2007; Nishikawa et al. 1998; Tavian et al. 2001; de Bruijn et al. 2002). Two detailed studies analyzed whether ECFCs are derived from this bi-potent stem cells during postnatal life. Isolation of circulating ECFCs from patients with clonal hematopoietic stem cell disorders showed that 3 of 89 ECFCs obtained from 11 patients carried a mutation identical to the hematopoietic stem cell, providing

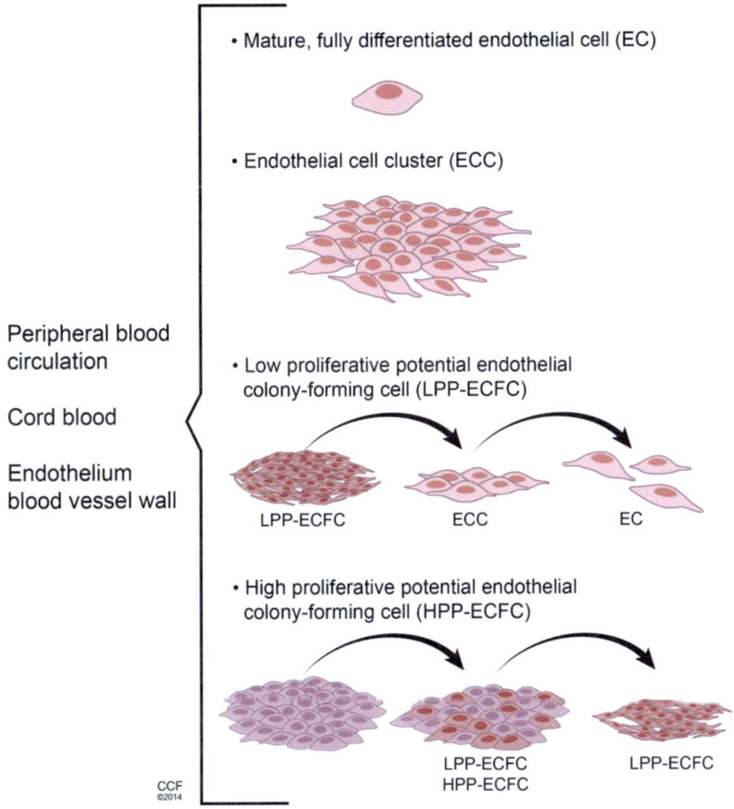

Fig. 9.1 Classification of endothelial cell colonies. Endothelial cell colonies are isolated from mononuclear cell fractions of cord blood, peripheral blood circulation, or derived from endothelial cell cultures after single cell sorting. Mature, fully differentiated endothelial cells (ECs) are non-proliferating individual cells. Endothelial cell clusters (ECCs) comprise colonies of 2–50 cells. Low proliferative potential endothelial colony forming cells (LPP-ECFC) contain more than 50 cells that form ECCs on replating. High proliferative potential endothelial colony forming cells (HPP-ECFC) are macroscopic colonies capable of forming secondary HPP-ECFC and LPP-ECFC. HPP-ECFCs maintain high telomerase activity. LPP-ECFCs and HPP-ECFCs typically contain cells that have a smaller size compared to EC or ECC

evidence for a common precursor for circulating ECFCs and hematopoietic stem cells for at least a minority of ECFCs in the peripheral blood (Yoder et al. 2007). On the other hand, in a larger cohort of 76 patients with clonal myeloproliferative disorders, all ECFCs were mutation negative (Piaggio et al. 2009). Study of circulating ECFCs in non-human primates over the lifespan from birth to elderly showed that clonogenic and proliferation capacity and in vivo vasculogenesis potential significantly decreased with age, suggesting that ECFC mobilization and function are adversely affected by aging (Shelley et al. 2012).

9.2 Mechanisms of Mobilization, Homing, and Angiogenic Function

ECFCs are mobilized and recruited to ischemic sites to contribute to neovasculogenesis and repair of injured tissues. However, little is known about the mechanisms via which ECFCs are released from the endothelium and become blood-borne. Whether mobilization of ECFCs occurs by passive shedding from the endothelium or an active migration out of the endothelium, and the molecular activators of this process have yet to be revealed.

Interestingly, in patients with acute myocardial infarction circulating ECFCs numbers were higher than cord blood, the richest healthy source of circulating ECFCs, suggesting that ischemia/injury governs the mobilization of these cells (Massa et al. 2009). Tracing of intracoronary-delivered lacZ-labeled ECFCs into acute myocardial infracted pigs showed specific incorporation of the cells into the ischemic/injured area and not in other organs, except for the splenic sinusoids (Dubois et al. 2010). C-X-C chemokine receptor type 4 (CXCR-4) expression on ECFCs played a critical role in ex vivo migration of the cells towards stromal cell-derived factor-1α (SDF-1α). Sorting of cord blood and human umbilical cord ECFCs based on the level of CXCR-4 expression showed that the cells displaying high numbers of receptors homed more efficiently to ischemic sites in mouse hindlimb injury model and exhibited increased vasculogenesis potential compared to ECFCs with low-CXCR4 expression (Oh et al. 2010). The expression of CXCR-4 on cord and adult blood ECFC is suppressed by thrombospondin-1, an endogenous angiogenesis inhibitor. Blocking of thrombospondin-1 in ECFCs resulted in increased migration in vitro and boosted the angiogenic potential of these cells in the hindlimb ischemia model (Smadja et al. 2011a).

More recently, synthetic peptide WKYMVm, a selective agonist of G-protein-coupled formyl peptide receptor-2, was identified as potent promoter of homing and incorporation of intravenously injected cord blood ECFCs into ischemic limb in a mouse hindleg ischemia model. Activation of the formyl peptide receptor-2 on the cell surface of ECFCs by WKYMVm augmented migration in scratch wound assays, cell proliferation, and in vitro tube formation in Matrigel assays. Intramuscular injection of WKYMVm at the site of injury increased ECFC homing and engraftment into newly formed blood vessels resulting in greater blood perfusion and limb salvage (Heo et al. 2014).

ECFC function is critically dependent on migration and luminalization. Vascular endothelial cell growth factor (VEGF)-induced migration and in vivo capillary formation by cord blood-derived ECFCs are critically regulated by urokinase-type plasminogen activator receptor (uPAR), which has an established role in extracellular matrix degradation and cell adhesion (Margheri et al. 2011). Lumen formation is a critical step in neovascularization. Angiopoietin-like protein 2 has recently been shown to control ECFC luminalization in three dimensional collagen assays by activation of c-Jun NH_2-terminal kinase and up-regulation of membrane type I matrix metalloproteinase (Richardson et al. 2014). There is conclusive evidence that ECFCs are able to form de novo endothelium in newly formed blood vessels.

ECFCs in collagen-1/fibronectin or Matrigel implanted subcutaneously into NODSCID mice promoted vascularization (Yoder et al. 2007; Melero-Martin et al. 2007). In addition to structural contribution to the nascent endothelium, ECFCs also stimulate new vessel formation in a paracrine fashion by activation of local ECFCs (Alphonse et al. 2014; Baker et al. 2013). The soluble factors driving this paracrine angiogenic activity remain to be determined.

9.3 ECFCs in Healthy Lungs and Pulmonary Arterial Hypertension

ECFCs with robust self-renewal and de novo vessel formation potential have been isolated from neonatal rat lungs and human fetal lungs at 17–22 weeks of gestation, indicating that these cells are critical in alveolar development (Alphonse et al. 2014). ECFC-like cells with high proliferation capacity, self-regeneration potential, and in vivo vasculogenesis activity were also found to be enriched in adult mouse or rat lung mircovascular endothelium as compared to the pulmonary artery (Alvarez et al. 2008; Schniedermann et al. 2010).

Pulmonary arterial hypertension (PAH) is characterized by severe angiogenic remodeling of the pulmonary circulation (Heath and Edwards 1958; Tuder et al. 2007). Emerging evidence suggests that ECFCs may play an active role in the angiogenic remodeling (Alphonse et al. 2014; Duong et al. 2011; Toshner et al. 2009). Both healthy and PAH pulmonary artery endothelial cells harbor similar numbers of ECFCs (Duong et al. 2011). However, ECFCs from PAH PAECs exhibited significantly higher cell numbers in culture and gave rise to more HPP-ECFCs (Duong et al. 2011) and increased vasculogenesis in vivo (Fig. 9.2). These findings demonstrate that ECFCs are enriched among adult PAECs. In PAH, ECFC frequency in the pulmonary endothelium is unaltered, but the expansion capacity of these ECFCs is amplified. In children with PAH, ECFCs have been isolated from the peripheral blood mononuclear cell fraction (Smadja et al. 2011b). Their numbers, cell proliferation, and vascular injury repair capacity in a nude mouse hindleg ischemia model increased after prostacyclin therapy, suggesting that this class of drug may provide clinical benefits to patients by enhancing ECFC function (Smadja et al. 2011b). Circulating ECFCs cultured from adult PAH patients had a higher proliferation rate compared to ECFCs obtained from healthy individuals and impaired angiogenic capacity in ex vivo Matrigel assays (Toshner et al. 2009). However, enthusiasm that peripheral blood circulation may be a useful source of pulmonary circulation-derived ECFCs was dampened by analysis of circulating ECFCs from a pair of monozygotic twins carrying bone morphogenetic protein type 2 receptor (BMPR2) mutation. One of the twins received a heart and double lung transplant for severe PAH. Five years after the transplantation of the afflicted twin, ECFCs from both twins were analyzed for host or donor origin by sequencing of the specific BMPR2 mutation. Circulating ECFCs from both twins after transplantation carried equal amounts of the mutated allele, demonstrating that they did not come from the donor lungs or heart. This case study

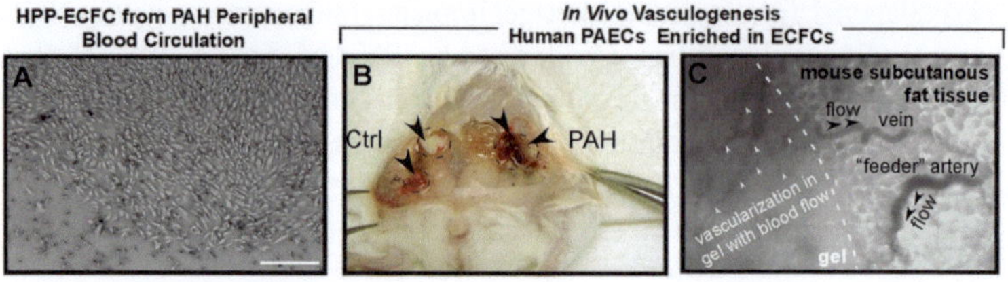

Fig. 9.2 ECFC in pulmonary arterial hypertension blood circulation and pulmonary artery endothelium. (**a**) HPP-ECFC obtained from PAH patient's peripheral blood. Bar = 200 μm. (**b**) Increased de novo vasculogenesis by PAH PAECs enriched in HPP-ECFCs. Healthy control PAECs and PAH PAECs, enriched in ECFCs, were inoculated in collagen-I/fibronectin extracellular matrix gel. After overnight incubation at 37 °C in EGM-2 medium, the gels were surgically implanted into the abdominal wall, between the dermis and peritoneum of humanized NODSCID mice (Yoder et al. 2007). In contrast to the traditional matrigel assay, this technique allows the study of vasculogenic capacity of ECFCs without invasion of the gel by mouse cells. Animals were analyzed after 1 week. (**c**) Sidestream dark field (SDF) imaging (MicroScanTM, MicroVisionMedical, Wallingford, CT) was used to visualize vasculogenesis in the gel. This is a newly available technique in which tissue-penetrating green light (530 nm) is emitted by a probe positioned on top of the collagen-I gel in an anesthetized mouse. The light is absorbed by the hemoglobin in the red blood cells. Stroboscopic illumination and capturing of the reflected light allows the observation of flowing cells. A captioned still image of this video is shown. Dark network (*white arrows*) indicates newly formed blood vessels in the gel connected to the mouse circulation

suggested that ECFCs in the peripheral blood circulation may not be derived from a pulmonary or cardiac origin (Ormiston et al. 2013).

Animal models support that ECFCs may play a role in PAH. Studies of chronic hypobaric hypoxic calves revealed increased frequency of HPP- and LPP-ECFCs in the adventitial *vasa vasorum* of the thickened pulmonary artery wall compared to normoxic calves. Of note, *vasa vasorum* ECFCs' number in hypoxic animals was considerably higher compared to the frequency among aortic and main pulmonary artery endothelial cells. Furthermore, in Matrigel tube formation assays, *vasa vasorum* ECFCs from hypoxic calves exhibited increased angiogenesis compared to control ECFCs (Nijmeh et al. 2014).

The greater presence of highly proliferation angiogenic ECFCs among pulmonary artery endothelial cells and pulmonary artery *vasa vasorum* endothelial cells suggests that these cells may play an active role in the pathological angiogenesis of PAH (Fig. 9.3).

9.4 Therapeutic Potential of ECFCs in Cardiovascular Diseases

Several lines of evidence substantiate that ECFCs may hold promise for cell therapy. Infusion of autologous ECFCs in pigs, 1 week after the induction of experimental acute myocardial infarction, resulted in attenuated pathological remodeling and

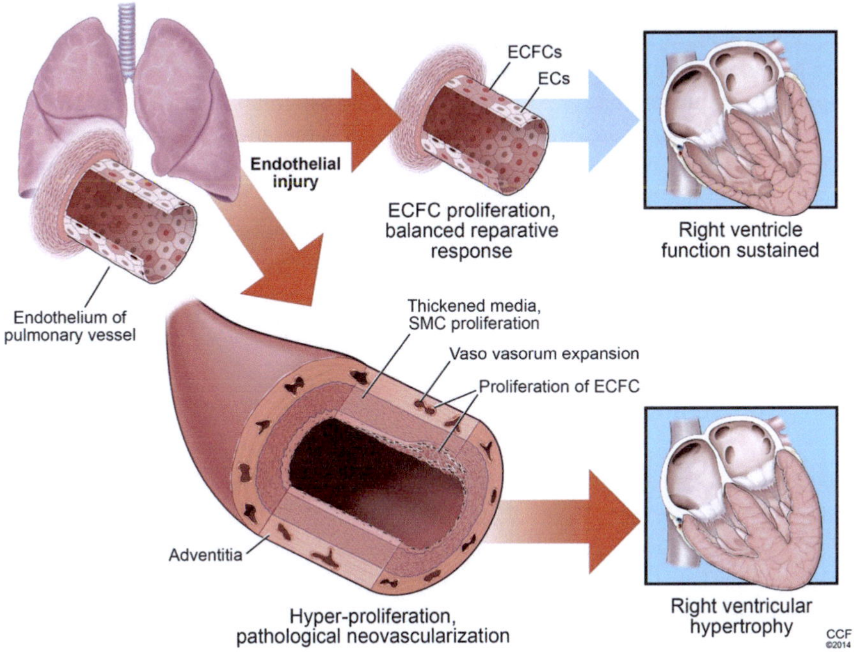

Fig. 9.3 ECFCs in pulmonary arterial hypertension. Our knowledge of ECFC biology in lung health or disease is at a primordial stage. Likewise, very little is known about the exact function of ECFCs in pulmonary arterial hypertension. A following working model could be proposed based on the current literature. ECFCs play both beneficial and pathological roles in pulmonary arterial hypertension. After an initial insult to the pulmonary artery endothelium, ECFCs are activated to repair the injury. Both, resident and circulating, ECFCs may participate in this process. In pulmonary arterial hypertension with intrinsic impaired ECFC function, prostacyclin therapy or administration soluble factors derived from healthy ECFCs (such as cord blood) could be used to restore ECFC activity to promote a reparative response. In situations with repetitive injury or uncontrolled propagation of the reparative response, proliferation of ECFCs in the pulmonary artery is increased contributing to pathological remodeling by expansion of *vasa vasorum* in the pulmonary adventitia wall and formation of angioproliferative lesions in the pulmonary endothelium

increased vascularization of the infracted zones with neoendothelium formation by the ECFCs and improvement of left ventricle function (Dubois et al. 2010). In a rabbit model of wire injury-induced intima hyperplasia, transfusion of autologous ECFCs inhibited neointima formation by facilitating restoration of the damaged endothelium (Liu et al. 2011). ECFCs isolated from hyperoxia-exposed neonatal rats or human fetal lungs subjected to hyperoxia ex vivo showed impaired proliferation, self-renewal, and reduced in vitro tube formation on Matrigel, all of which were restored after intra-jugular administration of cord blood ECFCs in the hyperoxic neonatal rats (Alphonse et al. 2014). Despite this therapeutic effect, a poor engraftment of the cord blood ECFCs in neonatal rats was observed and similar beneficial restoration of alveolar growth was obtained by infusion of

ECFC-conditioned medium, indicating a potential *paracrine* mechanism (Alphonse et al. 2014). Interestingly, bronchopulmonary dysplasia-induced pulmonary hypertension in the hyperoxic newborn rats, including right ventricular hypertrophy and pulmonary artery acceleration time, a sonographic surrogate measure of pulmonary artery pressure, was attenuated by ECFC therapy (Alphonse et al. 2014). Similar restorative benefits were observed with cord blood ECFCs and cord blood ECFC-conditioned medium in bleomycin-exposed newborn rats, another experimental model for bronchopulmonary dysplasia (Baker et al. 2013). In this study, cord blood ECFC-conditioned medium inhibited right ventricular hypertrophy, but did not contribute to alveolarization

9.5 Conclusion

The identification of ECFCs has provided deeper understanding of pulmonary endothelial biology, but much work is needed to decipher the precise role(s) of ECFCs in lung physiology and disease. In vivo studies have been hampered by the lack of biomarkers to distinguish vasculogenic ECFCs from mature, fully differentiated endothelial cells. Further studies are necessary to analyze how ECFC function evolves during lung vascular disease and what are the molecular determinants that control the balance between a reparative response and pathological remodeling.

Acknowledgments Illustration by David Schumick, B.S., CMI. Reprinted with the permission of the Cleveland Clinic Center for Medical Art and Photography © 2014. All Rights Reserved. Kewal Asosingh is a Scholar of the International Society for Advancement of Cytometry (ISAC). This work was supported by grants HL60917, HL115008, and M01 RR018390 from the National Institutes of Health, American Thoracic Society/Pulmonary Association Research grant (PH-07-003), and the Hematopoietic Stem Cell Core Facility of the Case Comprehensive Cancer Center (P30 CA43703).

References

Alphonse RS, Vadivel A, Fung M, Shelley WC, Critser PJ, Ionescu L, O'Reilly M, Ohls RK, McConaghy S, Eaton F, Zhong S, Yoder M, Thebaud B (2014) Existence, functional impairment and lung repair potential of endothelial colony forming cells in oxygen-induced arrested alveolar growth. Circulation 129(21):2144–2157. doi:10.1161/CIRCULATIONAHA.114.009124

Alvarez DF, Huang L, King JA, ElZarrad MK, Yoder MC, Stevens T (2008) Lung microvascular endothelium is enriched with progenitor cells that exhibit vasculogenic capacity. Am J Physiol Lung Cell Mol Physiol 294(3):L419–L430. doi:10.1152/ajplung.00314.2007

Baker CD, Seedorf GJ, Wisniewski BL, Black CP, Ryan SL, Balasubramaniam V, Abman SH (2013) Endothelial colony-forming cell conditioned media promote angiogenesis in vitro and prevent pulmonary hypertension in experimental bronchopulmonary dysplasia. Am J Physiol Lung Cell Mol Physiol 305(1):L73–L81. doi:10.1152/ajplung.00400.2012

de Bruijn MF, Ma X, Robin C, Ottersbach K, Sanchez MJ, Dzierzak E (2002) Hematopoietic stem cells localize to the endothelial cell layer in the midgestation mouse aorta. Immunity 16(5):673–683

Dubois C, Liu X, Claus P, Marsboom G, Pokreisz P, Vandenwijngaert S, Depelteau H, Streb W, Chaothawee L, Maes F, Gheysens O, Debyser Z, Gillijns H, Pellens M, Vandendriessche T, Chuah M, Collen D, Verbeken E, Belmans A, Van de Werf F, Bogaert J, Janssens S (2010) Differential effects of progenitor cell populations on left ventricular remodeling and myocardial neovascularization after myocardial infarction. J Am Coll Cardiol 55(20):2232–2243. doi:10.1016/j.jacc.2009.10.081

Duong HT, Comhair SA, Aldred MA, Mavrakis L, Savasky BM, Erzurum SC, Asosingh K (2011) Pulmonary artery endothelium resident endothelial colony-forming cells in pulmonary arterial hypertension. Pulm Circ 1(4):475–486. doi:10.4103/2045-8932.93547PC-1-475

Heath D, Edwards JE (1958) The pathology of hypertensive pulmonary vascular disease; a description of six grades of structural changes in the pulmonary arteries with special reference to congenital cardiac septal defects. Circulation 18(4 Part 1):533–547

Heo SC, Kwon YW, Jang IH, Jeong GO, Yoon JW, Kim CD, Kwon SM, Bae YS, Kim JH (2014) WKYMVm-induced activation of formyl peptide receptor 2 stimulates ischemic neovasculogenesis by promoting homing of endothelial colony-forming cells. Stem Cells 32(3):779–790. doi:10.1002/stem.1578

Hur J, Yoon CH, Kim HS, Choi JH, Kang HJ, Hwang KK, Oh BH, Lee MM, Park YB (2004) Characterization of two types of endothelial progenitor cells and their different contributions to neovasculogenesis. Arterioscler Thromb Vasc Biol 24(2):288–293. doi:10.1161/01.ATV.0000114236.77009.06

Ingram DA, Mead LE, Tanaka H, Meade V, Fenoglio A, Mortell K, Pollok K, Ferkowicz MJ, Gilley D, Yoder MC (2004) Identification of a novel hierarchy of endothelial progenitor cells using human peripheral and umbilical cord blood. Blood 104(9):2752–2760. doi:10.1182/blood-2004-04-1396

Ingram DA, Mead LE, Moore DB, Woodard W, Fenoglio A, Yoder MC (2005) Vessel wall-derived endothelial cells rapidly proliferate because they contain a complete hierarchy of endothelial progenitor cells. Blood 105(7):2783–2786. doi:10.1182/blood-2004-08-3057

Liu SQ, Li ZL, Cao YX, Li L, Ma X, Zhao XG, Kang AQ, Liu CH, Yuan BX (2011) Transfusion of autologous late-outgrowth endothelial cells reduces arterial neointima formation after injury. Cardiovasc Res 90(1):171–181. doi:10.1093/cvr/cvq395

Margheri F, Chilla A, Laurenzana A, Serrati S, Mazzanti B, Saccardi R, Santosuosso M, Danza G, Sturli N, Rosati F, Magnelli L, Papucci L, Calorini L, Bianchini F, Del Rosso M, Fibbi G (2011) Endothelial progenitor cell-dependent angiogenesis requires localization of the full-length form of uPAR in caveolae. Blood 118(13):3743–3755. doi:10.1182/blood-2011-02-338681

Massa M, Campanelli R, Bonetti E, Ferrario M, Marinoni B, Rosti V (2009) Rapid and large increase of the frequency of circulating endothelial colony-forming cells (ECFCs) generating late outgrowth endothelial cells in patients with acute myocardial infarction. Exp Hematol 37(1):8–9. doi:10.1016/j.exphem.2008.09.007

Melero-Martin JM, Khan ZA, Picard A, Wu X, Paruchuri S, Bischoff J (2007) In vivo vasculogenic potential of human blood-derived endothelial progenitor cells. Blood 109(11):4761–4768. doi:10.1182/blood-2006-12-062471

Nijmeh H, Balasubramaniam V, Burns N, Ahmad A, Stenmark KR, Gerasimovskaya EV (2014) High proliferative potential endothelial colony-forming cells contribute to hypoxia-induced pulmonary artery vasa vasorum neovascularization. Am J Physiol Lung Cell Mol Physiol 306(7):L661–L671. doi:10.1152/ajplung.00244.2013

Nishikawa SI, Nishikawa S, Kawamoto H, Yoshida H, Kizumoto M, Kataoka H, Katsura Y (1998) In vitro generation of lymphohematopoietic cells from endothelial cells purified from murine embryos. Immunity 8(6):761–769. doi:10.1016/S1074-7613(00)80581-6

Oh BJ, Kim DK, Kim BJ, Yoon KS, Park SG, Park KS, Lee MS, Kim KW, Kim JH (2010) Differences in donor CXCR4 expression levels are correlated with functional capacity and therapeutic outcome of angiogenic treatment with endothelial colony forming cells. Biochem Biophys Res Commun 398(4):627–633. doi:10.1016/j.bbrc.2010.06.108

Ormiston ML, Southgate L, Treacy C, Pepke-Zaba J, Trembath RC, Machado RD, Morrell NW (2013) Assessment of a pulmonary origin for blood outgrowth endothelial cells by examination of identical twins harboring a BMPR2 mutation. Am J Respir Crit Care Med 188(2):258–260. doi:10.1164/rccm.201301-0078LE

Piaggio G, Rosti V, Corselli M, Bertolotti F, Bergamaschi G, Pozzi S, Imperiale D, Chiavarina B, Bonetti E, Novara F, Sessarego M, Villani L, Garuti A, Massa M, Ghio R, Campanelli R, Bacigalupo A, Pecci A, Viarengo G, Zuffardi O, Frassoni F, Barosi G (2009) Endothelial colony-forming cells from patients with chronic myeloproliferative disorders lack the disease-specific molecular clonality marker. Blood 114(14):3127–3130. doi:10.1182/blood-2008-12-190991

Richardson MR, Robbins EP, Vemula S, Critser PJ, Whittington C, Voytik-Harbin SL, Yoder MC (2014) Angiopoietin-like protein 2 regulates endothelial colony forming cell vasculogenesis. Angiogenesis 17(3):675–683. doi:10.1007/s10456-014-9423-8

Samokhvalov IM, Samokhvalova NI, Nishikawa S (2007) Cell tracing shows the contribution of the yolk sac to adult haematopoiesis. Nature 446(7139):1056–1061. doi:10.1038/nature05725

Schniedermann J, Rennecke M, Buttler K, Richter G, Stadtler AM, Norgall S, Badar M, Barleon B, May T, Wilting J, Weich HA (2010) Mouse lung contains endothelial progenitors with high capacity to form blood and lymphatic vessels. BMC Cell Biol 11:50. doi:10.1186/1471-2121-11-50

Shelley WC, Leapley AC, Huang L, Critser PJ, Zeng P, Prater D, Ingram DA, Tarantal AF, Yoder MC (2012) Changes in the frequency and in vivo vessel-forming ability of rhesus monkey circulating endothelial colony-forming cells across the lifespan (birth to aged). Pediatr Res 71(2):156–161. doi:10.1038/pr.2011.22

Smadja DM, d'Audigier C, Bieche I, Evrard S, Mauge L, Dias JV, Labreuche J, Laurendeau I, Marsac B, Dizier B, Wagner-Ballon O, Boisson-Vidal C, Morandi V, Duong-Van-Huyen JP, Bruneval P, Dignat-George F, Emmerich J, Gaussem P (2011a) Thrombospondin-1 is a plasmatic marker of peripheral arterial disease that modulates endothelial progenitor cell angiogenic properties. Arterioscler Thromb Vasc Biol 31(3):551–559. doi:10.1161/ATVBAHA.110.220624

Smadja DM, Mauge L, Gaussem P, d'Audigier C, Israel-Biet D, Celermajer DS, Bonnet D, Levy M (2011b) Treprostinil increases the number and angiogenic potential of endothelial progenitor cells in children with pulmonary hypertension. Angiogenesis 14(1):17–27. doi:10.1007/s10456-010-9192-y

Tavian M, Robin C, Coulombel L, Peault B (2001) The human embryo, but not its yolk sac, generates lympho-myeloid stem cells: mapping multipotent hematopoietic cell fate in intraembryonic mesoderm. Immunity 15(3):487–495. doi:10.1016/S1074-7613(01)00193-5

Toshner M, Voswinckel R, Southwood M, Al-Lamki R, Howard LS, Marchesan D, Yang J, Suntharalingam J, Soon E, Exley A, Stewart S, Hecker M, Zhu Z, Gehling U, Seeger W, Pepke-Zaba J, Morrell NW (2009) Evidence of dysfunction of endothelial progenitors in pulmonary arterial hypertension. Am J Respir Crit Care Med 180(8):780–787. doi:10.1164/rccm.200810-1662OC

Tuder RM, Marecki JC, Richter A, Fijalkowska I, Flores S (2007) Pathology of pulmonary hypertension. Clin Chest Med 28(1):23–42, vii. doi:10.1016/j.ccm.2006.11.010

Yoder MC, Mead LE, Prater D, Krier TR, Mroueh KN, Li F, Krasich R, Temm CJ, Prchal JT, Ingram DA (2007) Redefining endothelial progenitor cells via clonal analysis and hematopoietic stem/progenitor cell principals. Blood 109(5):1801–1809. doi:10.1182/blood-2006-08-043471

Chapter 10
Mesothelial Progenitors in Development, Lung Homeostasis, and Tissue Repair

Radhika Dixit, Xingbin Ai, and Alan Fine

Abbreviations

CD34	Cluster of differentiation molecule 34
EMT	Epithelial–mesenchymal transition
HGF	Hepatocyte growth factor
IL-8, -10, -1	Interleukin-8, -10, -1
RALDH2	Retinaldehyde dehydrogenase 2
TGFβ	Transforming growth factor beta
TNFα	Tumor necrosis factor alpha
WT1	Wilm's tumor1

10.1 Mesothelia: Definition and Origin

The term mesothelium was coined by Minot in 1890, following a detailed characterization of the "epithelial cells lining mammalian mesodermic cavities" (Minot 1890). Mesothelial cells arise from the primitive mesoderm and line serosal cavities. Serous membranes have two layers; parietal (somatic) that forms the embryonic gut wall and visceral (splanchnic) that surrounds internal organs within the thoracic and abdominal cavities. The mesothelium lining the lung and heart is referred to as pleura and pericardium, respectively, whereas the mesothelium in the abdominopelvic cavity is referred to as peritoneum.

R. Dixit • X. Ai • A. Fine, M.D. (✉)
Division of Pulmonary, Allergy, Sleep and Critical Care Medicine, Boston University School of Medicine, 72 E. Concord St Housman (R), Boston, MA 02118, USA
e-mail: afine@bu.edu

A. Firth, J.X.-J. Yuan (eds.), *Lung Stem Cells in the Epithelium and Vasculature*, Stem Cell Biology and Regenerative Medicine, DOI 10.1007/978-3-319-16232-4_10

10.2 Mesothelium: Biology and Function

Mesothelial cells are 25 μm in diameter and contain a central round nucleus surrounded by a thin volume of cytoplasm (Slater et al. 1989). The luminal surface of mesothelial cells displays well-developed microvilli with an occasional primary cilium. The cells are closely packed and are joined by tight junctions and desmosomes. They are supported by a basement membrane and invested by connective tissue and elastic fibers that aid in epithelial–mesenchymal transition (EMT) and migration of cells into injured tissue (Mutsaers 2002).

Morphologically, mesothelial cells are squamous-like epithelial cells at most anatomical sites, except in septal folds of the pleura and the peritoneal side of diaphragm where they are cuboidal. The squamous-like and cuboidal mesothelia have differences in mitochondrial and microfilament density (Herrick and Mutsaers 2004). Cuboidal mesothelial cells, which are more metabolically active, are localized to regions around spleen, liver, diaphragm, and in areas of mesothelial injury (Tsilibary and Wissing 1977; Mutsaers 2002). Interestingly, a genome-wide profile study identified intrinsic genetic differences between visceral and parietal pleural cells (Røe et al. 2009).

10.2.1 Function of Mesothelial Cells

Mesothelial cells maintain organ integrity by protecting against the "wear and tear" associated with intracoelomic organ movement (Pfeiffer et al. 1987; Mutsaers 2004). They provide structural support and a protective barrier against pathogens by secreting glycosaminoglycans, predominantly hyaluron (Mutsaers 2002). Mesothelial cells mediate transport of fluid and cells across serosal layers (Mutsaers 2002) and release immunomodulatory substances including chemokines, cytokines, and growth factors that regulate inflammation and tissue repair (Slater et al. 1989). Human mesothelial cells produce IL-8 and IL-10 in response to IL-1, TNF-alpha, and *Staphylococci* (Visser et al. 1995; Topley et al. 1993). Mesothelial cells also have the capacity for antigen presentation and modulate coagulation cascades (Mutsaers 2004).

10.3 Markers of Mesothelial Cells

Mesothelial cells express mesenchymal markers such as Vimentin and Desmin as well as epithelial markers including Cytokeratins (Mutsaers 2002; Herrick and Mutsaers 2004). Other proteins that are relatively specific for mesothelial cells include Wilm's tumor1 (WT1), Mesothelin, MCP130, Calretinin, Sulfatase1, Uroplakin3B, and RALDH2. Genetically modified mice that take advantage of the relatively specific expression of WT1 and Mesothelin have been used for lineage fate analysis during embryogenesis. Whether mesothelial cells lining the adult lung surface express the same markers as the fetal mesothelium is unclear. Table 10.1 summarizes the status of the markers thought to be relatively specific for mesothelial cells.

Table 10.1 Markers of mesothelial cells

Marker	Expression	Function
Wilm's tumor1 (*WT1*) is a zinc finger transcription factor	WT1 is expressed in visceral and parietal mesothelial cells, kidney podocytes, and glomerular capillaries	WT1 controls organ formation (Kreidberg et al. 1993). Loss of WT1 results in kidney and spleen agenesis as well as reduced cardiomyocyte mass and lung malformation (Scholz and Kirschner 2005; Moore et al. 1999). In human lungs, bi-allelic *Wt1* mutations are associated with pulmonary dysplasia and mesothelioma (Dharnidharka et al. 2001; Loo et al. 2012)
Mesothelin is a 40 kDa membrane glycoprotein	Mesothelin is expressed in pleura, pericardium, and peritoneum	To date, the biological function of mesothelin is not known; however, the cleaved fragment that is released from the mesothelin transmembrane protein, called megakaryocyte-potentiating factor, is currently being evaluated as a marker for mesothelial malignancies (Chang and Pastan 1996; Hassan et al. 2004)
Leucine-rich repeat transmembrane neuronal-4 protein (*LRRN4*) is a 67 kDa transmembrane protein	LRRN4 is expressed in primary mesothelial cells and is significantly down-regulated in mesotheliomas	To date, its function is not known. LRRN4 is up-regulated in acute lung injury and in chronic obstructive pulmonary disease (Kanamori-Katayama et al. 2011)
Uroplakin 3B is a membrane glycoprotein	Uroplakin 3B is expressed in urothelial cells, retinal pigment cells, and mesothelial cells of the peritoneum, pleura, and pericardium	Uroplakins are cell-surface proteins that regulate signal transduction events. Uroplakin 3B expression is significantly down-regulated in mesothelioma, suggesting a role as a tumor suppressor gene (Kanamori-Katayama et al. 2011; Hu et al. 2002)
Sulfatase 1 (*Sulf1*) is a heparan sulfate 6-O-endosulfatase enzyme	*Sulfatase1* is expressed in mesothelial cells lining the lung surface and Sertoli cells of testes	This enzyme removes 6-O sulfate groups from heparan sulfate. *Sulfatase1* modulates heparan sulfate by regulating its binding partners. In Sertoli cells of the testes, *Sulfatase1* is regulated by WT1. (Langsdorf et al. 2011)
(*Raldh2*) *Retinaldehyde dehydrogenase 2* mRNA encodes for an enzyme that belongs to the aldehyde dehydrogenase family	*Raldh2* mRNA is expressed in mesodermal and mesothelial cells	This enzyme is required for the production of retinoic acid. Loss of *Raldh2* causes cardiovascular defects and embryonic lethality in mice. It plays a crucial role in lung bud formation (Desai et al. 2006)
Calretinin is a 29 kDa intracellular calcium-binding protein	Calretinin is expressed in the central and peripheral nervous system, retina, mesothelium, adipocytes, and Sertoli cells of testes	Calretinin is a calcium-binding protein involved in many signal transduction pathways. It is up-regulated in mesotheliomas and adenocarcinomas (Barberis et al. 1997)

10.4 Mesothelial Progenitor Cell Function

Evidence that the mesothelium gives rise to other cell lineages was originally derived from in vitro observation of cultured cells. In these studies, it was found that mesothelial cells "trans-differentiate" to distinct mesenchymal phenotypes such as fibroblasts and smooth muscle when exposed to defined growth factors (Yang et al. 2003). Dye labeling studies and quail-chick chimera experiments showed that the epicardial mesothelium differentiates into endothelium and smooth muscle through an EMT process during embryonic development (Pérez-Pomares et al. 2002). Examination of WT1 genetically deficient mice provided further evidence that mesothelial cells are progenitors required for organ formation (Kreidberg et al. 1993).

Mice expressing Cre-recombinase under mesothelial-specific promoters are powerful tools that have been utilized to study mesothelial progenitor function. Table 10.2 lists the currently available mesothelial-Cre mouse strains that can be

Table 10.2 Mouse lines used to lineage trace mesothelial cells

Mouse strain	Mesothelium-derived lineages
WT280Cre YAC (*Wt1*^{Cre}) Transgenic mouse line (Wilm et al. 2005)	Gut—vascular smooth muscle and endothelial cells (Wilm et al. 2005)
	Lung—vascular smooth muscle (Que et al. 2008)
	Heart—smooth muscle cells and endothelial cells (Wilm et al. 2005)
Wt1$^{CreERT2/+}$ Loss-of-function *WT1* allele resulting from a knock-in TAM-inducible *Cre recombinase* gene in exon 1 of WT1 (Zhou et al. 2008)	Heart—epicardium, cardiomyocytes, vascular smooth muscle cells, endothelium (Zhou et al. 2008)
	Liver—hepatic stellate cells, myofibroblasts, perivascular mesenchyme (Asahina et al. 2011)
	Lung—vascular smooth muscle, bronchial smooth muscle, desmin+ fibroblasts (Dixit et al. 2013)
Wt1$^{GFP\text{-}Cre}$ Loss-of-function *Wt1* allele resulting from a knock-in *eGFP-Cre* fusion construct inserted in exon 1 of WT1 (Zhou et al. 2008)	Off-target Cre recombination throughout embryo, limits use as a tool for lineage analysis. However, fidelity of GFP supports use for cell sorting of mesothelial cells
Wt1/IRES/GFP-Cre Transgenic mouse line IRES/EGFP-Cre cassette is inserted 17 bp downstream of the translation stop codon of *Wt1* gene (del Monte et al. 2011; Wessels et al. 2012)	Intestine and gut—endothelium, smooth muscle, visceral musculature, pericytes, Cajal cells (Carmona et al. 2013)
	Heart—parietal leaflets of atrioventricular valves (Wessels et al. 2012), coronary vessels (del Monte et al. 2011)
	Lung—pulmonary endothelial, smooth muscle cells, bronchial musculature, tracheal and bronchial cartilage, CD34+ fibroblast-like interstitial cells (Cano et al. 2013)
Msln$^{CLN\text{-}CreERT2/+}$ TAM-inducible *Cre recombinase-IRES-LacZ* knocked into *Mesothelin* (Msln) locus (Rinkevich et al. 2012)	Msln+ cells give rise to fibroblasts and vascular smooth muscle in the lung, intestine, liver, gut, and thymus (Rinkevich et al. 2012)

used for lineage analysis. Using a mouse that expresses an inducible Cre-recombinase from the endogenous *Wt1* locus (Zhou et al. 2008), mesothelial cells lining fetal heart, gut, and liver were shown to migrate inward and contribute to differentiated mesenchymal cells during organogenesis. To date, mesothelial-derived cell types identified by this methodology include cardiomyocytes, airway smooth muscle, vascular smooth muscle, endothelium, hepatic stellate cells, fibroblasts, perivascular mesenchyme, and gut tube anlage (Zhou et al. 2008; Que et al. 2008; Wilm et al. 2005; Asahina et al. 2011; Winters et al. 2012; Dixit et al. 2013). More recently, mesothelial cells were shown to give rise to visceral adipocytes (Chau et al. 2014).

10.4.1 Mesothelial Progenitor Cells Lining the Lung

In the lung, independent studies using different Cre-lox lineage tracing system found conflicting results regarding the contribution of fetal lung mesothelial progenitors to differentiated lung cell compartments (Que et al. 2008; Greif et al. 2012, Rinkevich et al. 2012; Dixit et al. 2013; Cano et al. 2013). One study used a non-inducible *Wt1-Cre* (WT280Cre YAC) transgenic mouse line and showed that the mesothelium gives rise to intra-pulmonary artery smooth muscle cells (Que et al. 2008). These results were confounded, however, by uncertainties regarding the strength, timing, and specificity of the cellular marking in this transgenic Cre line. The second study, which focused on lineages in the main pulmonary artery, used an inducible knock-in $Wt1^{CreERT2/+}$ line and showed that the mesothelium is not a significant source of smooth muscle cells for this structure (Greif et al. 2012). Utilization of *Wt1/IRES/GFP-Cre* transgenic mouse line (Table 10.2) showed that mesothelial cells lining the lung surface were found to contribute to endothelial cells, smooth muscle, bronchial musculature, tracheal and bronchial cartilage, as well as CD34+ fibroblast-like interstitial cells (Cano et al. 2013). Like previous studies, these findings may be similarly limited by uncertainties in the fidelity of Cre activation. Finally, a study employing the $Msln^{CLN}$ mouse that contains an inducible *CreERT2* knocked into the *mesothelin* locus showed that fibroblasts and smooth muscle are derived from mesothelin+ cells. One potential concern of this study is that mesothelin expression in the embryonic lung may not be mesothelial-specific (Rinkevich et al. 2012; Dixit et al. 2013).

In order to clarify these issues, we employed a $Wt1^{CreERT2/+}$ mouse line and conducted a detailed lineage analysis of the early fetal mesothelium during lung development (Table 10.2). The strength of this tool is that, Cre expression can be controlled temporally and parallels endogenous WT1 expression. Before embarking on these studies we first confirmed that WT1 is restricted to the mesothelium only in the developing lung (Dixit et al. 2013). The lineage data showed that mesothelium is a source of a distinct subpopulation of bronchial smooth muscle, vascular smooth muscle, and desmin^{+} fibroblasts during embryonic and early post-natal lung development. Further, using genetically modified mice and ex-vivo organ cultures, we mechanistically demonstrated that active hedgehog signaling is required for mesothelial cell entry into the fetal lung parenchyma. In contrast, hedgehog function had no apparent role in mesothelial cell entry into the developing heart. Notably, Wnt and retinoic acid signaling were found to control EMT and migration of epicardial cells (von Gise and Pu 2012). In addition, TGF-β was found to induce EMT and migration of mesothelial cells in the developing liver (Li et al. 2013). In culture, hepatocyte growth factor/scatter factor (HGF/SF) was found to regulate migration and proliferation of human mesothelial cells isolated from pleural effusions (Warn et al. 2001). Taken together, these findings further underscore a paradigm in which fetal mesothelial cells are an important source of progenitors for development of mesenchymal structures and show that key signals that underlay the entry and migration of mesothelial cells are organ-specific (Dixit et al. 2013).

10.5 Mesothelial Cells in Tissue Repair and Remodeling

Injury responses in mesothelial cells occur following direct injury to the mesothelium and during injury to the underlying sub-mesothelial tissue. During homeostasis, mesothelia are quiescent; however, they proliferate after direct damage from air, water, foreign substances, intra-coelomic movement of organs, surgical procedures, and peritoneal dialysis (Mutsaers et al. 2000). These injuries cause swelling and sloughing from the surface into the surrounding anatomical space (Herrick and Mutsaers 2004). Mutsaers et al. reported that within 48 h of injury, 30 % of mesothelial cells begin to proliferate and subsequently restore the damaged area completely by 7–10 days (Mutsaers et al. 2000; Ryan et al. 1973). It has also been suggested that free-floating mesothelial cells migrate from the surrounding fluid to the wound site to reestablish tissue integrity (Herrick and Mutsaers 2004). Notably, a prominent pro-inflammatory role for mesothelial cells has been postulated for acute and chronic peritonitis in humans undergoing peritoneal dialysis (Hekking and van den Born 2007; Nocoladi et al. 1994; Devuyst et al. 2010; Yung and Chan 2009).

Several studies suggest a role for mesothelial cells in fibrotic reactions in the lung. In this regard, cultured pleura cells undergo EMT to a fibroblast-like state, produce collagens, and elaborate TGF-β (Lee et al. 2003). Nasreen et al. demonstrated that TGF-β1 stimulated in vitro differentiation of pleural mesothelial cells to myofibroblasts (Nasreen et al. 2009). This was also observed after delivery of *Wt1 shRNA* to isolated murine pleural cells (Karki et al. 2014). Furthermore, active TGF-β1 administration to $Wt1^{GFP\text{-}Cre}$ mice was found to stimulate mesothelial cell migration into the underlying lung (Karki et al. 2014). The relative contribution of mesothelial-derived cells in other diseases characterized by accumulation of myofibroblasts such as asthma and sarcoidosis requires further study.

Injury models in genetically modified mice demonstrated that WT1+ mesothelial cells can regenerate differentiated mesenchymal parenchymal cells (Smart et al. 2011; Li et al. 2013). In the adult heart, epicardial mesothelial cells are quiescent; however, after myocardial infarction, mesothelial cells proliferate, migrate, and subsequently differentiate into cardiac fibroblasts, vascular smooth muscle, a subset of endothelial cells, and possibly cardiomyocytes (Smart et al. 2011; von Gise and Pu 2012; Zhou et al. 2008). In the liver, WT1+ mesothelial cells undergo an epithelial-to-mesenchymal transition during liver fibrosis, differentiating into hepatic stellate cells and fibroblasts (Li et al. 2013). Mesothelial cells lining the diaphragm were shown to give rise to skeletal muscle fibers during chemical-induced peritonitis (Levine and Saltzman 1994). More recently, engrafted omental cells (mesothelial cells of the peritoneum) in a carotid artery injury model in mice were found to differentiate into vascular smooth muscle and endothelial cells (Shelton et al. 2013). Taken together, these studies indicate that regeneration of organ function involves the progenitor activity of mesothelial cells.

In summary, current data reveal the dynamic behavior of mesothelia in embryogenesis and in adult organs after injury. Collectively, these observations suggest a new biology not previously recognized. Thus, the detailed characterization of mesothelial cell diversity and function is crucial to our understanding of normal organ development, injury repair, and tissue regeneration.

References

Asahina K, Zhou B, Pu WT, Tsukamoto H (2011) Septum transversum-derived mesothelium gives rise to hepatic stellate cells and perivascular mesenchymal cells in developing mouse liver. Hepatology 53:983–995

Barberis MC, Faleri M, Veronese S, Casadio C, Viale G (1997) Calretinin: a selective marker of normal and neoplastic mesothelial cells in serous effusions. Acta Cytol 6:1757–1761

Cano E, Carmona R, Munoz-Chapuli R (2013) Wt1-expressing progenitors contribute to multiple tissues in the developing lung. Am J Physiol Lung Cell Mol Physiol 305:322–332

Carmona R, Cano E, Mattiotti A, Gaztambide J, Muñoz-Chápuli R (2013) Cells derived from the coelomic epithelium contribute to multiple gastrointestinal tissues in mouse embryos. PLoS One 8:e55890

Chang K, Pastan I (1996) Molecular cloning of mesothelin, a differentiation antigen present on mesothelium, mesotheliomas, and ovarian cancers. Proc Natl Acad Sci U S A 93:136–140

Chau YY, Bandiera R, Serrels A, Martínez-Estrada OM, Qing W, Lee M, Slight J, Thornburn A, Berry R, McHaffie S, Stimson RH, Walker BR, Chapuli RM, Schedl A, Hastie N (2014) Visceral and subcutaneous fat have different origins and evidence supports a mesothelial source. Nat Cell Biol 16:367–375

del Monte G, Casanova JC, Guadix JA, MacGrogan D, Burch JB, Pérez-Pomares JM, de la Pompa JL (2011) Differential Notch signaling in the epicardium is required for cardiac inflow development and coronary vessel morphogenesis. Circ Res 108:824–836

Desai TJ, Chen F, Lü J, Qian J, Niederreither K, Dollé P, Pierre C, Cardoso WV (2006) Distinct roles for retinoic acid receptors alpha and beta in early lung morphogenesis. Dev Biol 291: 12–24

Devuyst O, Margetts PJ, Topley N (2010) The pathophysiology of the peritoneal membrane. J Am Soc Nephrol 21:1077–1085

Dharnidharka VR, Ruteshouser EC, Rosen S, Kozakewich H, Harris HW Jr, Herrin JT, Huff V (2001) Pulmonary dysplasia, Denys-Drash syndrome and Wilms tumor 1 gene mutation in twins. Pediatr Nephrol 16:227–231

Dixit R, Ai X, Fine A (2013) Derivation of lung mesenchymal lineages from the fetal mesothelium requires hedgehog signaling for mesothelial cell entry. Development 140:4398–4406

Greif DM, Kumar M, Lighthouse JK, Hum J, An A, Ding L, Red-Horse K, Espinoza FH, Olson L, Offermanns S, Krasnow MA (2012) Radial construction of an arterial wall. Dev Cell 23:482–493

Hassan R, Bera T, Pastan I (2004) Mesothelin: a new target for immunotherapy. Clin Cancer Res 10:3937–3942

Hekking LH, van den Born J (2007) Feasibility of mesothelial transplantation during experimental peritoneal dialysis and peritonitis. Int J Artif Organs 30:513–519

Herrick SE, Mutsaers SE (2004) Mesothelial progenitor cells and their potential in tissue engineering. Int J Biochem Cell Biol 36:621–642

Hu P, Meyers S, Liang FX, Deng FM, Kachar B, Zeidel ML, Sun TT (2002) Role of membrane proteins in permeability barrier function: uroplakin ablation elevates urothelial permeability. Am J Physiol Renal Physiol 283:1200–1207

Kanamori-Katayama M, Kaiho A, Ishizu Y, Okamura-Oho Y, Hino O, Abe M, Kishimoto T, Sekihara H, Nakamura Y, Suzuki H, Forrest AR, Hayashizaki Y (2011) LRRN4 and UPK3B are markers of primary mesothelial cells. PLoS One 6:e25391

Karki S, Surolia R, Hock TD, Guroji P, Zolak JS, Duggal R, Ye T, Thannickal VJ, Antony VB (2014) Wilms' tumor 1 (Wt1) regulates pleural mesothelial cell plasticity and transition into myofibroblasts in idiopathic pulmonary fibrosis. FASEB J 28:1122–1131

Kreidberg JA, Sariola H, Loring JM, Maeda M, Pelletier J, Housman D, Jaenisch R (1993) WT-1 is required for early kidney development. Cell 74:679–691

Langsdorf A, Schumacher V, Shi X, Tran T, Zaia J, Jain S, Taglienti M, Kreidberg J, Fine A, Ai X (2011) Expression regulation and function of Sulfs in the spermatogonial stem cell niche. Glycobiology 21:152–161

Lee YC, Lane KB, Zoia O, Thompson PJ, Light RW, Blackwell TS (2003) Transforming growth factor-beta induces collagen synthesis without inducing IL-8 production in mesothelial cells. Eur Respir 22:197–202

Levine S, Saltzman A (1994) Neogenesis of skeletal rnuscle in the postinflammatory rat peritoneum. Exp Mol Pathol 60:60–69

Li Y, Wang J, Asahina K (2013) Mesothelial cells give rise to hepatic stellate cells and myofibroblasts via mesothelial-mesenchymal transition in liver injury. Proc Natl Acad Sci U S A 110:2324–2329

Loo C, Algar E, Payton D, Perry-Keene J, Pereira T, Ramm G (2012) Possible role of WT1 in a human fetus with bronchial atresia, lung malformation and renal agenesis. Pediatr Dev Pathol 15:39–44

Minot CS (1890) The mesoderm and the coelom of vertebrates. Am Nat 24:877–898

Moore AW, McInnes L, Kreidberg J, Hastie ND, Schedl A (1999) YAC complementation shows a requirement for Wt1 in the development of epicardium, adrenal gland and throughout nephrogenesis. Development 126:1845–1857

Mutsaers SE (2002) Mesothelial cells: their structure, function and role in serosal repair. Respirology 7:171–191

Mutsaers SE (2004) The mesothelial cell. Int J Biochem Cell Biol 36:9–16

Mutsaers SE, Whitaker D, Papadimitriou JM (2000) Mesothelial regeneration is not dependent on subserosal cells. J Pathol 190:86–92

Nasreen N, Mohammed KA, Mubarak KK, Baz MA, Akindipe OA, Fernandez-Bussy S, Antony VB (2009) Pleural mesothelial cell transformation into myofibroblasts and haptotactic migration in response to TGF-beta1 in vitro. Am J Physiol Lung Cell Mol Physiol 297:115–124

Nocoladi P, Garosi G, Petrini G, Mnaci G (1994) Morphology and morphometric changes in mesothelial cells during peritoneal dialysis in rabbit. Nephron 74:594–599

Pérez-Pomares J-M, Carmona R, González-Iriarte M, Atencia G, Wessels A, Muñoz-Chápuli R (2002) Origin of coronary endothelial cells from epicardial mesothelium in avian embryos. Int J Dev Biol 46:1005–1013

Pfeiffer CJ, Pfeiffer DC, Misra HP (1987) Enteric serosal surface in the piglet: a scanning and transmission electron microscopic study of the mesothelium. J Submicrosc Cytol 19:237–246

Que J, Wilm B, Hasegawa H, Wang F, Bader D, Hogan BL (2008) Mesothelium contributes to vascular smooth muscle and mesenchyme during lung development. Proc Natl Acad Sci U S A 105:16626–16630

Rinkevich Y, Mori T, Sahoo D, Xu PX, Jr B, Weissman IL (2012) Identification and prospective isolation of a mesothelial precursor lineage giving rise to smooth muscle cells and fibroblasts for mammalian internal organs, and their vasculature. Nat Cell Biol 12:1251–1260

Røe DO, Anderssen E, Helge E, Pettersen HC, Olsen SK, Sandeck H, Haaverstad R, Lundgren S, Larsson E (2009) Genome-wide profile of pleural mesothelioma versus parietal and visceral pleura: the emerging gene portrait of the mesothelioma phenotype. PLoS One 4:e6554

Ryan GB, Grobety J, Majno G (1973) Mesothelial injury and recovery. Am J Pathol 71:93–112

Scholz H, Kirschner KM (2005) A role for the Wilms' tumor protein WT1 in organ development. Physiology 20:54–59

Shelton EL, Poole SD, Reese J, Bader DM (2013) Omental grafting: a cell-based therapy for blood vessel repair. J Tissue Eng Regen Med 7:421–433

Slater NJ, Raftery AT, Cope GH (1989) The ultrastructure of human abdominal mesothelium. J Anat 167:47–56

Smart N, Bollini S, Dubé KN, Vieira JM, Zhou B, Davidson S, Yellon D, Riegler J, Price AN, Lythgoe MF, Pu WT, Riley PR (2011) De novo cardiomyocytes from within the activated adult heart after injury. Nature 474:640–644

Topley N, Jorres A, Luttmann W, Petersen MM, Lang MJ, Thierauch KH, Müller C, Coles GA, Davies M, Williams JD (1993) Human peritoneal mesothelial cells synthesize interleukin-6: induction by IL-1 beta and TNF alpha. Kidney Int 43:226–233

Tsilibary EC, Wissing SL (1977) Absorption from the peritoneal cavity: SEM study of the mesothelium covering the peritoneal surface of the muscular portion of the diaphragm. Am J Anat 149:127–133

Visser CE, Steenbergen JJ, Betjes MG, Meijer S, Arisz L, Hoefsmit EC, Krediet RT, Beelen RH (1995) IL-8 production by human mesothelial cells after direct stimulation with Staphylococci. Infect Immun 63:4206–4209

von Gise A, Pu WT (2012) Endocardial and epicardial epithelial to mesenchymal transitions in heart development and disease. Circ Res 110:1628–1645

Warn R, Harvey P, Warn A, Foley-Comer A, Heldin P, Versnel M, Arakaki N, Daikuhara Y, Laurent GJ, Herrick SE, Mutsaers SE (2001) HGF/SF induces mesothelial cell migration and proliferation by autocrine and paracrine pathways. Exp Cell Res 267:258–266

Wessels A, van den Hoff MJ, Adamo RF, Phelps AL, Lockhart MM, Sauls K, Briggs LE, Norris RA, van Wijk B, Perez-Pomares MJ, Dettman RW, Burch JB (2012) Epicardially derived fibroblasts preferentially contribute to the parietal leaflets of the atrioventricular valves in the murine heart. Dev Biol 366:111–124

Wilm B, Ipenberg A, Hastie ND, Burch JB, Bader DM (2005) The serosal mesothelium is a major source of smooth muscle cells of the gut vasculature. Development 132:5317–5328

Winters NI, Thomason RT, Bader DM (2012) Identification of a novel developmental mechanism in the generation of mesothelia. Development 139:2926–2934

Yang AH, Chen JY, Lin JK (2003) Myofibroblastic conversion of mesothelial cells. Kidney Int 63:1530–1539

Yung S, Chan TM (2009) Intrinsic cells: mesothelial cells—central players in regulating inflammation and resolution. Perit Dial Int 29:21–27

Zhou B, Ma Q, Rajagopal S, Wu SM, Domian I, Rivera-Feliciano J, Jiang D, Von Gise A, Ikeda S, Chien KR, Pu WT (2008) Epicardial progenitors contribute to the cardiomyocyte lineage in the developing heart. Nature 454:109–113

Chapter 11
Lung Microvascular Endothelium as a Putative Progenitor Cell Niche

Lauren Hartman and Troy Stevens

11.1 Introduction

The pulmonary circulation is a unique vascular bed; it receives 100 % of the cardiac output from the right ventricle while maintaining low vascular pressures. Even with intense exercise, when cardiac output increases 5–8-fold, pulmonary artery pressure increases only modestly. The ability to maintain low pressures even in the face of high cardiac outputs is partly attributable to the vast surface area of the lung capillaries, their ability to distend, and the ability to recruit blood into capillaries that are not otherwise continuously perfused. Indeed, blood flow through individual capillaries is intermittent and subject to active regulation in order to match ventilation with perfusion to optimize gas exchange. Such specialization of the lung's capillaries has fascinated investigators since their original discovery in 1661 by Marcello Malpighi (West 2013).

We now know that endothelium contributes to the alveolar-capillary membrane, where capillary endothelial cells and type I pneumocytes border one another through

L. Hartman
Department of Physiology and Cell Biology, University of South Alabama, Mobile, AL 36688, USA

Center for Lung Biology, College of Medicine, University of South Alabama, Mobile, AL 36688, USA

T. Stevens, Ph.D. (✉)
Department of Physiology and Cell Biology, University of South Alabama, Mobile, AL 36688, USA

Department of Medicine, University of South Alabama, Mobile, AL 36688, USA

Center for Lung Biology, College of Medicine, University of South Alabama, Mobile, AL 36688, USA
e-mail: tstevens@southalabama.edu

A. Firth, J.X.-J. Yuan (eds.), *Lung Stem Cells in the Epithelium and Vasculature*, Stem Cell Biology and Regenerative Medicine, DOI 10.1007/978-3-319-16232-4_11

a fused basement membrane. However, this association was impossible to fully appreciate until development of the electron microscope allowed for high-resolution analysis of intact anatomy. Cell biology and physiology studies of lung capillary endothelium have only recently been advanced. This ongoing work indicates that the lung capillary endothelium represents a highly specialized, differentiated cell phenotype (Stevens 2005; Gebb and Stevens 2004). One of the unique features of this cell is its high replication competence and the ability to self-renew (Alvarez et al. 2008). Highly replication competent cells that self-renew are features of progenitor cells, meaning that lung capillary endothelium constitutes an important progenitor cell niche. Here, we highlight evidence that lung capillary endothelium represents a progenitor cell niche, and we discuss the physiological demands for rapid neoangiogenesis in the alveolar-capillary compartment.

11.2 Stem Cells

Genetic evidence for the existence of stem cells arose from studies examining hematopoietic cells (Becker et al. 1963; Till 1961; Wu et al. 1968a, b). During these experiments, investigators discovered colonies of progenitor cells, each derived from a single clonogenic precursor, in the spleens of conditioned hosts. The colonies were composed not only of differentiated cells, but also cells that could be used to reconstitute all blood cell lineages. From these initial studies, numerous investigations ensued revealing that stem cells are very unique cells that have the potential to self-renew, that is, to divide and create additional stem cells or to differentiate into mature cells of any particular cell lineage. Stem cells will either go through symmetric cell division giving rise to two daughter cells that remain undifferentiated or perform asymmetric cell division to become differentiated. This unique characteristic of stem cells is vital in the generation of specialized cells forming different types of tissues, in addition to continually renewing normal tissue and repairing injured tissue (Reya et al. 2001).

There are two major types of stem cells. The first type is an embryonic stem cell, which is found in the inner cell mass of the mammalian blastocyst. Embryonic stem cells are able to differentiate into the three types of germ layers which are the ectoderm (gives rise to the skin and neural lineages), mesoderm (generates blood, bone, muscle, cartilage, and fat), and endoderm (contribute to tissues of the respiratory and digestive tracts) along with having the capability to infinitely grow (Wagers and Weissman 2004). Specific signaling mechanisms between stem cells, their progeny, and the surrounding tissues guarantee that the correct amount of new cells will be formed within the proper locations. This is especially important during development when the embryonic stem cells are induced down a specific cell lineage pathway as opposed to all the other lineage pathways through intercellular cross-talk (Fuchs and Segre 2000).

Stem cell hierarchy determines the potential of the stem cell to differentiate into varying cell types. The earliest stem cells in the hierarchy are totipotent stem cells, which are capable of differentiating into all embryonic and extra-embryonic cell types. Totipotent stem cells give rise to pluripotent stem cells that are able to give rise to all cell types of the embryo proper. Pluripotent stem cells generate multipotent stem cells, which are progenitor cells within the same embryonic layer. Multipotent stem cells further differentiate into stem cells that are incapable of self-renewal, called oligolineage progenitors. Oligolineage progenitors only have the capacity to divide into one particular cell lineage and give rise to progeny that are more restricted in their differentiating potential. A hallmark of both stem and progenitor cells is their ability to proliferate and give rise to functional progeny; however, progenitor cells are unable to infinitely divide like stem cells (Weissman 2000). Finally, unipotent cells are produced that are only able to contribute one mature cell type.

In contrast to embryonic stem cells, the second type of stem cell, the adult stem cell, resides within an adult organ or tissue and is only able to select a differentiation program from a few possible pathways (Fuchs and Segre 2000). Adult stem cells include the somatic stem cell and the germ stem cell (Takahashi and Yamanaka 2006). Although the majority of the diversification of various cell types is carried out at or shortly after birth, adult tissues must still be regenerated during adult life from stem cell to progenitor cell to functional progeny. Therefore, populations of stem cells must be present to undergo self-renewal as well as select for a particular differentiation program to replenish dying cells and regenerate damaged tissues (Fuchs and Segre 2000; Takahashi and Yamanaka 2006; Weissman et al. 2001). For this reason, adult stem cells are frequently localized to specific microenvironments called stem cell niches. External and intrinsic cues within these niches act on the adult stem cells to alter gene expression and induce self-renewal or terminal differentiation (Fuchs and Segre 2000).

Pioneering studies performed by Yamanaka and Takahashi (2006) have shown that differentiated somatic cells can be reprogrammed to an undifferentiated state leading to the formation of pluripotent stem cells. Pluripotent stem cells are able to differentiate into any of the three types of embryonic layers. Through the introduction of Oct3/4, Sox2, c-Myc, and Klf4 under embryonic stem cell conditions, pluripotent stem cells can be induced, which exhibit normal embryonic stem cell morphology. They proliferate indefinitely in culture, as well as express genes characteristic of embryonic stem cells. Because of these qualities, inducible pluripotent cells are useful in disease models along with drug development and transplantation medicine (Yamanaka and Takahashi 2006).

11.3 Endothelial Progenitor Cells

A landmark study was performed in 1997 by Asahara et al. (1997) examining adult human peripheral blood for a circulating progenitor cell capable of differentiating into endothelial cells. Asahara et al. (1997) documented a particular group of human-circulating cells that had become endothelial cell-like in culture and were able to engraft into injured vessels. These particular cells were then termed "endothelial progenitor cells." Therefore, the authors concluded that some circulating cells in peripheral blood have the potential to serve as progenitors of the endothelial lineage. This paper defined endothelial progenitor cells as cells having key criteria, such as displaying CD34, Tie-2, CD31, UEA-1, ac-LDL, and expressing CD45 in certain instances (Asahara et al. 1997).

For nearly two decades, the term "endothelial progenitor cell" was used to broadly describe a heterogeneous group of circulating cells that putatively give rise to endothelial cells in vitro or in vivo. Studies have shown over the years that there are two main categories of endothelial progenitor cells. The first category is a heterogeneous population of hematopoietic cells that display crucial paracrine angiogenic activity called pro-angiogenic hematopoietic progenitor cells (Chao and Hirschi 2010; Duong et al. 2011; Richardson and Yoder 2011). The second category is endothelial colony-forming cells, which are capable of generating blood vessels de novo.

Recent evidence has revealed that the original endothelial progenitor cells identified by Asahara et al. (1997) were most likely circulating pro-angiogenic hematopoietic progenitor cells (Medina et al. 2010). These cells have been shown to correlate with the degree of diseases such as cardiovascular disease and pulmonary hypertension (Asosingh et al. 2008; Hill et al. 2003; Kissel et al. 2007). Studies have also demonstrated that pro-angiogenic hematopoietic progenitor cells originate from the bone marrow, circulate within the peripheral blood, and temporarily engraft themselves into injured host vessels (Asahara et al. 1997; Hristov and Weber 2004; Lin et al. 2000). Although pro-angiogenic hematopoietic cells exhibit some properties of endothelial cells, they are not capable of vessel formation (Yoder et al. 2007).

Numerous methods have been developed to properly identify endothelial progenitor cells, including various combinations of cell surface markers, functional assays, and colony-forming assays; however, there is not a uniform definition of an endothelial progenitor cell (Hirschi et al. 2008). In 2004, Ingram et al. (2004) described a novel approach that identified endothelial progenitor cells based on their clonogenic and proliferative capacities rather than on the expression of cell surface antigens. After single-cell in vitro clonogenic assays were performed on both adult peripheral blood and umbilical cord blood, a hierarchy was discovered in both blood samples (Ingram et al. 2004). This growth hierarchy was similar to hematopoietic stem cells, and a second category of endothelial progenitor cells was discovered—cells which proliferate to form new blood vessels, called endothelial colony-forming cells (Chao and Hirschi 2010; Duong et al. 2011; Richardson and Yoder 2011).

Endothelial colony-forming cells possess true endothelial progenitor cell characteristics because they are highly proliferative, form blood vessels de novo, and contribute to endothelialization and angiogenesis (Basile and Yoder 2014). The endothelial colony-forming cells found in the umbilical cord blood were capable of achieving at least 100 population doublings, form re-platable secondary and tertiary colonies, and retain high levels of telomerase activity. Overall, the endothelial colony-forming cells found within umbilical cord blood have a significantly higher proliferative potential compared to those isolated from the adult peripheral blood (Ingram et al. 2004; Yoder 2012).

Currently, it is unknown whether endothelial colony-forming cells are unipotent, multipotent, or a fully differentiated endothelial cell possessing a high proliferative potential (Duong et al. 2011). Similar to pro-angiogenic hematopoietic progenitor cells, endothelial colony-forming cells cannot be identified through specific cell markers, but they have been shown to express CD34, CD146, CD31, Flk-1, and CD105. Importantly, endothelial colony-forming cells do not express CD133 or CD45, which indicates that they are not hematopoietic cells (Ingram et al. 2004).

Along with peripheral and umbilical cord blood, endothelial colony-forming cells have been found within the endothelium of the umbilical vein, human aorta, and the pulmonary microvasculature (Alvarez et al. 2008; Ingram et al. 2005). In 2008, Alvarez et al. (2008) found that the pulmonary microvasculature is enriched with endothelial colony-forming cells that have a high proliferative potential, are able to reconstitute the entire hierarchy of growth potentials, and perform postnatal vasculogenesis while retaining their endothelial microvascular phenotype (Alvarez et al. 2008) (see Sect. 11.6). These findings suggest a progenitor cell niche within the pulmonary microvasculature that regenerates and repairs the pulmonary endothelium in order to maintain vascular homeostasis.

Recently, Prasain et al. (2014) derived cells possessing endothelial colony-forming cell-like properties from human iPS cells and human embryonic stem cells. These cells were NRP-1^{+}CD31^{+} and displayed high clonal proliferative potential, angiogenic capacity, and significantly contributed to vascular repair of ischemic tissue. Additionally, Yoon et al. (2005) revealed that pro-angiogenic cells and endothelial colony-forming cells synergistically interact to perform vascular repair and neovascularization. Individually, both types of endothelial progenitor cells enhance angiogenesis; however, a combination of both cells has been found to yield the largest angiogenic response in both the mouse hind limb ischemia model and the Matrigel plug model (Yoon et al. 2005). Due to pro-angiogenic hematopoietic cells expressing greater amounts of pro-angiogenic cytokines, it has been suggested that these cells exert their pro-angiogenic effects via paracrine mechanisms interacting with endothelial colony-forming cells distributed throughout the tissue. In turn, endothelial colony-forming cells are activated by the pro-angiogenic chemokines to fulfill the reparative needs of the endothelium (Fig. 11.1) (Asosingh et al. 2008; Basile and Yoder 2014).

Questions remain as to the location (niche) of resident endothelial colony-forming cells within the vessel wall, and how these cells interpret tissue-specific cues to control development, neo-angiogenesis, and the response to injury. These issues are highly relevant to the pulmonary circulation, which possesses the largest vascular surface area and the highest number of endothelial cells among all internal organs. Even within the lung, endothelium changes structurally and functionally along the arterial-capillary-venous axis. The richest abundance of endothelial progenitors cells is found within the lung's capillary segment.

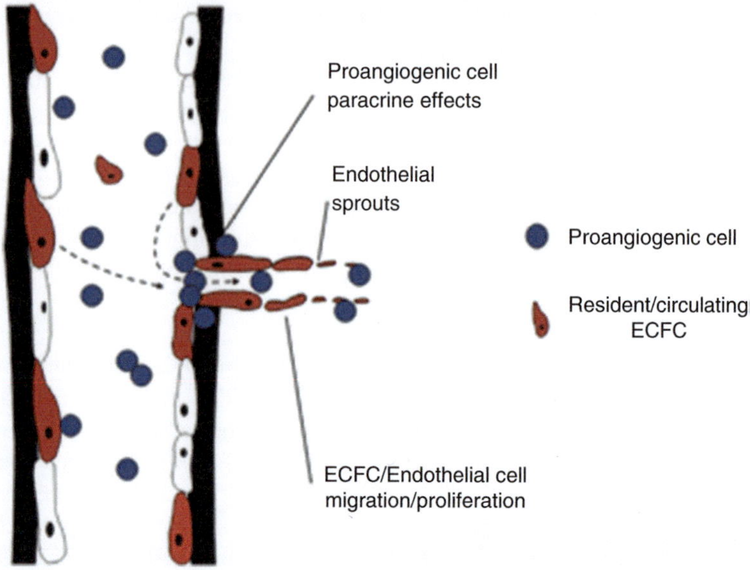

Fig. 11.1 Proangiogenic hematopoietic cells may interact with resident endothelial progenitor cells to promote neoangiogenesis. Circulating hematopoietic (bone marrow-derived) cells are proposed to provide paracrine signals that stimulate resident endothelial cell colony-forming cells to undergo neoangiogenesis. Schematic adapted from Basile and Yoder (2014)

11.4 Endothelial Heterogeneity in the Pulmonary Circulation

Although anatomical differences in endothelium lining pulmonary arteries, capillaries, and veins were widely acknowledged in the twentieth century (Crapo et al. 1982; Haies et al. 1981; Weibel 1973), in general, lung endothelium was considered to be a functionally inert, homogeneous layer functioning primarily as a semi-permeable barrier separating blood from the underlying tissue (Gebb and Stevens 2004). Evidence that the endothelium contributes to vasoregulation and metabolism, such as the conversion of angiotensin I to angiotensin II, directed a research focus on the "non-respiratory" functions of the lung (Pfannkuch and Blumcke 1985). These studies contributed to our understanding of functional heterogeneity among lung vascular compartments; for example, lung capillaries possess extensive angiotensin converting enzyme activity (Ryan et al. 1975, 1976; Ryan and Ryan 1984a). Recognition that certain lectins discriminate between pulmonary artery and microvascular endothelial cells provided a way to test whether these cells were phenotypically different in vivo, and ultimately led to a way to purify cell phenotypes in vitro (Gebb and Stevens 2004). Lectins are plant and animal proteins that recognize sugar moieties with nominal specificity, and therefore, differential lectin binding is indicative of distinctive cell surface molecular signatures.

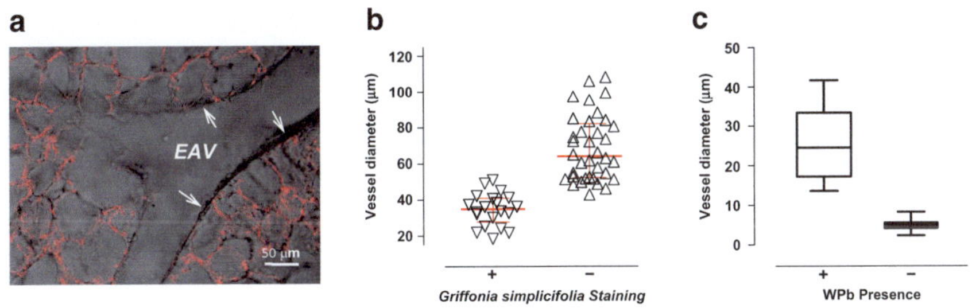

Fig. 11.2 Lung capillary endothelium is discriminated by Griffonia lectin, whereas precapillary endothelium possesses Weibel–Palade bodies. (**a**) The lung's circulation was gelatin-filled and the airways were agarose-filled, thick sections were cut, and the lung slices were incubated with Griffonia lectin. Whereas endothelium lining extra-alveolar blood vessels does not interact with the lectin, capillary endothelium uniformly interacts with *Griffonia simplicifolia*. (**b**) Griffonia lectin reveals a zone, in blood vessels ranging from 38 to 60 μm, where the macrovascular endothelial cell phenotype transitions to a microvascular endothelial cell phenotype. Endothelium lining blood vessels smaller than 38 μm in diameter are uniformly Griffonia-positive, and endothelium in blood vessels greater than 60 μm in diameter is Griffonia-negative. (**c**) Weibel–Palade bodies are seen in precapillary endothelial cells, but are not seen in capillary endothelial cells. Adapted from Wu et al. (2014)

Simultaneous infusion of labeled *Griffonia simplicifolia* and *Helix pomatia* lectins revealed a distinctive border between endothelial cell phenotypes, where pulmonary artery endothelial cells preferentially interacted with *Helix pomatia* and capillary endothelial cells preferentially interacted with *Griffonia simplicifolia*. No apparent overlap in lectin binding was noted. Recent studies seeking to resolve the transition zone found that all endothelial cells in blood vessels less than 38 μm in diameter interacted with Griffonia lectin, whereas half of the endothelial cells in blood vessels 38–60 μm in diameter interacted with Griffonia lectin, and none of the endothelial cells in blood vessels >60 μm in diameter interacted with *Griffonia simplicifolia* (Wu et al. 2014). These data illustrated a stark transition zone between 38 and 60 μm diameter pulmonary arterioles where a shift in endothelial cell phenotype is observed (Fig. 11.2). This transition zone resides immediately before arterioles enter the capillary plexus.

Weibel–Palade bodies are endothelial-specific organelles that contain many proteins important to the stimulated immune response, including P-selectin and von Willebrand factor (vWf). Interestingly, not all endothelial cells possess Weibel–Palade bodies. Fuchs and Weibel first recognized that lung capillary endothelial cells do not possess Weibel–Palade bodies, and more specifically, they found that the transition zone was in blood vessels with an approximately 20 μm internal diameter (Fuchs and Weibel 1966; Weibel 2012). The relationship between Weibel–Palade bodies and interaction with Griffonia lectin was recently compared in lung endothelium (Wu et al. 2014). In agreement with the results of Fuchs and Weibel (Fuchs and Weibel 1966; Weibel 2012), Weibel–Palade bodies were present in endothelium in small precapillary vessels, with internal diameters of approximately 18 μm. All of the endothelial cells in these blood vessels interacted with Griffonia lectin (Fig. 11.2). Thus, it appears that a macroheterogeneity among lung endothelial cell phenotypes is defined by recognition of the Griffonia lectin, although this molecular pattern does not designate the formation of Weibel–Palade bodies. These findings also resolve an important precapillary niche in lung endothelium, where a transition in phenotype occurs.

Recognition that changes in the structure of pulmonary artery and capillary endothelial cells are paralleled by distinct molecular signatures gave rise to the idea that not all endothelial cells within the pulmonary circulation are homogeneous in their function. The study of highly purified macro- and microvascular endothelial cell populations in culture has supported this view and provides insight into the rich source of endothelial progenitor cells seeded within lung capillaries.

11.5 Endothelial Heterogeneity: Insights Gained by the Study of Lung Endothelium in Culture

Mammalian cell isolation and culture has represented a breakthrough technology that is fundamental to identifying the molecular basis of cell physiology and pathophysiology. Endothelial cell cultures were not developed until the 1970s (Nachman and Jaffe 2004). Early endothelial cultures were generated from conduit-derived vessels, mostly from larger animals, because larger vessel segments were more easily accessible. What is considered normal endothelial cell identity in culture primarily arose from studies using aortic, pulmonary artery, and umbilical vein endothelial cells. In these primary cell lines, endothelial cells were shown to take up low-density lipoprotein (LDL) and to express proteins such as vWf, factor VIII, and platelet-endothelial cell adhesion molecule-1 (PECAM-1). More recently, expressions of the endothelial cell nitric oxide synthase (eNOS) and vascular endothelial cell cadherin (VE-cadherin) have been found useful in resolving endothelial identity.

Pioneering work by Ryan and colleagues paved the way for isolating capillary endothelial cells from the lungs of small animals (Habliston et al. 1979; Ryan and Ryan 1977, 1984b; Ryan et al. 1982). Using a peripheral lung cut technique and a bead retro-perfusion approach, Ryan's group was able to obtain lung microvascular endothelial cells suitable for study in vitro. They tested the metabolic activity of these cells, including their ability to generate angiotensin II, since angiotensin-converting enzyme is largely expressed in lung capillary endothelium. These approaches have been widely adapted by investigators over the past four decades.

Comparison of lung "microvascular" (i.e., PMVECs) and "macrovascular" (PAECs) endothelial cells led to new insight in vascular biology (Stevens 2005; Gebb and Stevens 2004; King et al. 2004; Ochoa et al. 2010). Key to this insight, however, was rigorous characterization of the cell phenotype. Both PMVECs and

PAECs display cobblestone morphology at confluence. Both cell types express markers characteristic of endothelium, including vWf, eNOS, VE-cadherin, and PECAM-1. However, PMVECs differ from PAECs with respect to their lectin-binding characteristics. PMVECs preferentially interact with *Griffonia simplicifolia* and *Glycine max*, whereas PAECs preferentially interact with *Helix pomatia*. Retention of these discriminating features is critical to the study of endothelial cell heterogeneity. It is noteworthy that most commercially available pulmonary microvascular endothelial cell lines do not report interaction with *Griffonia simplicifolia* in cell characterization. However, this feature of cell phenotyping is critical to resolving PMVEC behavior in culture that is reflective of its in vivo behavior. Thus, lectin-binding criteria is a critical component of endothelial cell phenotyping, especially when considering the heterogeneity of cell phenotypes.

Study of PMVECs and PAECs characterized using the aforementioned approaches has revealed that these cells differ with regard to many physiological properties, including their semi-permeable barrier function, mechanotransduction, migration, proliferation, and angiogenic capacity [reviewed in (Stevens 2005; Gebb and Stevens 2004; Ochoa and Stevens 2012; King et al. 2004; Ochoa et al. 2010)]. Rapid migration, proliferation, and angiogenic capacity is consistent with the presence of progenitor cells. Rapid PMVEC migration was noted in studies designed to evaluate mechanisms by which thrombin induces gap formation (Cioffi et al. 2002). Thrombin stimulated PMVEC gap formation within minutes of its application. However, even large micron-sized gaps were resealed within 2 h. In contrast, gap formation was slow and progressive in PAECs, and gap size was at its greatest 2 h post-thrombin challenge. Time-lapse movies revealed rapid PMVEC movement with extensive lamellipodia as cells resealed the barrier. Subsequent scratch wound assays substantiated these initial observations. From this work, it was surmised that PMVECs rapidly migrate, perhaps to sustain alveolar capillary integrity especially following injury.

If PMVECs migrate rapidly as a cellular feature designed to protect the alveolar-capillary membrane, then they may also proliferate rapidly to repair the post-natal lung following injury. Consistent with this idea, serum-stimulated growth is greater in PMVECs than it is in PAECs, and accordingly, population doubling time during log phase growth is much shorter in the microvascular cells (King et al. 2004). In contrast to PAECs, PMVEC growth, while slowed, is not abolished in low serum conditions (0.1 %) (Solodushko and Fouty 2007). The microvascular cells do not undergo G0/G1 arrest and progress through S phase. They possess increased cdk4 and cdk2 kinase activity along with hyperphosphorylated and inactive retinoblastoma and increased cyclin D1 protein. Despite maintaining growth in low serum conditions, the establishment of a monolayer—confluence—causes G0/G1 arrest, with concomitant retinoblastoma hypophosphorylation and $p27^{Kip1}$ upregulation. Thus, PMVECs possess an intrinsic growth advantage, but their growth advantage remains regulated by cell adhesion, e.g., formation of a monolayer.

11.6 Endothelial Microheterogeneity: Progenitor Cells Within a Population

The studies of Solodushko and Fouty (2007) provided mechanistic insight into what controls PMVEC progression through the cell cycle. However, their work focused on the growth characteristics of cell populations and did not seek ways to evaluate the presence, or function, of endothelial cell progenitor cells within the population. As discussed above (see Sect. 11.3), Ingram and colleagues (2004) utilized single cell cloning approaches to establish growth hierarchies within endothelial cell populations, and thereby developed functional assays to identify endothelial progenitors. Alvarez and colleagues (2008) adapted this approach for evaluation of PMVEC and PAEC progenitor growth characteristics.

Using PMVEC and PAEC populations, characterized as described (see Sect. 11.5), Alvarez and colleagues (2008) subjected cells to the clonogenic assay and established a hierarchy of growth potentials among single cells. Whereas the hierarchy of PAEC growth potentials was similar to previous descriptions in HUVECs, with <10 % of cells displaying a high proliferative potential, nearly 50 % of single PMVECs were highly replication competent (Fig. 11.3). Similar results were obtained in endothelial cells isolated from mice (Schniedermann et al. 2010). These data suggest that endothelial cells isolated from lung capillaries possess an intrinsically high proliferative potential with the capacity to self-renew, distinguishing characteristics of progenitor cells.

PMVECs display a restrictive barrier property, limiting trans-endothelial water, solute and protein flux. However, it is generally believed that cell adhesion limits cell proliferation. High proliferative potential cells isolated from PMVECs were tested to determine whether they establish a restrictive barrier (Alvarez et al. 2008). Similar to the parent cell population, high proliferative potential cells generated a

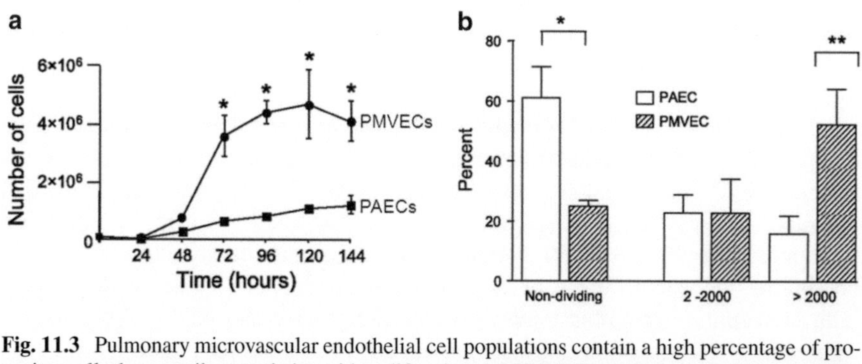

Fig. 11.3 Pulmonary microvascular endothelial cell populations contain a high percentage of progenitor cells that contribute to their rapid proliferation. (**a**) PMVEC populations proliferate rapidly when compared to PAEC populations. Cells were seeded at 10^5 cells in 35 mm dishes, and serum (10 %)-stimulated growth evaluated over 6 days. (**b**) Single cell clonogenic assay reveals a high percentage of replication competent PMVECs; as many at 50 % of PMVECs display a high proliferative capacity. Adapted from Alvarez et al. (2008)

restrictive barrier. These cells also retained expression of usual endothelial "marker" proteins, such as VE-cadherin, vWf, PECAM-1, eNOS, and N-cadherin. They retained an ability to interact with *Griffonia simplicifolia* and did not readily interact with *Helix pomatia*, similar to the parent cells. High proliferative potential cells expressed proteins characteristic of circulating endothelial progenitor cells, including vascular endothelial cell growth factor receptor-2 and CD105, but they did not express CD133 or CD45. Thus, high proliferative potential cells were selected from a population of lung microvascular endothelial cells, PMVECs, and while these individual cells grew rapidly and fulfilled the criterion of progenitor cells, they maintained their "microvascular" identity.

High proliferative potential endothelial cells were tested to determine whether they displayed evidence of replicative senescence (Alvarez et al. 2008). Short telomeres are associated with replicative senescence. Telomeres were fluorescently labeled, and their length evaluated against control cell populations. PAECs displayed the shortest telomere lengths and high proliferative potential PMVECs possessed the longest telomere lengths, consistent with their rapid proliferative capacity. Growth in soft agar matrix was measured to evaluate possible cell transformation. Whereas breast cancer cells grew into large, anchorage-independent clumps, growth was suppressed in PAECs, PMVECs, and high proliferative potential PMVECs, suggesting high proliferative potential cells displayed regulated cell cycle progression. Consistent with this idea, as high proliferative potential PMVECs grew to confluence in a serum (10 %)-stimulated growth curve, the percentage of S phase cells decreased substantially (e.g., cells came out of the cell cycle), similar to both PAECs and PMVECs. Both PMVECs and high proliferative potential PMVECs were rapidly neo-angiogenic as well, when compared to PAECs. However, selection of high proliferative potential cells did not enrich for a cell population with even greater angiogenic benefit, as both the parent PMVEC population and the high proliferative potential PMVECs generated similar numbers of blood vessels.

Studies seeking to resolve a molecular basis for rapid endothelial growth identified increased expression of nucleosome assembly protein 1 in PMVECs and in high proliferative potential PMVECs (Clark et al. 2008). Nucleosome assembly protein 1 is an epigenetic, pro-proliferative factor that is conserved from yeast to mammals. This protein was detected in a comparative mRNA profiling transcript analysis in PAECs and PMVECs. In the initial screen, nucleosome assembly protein 1 was highly expressed in PMVECs, even at confluence. Nucleosome assembly protein 1 downregulation decreased the proliferative and neo-angiogenic capacity of PMVECs, and its overexpression increased the proliferative and neo-angiogenic capacity of PAECs. In both experimental cases, however, changing nucleosome assembly protein 1 expression did not impact the expression of endothelial cell markers, including PECAM-1, VE-cadherin, and vWf. It also had no effect on lectin-binding criteria; PAECs recognized *Helix pomatia* even following nucleosome assembly protein 1 overexpression, and PMVECs recognized *Griffonia simplicifolia* even following nucleosome assembly protein 1 silencing. These findings implicate nucleosome assembly protein 1 as an epigenetic determinant of endothelial growth, but not in either endothelial or "microvascular" versus "macrovascular" phenotype specifications. Studies have not yet been completed to evaluate how nucleosome assembly protein 1 impacts the clonogenic potential, or the hierarchy of growth potentials, among single endothelial cells.

Endothelial cells that display rapid growth characteristics, such as PMVECs, can easily be identified in culture systems by the color of their medium at post-confluence. Mammalian cell culture medium contains phenol red, which is red at normal pH, light red-to-pink with alkalotic pH, and yellow with acidotic pH. The medium of post-confluent PMVECs becomes increasingly yellow, suggesting development of a lactic acidosis. This observation prompted investigation into the means by which PMVECs sustain bioenergetic demands during rapid growth (Parra-Bonilla et al. 2010, 2013). Microarray, RT-PCR, and western blot analysis revealed increased expression of glycolytic enzymes in PMVECs when compared with PAECs. Moreover, as cells grew to confluence in a standard serum-stimulated growth curve, PMVECs consumed glucose from the medium, generated lactate, and produced acidosis. Oxygen consumption was lower in PMVECs than it was in PAECs, and yet total ATP concentrations were higher in the microvascular cells. These data suggest that PMVECs, and likely high proliferative potential PMVECs, rely on glucose fermentation to sustain their rapid proliferation.

To test this idea more directly, glucose was restricted from the PMVEC medium during growth (Fig. 11.4). Glucose restriction produced a dose-dependent decrease in cell growth and attenuated the lactate accumulation and acidosis.

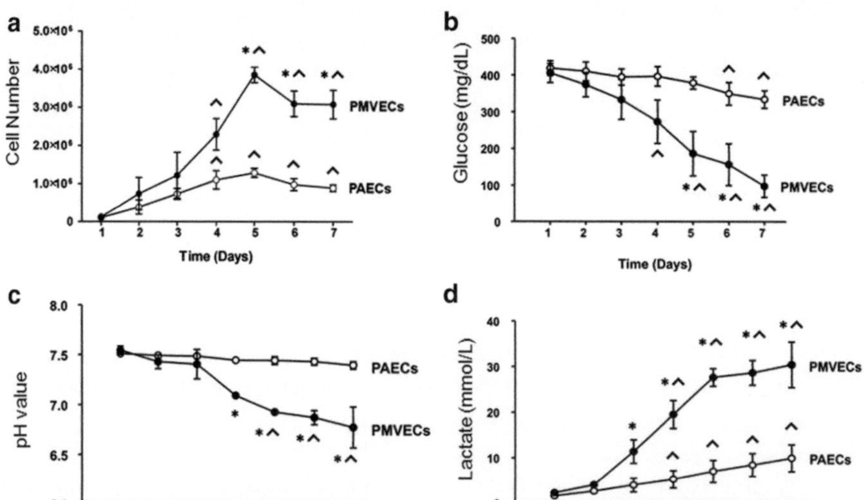

Fig. 11.4 Pulmonary microvascular endothelial cells utilize aerobic glycolysis to sustain their bioenergetic demands during proliferation. As PMVECs undergo rapid proliferation (**a**), they consume glucose from the media (**b**), produce an acidosis (**c**), and generate lactate (**d**). These features, including glucose consumption and production of lactic acidosis in the presence of sufficient oxygen, are characteristic of aerobic glycolysis. Adapted from Parra-Bonilla and coworkers (2010)

Moreover, galactose substitution for glucose abolished the lactic acidosis and greatly reduced cell proliferation, while reducing PMVEC ATP concentrations. Interestingly, supplying extracellular lactate to the galactose-treated PMVECs rescued ATP concentrations. Glucose consumption and the reliance on glycolysis to sustain ATP concentrations and cell proliferation are consistent with the use of aerobic glycolysis to sustain the bioenergetic demands of PMVEC growth.

Lactate dehydrogenase is the enzyme responsible for converting pyruvate to lactate in a reaction that oxidizes NADH to NAD^+ and H^+. The functional lactate dehydrogenase enzyme is a tetramer comprised of different combinations of “A” (or M, predominates in muscle and liver) and “B” (or H, predominates in heart) proteins (Jungmann et al. 1998; Zhong and Howard 1990). The different combinations of A and B subunits produce five separate enzymes, lactate dehydrogenase 1–5. Lactate dehydrogenase 5 is comprised of four A subunits. Lactate dehydrogenase A activity had been linked to cell proliferation and neoangiogenesis, prompting investigation into the function of this protein in maintaining PMVEC rapid growth (Parra-Bonilla et al. 2010, 2013). Pharmacological inhibition, and genetic disruption, of lactate dehydrogenase A reversibly impaired aerobic glycolysis, decreased proliferation, and decreased neo-angiogenesis. Thus, PMVECs utilize aerobic glycolysis to meet the bioenergetic demands of rapid angiogenesis.

11.7 Lung Capillaries: A Progenitor Endothelial Cell Niche?

Considerable work in the past decade has validated the anatomical, biochemical, and physiological heterogeneity of lung macrovascular and microvascular endothelium. A specialized transition zone has been identified in small precapillary blood vessels, where the phenotype switches from one characteristic of a “macrovascular” to a “microvascular” cell type (Wu et al. 2014). It is interesting to note that cell–cell junctions between these PAECs and PMVECs is quite distinct (Ofori-Acquah et al. 2008), leading to the hypothesis that this vascular site—where phenotypically different cell types border one another—might contribute to pulmonary vasculopathy (Stevens 2005). Indeed, this transition zone is recognized as the site where complex, lumen-occluding lesions form in pulmonary arterial hypertension. These lesions are at least partly due to the exuberant overgrowth of endothelial cells that lose the “law of the monolayer,” perhaps beginning with the overgrowth of apoptosis-resistant cells (Tuder et al. 2001). Preliminary results suggest that cells in the occlusive lesion interact with *Griffonia simplicifolia* and overexpress both nucleosome assembly protein 1 and lactate dehydrogenase A (Stevens 2005). It will be important to better resolve whether the apoptosis-resistant, lumen-occluding cells represent endothelial progenitor cells, and whether they have a “macrovascular” or “microvascular” specification.

Evidence supporting the idea that lung capillary endothelial cells represent a progenitor cell niche comes from the study of highly characterized cells in culture. PMVECs grow rapidly in cell population studies. An abundance of individual

PMVECs display high proliferative potential, and these cells reconstitute the entire growth hierarchy of growth potentials. These cells express circulating endothelial cell progenitor cell markers CD34 and VEGFR2, but not CD113, and they express a high level of nucleosome assembly protein 1 and glycolytic enzymes, including lactate dehydrogenase A. They require aerobic glycolysis to sustain rapid proliferation and neo-angiogenesis. All of these cellular characteristics support the notion that lung capillaries are enriched with resident progenitor cells.

However, resident endothelial progenitor cells have not been specifically distinguished in the intact capillary. Currently, there is not a universally accepted approach to visualize progenitor cells within the vessel wall and discriminate their location and function among neighboring cells. Clear evidence for endothelial microheterogeneity—a heterogeneity between adjacent cells within any specific vascular segment—has been seen in intact vessels (Majno and Palade 1961; Majno et al. 1961; Aird 2005). Perhaps this concept is best known from dynamic measurements of cytosolic calcium, where endothelial cells display dynamic transitions in cytosolic calcium even at rest. Here, the vessel-averaged endothelial cell cytosolic calcium signals are not synonymous with individual cell cytosolic calcium signals. How these different calcium signals translate into cellular physiology, within the consortium of cells comprising the vessel wall, remains unclear. Decoding complex cytosolic calcium signatures may be useful in identifying endothelial progenitor cells within a vessel wall, to match the signature with an important cellular physiology. Such functional assays may complement ongoing efforts to identify protein matrices that define cellular function (Li et al. 2011; Nolan et al. 2013). While we recognize that no single protein marker is sufficient to define endothelial cell progenitor cells, in time, a protein matrix can aid the search for their location within the vascular wall.

Recognition that endothelial populations possess cells with a hierarchy of growth potentials, including highly replication competent cells that fulfill the definition of progenitor cells, represents a major advance in vascular biology. As the field moves forward, it will be essential to identify the in vivo location of these progenitor cells in the post-natal lung with precision, to understand their role in vascular maintenance and to appreciate how they contribute to vascular repair following injury. Advancing our fundamental knowledge of resident endothelial progenitor cells may reveal their therapeutic potential.

References

Aird WC (2005) Spatial and temporal dynamics of the endothelium. J Thromb Haemost 3(7):1392–1406

Alvarez DF, Huang L, King JA, ElZarrad MK, Yoder MC, Stevens T (2008) Lung microvascular endothelium is enriched with progenitor cells that exhibit vasculogenic capacity. Am J Physiol Lung Cell Mol Physiol 294(3):L419–L430

Asahara T, Murohara T, Sullivan A, Silver M, van der Zee R, Li T, Witzenbichler B, Schatteman G, Isner JM (1997) Isolation of putative progenitor endothelial cells for angiogenesis. Science 275(5302):964–967

Asosingh K, Aldred MA, Vasanji A, Drazba J, Sharp J, Farver C, Comhair SA, Xu W, Licina L, Huang L, Anand-Apte B, Yoder MC, Tuder RM, Erzurum SC (2008) Circulating angiogenic precursors in idiopathic pulmonary arterial hypertension. Am J Pathol 172(3):615–627

Basile DP, Yoder MC (2014) Circulating and tissue resident endothelial progenitor cells. J Cell Physiol 229(1):10–16

Becker AJ, Mc CE, Till JE (1963) Cytological demonstration of the clonal nature of spleen colonies derived from transplanted mouse marrow cells. Nature 197:452–454

Chao H, Hirschi KK (2010) Hemato-vascular origins of endothelial progenitor cells? Microvasc Res 79(3):169–173

Cioffi DL, Moore TM, Schaack J, Creighton JR, Cooper DM, Stevens T (2002) Dominant regulation of interendothelial cell gap formation by calcium-inhibited type 6 adenylyl cyclase. J Cell Biol 157(7):1267–1278

Clark J, Alvarez DF, Alexeyev M, King JA, Huang L, Yoder MC, Stevens T (2008) Regulatory role for nucleosome assembly protein-1 in the proliferative and vasculogenic phenotype of pulmonary endothelium. Am J Physiol Lung Cell Mol Physiol 294(3):L431–L439

Crapo JD, Barry BE, Gehr P, Bachofen M, Weibel ER (1982) Cell number and cell characteristics of the normal human lung. Am Rev Respir Dis 125(6):740–745

Duong HT, Erzurum SC, Asosingh K (2011) Pro-angiogenic hematopoietic progenitor cells and endothelial colony-forming cells in pathological angiogenesis of bronchial and pulmonary circulation. Angiogenesis 14(4):411–422

Fuchs E, Segre JA (2000) Stem cells: a new lease on life. Cell 100(1):143–155

Fuchs A, Weibel ER (1966) Morphometric study of the distribution of a specific cytoplasmatic organoid in the rat's endothelial cells. Z Zellforsch Mikrosk Anat 73:1–9

Gebb S, Stevens T (2004) On lung endothelial cell heterogeneity. Microvasc Res 68(1):1–12

Habliston DL, Whitaker C, Hart MA, Ryan US, Ryan JW (1979) Isolation and culture of endothelial cells from the lungs of small animals. Am Rev Respir Dis 119(6):853–868

Haies DM, Gil J, Weibel ER (1981) Morphometric study of rat lung cells. I. Numerical and dimensional characteristics of parenchymal cell population. Am Rev Respir Dis 123(5):533–541

Hill JM, Zalos G, Halcox JP, Schenke WH, Waclawiw MA, Quyyumi AA, Finkel T (2003) Circulating endothelial progenitor cells, vascular function, and cardiovascular risk. N Engl J Med 348(7):593–600

Hirschi KK, Ingram DA, Yoder MC (2008) Assessing identity, phenotype, and fate of endothelial progenitor cells. Arterioscler Thromb Vasc Biol 28(9):1584–1595

Hristov M, Weber C (2004) Endothelial progenitor cells: characterization, pathophysiology, and possible clinical relevance. J Cell Mol Med 8(4):498–508

Ingram DA, Mead LE, Tanaka H, Meade V, Fenoglio A, Mortell K, Pollok K, Ferkowicz MJ, Gilley D, Yoder MC (2004) Identification of a novel hierarchy of endothelial progenitor cells using human peripheral and umbilical cord blood. Blood 104(9):2752–2760

Ingram DA, Mead LE, Moore DB, Woodard W, Fenoglio A, Yoder MC (2005) Vessel wall-derived endothelial cells rapidly proliferate because they contain a complete hierarchy of endothelial progenitor cells. Blood 105(7):2783–2786

Jungmann RA, Huang D, Tian D (1998) Regulation of LDH-A gene expression by transcriptional and posttranscriptional signal transduction mechanisms. J Exp Zool 282(1–2):188–195

King J, Hamil T, Creighton J, Wu S, Bhat P, McDonald F, Stevens T (2004) Structural and functional characteristics of lung macro- and microvascular endothelial cell phenotypes. Microvasc Res 67(2):139–151

Kissel CK, Lehmann R, Assmus B, Aicher A, Honold J, Fischer-Rasokat U, Heeschen C, Spyridopoulos I, Dimmeler S, Zeiher AM (2007) Selective functional exhaustion of hematopoietic progenitor cells in the bone marrow of patients with postinfarction heart failure. J Am Coll Cardiol 49(24):2341–2349

Li Y, Massey K, Witkiewicz H, Schnitzer JE (2011) Systems analysis of endothelial cell plasma membrane proteome of rat lung microvasculature. Proteome Sci 9(1):15

Lin Y, Weisdorf DJ, Solovey A, Hebbel RP (2000) Origins of circulating endothelial cells and endothelial outgrowth from blood. J Clin Invest 105(1):71–77

Majno G, Palade GE (1961) Studies on inflammation. 1. The effect of histamine and serotonin on vascular permeability: an electron microscopic study. J Biophys Biochem Cytol 11:571–605

Majno G, Palade GE, Schoefl GI (1961) Studies on inflammation. II. The site of action of histamine and serotonin along the vascular tree: a topographic study. J Biophys Biochem Cytol 11:607–626

Medina RJ, O'Neill CL, Sweeney M, Guduric-Fuchs J, Gardiner TA, Simpson DA, Stitt AW (2010) Molecular analysis of endothelial progenitor cell (EPC) subtypes reveals two distinct cell populations with different identities. BMC Med Genomics 3:18

Nachman RL, Jaffe EA (2004) Endothelial cell culture: beginnings of modern vascular biology. J Clin Invest 114(8):1037–1040

Nolan DJ, Ginsberg M, Israely E, Palikuqi B, Poulos MG, James D, Ding BS, Schachterle W, Liu Y, Rosenwaks Z, Butler JM, Xiang J, Rafii A, Shido K, Rabbany SY, Elemento O, Rafii S (2013) Molecular signatures of tissue-specific microvascular endothelial cell heterogeneity in organ maintenance and regeneration. Dev Cell 26(2):204–219

Ochoa CD, Stevens T (2012) Studies on the cell biology of interendothelial cell gaps. Am J Physiol Lung Cell Mol Physiol 302(3):L275–L286

Ochoa CD, Wu S, Stevens T (2010) New developments in lung endothelial heterogeneity: von Willebrand factor, P-selectin, and the Weibel-Palade body. Semin Thromb Hemost 36(3): 301–308

Ofori-Acquah SF, King J, Voelkel N, Schaphorst KL, Stevens T (2008) Heterogeneity of barrier function in the lung reflects diversity in endothelial cell junctions. Microvasc Res 75(3): 391–402

Parra-Bonilla G, Alvarez DF, Al-Mehdi AB, Alexeyev M, Stevens T (2010) Critical role for lactate dehydrogenase A in aerobic glycolysis that sustains pulmonary microvascular endothelial cell proliferation. Am J Physiol Lung Cell Mol Physiol 299(4):L513–L522

Parra-Bonilla G, Alvarez DF, Alexeyev M, Vasauskas A, Stevens T (2013) Lactate dehydrogenase a expression is necessary to sustain rapid angiogenesis of pulmonary microvascular endothelium. PLoS One 8(9):e75984

Pfannkuch F, Blumcke S (1985) What's new in lung physiology? Pulmonary vessel regulation/non-respiratory metabolic lung functions. Pathol Res Pract 180(6):718–720

Prasain N, Lee MR, Vemula S, Meador JL, Yoshimoto M, Ferkowicz MJ, Fett A, Gupta M, Rapp BM, Saadatzadeh MR, Ginsberg M, Elemento O, Lee Y, Voytik-Harbin SL, Chung HM, Hong KS, Reid E, O'Neill CL, Medina RJ, Stitt AW, Murphy MP, Rafii S, Broxmeyer HE, Yoder MC (2014) Differentiation of human pluripotent stem cells to cells similar to cord-blood endothelial colony-forming cells. Nat Biotechnol 32(11):1151–1157

Reya T, Morrison SJ, Clarke MF, Weissman IL (2001) Stem cells, cancer, and cancer stem cells. Nature 414(6859):105–111

Richardson MR, Yoder MC (2011) Endothelial progenitor cells: quo vadis? J Mol Cell Cardiol 50(2):266–272

Ryan JW, Ryan US (1977) Pulmonary endothelial cells. Fed Proc 36(13):2683–2691

Ryan US, Ryan JW (1984a) The ultrastructural basis of endothelial cell surface functions. Biorheology 21(1–2):155–170

Ryan US, Ryan JW (1984b) Cell biology of pulmonary endothelium. Circulation 70(5 Pt 2): III46–III62

Ryan JW, Ryan US, Schultz DR, Whitaker C, Chung A (1975) Subcellular localization of pulmonary antiotensin-converting enzyme (kininase II). Biochem J 146(2):497–499

Ryan US, Ryan JW, Whitaker C, Chiu A (1976) Localization of angiotensin converting enzyme (kininase II). II. Immunocytochemistry and immunofluorescence. Tissue Cell 8(1):125–145

Ryan US, White LA, Lopez M, Ryan JW (1982) Use of microcarriers to isolate and culture pulmonary microvascular endothelium. Tissue Cell 14(3):597–606

Schniedermann J, Rennecke M, Buttler K, Richter G, Stadtler AM, Norgall S, Badar M, Barleon B, May T, Wilting J, Weich HA (2010) Mouse lung contains endothelial progenitors with high capacity to form blood and lymphatic vessels. BMC Cell Biol 11:50

Solodushko V, Fouty B (2007) Proproliferative phenotype of pulmonary microvascular endothelial cells. Am J Physiol Lung Cell Mol Physiol 292(3):L671–L677

Stevens T (2005) Molecular and cellular determinants of lung endothelial cell heterogeneity. Chest 128(6 Suppl):558S–564S

Takahashi K, Yamanaka S (2006) Induction of pluripotent stem cells from mouse embryonic and adult fibroblast cultures by defined factors. Cell 126(4):663–676

Till JE (1961) Radiation effects on the division cycle of mammalian cells in vitro. Ann N Y Acad Sci 95:911–919

Tuder RM, Cool CD, Yeager M, Taraseviciene-Stewart L, Bull TM, Voelkel NF (2001) The pathobiology of pulmonary hypertension. Endothelium. Clin Chest Med 22(3):405–418

Wagers AJ, Weissman IL (2004) Plasticity of adult stem cells. Cell 116(5):639–648

Weibel ER (1973) Morphological basis of alveolar-capillary gas exchange. Physiol Rev 53(2):419–495

Weibel ER (2012) Fifty years of Weibel-Palade bodies: the discovery and early history of an enigmatic organelle of endothelial cells. J Thromb Haemost 10(6):979–984

Weissman IL (2000) Stem cells: units of development, units of regeneration, and units in evolution. Cell 100(1):157–168

Weissman IL, Anderson DJ, Gage F (2001) Stem and progenitor cells: origins, phenotypes, lineage commitments, and transdifferentiations. Annu Rev Cell Dev Biol 17:387–403

West JB (2013) Marcello Malpighi and the discovery of the pulmonary capillaries and alveoli. Am J Physiol Lung Cell Mol Physiol 304(6):L383–L390

Wu AM, Siminovitch L, Till JE, McCulloch EA (1968a) Evidence for a relationship between mouse hemopoietic stem cells and cells forming colonies in culture. Proc Natl Acad Sci U S A 59(4):1209–1215

Wu AM, Till JE, Siminovitch L, McCulloch EA (1968b) Cytological evidence for a relationship between normal hematopoietic colony-forming cells and cells of the lymphoid system. J Exp Med 127(3):455–464

Wu S, Zhou C, King JA, Stevens T (2014) A unique pulmonary microvascular endothelial cell niche revealed by Weibel-Palade bodies and Griffonia simplicifolia. Pulm Circ 4(1):110–115

Yamanaka S, Takahashi K (2006) Induction of pluripotent stem cells from mouse fibroblast cultures. Tanpakushitsu Kakusan Koso 51(15):2346–2351

Yoder MC (2012) Human endothelial progenitor cells. Cold Spring Harb Perspect Med 2(7):a006692

Yoder MC, Mead LE, Prater D, Krier TR, Mroueh KN, Li F, Krasich R, Temm CJ, Prchal JT, Ingram DA (2007) Redefining endothelial progenitor cells via clonal analysis and hematopoietic stem/progenitor cell principals. Blood 109(5):1801–1809

Yoon CH, Hur J, Park KW, Kim JH, Lee CS, Oh IY, Kim TY, Cho HJ, Kang HJ, Chae IH, Yang HK, Oh BH, Park YB, Kim HS (2005) Synergistic neovascularization by mixed transplantation of early endothelial progenitor cells and late outgrowth endothelial cells: the role of angiogenic cytokines and matrix metalloproteinases. Circulation 112(11):1618–1627

Zhong XH, Howard BD (1990) Phosphotyrosine-containing lactate dehydrogenase is restricted to the nuclei of PC12 pheochromocytoma cells. Mol Cell Biol 10(2):770–776

Chapter 12
Pulmonary Vascular Remodeling by Resident Lung Stem and Progenitor Cells

Rubin Baskir and Susan Majka

Abbreviations

ABCG2	ATP-binding cassette sub-family G member 2
BAL	Bronchiolar lavage
BM-MSC	Bone marrow-derived mesenchymal stem cells
BMP4	Bone morphogenic protein 4
BPD	Bronchopulmonary dysplasia
CD31	Cluster of differentiation molecule 31, also known as PECAM
CD45	Cluster of differentiation molecule 45, also known as leukocyte common antigen
COPD	Chronic obstructive lung disease
CVC	Calcifying vascular cells
EC	Endothelial cell
EndMT	Endothelial-mesenchymal transition
EPC	Endothelial progenitor cells
Fb	Fibroblast
FGF (R)	Fibroblast growth factor (receptor)
FLK1	Fetal liver kinase 1, also known as CD309, KDR, and VEGFR2
FOXF1	Forkhead box protein F1
GFP	Green fluorescent protein
hESC	Human embryonic stem cell
HOXB5/N5	Homeobox protein Hox-B5/N5
ILD	Interstitial lung diseases

R. Baskir • S. Majka, Ph.D. (✉)
Department of Medicine, Division of Allergy, Pulmonary and Critical Care Medicine, Vanderbilt Center for Stem Cell Biology, Cell and Developmental Biology, Vanderbilt University, P475 MRBIV/Langford, 2213 Garland Ave, Nashville, TN 37232, USA
e-mail: Susan.M.Majka@Vanderbilt.Edu

A. Firth, J.X.-J. Yuan (eds.), *Lung Stem Cells in the Epithelium and Vasculature*, Stem Cell Biology and Regenerative Medicine, DOI 10.1007/978-3-319-16232-4_12

IPF	Idiopathic pulmonary fibrosis
LAM	Lymphangioleiomyomatosis
MEF2D	Myocyte-specific enhancer factor 2D
MSC	Mesenchymal stem cell
MyoFb	Myofibroblast
PAH	Pulmonary arterial hypertension
PDGF (R)	Platelet-derived growth factor (receptor)
PH	Pulmonary hypertension
RGS5	Regulator of G-protein signaling 5
Shh	Sonic hedgehog
SMC	Smooth muscle cell
TGFβ	Transforming growth factor beta
Thy1	Thymocyte antigen 1, also known as CD90
VAFs	Vascular adventitial fibroblasts
VeCad	Vascular endothelial cadherin
VSMC	Vascular smooth muscle cell
WT1	Wilms' tumor protein

12.1 Introduction

Any attempt to determine the reparative capacity of stem cell populations in the lung vasculature must first begin with a precise understanding of the different types and locations of stem and progenitor cells in the lung. This chapter will describe known stem and progenitor cell populations associated with the lung vasculature in terms of their distinguishing cell makers and responses to injury and remodeling.

There are several stem cell types that reside in the proximal vasculature of the lung, such as the major arteries, versus the more distal areas of the lung, such as alveoli-associated microvessels. The distinction between proximal and distal vasculature is especially apparent in how both areas respond to the remodeling that takes place after injury: Remodeling of arteries includes adventitial thickening, and medial and intimal hypertrophy whereas the remodeling of alveoli-associated microvessels features muscularization and loss of microvessels. While we know that resident pulmonary stem and progenitor cells exist throughout the vascular tree, our current understanding of the role resident pulmonary stem and progenitor cells play during vascular remodeling and lung disease remains limited.

Pulmonary hypertension (PH) is a lung disease that is characterized by vasoconstriction, remodeling of the large and small pulmonary arteries and occlusion or rarefaction of microvessels in the lung. These functional and structural changes lead to increased pulmonary vascular resistance and subsequent right heart failure (Morrell et al. 2009). Types of PH include: familial or idiopathic pulmonary arterial hypertension (PAH) and PH as a secondary complication of many lung diseases (Simonneau et al. 2004).

Chronic adult lung diseases are associated with a high incidence of PH or vascular remodeling. These diseases include chronic obstructive lung disease (COPD), interstitial lung diseases (ILD) such as idiopathic pulmonary fibrosis (IPF), sarcoidosis or asthma, and Lymphangioleiomyomatosis (LAM) (Seeger et al. 2013; Alagappan et al. 2013). A recent study by Patel and colleagues demonstrated that IPF patients share common gene expression profiles in remodeled arterioles with patients with diagnosed PH (Patel et al. 2013). LAM is a rare lung disease that can occur sporadically or in association with tuberous sclerosis and is characterized by disorderly proliferation and accumulation of smooth muscle throughout the lungs, around pulmonary arteries, venules and in the perivascular spaces and airways (Ferri et al. 2004).

PH also complicates neonatal respiratory diseases such as bronchopulmonary dysplasia (BPD), Adams–Oliver syndrome, and infantile scleroderma. BPD is characterized by impaired alveolarization and vasculogenesis and features vascular remodeling, increased vascular tone, and altered vasoreactivity (Berkelhamer et al. 2013). Vascular pathology and PH associated with Adams–Oliver syndrome includes capillary rarefaction or loss as well as vessel stenosis (Patel et al. 2004). Loss of capillaries and vasculopathy are also major contributors to clinical manifestations of scleroderma (Fleming et al. 2009).

Understanding the roles pulmonary vascular stem and progenitor cells play in tissue homeostasis and remodeling may lead to new strategies that harness the reparative capacity of these cell types while preventing their participation in dysfunctional remodeling processes.

12.2 Putative Stem and Progenitor Niches in the Pulmonary Vascular Tree: Potential for Vascular Remodeling

12.2.1 Overview of Vascular Remodeling During Lung Disease

The progenitor and stem cell niches present in the pulmonary vascular tree are not as well-described as their corresponding airway epithelial counterparts. Novel primitive cell types have been identified in vitro; however, their study and localization in vivo is complicated by the lack of specific markers. The resident lung progenitor and stem cells that have the potential to contribute to vascular remodeling during disease are presented. The location, characteristics, and function of these cells are summarized.

The arterial vascular network has been characterized in detail due to the prevalence of remodeling during adult lung disease. The arteries parallel the epithelial respiratory tree, and similar to the airways, one might assume that the local populations of stem and progenitor cells are likely specialized based on the nature of the vessel branch with which they associate. There is gradual loss of muscularization over the branches of the vascular tree (Townsley 2012) (Fig. 12.1; Table 12.1). The pulmonary artery is the largest and most muscularized portion of the arterial circulation, proximal and closest to the trachea. The branching smaller arteries, also referred to as arterioles, are defined by size and muscularization. The small artery/arteriole is composed of 1–2 layers of smooth muscle which completely

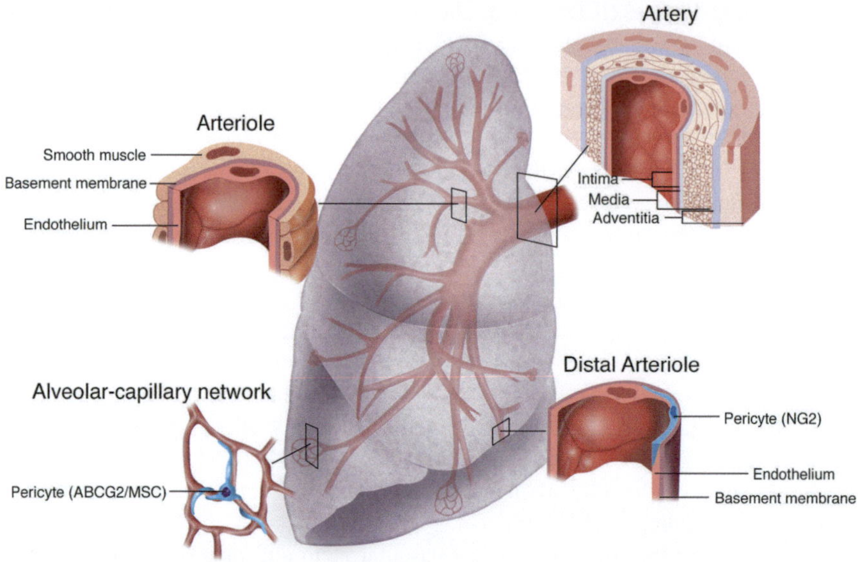

Fig. 12.1 The pulmonary arterial vascular tree. The pulmonary arterial circulation is characterized by a gradual loss of complexity and muscularization from proximal to distal branches of the vascular tree (Townsley 2012). The pulmonary artery is the most proximal and muscularized portion of the arterial circulation. The pulmonary artery consists of defined layers, including the intima (endothelium), the media (smooth muscle), and the adventitia (fibroblasts, microvessels, and vascular precursors). The branching smaller arteries, also referred to as arterioles, are defined by size and muscularization. The small artery/arteriole is composed of 1–2 layers of smooth muscle which completely encircle the endothelium of the vessels. With further branching, smooth muscle becomes less continuous and the most distal vessels and capillaries, consisting of endothelial cells forming a small diameter conduit, are stabilized by a non-continuous layer of pericytes. Pericytes are directly in contact with the endothelium and reside under a continuous basement membrane

Table 12.1 Putative location of stem and progenitor cells in the adult pulmonary vasculature

	Artery	Small artery	Distal artery	Alveolar—capillary network
Characteristics	Muscularized	Muscular arteriole	Non-muscular arteriole	Not muscularized
		1–2 SMC layers	Non-continuous pericyte coverage	Few pericytes with extended processes
Function	Contractile	Contractile	Contractile	Non-contractile
Associated stem and progenitor cells	Mesoangioblast/EPC, SMC progenitors	SMC progenitors	Pericytes (NG2)	MSC pericytes (ABCG2), EPC

encircle the vessels. With further branching, smooth muscle becomes less continuous and the most distal vessels and capillaries, consisting of 2–4 endothelial cells, are stabilized by a non-continuous layer of pericytes. The capillary endothelium consists of 30–50 % of the total cell population in the alveolar septal wall of both rodent and human (Townsley 2012).

12.2.2 *Mesoangioblasts*

The mesoangioblast is a vascular-associated progenitor cell, typically residing in artery wall, that can give rise to vascular cells as well as other mesodermal lineages including skeletal and cardiac muscle. The name "mesoangioblast" distinguishes this cell type from the "hemangioblast," another progenitor cell that can give rise to vascular as well as hematopoetic lineages. Minasi and colleagues first defined the mesoangioblast by transplanting cells derived from embryonic mouse or quail dorsal aorta into chick embryos (Minasi et al. 2002). Donor cells were found in numerous mesodermal tissues, including bone, smooth, and cardiac muscle. These cells could be cultured and differentiated into osteoblasts, adipocytes, and skeletal myotubes.

Although mesoangioblasts possess a number of cell markers similar to mesenchymal stem cells (MSCs), they cannot be classified as MSCs. Embryonic mesoangioblasts express CD34, while MSCs and pericytes do not (Wang et al. 2012). Mesoangioblasts may also express Flk1, MEF2D, Sca1, and Thy1, with varying reports of cKit (Passman et al. 2008; Campagnolo et al. 2010; Zengin et al. 2006). Transcriptome analysis of mesoangioblasts also determined that they express receptors for FGF, PDGF, TGFβ, and Wnt proteins (Tagliafico et al. 2004).

While mesoangioblasts were first isolated from dorsal aorta, the developmental origin of mesoangioblasts remains unknown. Flk1 lineage tracing determined that Flk1+ positive cells can give rise to skeletal and cardiac muscle, similar to Minasi's chimeric studies (Motoike et al. 2003). However, Flk1 is not a unique marker for mesoangioblasts. A recent study by Vodyanik and colleagues determined that hESC differentiated to an MSC lineage express genes specific to lateral plate mesoderm during differentiation, indicating that MSC and MSC-like cells have a lateral plate mesodermal origin (Vodyanik et al. 2010). However, the MSC subset Vodyanik studied appeared to possess a more restricted differentiation potential than mesoangioblasts, making it inaccurate to assume that mesoangioblasts arise from lateral plate mesoderm as well.

Mesoangioblasts have been studied as a potential cell therapy source for muscular regeneration, but have not been used to treat pulmonary diseases (Galli et al. 2005; Galvez et al. 2006; Sampaolesi et al. 2003, 2006). Mesoangioblasts were used to treat mice with X-chromosome-linked muscular dystrophy (mdx), alpha-SG-null mice, and golden retrievers with heredity muscular dystrophy. Animals that are administered mesoangioblasts showed some improvement in muscular function.

Interestingly, when GFP-labeled mesoangioblasts were injected into the femoral arteries of mice with cardiotoxin-induced muscular injury, 5 % of the injected mesoangioblast population migrated to the lung (Galvez et al. 2006). Whether mesoangioblast treatment could alleviate the pathology of pulmonary injury remains unknown. The role of these cells in remodeling during lung disease has not been evaluated and a precise understanding of their biology is confounded by the lack of a specific marker to define their origin in vivo.

12.2.3 Smooth Muscle Progenitors

Smooth muscle proliferation and hypertrophy are characteristic of vascular remodeling in the pulmonary artery and arterioles. A population of smooth muscle cell (SMC) progenitors is responsible for SMC expansion. These progenitors are hypothesized to reside in tissue interstitium, arterial adventitia as well as the mesothelium. The characterization of smooth muscle progenitors to date includes their identification by cell marker analysis including, but not limited to, smooth muscle alpha actin (*acta2*), SM22a (t*agln*), Tie-1, and PDGFRα/β (Majesky et al. 2011b; Minasi et al. 2002). The first smooth muscle progenitors described were defined by their expression of PDGFR α and β during lung development and terminal differentiation and their recruitment to newly forming endothelial tubes by PDGF ligands (Lindahl et al. 1997; Hellstrom et al. 1999).

The mesothelium is an extra-pulmonary tissue that may provide a source of SMC progenitors in the adult lung during tissue homeostasis as well as disease-associated remodeling. The mesothelium is a component of the pleura that encases the adult lung. Mesothelial expression of the Wilms tumor 1 gene (WT1) was exploited to perform lineage tracing of mesothelium to both mesenchymal and smooth muscle lineages within the lung. However, WT-1 expression is not strictly limited to the mesothelial cells in the adult lung, complicating lineage-tracing studies to determine the role these cells play during adult tissue homeostasis and disease.

SMC progenitors in homeostasis, development, and disease are controlled by a variety of cell signaling pathways, including FGF, Shh, Wnt, PDGF, and Notch. FGF10-FGFR2b signaling during lung development prevents SMC differentiation and limits matrix deposition (De Langhe et al. 2006). Studies by Passman et al. demonstrated that a gradient of Shh in the adventitia regulates the number of Sca1 SMC progenitors lacking differentiated markers, but expressing transcription factors of SMC lineage potential, including serum response factor, Klf-4, Msx1, and fox04 (Passman et al. 2008). Regulation of proliferation by the Wnt and PDGFR signaling pathways in SM22-labeled SMC progenitors was also reported by Cohen et al. (2009). A temporal increase in Notch signaling was required for Ve-cad^{neg}/CD45^{neg}/Tie1^{pos}/CD31^{dim} terminal differentiation to SMC, but not the maintenance of a stable vessel (Chang et al. 2012). These complex interactions during development involve coordinated signaling between the epithelium, endothelium, and mesenchyme to regulate appropriate tissue patterning and development.

12.2.4 Pericytes

A pericyte is a perivascular-associated progenitor cell found within the basement membrane of the vasculature. Pericytes were first described by Charles-Marie Benjamen Rouget (Armulik et al. 2011) and named "Rouget cells," until being renamed for their proximity to endothelial cells by Zimmerman in a 1923 publication. While pericytes are thought to derive from progenitors found in the neurocrest in the forebrain or mesenchyme, lung-resident pericytes also derive from the mesothelium (Que et al. 2008; Bergers and Song 2005). Pericytes, like many mesenchymal progenitor cells, exist in a heterogeneous population with no single cell marker. Also, because many other stromal populations exist in the periendothelial space, they are often confused with other cell types, such as MSC, vascular SMC (vSMC), Fb, and macrophages. Thus, the "gold standard" for pericyte identification is a combination of analysis using multiple markers as well as ultrastructural analysis.

Pericytes have been found to express SMA, Desmin, RGS5, PDGFβ, and NG2 among other markers (Armulik et al. 2011; Crisan et al. 2008; Feng et al. 2010; Rock et al. 2011; Bergers and Song 2005). These makers are not specific to pericytes. For example, all of the markers mentioned are also found on vSMC (Armulik et al. 2011). To correctly identify a pericyte, high-resolution histological examination is needed. Pericytes are vascular associated cells with a large nucleus, small cytoplasm, and long processes encircling the capillary wall (Bergers and Song 2005). Pericytes communicate with endothelial cells through paracrine signaling as well as direct contact in the form of gap and adherens junctions as well as "peg-and-socket" contacts in which pericyte processes are inserted into endothelial invaginations (Armulik et al. 2011; Feng et al. 2010).

Pericytes have been isolated from multiple human organs, including skeletal muscle, placenta, skin, pancreas, and bone marrow (Crisan et al. 2008). Crisan isolated pericytes using flow cytometry based on the expression of CD146 and the exclusion of CD34-, CD45-, and CD56-positive cells. These CD146-positive cells could give rise to adipocytes, chondrocytes, and osteoblasts both in vitro as well as express myofibers when administered to cardioxin-treated mouse skeletal muscle and become ectopic bone when implanted into mouse hind limb (Crisan et al. 2008). Based on marker expression and cell lineage potential, pericytes are similar to MSCs and to date have typically been grouped as the same population due to their perivascular localization.

12.2.5 Mesenchymal Stem Cell

Another potential source of multipotent cells in the lung are the resident MSCs. The first MSCs defined were bone marrow-derived mesenchymal stem cells (BM-MSC) identified by Friedenstein (Friedenstein et al. 1968; Lee et al. 2011). He established several foundational concepts of MSC biology, determining that isolated BM-MSCs grew in tissue culture plates in formations termed colony-forming unit-fibroblast (CFU-F) and that BM-MSCs could differentiate into a host of different mesenchymal lineages. Subsequent work discovered MSC populations outside the BM-MSC, prompting the International Society of Cellular Therapy to define an MSC by three conditions: (1) MSCs be adherent to plastic (2) MSCs must express specific cell markers (CD73, CD90, CD105) but not express other markers (CD45, CD34, CD14, CD11b), and (3) MSCs must differentiate into mesenchymal lineages including osteoblasts, adipocytes, and chondroblasts (Dominici et al. 2006). While it is common to differentiate potential MSC populations to multiple mesenchymal lineages in vitro, such an experiment is difficult to recapitulate in vivo. MSC are distinct from fibroblasts in their ability to form colonies in a CFU-F assay and their previously mentioned multilineage differentiation potential.

MSCs are also prominent in remodeling due to injury. Remodeling in arteries during disease or in response to injury is often coupled with the formation of ectopic tissue structures such as bone, fat, or cartilage as well as microvascularization of the outer most layers (Majesky et al. 2011a). Primitive mesenchymal cells associated with arteries have been termed calcifying vascular cells (CVC) and vascular adventitial fibroblasts (VAFs) and represent an additional group of multipotent mesenchymal stromal cells (MSC) (Tintut et al. 2003; Hoshino et al. 2008). These cells were similar in surface phenotype to bone marrow MSC including the expression of CD44, CD105, and CD29 and lack of all hematopoietic markers as well as CD34, distinguishing these populations from previously described mesoangioblasts (Tintut et al. 2003; Hoshino et al. 2008).

Resident MSC populations identified in the neonate and adult lung described to date differ both in their origin and function (Foronjy and Majka 2012). Lung MSCs have been identified in bronchoalveolar lavage fluid from patient allograft tissue or tracheal aspirates from ventilated neonates, respectively, through their ability to adhere to plastic (Lama et al. 2007; Hennrick et al. 2007). Concurrently, MSC have been identified and isolated by flow cytometry via visualization of a side population phenotype by Hoechst 33342 vital dye staining in combination with the absence of the hematopoietic marker, CD45 (Summer et al. 2007; Martin et al. 2008; Irwin et al. 2007). Lama et al. showed conclusively that MSCs reside in an adult lung in humans by demonstrating donor origin of BAL MSC up to 11 years after transplant (Lama et al. 2007). Jun et al. also showed that MSCs reside in the adult lung by transplanting genetically labeled bone marrow and observing the composition of the MSC following injury in murine models (Jun et al. 2011). Sca1 has also been reported to enrich for a population of mesenchymal/fibroblast progenitor cells resident in lung tissue (McQualter et al. 2009). Multipotent MSC have also been isolated from endarterectomized tissues from patients with chronic thromboembolic pulmonary hypertension (CTEPH) (Firth et al. 2010).

These MSC populations, independent of their origin, demonstrate multilineage mesenchymal differentiation potential to osteocyte, adipocyte, and chondrocyte lineages, and in varying combinations, express the characteristic mesenchymal cell surface determinants ABCG2, CD90, CD105, CD106, CD73, CD44, Stro-1, and Sca1.

In addition, they lack the hematopoietic markers c-kit and CD34 (Jun et al. 2011; Martin et al. 2008; Chateauvieux et al. 2007; Summer et al. 2007; McQualter et al. 2009). Both MSC isolated from BAL and via ABCG2 expression are capable of suppressing T-cell proliferation (Jarvinen et al. 2008; Jun et al. 2011; Lee et al. 2009). Analysis of gene expression of lung MSCs compared to BM-MSCs revealed that only lung MSCs express FOXF1, HOXA5, and HOXB5, transcription factors unique to the developing lung (Walker et al. 2011). In addition to delineating BM-MSC from organ-resident MSCs, this finding may indicate that MSCs are inherently organ-specific.

Pericytes and MSC have very similar cell surface characteristics and differentiation potential. Due to these similarities, pericytes have been hypothesized to be MSC in adult tissues. However, the limitations to address this hypothesis included the lack of a specific marker that could be used to distinguish MSC. Pericytes function to stabilize blood vessels, whereas the function of MSC during adult tissue homeostasis, other than as a supporting or stromal cell, is unknown. Recently, studies by Jun and Chow et al. reported the identification of ABCG2 as a marker to label lung MSC, which lack expression of SMA and NG2 protein (Chow et al. 2013; Jun et al. 2011). Further global gene profiling analyses demonstrated that these ABCG2^{pos} lung MSC were distinct from NG2 pericytes as well as lung fibroblasts. The differences between NG2^{pos} pericytes and ABCG2^{pos} MSC likely represent pericyte heterogeneity within the lung. Interestingly, these ABCG2^{pos} MSC exhibit vascular endothelial, smooth muscle, and myofibroblast differentiation potential, additionally qualifying them as endothelial and smooth muscle progenitors (Chow et al. 2013).

To distinguish differences between generic MSC, fibroblasts, side population MSC, pericytes, and ABCG2^{pos}MSC, we have formulated a comparison of defined "mesenchymal" cell populations present in the distal lung (Fig. 12.2). Here we compare the initial lung MSC described by the efflux of Hoechst 33342 dye and the appearance of the side population phenotype (CD45^{neg} SP (Summer et al. 2007; Martin et al. 2008; Irwin et al. 2007; Jun et al. 2011)), to lung fibroblasts, pericytes, ABCG2^{pos}MSC. All of the populations share common traditional MSC cell surface markers. However, only fibroblasts lack CFU-F ability. Although there is significant overlap, based on global gene expression analysis, the CD45^{neg} SP is a bit broader than the ABCG2^{pos}MSC. This is likely due the expression of transporters and additional multidrug-resistant transporters on the cell surfaces of other lung cell populations which can efflux Hoechst dye. This was confirmed by Jun et al. who demonstrated that ABCG2 marked ~80 % of the CD45^{neg} SP (Jun et al. 2011). These findings illustrate that ABCG2 is a more specific marker than the Hoechst 33342 staining. There is significant similarity between the ABCG2^{pos}MSC and NG2^{pos} pericytes. ABCG2^{pos}MSC are precursors of NG2 pericytes. The differences in gene expression are likely due to function. However, it is clear that in their naïve states, they represent two distinct subpopulations of cells. All of these cell populations share characteristics of the "classically defined MSC."

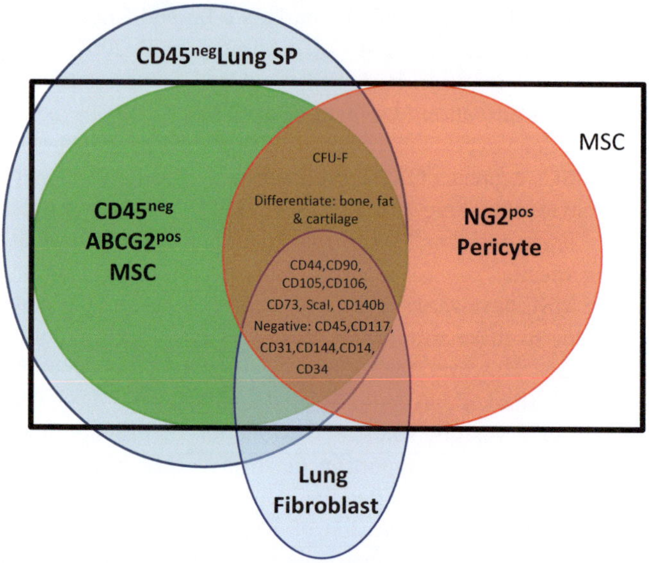

Fig. 12.2 Summary of lung mesenchymal cell subpopulations. Classically defined MSC (*black box*) have traditionally been isolated based on their ability to adhere to plastic and a panel of cell surface antigens. This limited their study in vivo because fibroblasts and/or endothelial cells express many of the same markers. However, more recently, lung mesenchymal subpopulations have been further delineated by genomic analyses, CFU-F, and multilineage differentiation potential (Chow et al. 2013; Firth et al. 2010; Summer et al. 2007; Jun et al. 2011; Lama et al. 2007; Martin et al. 2008; Irwin et al. 2007). These distinct subpopulations include the CD45^{neg} side population (SP), lung fibroblasts, ABCG2^{pos}MSC, and NG2^{pos} pericytes

12.2.6 Endothelial Progenitor Cells and Endothelial Mesenchymal Transition

The endothelial progenitor cell (EPC) is a cell type that generates endothelial cells in response to injury. The first indications of the existence of EPCs came from the spontaneous generation of microvasculature on implanted prosthetics in animals (Scott et al. 1994). Asahara and colleagues were the first to isolate and characterize EPCs from human peripheral blood (Asahara et al. 1997). While Asahara's initial isolation protocol has been adapted by other researchers (reviewed in Hirschi et al. 2008) (Yoder 2012), the most common method of EPC isolation is to plate peripheral blood on fibronectin-coated dishes, replate non-adherent cells onto separate fibronectin plates to deplete other hematopoietic lineages (such as macrophages and monocytes), and culture cells until clusters emerge at 7 days. Tissue-resident EPCs have also been isolated from artery walls in a location termed "the vasculogenic zone" between the smooth muscle and adventitial layers (Zengin et al. 2006).

While there is no individual accepted marker that defines an EPC, the cells that make up these clusters express CD34, CD133, and VEGFR2/KDR (Hirschi et al. 2008; Mao et al. 2013). Several functional assays define EPCs both in vitro and in vivo.

In vitro, EPCs must have the capacity to uptake Dil-Ac-LDL, stain with lectin dye, as well as form tubes when plated in Matrigel (Firth and Yuan 2012). In vivo, EPCs must be able to form tubes when injected into animals in matrigel "plugs." Several investigators have verified that EPCs contribute to neovascularization in vivo (Asahara et al. 1997; Lam et al. 2008, 2011), and therefore likely have the potential to contribute to pathological remodeling during disease. Studies of the origin, localization, and function of EPC during vascular homeostasis and remodeling are complicated by the lack of tissue and cell-specific markers of this cell type.

Endothelial cells (EC) also represent a potential progenitor population in the vasculature. EC can be induced to become migratory mesenchymal cells through the loss of endothelial markers and the progressive gain of mesenchymal characteristics (Kalluri and Weinberg 2009). Endothelial-to-mesenchymal transition (or EndoMT) has mainly been studied in the context of endocardial cushion and heart valve formation, but has become the focus of many investigators in research into sources of myofibroblasts during fibrosis. EndMT can be induced after bleomycin injury to the lungs (Hashimoto et al. 2010). Sixteen percent of lung fibroblasts isolated from transgenic animals expressing beta-galactosidase in $TIE2^+$ cells treated with bleomycin were $lacZ^+$, indicating these cells were likely derived from a vascular origin, the caveat of these studies being that Tie-2 expression is not restricted to EC.

Endothelial cells may also be induced to exhibit MSC characteristics. Medici and colleagues determined that a mouse model with constitutive activation of Alk2 had endothelially derived chondrocytes and osteoblasts (Medici et al. 2010). Injection of BMP4, an Alk2-activating ligand, could also generate endothelially derived chondrocytes and osteoblasts. Endothelial cell lines transfected with an Alk4-inducing adenovirus also began to show expression of the MSC markers STRO-1, CD10, CD44, CD71, and CD90. In addition, endothelial cells treated with TGFβ2 and BMP4 (both activators of Alk4) and cultured in osteogenic, chondrogenic, and adipogenic medium could differentiate into these three lineages. Endothelial cells treated with TGFβ2 and BMP4, placed in sponges and implanted into nude mice, could also differentiate into bone cartilage or fat. Taken together, these studies demonstrate that endothelial cells can acquire MSC phenotypes both in vivo and in vitro.

12.3 The Role of Resident Lung Stem Cells and Progenitors in Vascular Remodeling

12.3.1 Cell-Based Mechanisms of Vascular Remodeling During Lung Disease

Many of the cellular changes associated with PH include an increase in myofibroblasts expressing SMA (Morrell et al. 2009). The resident lung origin of these cells is hypothesized to be vascular SMC or adventitial fibroblasts. More recently, the possibility of resident progenitor cells or the process of endothelial-to-mesenchymal transition (EndoMT) is considered. Existing studies supporting a role for the contribution of resident lung progenitor and stem cells to pulmonary remodeling are summarized. To our knowledge, mesoangioblasts have not been studied in pulmonary remodeling and so are not included.

12.3.2 *Smooth Muscle Progenitors*

Excessive accumulation of SMCs and collagen in the vasculature is common to many adult lung diseases including PAH, PH associated with ILD or COPD, and lymphangioleiomyomatosis (LAM) (Ferri et al. 2004; Benedict et al. 2007; Hansmann et al. 2008; Harrison et al. 2005; Hemnes et al. 2011; Hong et al. 2008; Humbert et al. 2004; Majka et al. 2008, 2011; Rubin and Galie 2004). That being said, studies assessing the contribution of smooth muscle progenitors to vascular remodeling during adult lung disease are rather limited. Cohen et al. defined Wnt regulation of SM22 expressing SMC progenitor proliferation and differentiation during development through a tenascin C-PDGFR pathway (Cohen et al. 2009). The PDGFR pathway was also upregulated in vascular smooth muscle from pulmonary hypertension patients relative to controls. These studies illustrate the reactivation of developmental pathways during adult disease.

12.3.3 *Pericytes*

Pericytes have been implicated in vascular pathology, but there is some debate into the role they play in the process of lung remodeling. Pericytes function to stabilize vasculature and embryos lacking pericyte coverage are not viable (Armulik et al. 2011; Bergers and Song 2005). Whether the proliferation by and increased muscularization of arterioles by pericytes causes adaptive or adverse remodeling is a topic of current debate. Pericytes are also known to play a substantial role in the genesis of myofibroblasts and adverse remodeling associated with kidney fibrosis (Kida and Duffield 2011; Lin et al. 2008).

Rock and colleagues employed lineage tracing of NG2-expressing pericytes and determined that cells positive for PDGFRβ, desmin, and NG2 proliferated 21 days after bleomycin injury, but that the $NG2^+$ cells do not substantially differentiate into αSMA^+ myofibroblasts. This data implies that while $NG2^+$ pericytes expand in response to injury, they do not become myoFB. However, a recent study by Ricard et al. demonstrated that not only do $NG2^+$ cells expand in numbers 21 days post-hypoxia injury, but that the population of $NG2^+/SMA^+$ cells doubles in number as well (Ricard et al. 2014). While the proliferative response of pericytes to lung injury is becoming recognized, the role they play in disease is still relatively unknown.

Pericytes have been shown to play a role in some vascular diseases. Abnormal pericyte coverage of vessels has been implicated as causative for pulmonary hypertension associated with Adams–Oliver syndrome. Adams–Oliver syndrome features capillary rarefaction, loss or hyperproliferation for capillaries, and vessel stenosis (Patel et al. 2004). Loss of capillaries and vasculopathy are also major contributors to clinical manifestations of scleroderma (Fleming et al. 2009). However, Fleming et al. demonstrated that high dose immunosuppressive and hematopoietic stem cell therapies reversed the loss of functional pericytes, decreased rgs5 expression, and induced capillary regeneration (Fleming et al. 2008). These data suggest that preserving the function of pericytes and vascular supporting cells may have implications for preventing the progression of PAH or pulmonary hypertension associated with other chronic lung disease.

12.3.4 Mesenchymal Stem Cell

Mesenchymal cell proliferation and apoptosis are factors that determine whether a fibroproliferative response resolves, as in wound healing, or progresses to a chronic pathologic condition, as in pulmonary hypertension and fibrosis (Bonner 2010).

Functional studies of the ABCG2^{pos} Hoechst33342^{dim} CD45^{neg}-resident lung MSCs demonstrate that they regulate the severity of bleomycin injury via modulation of the T-cell response (Jun et al. 2011). Jun et al. elegantly documented that bleomycin treatment of mice induced the loss of these endogenous lung MSCs and elicited fibrosis (Cui et al.), inflammation, and PAH. Replacement of resident stem cells by administration of isolated lung MSCs attenuated the bleomycin-associated pathology and mitigated the development of PAH. These data suggest that lung MSCs function to protect lung integrity following injury; however, when endogenous MSCs are lost, this function is compromised. Lung MSCs may therefore regulate homeostasis and repair of their native tissue, the distal lung.

In addition to their reparative properties, several studies indicate that lung MSCs under certain conditions mediate pathogenic changes within the lung (Bonner 2010; Pierro and Thebaud 2010). Firth et al. described the isolation of MSC from endarterectomized tissues from patients with CTEPH (Firth et al. 2010). The isolation of both myofibroblasts and functional MSC fibrotic coagulant tissue suggested that multipotent lung MSC might participate in the vascular remodeling associated with pulmonary hypertension. More recently, lineage-tracing analyses performed by Chow and colleagues confirmed the lineage potential of lung MSC in vivo. Using an inducible ABCG2-driven expression of Cre and an eGFP label, these studies demonstrated that multipotent MSC contributed to muscularization of distal lung microvessels in a murine model of hypobaric hypoxia-induced PAH (Chow et al. 2013). When these cells were genetically depleted of superoxide dismutase, a potent antioxidant, their contribution to muscularization was significantly increased in the murine model of PAH. In vitro, increased oxidant stress resulted in transition of the MSC to a contractile myofibroblast phenotype. This transition from a MSC to a contractile myofibroblast was correlated to alterations in Wnt pathway signaling.

Indeed, the behavior of MSCs is highly sensitive to the microenvironment to which these cells are exposed (Yan et al. 2007). As an example, it was recently shown that TGF-β expression within the lungs of premature infants stimulates MSCs to differentiate into myofibroblasts (Popova et al. 2010). Myofibroblasts induce abnormal matrix remodeling that results in the development of the chronic lung disease, BPD. BPD is characterized by decreased functionality of the alveolar-capillary surfaced for gas exchange including both the vasculature and airways (Toti et al. 1997). Similar findings were observed in lung allografts from transplanted patients (Walker et al. 2011). In this study, lung-derived MSCs isolated from the airways had increased expression of type I collagen and α-smooth muscle actin and readily differentiated into myofibroblasts upon treatment with IL-13 or TGF-β (Walker et al. 2011). Allograft survival is limited by bronchiolitis obliterans (BOS), a fibrotic constriction of the airways (Todd and Palmer 2011). MSCs from BOS subjects had a profibrotic phenotype, indicating that this cell type was an important mediator of the fibrotic changes and associated remodeling. Thus, these findings indicate that MSCs are a critical factor in the development of dysfunctional lung vascular and interstitial remodeling in these adult lung diseases.

12.3.5 EPC and EndoMT

EPCs, while difficult to define conclusively, have been studied as an indicator of disease as well as a potential treatment for neonatal lung disease. Heterogeneity within EC populations and lack of a specific definitive marker (Rajotte et al. 1998) has limited our understanding of the true differentiation potential of resident lung progenitor populations. Typically, tissue-resident EPC share markers with differentiated ECs and in some cases MSC (along with hematopoietic and circulating cells). These common markers include VE-cadherin (CD144), Tie-1/2, VEGFR1/2, and PECAM (CD31). EPC levels in the peripheral circulation of patients with pulmonary disease have been correlated to their prognosis and severity (Burnham et al. 2005). Recently, Alphonse and colleagues were able to isolate a subset of EPCs known as endothelial colony-forming cells (ECFCs) from human fetal lung (Alphonse et al. 2014). These EPCs could be altered by hypoxic injury to form fewer colonies in vitro as well as fewer capillary networks in capillary assays. In addition, ECFCs isolated from a rat model of BPD had similar reductions of functionality. However, human umbilical cord ECFC delivered through intra-jugular injections to the BPD rat-model preserved alveolar architecture, vasculature, and reduced hypoxic-related PH, showing for the first time that EPCs can be used to treat the alveolar as well as vascular abnormalities associated with BPD.

EndoMT occurs in remodeling associated with cardiac, kidney, and liver fibrosis (Piera-Velazquez et al. 2011). To date, evidence for EndoMT during lung disease is more limited. The studies presented link EPC and differentiated EC to remodeling associated with fibrosis and pulmonary hypertension through EndoMT. To model EndoMT pulmonary hypertension, Zhu et al. performed a limited in vitro analysis of porcine pulmonary artery endothelial cells cultured under relative hypoxic conditions (Zhu et al. 2006). Zhu measured mesenchymal transition by the increased expression of smooth muscle alpha actin (Zhu et al. 2006). Hashimoto et al. performed lineage-tracing analysis of Tie-2-positive cells in vivo during bleomycin-induced fibrosis in the murine lung (Hashimoto et al. 2010). In this study, Hashimoto crossed a Tie-2 Cre driver strain to a lacZ reporter to label capillary endothelium in the lung. Resulting Tie-2-derived lacZ populations expressed collagen I and SMA. A limitation of this study was that in using a non-inducible Cre system, Tie-2 expression may have been upregulated in non-ECs following bleomycin injury, confounding the exact origin of the lacZ-positive cells. Hashimoto addressed this limitation by performing in vitro analyses of EndoMT. TGFβ treatment of EC reduced endothelial marker expression and increased the mesenchymal markers collagen 1, fibronectin, snail, and twist. The results of these studies suggest that EC may be a source of contractile cells that contribute to pulmonary vascular remodeling.

12.4 Summary

In summary, studies are needed to broaden our understanding of resident lung stem and progenitors during pulmonary vascular tissue homeostasis as well as how their functions are impaired during disease. This knowledge will lead to understanding their role during the development of disease and will be useful for designing further intervention.

Acknowledgment This work was funded by grants to S.M. Majka from the NIH NHLBI R01HL091105 and R01HL11659701.

References

Alagappan VT, Boer W, Misra V, Mooi W, Sharma H (2013) Angiogenesis and vascular remodeling in chronic airway diseases. Cell Biochem Biophys 67(2):219–234. doi:10.1007/s12013-013-9713-6

Alphonse RS, Vadivel A, Fung M, Shelley WC, Critser PJ, Ionescu L, O'Reilly M, Ohls RK, McConaghy S, Eaton F, Zhong S, Yoder M, Thebaud B (2014) Existence, functional impairment and lung repair potential of endothelial colony forming cells in oxygen-induced arrested alveolar growth. Circulation 129(21):2144–2157. doi:10.1161/CIRCULATIONAHA.114.009124

Armulik A, Genove G, Betsholtz C (2011) Pericytes: developmental, physiological, and pathological perspectives, problems, and promises. Dev Cell 21(2):193–215. doi:10.1016/j.devcel.2011.07.001

Asahara T, Murohara T, Sullivan A, Silver M, van der Zee R, Li T, Witzenbichler B, Schatteman G, Isner JM (1997) Isolation of putative progenitor endothelial cells for angiogenesis. Science 275(5302):964–967

Benedict N, Seybert A, Mathier MA (2007) Evidence-based pharmacologic management of pulmonary arterial hypertension. Clin Ther 29(10):2134–2153

Bergers G, Song S (2005) The role of pericytes in blood-vessel formation and maintenance. Neuro Oncol 7(4):452–464. doi:10.1215/S1152851705000232

Berkelhamer SK, Mestan KK, Steinhorn RH (2013) Pulmonary hypertension in bronchopulmonary dysplasia. Semin Perinatol 37(2):124–131. doi:10.1053/j.semperi.2013.01.009

Bonner JC (2010) Mesenchymal cell survival in airway and interstitial pulmonary fibrosis. Fibrogenesis Tissue Repair 3:15. doi:10.1186/1755-1536-3-15

Burnham EL, Taylor WR, Quyyumi AA, Rojas M, Brigham KL, Moss M (2005) Increased circulating endothelial progenitor cells are associated with survival in acute lung injury. Am J Respir Crit Care Med 172(7):854–860. doi:10.1164/rccm.200410-1325OC

Campagnolo P, Cesselli D, Al Haj Zen A, Beltrami AP, Kränkel N, Katare R, Angelini G, Emanueli C, Madeddu P (2010) Human adult vena saphena contains perivascular progenitor cells endowed with clonogenic and proangiogenic potential. Circulation 121(15):1735–1745

Chang L, Noseda M, Higginson M, Ly M, Patenaude A, Fuller M, Kyle AH, Minchinton AI, Puri MC, Dumont DJ, Karsan A (2012) Differentiation of vascular smooth muscle cells from local precursors during embryonic and adult arteriogenesis requires Notch signaling. Proc Natl Acad Sci U S A 109(18):6993–6998

Chateauvieux S, Ichanté J-L, Delorme B, Frouin V, Piétu G, Langonné A, Gallay N, Sensebé L, Martin MT, Moore KA, Charbord P (2007) Molecular profile of mouse stromal mesenchymal stem cells. Physiol Genomics 29(2):128–138. doi:10.1152/physiolgenomics.00197.2006

Chow K, Fessel JP, Ihida-Stansbury K, Schmidt EP, Gaskill C, Alvarez D, Graham B, Harrison DG, Wagner DH Jr, Nozik-Grayck E, West JD, Klemm DJ, Majka SM (2013) Dysfunctional resident lung mesenchymal stem cells contribute to pulmonary microvascular remodeling. Pulm Circ 3(1):31–49. doi:10.4103/2045-8932.109912

Cohen ED, Ihida-Stansbury K, Lu MM, Panettieri RA, Jones PL, Morrisey EE (2009) Wnt signaling regulates smooth muscle precursor development in the mouse lung via a tenascin C/PDGFR pathway. J Clin Invest 119(9):2538–2549. doi:10.1172/jci38079

Crisan M, Yap S, Casteilla L, Chen CW, Corselli M, Park TS, Andriolo G, Sun B, Zheng B, Zhang L, Norotte C, Teng PN, Traas J, Schugar R, Deasy BM, Badylak S, Buhring HJ, Giacobino JP, Lazzari L, Huard J, Peault B (2008) A perivascular origin for mesenchymal stem cells in multiple human organs. Cell Stem Cell 3(3):301–313. doi:10.1016/j.stem.2008.07.003

Cui B, Zhang S, Chen L, Yu J, Widhopf GF II, Fecteau JF, Rassenti LZ, Kipps TJ (2013) Targeting ROR1 inhibits epithelial-mesenchymal transition and metastasis. Cancer Res 73(12):3649–3660. doi:10.1158/0008-5472.CAN-12-3832

De Langhe SP, Carraro G, Warburton D, Hajihosseini MK, Bellusci S (2006) Levels of mesenchymal FGFR2 signaling modulate smooth muscle progenitor cell commitment in the lung. Dev Biol 299(1):52–62

Dominici M, Le Blanc K, Mueller I, Slaper-Cortenbach I, Marini F, Krause D, Deans R, Keating A, Prockop DJ, Horwitz E (2006) Minimal criteria for defining multipotent mesenchymal stromal cells. The International Society for Cellular Therapy position statement. Cytotherapy 8(4):315–317

Feng J, Mantesso A, Sharpe PT (2010) Perivascular cells as mesenchymal stem cells. Expert Opin Biol Ther 10(10):1441–1451. doi:10.1517/14712598.2010.517191

Ferri N, Carragher NO, Raines EW (2004) Role of discoidin domain receptors 1 and 2 in human smooth muscle cell-mediated collagen remodeling: potential implications in atherosclerosis and lymphangioleiomyomatosis. Am J Pathol 164(5):1575–1585. doi:10.1016/S0002-9440(10)63716-9

Firth AL, Yuan JX (2012) Identification of functional progenitor cells in the pulmonary vasculature. Pulm Circ 2(1):84–100. doi:10.4103/2045-8932.94841

Firth AL, Yao W, Ogawa A, Madani MM, Lin GY, Yuan JXJ (2010) Multipotent mesenchymal progenitor cells are present in endarterectomized tissues from patients with chronic thromboembolic pulmonary hypertension. Am J Physiol Cell Physiol 298(5):C1217–C1225

Fleming JN, Nash RA, McLeod DO, Fiorentino DF, Shulman HM, Connolly MK, Molitor JA, Henstorf G, Lafyatis R, Pritchard DK, Adams LD, Furst DE, Schwartz SM (2008) Capillary regeneration in scleroderma: stem cell therapy reverses phenotype? PLoS One 3(1):e1452

Fleming J, Nash RA, Mahoney WM Jr, Schwartz SM (2009) Is scleroderma a vasculopathy? Curr Rheumatol Rep 11(2):103–110

Foronjy R, Majka SM (2012) The potential for resident lung mesenchymal stem cells to promote functional tissue regeneration: understanding microenvironmental cues. Cells 4:874

Friedenstein AJ, Petrakova KV, Kurolesova AI, Frolova GP (1968) Heterotopic of bone marrow. Analysis of precursor cells for osteogenic and hematopoietic tissues. Transplantation 6(2):230–247

Galli D, Innocenzi A, Staszewsky L, Zanetta L, Sampaolesi M, Bai A, Martinoli E, Carlo E, Balconi G, Fiordaliso F, Chimenti S, Cusella G, Dejana E, Cossu G, Latini R (2005) Mesoangioblasts, vessel-associated multipotent stem cells, repair the infarcted heart by multiple cellular mechanisms: a comparison with bone marrow progenitors, fibroblasts, and endothelial cells. Arterioscler Thromb Vasc Biol 25(4):692–697. doi:10.1161/01.ATV.0000156402.52029.ce

Galvez BG, Sampaolesi M, Brunelli S, Covarello D, Gavina M, Rossi B, Constantin G, Torrente Y, Cossu G (2006) Complete repair of dystrophic skeletal muscle by mesoangioblasts with enhanced migration ability. J Cell Biol 174(2):231–243. doi:10.1083/jcb.200512085

Hansmann G, de Jesus Perez VA, Alastalo TP, Alvira CM, Guignabert C, Bekker JM, Schellong S, Urashima T, Wang L, Morrell NW, Rabinovitch M (2008) An antiproliferative BMP-2/PPARgamma/apoE axis in human and murine SMCs and its role in pulmonary hypertension. J Clin Invest 118(5):1846–1857. doi:10.1172/JCI32503

Harrison RE, Berger R, Haworth SG, Tulloh R, Mache CJ, Morrell NW, Aldred MA, Trembath RC (2005) Transforming growth factor-beta receptor mutations and pulmonary arterial hypertension in childhood. Circulation 111(4):435–441

Hashimoto N, Phan SH, Imaizumi K, Matsuo M, Nakashima H, Kawabe T, Shimokata K, Hasegawa Y (2010) Endothelial-mesenchymal transition in bleomycin-induced pulmonary fibrosis. Am J Respir Cell Mol Biol 43(2):161–172. doi:10.1165/rcmb.2009-0031OC

Hellstrom M, Kalén M, Lindahl P, Abramsson A, Betsholtz C (1999) Role of PDGF-B and PDGFR-beta in recruitment of vascular smooth muscle cells and pericytes during embryonic blood vessel formation in the mouse. Development 126(14):3047–3055

Hemnes A, Austin E, Robbins I, Loyd J, West J, Newman J, Cogan J, Fox K, Lane K, Robinson L, Hedges L, Talati M, Hamid R, Menon S (2011) Idiopathic and heritable PAH perturb common molecular pathways, correlated with increased MSX1 expression. Pulm Circ 1(3):389–398

Hennrick KT, Keeton AG, Nanua S, Kijek TG, Goldsmith AM, Sajjan US, Bentley JK, Lama VN, Moore BB, Schumacher RE, Thannickal VJ, Hershenson MB (2007) Lung cells from neonates show a mesenchymal stem cell phenotype. Am J Respir Crit Care Med 175(11):1158–1164. doi:10.1164/rccm.200607-941OC

Hirschi KK, Ingram DA, Yoder MC (2008) Assessing identity, phenotype, and fate of endothelial progenitor cells. Arterioscler Thromb Vasc Biol 28(9):1584–1595. doi:10.1161/ATVBAHA.107.155960

Hong KH, Lee YJ, Lee E, Park SO, Han C, Beppu H, Li E, Raizada MK, Bloch KD, Oh SP (2008) Genetic ablation of the BMPR2 gene in pulmonary endothelium is sufficient to predispose to pulmonary arterial hypertension. Circulation 118(7):722–730

Hoshino A, Chiba H, Nagai K, Ishii G, Ochiai A (2008) Human vascular adventitial fibroblasts contain mesenchymal stem/progenitor cells. Biochem Biophys Res Commun 368(2):305–310. doi:10.1016/j.bbrc.2008.01.090

Humbert M, Morrell NW, Archer SL, Stenmark KR, MacLean MR, Lang IM, Christman BW, Weir EK, Eickelberg O, Voelkel NF, Rabinovitch M (2004) Cellular and molecular pathobiology of pulmonary arterial hypertension. J Am Coll Cardiol 43(12 Suppl S):13S–24S

Irwin D, Helm K, Campbell N, Imamura M, Fagan K, Harral J, Carr M, Young KA, Klemm D, Gebb S, Dempsey EC, West J, Majka S (2007) Neonatal lung side population cells demonstrate endothelial potential and are altered in response to hyperoxia-induced lung simplification. Am J Physiol Lung Cell Mol Physiol 293:L941–L951

Jarvinen L, Badri L, Wettlaufer S, Ohtsuka T, Standiford TJ, Toews GB, Pinsky DJ, Peters-Golden M, Lama VN (2008) Lung resident mesenchymal stem cells isolated from human lung allografts inhibit T cell proliferation via a soluble mediator. J Immunol 181(6):4389–4396

Jun D, Garat C, West J, Thorn N, Chow K, Cleaver T, Sullivan T, Torchia EC, Childs C, Shade T, Tadjali M, Lara A, Nozik-Grayck E, Malkoski S, Sorrentino B, Meyrick B, Klemm D,

Rojas M, Wagner DH, Majka SM (2011) The pathology of bleomycin-induced fibrosis is associated with loss of resident lung mesenchymal stem cells that regulate effector T-cell proliferation. Stem Cells 29(4):725–735. doi:10.1002/stem.604

Kalluri R, Weinberg RA (2009) The basics of epithelial-mesenchymal transition. J Clin Invest 119(6):1420–1428. doi:10.1172/JCI39104

Kida Y, Duffield JS (2011) Pivotal role of pericytes in kidney fibrosis. Clin Exp Pharmacol Physiol 38(7):467–473

Lam CF, Liu YC, Hsu JK, Yeh PA, Su TY, Huang CC, Lin MW, Wu PC, Chang PJ, Tsai YC (2008) Autologous transplantation of endothelial progenitor cells attenuates acute lung injury in rabbits. Anesthesiology 108(3):392–401. doi:10.1097/ALN.0b013e318164ca64

Lam CF, Roan JN, Lee CH, Chang PJ, Huang CC, Liu YC, Jiang MJ, Tsai YC (2011) Transplantation of endothelial progenitor cells improves pulmonary endothelial function and gas exchange in rabbits with endotoxin-induced acute lung injury. Anesth Analg 112(3):620–627. doi:10.1213/ANE.0b013e3182075da4

Lama VN, Smith L, Badri L, Flint A, Andrei A-C, Murray S, Wang Z, Liao H, Toews GB, Krebsbach PH, Peters-Golden M, Pinsky DJ, Martinez FJ, Thannickal VJ (2007) Evidence for tissue-resident mesenchymal stem cells in human adult lung from studies of transplanted allografts. J Clin Invest 117(4):989–996

Lee JW, Gupta N, Serikov V, Matthay MA (2009) Potential application of mesenchymal stem cells in acute lung injury. Expert Opin Biol Ther 9(10):1259–1270. doi:10.1517/14712590903213651

Lee JW, Fang X, Krasnodembskaya A, Howard JP, Matthay MA (2011) Concise review: mesenchymal stem cells for acute lung injury: role of paracrine soluble factors. Stem Cells 29(6):913–919. doi:10.1002/stem.643

Lin S-L, Kisseleva T, Brenner DA, Duffield JS (2008) Pericytes and perivascular fibroblasts are the primary source of collagen-producing cells in obstructive fibrosis of the kidney. Am J Pathol 173(6):1617–1627. doi:10.2353/ajpath.2008.080433

Lindahl P, Karlsson L, Hellstrom M, Gebre-Medhin S, Willetts K, Heath JK, Betsholtz C (1997) Alveogenesis failure in PDGF-A-deficient mice is coupled to lack of distal spreading of alveolar smooth muscle cell progenitors during lung development. Development 124(20):3943–3953

Majesky MW, Dong XR, Hoglund V, Mahoney WM, Daum G (2011a) The adventitia: a dynamic interface containing resident progenitor cells. Arterioscler Thromb Vasc Biol 31(7):1530–1539

Majesky MW, Dong XR, Regan JN, Hoglund VJ (2011b) Vascular smooth muscle progenitor cells: building and repairing blood vessels. Circ Res 108(3):365–377

Majka SM, Skokan M, Wheeler L, Harral J, Gladson S, Burnham E, Loyd JE, Stenmark KR, Varella-Garcia M, West J (2008) Evidence for cell fusion is absent in vascular lesions associated with pulmonary arterial hypertension. Am J Physiol Lung Cell Mol Physiol 295(6):L1028–L1039. doi:10.1152/ajplung.90449.2008

Majka S, Hagen M, Blackwell T, Harral J, Johnson J, Gendron R, Paradis H, Crona D, Loyd J, Nozik-Grayck E, Stenmark K, West J (2011) Physiologic and molecular consequences of endothelial Bmpr2 mutation. Respir Res 12(1):84

Mao M, Xu X, Zhang Y, Zhang B, Fu ZH (2013) Endothelial progenitor cells: the promise of cell-based therapies for acute lung injury. Inflamm Res 62(1):3–8. doi:10.1007/s00011-012-0570-3

Martin J, Helm K, Ruegg P, Varella-Garcia M, Burnham E, Majka S (2008) Adult lung side population cells have mesenchymal stem cell potential. Cytotherapy 10(2):140–151. doi:10.1080/14653240801895296

McQualter JL, Brouard N, Williams B, Baird BN, Sims-Lucas S, Yuen K, Nilsson SK, Simmons PJ, Bertoncello I (2009) Endogenous fibroblastic progenitor cells in the adult mouse lung are highly enriched in the sca-1 positive cell fraction. Stem Cells 27(3):623–633. doi:10.1634/stemcells. 2008-0866

Medici D, Shore EM, Lounev VY, Kaplan FS, Kalluri R, Olsen BR (2010) Conversion of vascular endothelial cells into multipotent stem-like cells. Nat Med 16(12):1400–1406. doi:10.1038/nm.2252

Minasi MG, Riminucci M, De Angelis L, Borello U, Berarducci B, Innocenzi A, Caprioli A, Sirabella D, Baiocchi M, De Maria R, Boratto R, Jaffredo T, Broccoli V, Bianco P, Cossu G (2002) The meso-angioblast: a multipotent, self-renewing cell that originates from the dorsal aorta and differentiates into most mesodermal tissues. Development 129(11):2773–2783

Morrell NW, Adnot S, Archer SL, Dupuis J, Jones PL, MacLean MR, McMurtry IF, Stenmark KR, Thistlethwaite PA, Weissmann N, Yuan JX, Weir EK (2009) Cellular and molecular basis of pulmonary arterial hypertension. J Am Coll Cardiol 54(1 Suppl):S20–S31. doi:10.1016/j.jacc.2009.04.018

Motoike T, Markham DW, Rossant J, Sato TN (2003) Evidence for novel fate of Flk1+ progenitor: contribution to muscle lineage. Genesis 35(3):153–159. doi:10.1002/gene.10175

Passman JN, Dong XR, Wu S-P, Maguire CT, Hogan KA, Bautch VL, Majesky MW (2008) A sonic hedgehog signaling domain in the arterial adventitia supports resident Sca1+ smooth muscle progenitor cells. Proc Natl Acad Sci U S A 105(27):9349–9354

Patel MS, Taylor GP, Bharya S, Al-Sanna'a N, Adatia I, Chitayat D, Suzanne Lewis ME, Human DG (2004) Abnormal pericyte recruitment as a cause for pulmonary hypertension in Adams–Oliver syndrome. Am J Med Genet A 129A(3):294–299. doi:10.1002/ajmg.a.30221

Patel NM, Kawut SM, Jelic S, Arcasoy SM, Lederer DJ, Borczuk AC (2013) Pulmonary arteriole gene expression signature in idiopathic pulmonary fibrosis. Eur Respir J 41(6):1324–1330

Piera-Velazquez S, Li Z, Jimenez SA (2011) Role of endothelial-mesenchymal transition (EndoMT) in the pathogenesis of fibrotic disorders. Am J Pathol 179(3):1074–1080. doi:10.1016/j.ajpath.2011.06.001

Pierro M, Thebaud B (2010) Mesenchymal stem cells in chronic lung disease: culprit or savior? Am J Physiol Lung Cell Mol Physiol 298(6):L732–L734. doi:10.1152/ajplung.00099.2010

Popova AP, Bozyk PD, Bentley JK, Linn MJ, Goldsmith AM, Schumacher RE, Weiner GM, Filbrun AG, Hershenson MB (2010) Isolation of tracheal aspirate mesenchymal stromal cells predicts bronchopulmonary dysplasia. Pediatrics 126(5):e1127–e1133. doi:10.1542/peds.2009-3445

Que J, Wilm B, Hasegawa H, Wang F, Bader D, Hogan BL (2008) Mesothelium contributes to vascular smooth muscle and mesenchyme during lung development. Proc Natl Acad Sci U S A 105(43):16626–16630. doi:10.1073/pnas.0808649105

Rajotte D, Arap W, Hagedorn M, Koivunen E, Pasqualini R, Ruoslahti E (1998) Molecular heterogeneity of the vascular endothelium revealed by in vivo phage display. J Clin Invest 102(2):430–437

Ricard N, Tu L, Le Hiress M, Huertas A, Phan C, Thuillet R, Sattler C, Fadel E, Seferian A, Montani D, Dorfmuller P, Humbert M, Guignabert C (2014) Increased pericyte coverage mediated by endothelial-derived fibroblast growth factor-2 and interleukin-6 is a source of smooth muscle-like cells in pulmonary hypertension. Circulation 129(15):1586–1597. doi:10.1161/CIRCULATIONAHA.113.007469

Rock JR, Barkauskas CE, Cronce MJ, Xue Y, Harris JR, Liang J, Noble PW, Hogan BL (2011) Multiple stromal populations contribute to pulmonary fibrosis without evidence for epithelial to mesenchymal transition. Proc Natl Acad Sci U S A 108(52):E1475–E1483. doi:10.1073/pnas.1117988108

Rubin LJ, Galie N (2004) Pulmonary arterial hypertension: a look to the future. J Am Coll Cardiol 43(12 Suppl S):89S–90S

Sampaolesi M, Torrente Y, Innocenzi A, Tonlorenzi R, D'Antona G, Pellegrino MA, Barresi R, Bresolin N, De Angelis MG, Campbell KP, Bottinelli R, Cossu G (2003) Cell therapy of alpha-sarcoglycan null dystrophic mice through intra-arterial delivery of mesoangioblasts. Science 301(5632):487–492. doi:10.1126/science.1082254

Sampaolesi M, Blot S, D'Antona G, Granger N, Tonlorenzi R, Innocenzi A, Mognol P, Thibaud JL, Galvez BG, Barthelemy I, Perani L, Mantero S, Guttinger M, Pansarasa O, Rinaldi C, Cusella De Angelis MG, Torrente Y, Bordignon C, Bottinelli R, Cossu G (2006) Mesoangioblast stem cells ameliorate muscle function in dystrophic dogs. Nature 444(7119):574–579. doi:10.1038/nature05282

Scott SM, Barth MG, Gaddy LR, Ahl ET Jr (1994) The role of circulating cells in the healing of vascular prostheses. J Vasc Surg 19(4):585–593

Seeger W, Adir Y, Barberà JA, Champion H, Coghlan JG, Cottin V, De Marco T, Galiè N, Ghio S, Gibbs S, Martinez FJ, Semigran MJ, Simonneau G, Wells AU, Vachièry J-L (2013) Pulmonary hypertension in chronic lung diseases. J Am Coll Cardiol 62(25 Suppl):D109–D116. doi:10.1016/j.jacc.2013.10.036

Simonneau G, Galie N, Rubin LJ, Langleben D, Seeger W, Domenighetti G, Gibbs S, Lebrec D, Speich R, Beghetti M, Rich S, Fishman A (2004) Clinical classification of pulmonary hypertension. J Am Coll Cardiol 43(12 Suppl S):5S–12S

Summer R, Fitzsimmons K, Dwyer D, Murphy J, Fine A (2007) Isolation of an adult mouse lung mesenchymal progenitor cell population. Am J Respir Cell Mol Biol 37(2):152–159. doi:10.1165/rcmb.2006-0386OC

Tagliafico E, Brunelli S, Bergamaschi A, De Angelis L, Scardigli R, Galli D, Battini R, Bianco P, Ferrari S, Cossu G, Ferrari S (2004) TGFbeta/BMP activate the smooth muscle/bone differentiation programs in mesoangioblasts. J Cell Sci 117(Pt 19):4377–4388. doi:10.1242/jcs.01291

Tintut Y, Alfonso Z, Saini T, Radcliff K, Watson K, Boström K, Demer LL (2003) Multilineage potential of cells from the artery wall. Circulation 108(20):2505–2510. doi:10.1161/01.cir.0000096485.64373.c5

Todd JL, Palmer SM (2011) Bronchiolitis obliterans syndrome: the final frontier for lung transplantation. Chest 140(2):502–508. doi:10.1378/chest.10-2838

Toti P, Buonocore G, Tanganelli P, Catella AM, Palmeri ML, Vatti R, Seemayer TA (1997) Bronchopulmonary dysplasia of the premature baby: an immunohistochemical study. Pediatr Pulmonol 24(1):22–28

Townsley MI (2012) Structure and composition of pulmonary arteries, capillaries, and veins. Compr Physiol 2:675–709. doi:10.1002/cphy.c100081

Vodyanik MA, Yu J, Zhang X, Tian S, Stewart R, Thomson JA, Slukvin II (2010) A mesoderm-derived precursor for mesenchymal stem and endothelial cells. Cell Stem Cell 7(6):718–729. doi:10.1016/j.stem.2010.11.011

Walker N, Badri L, Wettlaufer S, Flint A, Sajjan U, Krebsbach PH, Keshamouni VG, Peters-Golden M, Lama VN (2011) Resident tissue-specific mesenchymal progenitor cells contribute to fibrogenesis in human lung allografts. Am J Pathol 178(6):2461–2469. doi:10.1016/j.ajpath.2011.01.058

Wang L, Kamath A, Frye J, Iwamoto GA, Chun JL, Berry SE (2012) Aorta-derived mesoangioblasts differentiate into the oligodendrocytes by inhibition of the Rho kinase signaling pathway. Stem Cells Dev 21(7):1069–1089. doi:10.1089/scd.2011.0124

Yan X, Liu Y, Han Q, Jia M, Liao L, Qi M, Zhao RC (2007) Injured microenvironment directly guides the differentiation of engrafted Flk-1(+) mesenchymal stem cell in lung. Exp Hematol 35(9):1466–1475. doi:10.1016/j.exphem.2007.05.012

Yoder MC (2012) Human endothelial progenitor cells. Cold Spring Harb Perspect Med 2(7):a006692

Zengin E, Chalajour F, Gehling UM, Ito WD, Treede H, Lauke H, Weil J, Reichenspurner H, Kilic N, Ergün S (2006) Vascular wall resident progenitor cells: a source for postnatal vasculogenesis. Development 133(8):1543–1551

Zhu P, Huang L, Ge X, Yan F, Wu R, Ao Q (2006) Transdifferentiation of pulmonary arteriolar endothelial cells into smooth muscle-like cells regulated by myocardin involved in hypoxia-induced pulmonary vascular remodelling. Int J Exp Pathol 87(6):463–474. doi:10.1111/j.1365-2613.2006.00503.x

Chapter 13
Hematopoietic Stem Cells and Chronic Hypoxia-Induced Pulmonary Vascular Remodelling

Alice Huertas, Marc Humbert, and Christophe Guignabert

Abbreviations

AP-1	Activating protein-1
C/EBPβ	CCAAT/enhancer binding protein β
CD	Cluster of differentiation
COPD	Chronic obstructive pulmonary disease
CPFE	Syndrome combining pulmonary fibrosis and emphysema
CREB	Cyclic AMP response element-binding protein
CXCL-12	C-X-C motif chemokine 12
CXCR	C-X-C chemokine receptor
EGF	Epidermal growth factor
FGF-2	Fibroblast growth factor-2 (basic)
G-CSF	Granulocyte colony-stimulating factor
GFP	Green fluorescence protein
HAPE	High altitude pulmonary edema
Hg	Mercury
HIF	Hypoxia inducible factor

A. Huertas • M. Humbert
INSERM UMR_S 999, LabEx LERMIT, Centre Chirurgical Marie Lannelongue, Le Plessis-Robinson, France

Univ Paris-Sud, School of medicine, Kremlin-Bicêtre, France

AP-HP, Service de Pneumologie, Centre de Référence de l'Hypertension Pulmonaire Sévère DHU Thorax Innovation, Hôpital de Bicêtre, France

C. Guignabert, Ph.D. (✉)
INSERM UMR_S 999, LabEx LERMIT, Centre Chirurgical Marie Lannelongue, Le Plessis-Robinson, France

Univ Paris-Sud, School of medicine, Kremlin-Bicêtre, France
e-mail: christophe.guignabert@inserm.fr

A. Firth, J.X.-J. Yuan (eds.), *Lung Stem Cells in the Epithelium and Vasculature*, Stem Cell Biology and Regenerative Medicine, DOI 10.1007/978-3-319-16232-4_13

HPV	Hypoxic pulmonary vasoconstriction
HSC	Hematopoietic stem cells
IL	Interleukin
IPF	Idiopathic pulmonary fibrosis
mPAP	Mean pulmonary arterial pressure
mRNA	Messenger RNA
NF-κB	Nuclear factor-kappa B
PAH	Pulmonary arterial hypertension
PH	Pulmonary hypertension
PHDs	Prolyl hydroxylase domain-containing proteins
SDF-1	Stromal-derived factor-1
TNF-α	Tumor necrosis factor alpha

13.1 Introduction

Under physiological conditions, hematopoietic stem cells (HSCs) represent a heterogeneous cell population that are primarily located in the bone marrow and that can generate all hematopoietic cell types and produce new HSCs through cell division and long-term self-renewal. Therefore, HSCs are characterized by multi-lineage potentials and are kept in a quiescent state by a special microenvironment called a *niche*, in which many cells and factors maintain HSCs in a dormant state with a very low mitochondrial activity. The factors involved in the conversion from quiescence to activated/mobilized state and vice versa are still unclear, but it is known that oxygen concentration can influence the metabolic state of the HSCs in the bone marrow *niches*.

CXCL12 (also known as stromal-derived factor-1 [SDF-1]) promoter contains hypoxia-inducible factor-1α (HIF-1α) binding sites and it is known that CXCL12-CXCR4 axis plays a role in stem cell homing in response to organ injury such as ischemia. However, it is still totally unknown whether the lung under chronic hypoxic conditions can mobilize HSCs and, if so, whether these recruited HSCs can contribute to pulmonary vascular remodelling. An improved understanding of HSC hypoxia-induced recruitment and role in the lung could help developing new therapeutic tools to either generate or eliminate HSCs in chronic hypoxia-induced pulmonary vascular remodelling.

This chapter aims to review the current knowledge of the role of HSCs in hypoxia-induced pulmonary vascular remodelling.

13.2 Hematopoietic Stem Cells and Hypoxia

13.2.1 Hematopoietic Stem Cells

HSCs are multipotent, self-renewing progenitor cells that develop from mesodermal hemangioblast cells. Under physiological conditions, numbers of HSCs in bone marrow and in peripheral blood are very low. However, it is possible to purify reconstituting HSC from mouse bone marrow with very high efficiency using different panels of specific surface markers (Yokota et al. 2012). Nevertheless, there are significant challenges with these methods for HSC isolation. First, none of these markers are exclusively specific for HSCs and their expression varies depending on whether the cells are proliferating or differentiating. Another limitation is the fact that these specific surface markers vary greatly between humans and animals, and even between different species. For example, Larochelle and colleagues have clearly shown that HSCs from human and macaque, in contrast to mice, do not express Sca1 or the SLAM family CD150 (Larochelle et al. 2011).

13.2.2 Role of Hypoxia on HSCs

In adults, HSCs are maintained in hypoxic *niches* with a physiological low oxygen concentration (0.1–5 %; similar to those in bone marrow) (Yin and Li 2006). This hypoxic environment plays a critical role in the regulation of their activities and can contribute to their egress from the bone marrow and mobilization (Parmar et al. 2007; Takubo et al. 2010; Imanirad and Dzierzak 2013; Lord and Murphy 1973; Murphy and Lord 1973; Serebrovskaya et al. 2011). However, the egress and mobilization of HSCs can be rapidly enhanced in the peripheral blood circulation during stress situations or with induced mobilization regimens (Hoggatt et al. 2013; Reeves 2014).

In vitro findings using human cells support the fact that human HSCs formed greater colonies under hypoxic (5–7 % oxygen) versus normoxic conditions (Ishikawa and Ito 1988; Villarruel et al. 2008). Under physiologic conditions, Takubo K. and colleagues have clearly demonstrated that HSCs over-stabilize HIF-1α protein. They also showed that the selective loss of HIF-1α affects the quiescence state of HSCs in an inducible HIF-1α deletion model in vivo. In addition, they showed that stabilization of HIF-1α induced cell cycle quiescence in HSCs and their progenitors and resulted in impairment in transplantation capacity (Takubo et al. 2010). Therefore, the precise regulation of HIF-1α levels is a critical element for the maintenance of appropriate HSC quiescence and limits their presence into the bloodstream. In addition to HIF-1α, activation states of other signalling pathways/transcriptional factors are also critical in maintaining undifferentiated HSCs and limiting their circulation. The hypoxia-related Egr1 transcription factor is another key master of HSC quiescence and localization in vivo (Min et al. 2008). Egr1 is abundantly expressed by quiescent HSCs and is downregulated when these cells are induced to divide and migrate in response to mobilizing stimuli. Furthermore, transgenic mice deficient in Egr1 exhibit an increased proportion of proliferating HSC populations in the bone marrow and increased numbers of circulating HSCs (Min et al. 2008).

Several factors have been described to mobilize HSCs and modulate HSC behaviour. Among them, some are modulated by chronic hypoxia and play a role in vascular remodelling such as fibroblast growth factor (FGF)-2, epidermal growth factor (EGF), and CXCL12/CXCR4 signalling. Indeed, systemic administration of FGF-2 in mice disrupts normal bone marrow hematopoiesis in part through reduced bone marrow levels of CXCL12 and c-kit activation (Itkin et al. 2012; Nakayama et al. 2007). In addition, an inhibitory role of EGF-receptor signaling in HSC mobilization has been demonstrated: inhibition of EGF-receptor signalling by both genetic (wa2/+ mice) and pharmacological (erlotinib) approaches enhance HSC mobilization, while administration of EGF dose-dependently reduces G-CSF-induced mobilization of HSCs (Ryan et al. 2010). On the other hand, the CXCL12 and its major receptor CXCR4 mediate both retention and mobilization of human and murine HSCs (Mohle et al. 1998; Peled et al. 1999). Interestingly, our group has demonstrated that remodelled vessels from chronically hypoxic mice exhibit a marked induction of CXCL12 and its receptors, CXCR4 and CXCR7 (Gambaryan et al. 2011) (Fig. 13.1). CXCL12 is highly expressed by various bone marrow stromal niche cells and has important roles in HSC homing, retention, survival, and quiescence. The conditional deletion

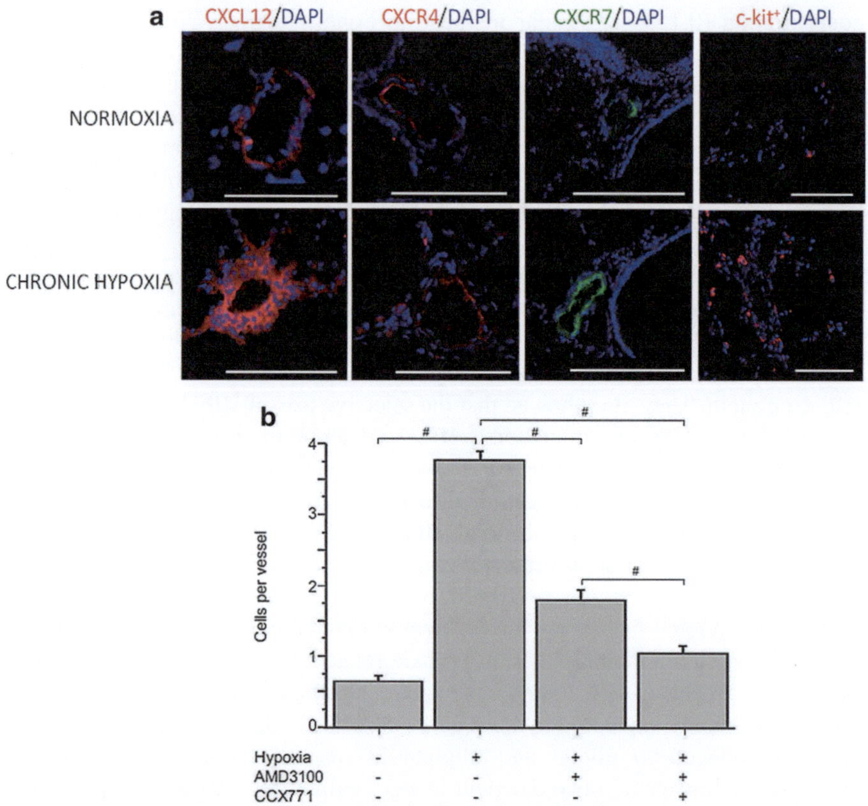

Fig. 13.1 (**a**) Immunofluorescence for CXCL12, CXCR4, CXCR7, and c-kit in mice lung exposed to chronic hypoxia (10 % oxygen for 3 weeks) or normoxia. Scale bars = 50 μm; (**b**) Pulmonary perivascular c-kit+ cell infiltration in mice exposed to chronic hypoxia (10 % oxygen for 3 weeks) treated or not with AMD3100 (a CXCR4 antagonist; intraperitoneal injection: 10 mg/kg/day) and CCX771 (a CXCR7 antagonist; intraperitoneal injection: 10 mg/kg/day) or both. #: $p<0.0001$. *Reproduced with permission of the European Respiratory Society. Eur Respir J 2011 37:1392–1399; published ahead of print 2010, doi:10.1183/09031936.00045710*

of CXCR4, CXCL12, or their inhibition leads to marked decrease in bone marrow cellularity and HSC numbers as well as impaired repopulation capacity (Sugiyama et al. 2006; Tzeng et al. 2011). Interestingly, administration of the small non-peptide CXCR4 inhibitor ADM3100 causes a rapid HSC mobilization within hours in mouse, human, and non-human primate (Broxmeyer et al. 2005; Liles et al. 2003) and synergistically augments G-CSF-induced mobilization (Abraham et al. 2007).

13.3 Chronic Hypoxia-Induced Pulmonary Vascular Remodelling

13.3.1 The Physiological Adaptive Response to Hypoxia

The pulmonary circulation is unique among vascular beds, characterized by a high flow, low-resistance, low-pressure system. Although all of the cardiac output is pumped through the pulmonary circulation, the measured values of mean pulmonary arterial pressure (mPAP) in pulmonary arteries for a resting, healthy adult human is about 14 ± 3 mm of mercury (Hg), when systemic arterial blood pressure is about 100 ± 20 mmHg. This unique feature of the pulmonary circulation is mainly explained by two important characteristics of the pulmonary vasculature: a high compliance of pulmonary pre-capillary arterioles characterized by a thin vessel wall and a high capacity to recruit vessels available to accommodate increase in flow. Indeed, the system can accommodate flow rates ranging from about 6 L/min under resting conditions to 25 L/min in strenuous exercise, with minimal increases in pulmonary pressures. In contrast to systemic arteries, the pulmonary artery carries oxygen-poor blood and the pulmonary vein carries oxygen-rich blood. In addition, oxygen deprivation causes vasoconstriction of the small pulmonary arteries, when systemic arteries dilate under hypoxic conditions. This phenomenon, known as hypoxic pulmonary vasoconstriction (HPV), is more pronounced as the vessel diameter decreases and represents an important physiological mechanism by which pulmonary arteries constrict in hypoxic lung areas in order to redirect blood flow to areas with greater oxygen supply.

Prolonged hypoxia, however, is a critical pathological stimulus that affects the expression/activity of the major regulators of vascular function necessary for maintenance of homeostasis in both pulmonary and systemic circulations. Such regulators include production of vasoconstrictors and vasodilators, activators and inhibitors of smooth muscle cell growth and migration, pro-thrombotic and anti-thrombotic mediators, as well as pro- and anti-inflammatory signals. Prolonged exposure to a hypoxic environment leads to increased pulmonary vascular resistance due to progressive structural and functional adaptations of the pulmonary vascular bed. There are age and/or sex differences and these pulmonary vascular remodelling can be variable between individuals, probably based on genetics and adaptive mechanisms. The magnitude of these responses is intense in pig, horse, and cow, moderate in rodent and human, and very low in dog, guinea pig, yak, and lama (Grover et al. 2011).

Importantly, chronic hypoxia-induced lesions, such as high altitude pulmonary edema (HAPE) in humans or chronic hypoxia-induced PH in animal models, are reversible. Interestingly, HAPE is mostly dependent on individual susceptibility as the acute and excessive rise of mPAP (35–55 mmHg) within few days after arrival at high altitude, which precedes the formation of pulmonary edema, occurs in less than 5 % of the mountaineers and climbers. Recent elegant studies demonstrated, in cattle susceptible to brisket disease, that this response is inherited, though the identification of the genetic basis is the subject of ongoing research (Newman et al. 2011). Beside individual susceptibility, the extent of the structural and functional changes of the vascular bed in humans depends on the type, the severity, and the duration of hypoxemia. The predominant cause of PH development is represented by alveolar hypoxia due to chronic lung diseases, impaired control of breathing, or residence at high altitude as reported above. Among chronic lung diseases, the diseases most commonly associated with PH, which is usually moderate, are chronic obstructive pulmonary disease (COPD), idiopathic pulmonary fibrosis (IPF), and a syndrome combining pulmonary fibrosis and emphysema (CPFE). The prevalence of PH varies depending on the associated disease and its severity. Clinical and epidemiological data highlight the notion that only a small percentage of patients with chronic lung diseases develop severe PH (defined as mPAP above 35–40 mmHg), clearly suggesting the need of additional "hit(s)" for the development of chronic hypoxia-induced severe PH.

13.3.2 Mechanisms Underlying Chronic Hypoxia-Induced Pulmonary Vascular Remodelling

The chronic hypoxia-induced remodelling of the pulmonary circulation takes place sequentially and includes: (a) endothelial dysfunction with release of vasoactive mediators, growth factors, cytokines, and chemokines; (b) smooth muscle cell hyperplasia and hypertrophy with resultant pulmonary arterial medial wall thickening; (c) adventitial fibroblast proliferation, accumulation of extracellular matrix components, and myofibroblast differentiation (Stenmark et al. 2000, 2002); and (d) changes in pulmonary pericyte behaviour including proliferation, migration, and differentiation (Ricard et al. 2014). These alterations in structure and function of the lung vasculature are accompanied by pronounced changes in cell metabolism, gene expression, and cytoskeletal organization. During hypoxic exposure, cells switch from oxidative metabolism to anaerobic glycolysis for energy production to sustain the same level of cell function. In addition, it is now well-recognized that prolonged hypoxia results in an inflammatory phenotype in the lung, both in humans and in animals (Carpenter and Stenmark 2001; Frid et al. 2006; Frohlich et al. 2013; Kubo et al. 1996; Madjdpour et al. 2003; Marsch et al. 2013; Minamino et al. 2001; Schoene et al. 1988; Vergadi et al. 2011; Zampetaki et al. 2003; West et al. 2004). Elevated levels of pro-inflammatory cytokines (i.e., IL- 1β, IL-6, IL-8, and TNF-α) and increased numbers of neutrophils and macrophages have been detected in bronchoalveolar lavage fluid of humans exposed to hypobaric hypoxia (Kubo et al. 1996; Schoene et al. 1988; West et al. 2004). These observations may explain the positive effects of corticosteroids prophylaxis on acute hypoxic hypertension in adults prone to HAPE (Fischler et al. 2009; Maggiorini et al. 2006). Finally, there is clear evidence that moderate hypoxia stimulates angiogenesis and increases vascular permeability in the pulmonary circulation, potentially explained by alterations of pulmonary pericyte functions or vessel coverage (Ricard et al. 2014). These phenomena may potentially contribute to the formation of alveolar edema in acutely hypoxic lung (Howell et al. 2003; Hyvelin et al. 2005).

The molecular networks underlying chronic hypoxia-induced pulmonary vascular remodelling are not fully understood; however, integrative genomic analyses have revealed a panel of key regulated genes and transcriptional factors that may contribute to hypoxia-induced phenotypic changes. HIF-1α and HIF-2α are among these master regulators of hypoxic signalling (Schofield and Ratcliffe 2004). Lowered oxygen tension inhibits hydroxylation of proline and asparagine residues by prolyl hydroxylase domain-containing proteins (PHDs) and stabilizes or induces a rapid nuclear translocation of cytoplasmic HIF-1α and/or HIF-2α that heterodimerizes with HIF-1β (also known as ARNT) and subsequently leads to induction of gene transcription and adaptive responses that accommodate cellular function in ischemia or hypoxia. In addition to HIF activation, hypoxia has been demonstrated to activate several other transcription factors including to varying degrees: Egr-1 (Bae et al. 1999; Jin et al. 2000; Nishi et al. 2002; Yan et al. 1998, 1999), nuclear factor-kB (NF-kB) (Chandel et al. 2000b; Koong et al. 1994; Leeper-Woodford and Detmer 1999; Liu et al. 2014; Matsui et al. 1999; Schmedtje et al. 1997; Scholz and Taylor 2013; Taylor et al. 2000), cyclic AMP response element-binding protein (CREB) (Beitner-Johnson and Millhorn 1998; Comerford et al. 2003), activating protein (AP)-1 (Laderoute 2005, Minet et al. 2001; Salnikow et al. 2002:53; Alarcon et al. 1999; Chen et al. 2003; Sermeus and Michiels 2011), Sp1, Sp3 (Deacon et al. 2012; Discher et al. 1998; Yang et al. 2014), and CCAAT/enhancer-binding protein β (C/EBPβ) (Semenza 2000; Yan et al. 1995; Matsui et al. 1999; Park and Park 2010; Larsen et al. 2008). In addition, reactive oxygen species (ROS) (Chandel et al. 1998, 2000a, b; Chandel and Schumacker 2000; Killilea et al. 2000; Zepeda et al. 2013), calcium (Pamenter and Haddad 2014; Peers and Kemp 2004; Zepeda et al. 2013), and intracellular cyclic AMP levels (Peers and Kemp 2004) or the response to extracellular adenosine have been demonstrated to participate in the modulation of the molecular networks underlying chronic hypoxia-induced pulmonary vascular remodelling. Low oxygen availability also controls mRNA concentrations by inducing changes in microRNA transcription and expression (Nallamshetty et al. 2013). All of these mediators lead to a cellular response when this balance between oxygen supply and demand is disturbed.

Acute (normobaric 10 % FiO_2 for 10–30 min) and chronic hypoxia (normobaric/hypobaric 10 % FiO_2 for 3–6 weeks) are among the most widely used models to study pulmonary vasoreactivity and PH, respectively (Ryan et al. 2011; Stenmark et al. 2009; Weir and Archer 1995). They are very valuable to test new targets and/or treatments and to give insights into the disease mechanisms. Animal PH models allow study of the early phases of disease process, at several different levels of cell function and also help to analyse precise mechanisms of action of potential agents for clinical use. However, chronic hypoxia-induced pulmonary vascular lesions are reversible when back to normoxic conditions, clearly demonstrating that such animal models do not reproduce the full spectrum of changes seen in lung specimens from severe PH patients with advanced pulmonary arterial hypertension (PAH). Therefore, efforts have been made to improve this chronic hypoxia-induced PH by treating rats with a single dose of VEGF-receptor antagonist (SU5416) plus hypoxia (de Raaf et al. 2014; Ryan et al. 2011; Stenmark et al. 2009; Taraseviciene-Stewart et al. 2001). However, pulmonary vascular lesions experimentally induced, even in this latter animal model, do not mimic human disease in particular regarding the plexogenic lesions. Furthermore, a low mortality rate is often reported, contrasting with the devastating prognosis of human PAH. In addition, interspecies-, age-, sex-related differences have been reported.

13.4 HSCs and Chronic Hypoxia-Induced Pulmonary Vascular Remodelling: Is There Any Link?

Our group has reported that idiopathic PAH patients display high serum CXCL12 levels and that pulmonary endothelium represents one local abnormal source (Montani et al. 2011) (Fig. 13.2). We have also demonstrated that remodelled vessels from chronically hypoxic mice exhibit a marked induction of CXCL12 and its receptors,

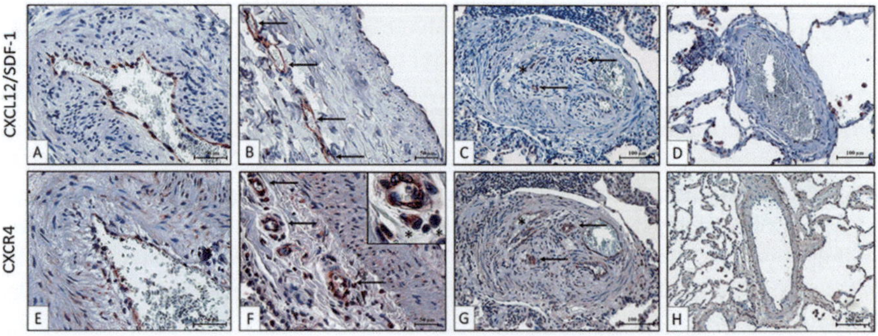

Fig. 13.2 Immunochemistry of CXCL12 and CXCR4 in pulmonary arterial lesions from patients with idiopathic pulmonary arterial hypertension (IPAH) (**a**–**c** and **e**–**g**) and controls (**d**, **h**). In panels **b** and **f**, *arrows* indicate the *vasa vasorum* and *asterisks* indicate inflammatory cells. In panels **c** and **g**, *arrows* and *asterisks* indicate endothelial cells. *Reprinted with permission of the American Thoracic Society. Copyright © 2014 American Thoracic Society. Montani D, Perros F, Gambaryan N, Girerd B, Dorfmuller P, Price LC, Huertas A, Hammad H, Lambrecht B, Simonneau G, Launay JM, Cohen-Kaminsky S, Humbert M (2011) C-kit-positive cells accumulate in remodeled vessels of idiopathic pulmonary arterial hypertension. American journal of respiratory and critical care medicine 184 (1):116–123. doi:10.1164/rccm.201006-0905OC. Official Journal of the American Thoracic Society*

CXCR4 and CXCR7 (Gambaryan et al. 2011) (Fig. 13.1). In the chronic hypoxia-induced PH mouse model, we found that administration of AMD3100, a CXCR4 antagonist, alone or combined with CCX771, a CXCR7 antagonist, prevented vascular remodelling, PH, and perivascular accumulation of c-kit$^+$/sca-1$^+$ progenitor cells, with a synergistic effect of these agents (Gambaryan et al. 2011) (Fig. 13.1). CXCL12 represents an important factor for recruitment and mobilization of progenitor cells, in particular HSCs expressing the transmembrane tyrosine kinase receptor c-kit (although c-kit may be present on other cell types such as mast cells). Interestingly, c-kit$^+$ cells can be mobilized from the bone marrow and may potentially differentiate into vascular cells to promote myocardial repair after infarction (Orlic et al. 2001) and to contribute to atherosclerosis (Sata et al. 2002). Furthermore, it is known that c-kit$^+$ HSCs can be mobilized from the bone marrow into the systemic circulation by hypoxia (Parmar et al. 2007; Takubo et al. 2010; Imanirad and Dzierzak 2013; Lord and Murphy 1973; Murphy and Lord 1973; Serebrovskaya et al. 2011). Interestingly, an increase in c-kit$^+$ cell number was found in remodelled pulmonary arterial wall in chronically hypoxic mice (Davie et al. 2004; Gambaryan et al. 2011) and in idiopathic PAH (Montani et al. 2011) (Figs. 13.1 and 13.3).

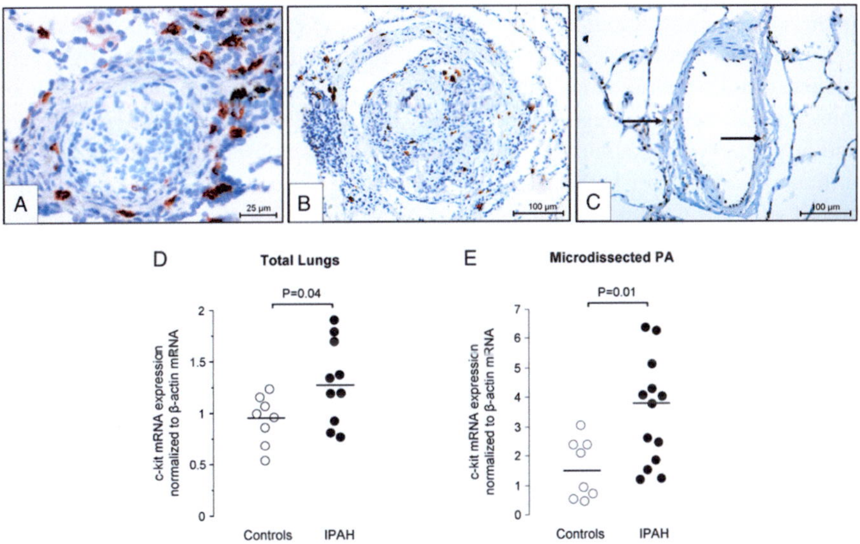

Fig. 13.3 Immunohistochemistry for c-kit (red staining) in paraffin-embedded lung sections: from patients with idiopathic pulmonary arterial hypertension (IPAH) (**a**, **b**) and controls (**c**). Real-time polymerase chain reaction quantifications of c-kit mRNA levels normalized to β-actin in total lungs from controls and IPAH patients (**d**) and in microdissected small pulmonary arteries (PA) from controls and IPAH patients (**e**). *Reprinted with permission of the American Thoracic Society. Copyright © 2014 American Thoracic Society. Montani D, Perros F, Gambaryan N, Girerd B, Dorfmuller P, Price LC, Huertas A, Hammad H, Lambrecht B, Simonneau G, Launay JM, Cohen-Kaminsky S, Humbert M (2011) C-kit-positive cells accumulate in remodeled vessels of idiopathic pulmonary arterial hypertension. American journal of respiratory and critical care medicine 184 (1):116–123. doi:10.1164/rccm.201006-0905OC. Official Journal of the American Thoracic Society*

Moreover, a work from Launay and colleagues shows a role of 5-HT2B receptors in the maturation of different bone marrow-derived cells and in their localization in the pulmonary vascular wall in the chronic hypoxia-induced PH mouse model (Launay et al. 2012).

By highlighting the CXCL12 signalling pathway dysregulation and the increased numbers of c-kit$^+$ cells in human PAH and experimental PH, we raise the question of the role of HSCs in pulmonary vascular remodelling and disease progression: whether HSCs can be recruited and mobilized in the lung and whether these cells can differentiate into vascular cells and actively participate to lesion formation and pulmonary vascular remodelling needs to be addressed.

Unfortunately, there is no robust data on the role of HSCs in chronic hypoxia-induced pulmonary vascular remodelling. The available evidences so far have been obtained using bone marrow cells without clear consensus about HSC markers, cell phenotypes, disease read-outs, and animal tools. In the first study, Hayashida and colleagues used mice transplanted with bone marrow from enhanced green fluorescence protein (GFP)-transgenic mice. After 8 weeks, they exposed the transplanted mice to chronic hypoxia, showing GFP$^+$ cells in pulmonary arterial wall, including into the adventitia. In addition, most of GFP$^+$ cells expressed α-smooth muscle cell actin, suggesting that these cells can differentiate in situ (Hayashida et al. 2005). In the second study, Raoul and coworkers showed no effect of intravenous administration of bone marrow cells on pulmonary hemodynamic measurements (right ventricular systolic pressure), on right ventricular hypertrophy, and on the percentage of muscularized vessels, suggesting that HSCs had no role in the disease development and/or progression (Raoul et al. 2007). Based on these observations, it is therefore difficult to conclude whether HSCs contribute to pulmonary vascular remodelling and, if so, by which underlying mechanisms. Additional studies are clearly needed to define and understand the potential role of HSCs in chronic hypoxia-induced pulmonary vascular remodelling.

13.5 Concluding Remarks

In this chapter, we have reviewed the current knowledge on the role of HSCs in chronic hypoxia-induced pulmonary vascular remodelling. Unfortunately, we still lack a clear understanding of HSC behaviour, activation, recruitment, and potential differentiation in the pulmonary vasculature, particularly under chronic hypoxia. A consensus on techniques such as bone marrow transplantation, animal models, and tools to follow HSCs in a time-dependent manner in the setting of chronic hypoxia would help defining whether these cells are mobilized in a chronic hypoxia-dependent manner, whether they are recruited in the pulmonary vasculature, and whether they are able to differentiate into pulmonary wall cells and contribute to and/or delay the progression of pulmonary vascular remodelling.

References

Abraham M, Biyder K, Begin M, Wald H, Weiss ID, Galun E, Nagler A, Peled A (2007) Enhanced unique pattern of hematopoietic cell mobilization induced by the CXCR4 antagonist 4F-benzoyl-TN14003. Stem Cells 25(9):2158–2166. doi:10.1634/stemcells. 2007-0161

Alarcon R, Koumenis C, Geyer RK, Maki CG, Giaccia AJ (1999) Hypoxia induces p53 accumulation through MDM2 down-regulation and inhibition of E6-mediated degradation. Cancer Res 59(24):6046–6051

Bae SK, Bae MH, Ahn MY, Son MJ, Lee YM, Bae MK, Lee OH, Park BC, Kim KW (1999) Egr-1 mediates transcriptional activation of IGF-II gene in response to hypoxia. Cancer Res 59(23): 5989–5994

Beitner-Johnson D, Millhorn DE (1998) Hypoxia induces phosphorylation of the cyclic AMP response element-binding protein by a novel signaling mechanism. J Biol Chem 273(31):19834–19839

Broxmeyer HE, Orschell CM, Clapp DW, Hangoc G, Cooper S, Plett PA, Liles WC, Li X, Graham-Evans B, Campbell TB, Calandra G, Bridger G, Dale DC, Srour EF (2005) Rapid mobilization of murine and human hematopoietic stem and progenitor cells with AMD3100, a CXCR4 antagonist. J Exp Med 201(8):1307–1318. doi:10.1084/jem.20041385

Carpenter TC, Stenmark KR (2001) Hypoxia decreases lung neprilysin expression and increases pulmonary vascular leak. Am J Physiol Lung Cell Mol Physiol 281(4):L941–L948

Chandel NS, Schumacker PT (2000) Cellular oxygen sensing by mitochondria: old questions, new insight. J Appl Physiol 88(5):1880–1889

Chandel NS, Maltepe E, Goldwasser E, Mathieu CE, Simon MC, Schumacker PT (1998) Mitochondrial reactive oxygen species trigger hypoxia-induced transcription. Proc Natl Acad Sci U S A 95(20):11715–11720

Chandel NS, McClintock DS, Feliciano CE, Wood TM, Melendez JA, Rodriguez AM, Schumacker PT (2000a) Reactive oxygen species generated at mitochondrial complex III stabilize hypoxia-inducible factor-1α during hypoxia: a mechanism of O_2 sensing. J Biol Chem 275(33):25130–25138. doi:10.1074/jbc.M001914200

Chandel NS, Trzyna WC, McClintock DS, Schumacker PT (2000b) Role of oxidants in NF-kappa B activation and TNF-alpha gene transcription induced by hypoxia and endotoxin. J Immunol 165(2):1013–1021

Chen D, Li M, Luo J, Gu W (2003) Direct interactions between HIF-1 alpha and Mdm2 modulate p53 function. J Biol Chem 278(16):13595–13598. doi:10.1074/jbc.C200694200

Comerford KM, Leonard MO, Karhausen J, Carey R, Colgan SP, Taylor CT (2003) Small ubiquitin-related modifier-1 modification mediates resolution of CREB-dependent responses to hypoxia. Proc Natl Acad Sci U S A 100(3):986–991. doi:10.1073/pnas.0337412100

Davie NJ, Crossno JT Jr, Frid MG, Hofmeister SE, Reeves JT, Hyde DM, Carpenter TC, Brunetti JA, McNiece IK, Stenmark KR (2004) Hypoxia-induced pulmonary artery adventitial remodelling and neovascularization: contribution of progenitor cells. Am J Physiol Lung Cell Mol Physiol 286(4):L668–L678. doi:10.1152/ajplung.00108.2003

de Raaf MA, Schalij I, Gomez-Arroyo JG, Rol N, Happe C, de Man FS, Vonk-Noordegraaf A, Westerhof N, Voelkel NF, Bogaard HJ (2014) SuHx rat model: partly reversible pulmonary hypertension and progressive intima obstruction. Eur Respir J. doi:10.1183/09031936.00204813

Deacon K, Onion D, Kumari R, Watson SA, Knox AJ (2012) Elevated SP-1 transcription factor expression and activity drives basal and hypoxia-induced vascular endothelial growth factor (VEGF) expression in non-small cell lung cancer. J Biol Chem 287(47):39967–39981. doi:10.1074/jbc.M112.397042

Discher DJ, Bishopric NH, Wu X, Peterson CA, Webster KA (1998) Hypoxia regulates beta-enolase and pyruvate kinase-M promoters by modulating Sp1/Sp3 binding to a conserved GC element. J Biol Chem 273(40):26087–26093

Fischler M, Maggiorini M, Dorschner L, Debrunner J, Bernheim A, Kiencke S, Mairbaurl H, Bloch KE, Naeije R, Brunner-La Rocca HP (2009) Dexamethasone but not tadalafil improves exercise capacity in adults prone to high-altitude pulmonary edema. Am J Respir Crit Care Med 180(4):346–352. doi:10.1164/rccm.200808-1348OC

Frid MG, Brunetti JA, Burke DL, Carpenter TC, Davie NJ, Reeves JT, Roedersheimer MT, van Rooijen N, Stenmark KR (2006) Hypoxia-induced pulmonary vascular remodelling requires recruitment of circulating mesenchymal precursors of a monocyte/macrophage lineage. Am J Pathol 168(2):659–669. doi:10.2353/ajpath.2006.050599

Frohlich S, Boylan J, McLoughlin P (2013) Hypoxia-induced inflammation in the lung: a potential therapeutic target in acute lung injury? Am J Respir Cell Mol Biol 48(3):271–279. doi:10.1165/rcmb.2012-0137TR

Gambaryan N, Perros F, Montani D, Cohen-Kaminsky S, Mazmanian M, Renaud JF, Simonneau G, Lombet A, Humbert M (2011) Targeting of c-kit + haematopoietic progenitor cells prevents hypoxic pulmonary hypertension. Eur Respir J 37(6):1392–1399. doi:10.1183/09031936.00045710

Grover RF, Wagner WW, McMurtry IF, Reeves JT (2011) Pulmonary circulation. In: Compr Physiol. Wiley. doi:10.1002/cphy.cp020304

Hayashida K, Fujita J, Miyake Y, Kawada H, Ando K, Ogawa S, Fukuda K (2005) Bone marrow-derived cells contribute to pulmonary vascular remodelling in hypoxia-induced pulmonary hypertension. Chest 127(5):1793–1798. doi:10.1378/chest.127.5.1793

Hoggatt J, Mohammad KS, Singh P, Hoggatt AF, Chitteti BR, Speth JM, Hu P, Poteat BA, Stilger KN, Ferraro F, Silberstein L, Wong FK, Farag SS, Czader M, Milne GL, Breyer RM, Serezani CH, Scadden DT, Guise TA, Srour EF, Pelus LM (2013) Differential stem- and progenitor-cell trafficking by prostaglandin E2. Nature 495(7441):365–369. doi:10.1038/nature11929

Howell K, Preston RJ, McLoughlin P (2003) Chronic hypoxia causes angiogenesis in addition to remodelling in the adult rat pulmonary circulation. J Physiol 547(Pt 1):133–145. doi:10.1113/jphysiol.2002.030676

Hyvelin JM, Howell K, Nichol A, Costello CM, Preston RJ, McLoughlin P (2005) Inhibition of Rho-kinase attenuates hypoxia-induced angiogenesis in the pulmonary circulation. Circ Res 97(2):185–191. doi:10.1161/01.RES.0000174287.17953.83

Imanirad P, Dzierzak E (2013) Hypoxia and HIFs in regulating the development of the hematopoietic system. Blood Cells Mol Dis 51(4):256–263. doi:10.1016/j.bcmd.2013.08.005

Ishikawa Y, Ito T (1988) Kinetics of hemopoietic stem cells in a hypoxic culture. Eur J Haematol 40(2):126–129

Itkin T, Ludin A, Gradus B, Gur-Cohen S, Kalinkovich A, Schajnovitz A, Ovadya Y, Kollet O, Canaani J, Shezen E, Coffin DJ, Enikolopov GN, Berg T, Piacibello W, Hornstein E, Lapidot T (2012) FGF-2 expands murine hematopoietic stem and progenitor cells via proliferation of stromal cells, c-Kit activation, and CXCL12 down-regulation. Blood 120(9):1843–1855. doi:10.1182/blood-2011-11-394692

Jin N, Hatton N, Swartz DR, Xia X, Harrington MA, Larsen SH, Rhoades RA (2000) Hypoxia activates jun-N-terminal kinase, extracellular signal-regulated protein kinase, and p38 kinase in pulmonary arteries. Am J Respir Cell Mol Biol 23(5):593–601. doi:10.1165/ajrcmb.23.5.3921

Killilea DW, Hester R, Balczon R, Babal P, Gillespie MN (2000) Free radical production in hypoxic pulmonary artery smooth muscle cells. Am J Physiol Lung Cell Mol Physiol 279(2):L408–L412

Koong AC, Chen EY, Giaccia AJ (1994) Hypoxia causes the activation of nuclear factor kappa B through the phosphorylation of I kappa B alpha on tyrosine residues. Cancer Res 54(6):1425–1430

Kubo K, Yamaguchi S, Fujimoto K, Hanaoka M, Hayasaka M, Honda T, Sodeyama T, Kiyosawa K (1996) Bronchoalveolar lavage fluid findings in patients with chronic hepatitis C virus infection. Thorax 51(3):312–314

Laderoute KR (2005) The interaction between HIF-1 and AP-1 transcription factors in response to low oxygen. Semin Cell Dev Biol 16(4–5):502–513. doi:10.1016/j.semcdb.2005.03.005

Larochelle A, Savona M, Wiggins M, Anderson S, Ichwan B, Keyvanfar K, Morrison SJ, Dunbar CE (2011) Human and rhesus macaque hematopoietic stem cells cannot be purified based only on SLAM family markers. Blood 117(5):1550–1554. doi:10.1182/blood-2009-03-212803

Larsen M, Tazzyman S, Lund EL, Junker N, Lewis CE, Kristjansen PE, Murdoch C (2008) Hypoxia-induced secretion of macrophage migration-inhibitory factor from MCF-7 breast cancer cells is regulated in a hypoxia-inducible factor-independent manner. Cancer Lett 265(2):239–249. doi:10.1016/j.canlet.2008.02.012

Launay JM, Herve P, Callebert J, Mallat Z, Collet C, Doly S, Belmer A, Diaz SL, Hatia S, Cote F, Humbert M, Maroteaux L (2012) Serotonin 5-HT2B receptors are required for bone-marrow contribution to pulmonary arterial hypertension. Blood 119(7):1772–1780. doi:10.1182/blood-2011-06-358374

Leeper-Woodford SK, Detmer K (1999) Acute hypoxia increases alveolar macrophage tumor necrosis factor activity and alters NF-kappaB expression. Am J Physiol 276(6 Pt 1):L909–L916

Liles WC, Broxmeyer HE, Rodger E, Wood B, Hubel K, Cooper S, Hangoc G, Bridger GJ, Henson GW, Calandra G, Dale DC (2003) Mobilization of hematopoietic progenitor cells in healthy volunteers by AMD3100, a CXCR4 antagonist. Blood 102(8):2728–2730. doi:10.1182/blood-2003-02-0663

Liu L, Salnikov AV, Bauer N, Aleksandrowicz E, Labsch S, Nwaeburu C, Mattern J, Gladkich J, Schemmer P, Werner J, Herr I (2014) Triptolide reverses hypoxia-induced epithelial-mesenchymal transition and stem-like features in pancreatic cancer by NF-kappaB downregulation. Int J Cancer 134(10):2489–2503. doi:10.1002/ijc.28583

Lord BI, Murphy MJ Jr (1973) Hematopoietic stem cell regulation. II. Chronic effects of hypoxic-hypoxia on CFU kinetics. Blood 42(1):89–98

Madjdpour C, Jewell UR, Kneller S, Ziegler U, Schwendener R, Booy C, Klausli L, Pasch T, Schimmer RC, Beck-Schimmer B (2003) Decreased alveolar oxygen induces lung inflammation. Am J Physiol Lung Cell Mol Physiol 284(2):L360–L367. doi:10.1152/ajplung.00158.2002

Maggiorini M, Brunner-La Rocca HP, Peth S, Fischler M, Bohm T, Bernheim A, Kiencke S, Bloch KE, Dehnert C, Naeije R, Lehmann T, Bartsch P, Mairbaurl H (2006) Both tadalafil and dexamethasone may reduce the incidence of high-altitude pulmonary edema: a randomized trial. Ann Intern Med 145(7):497–506

Marsch E, Sluimer JC, Daemen MJ (2013) Hypoxia in atherosclerosis and inflammation. Curr Opin Lipidol 24(5):393–400. doi:10.1097/MOL.0b013e32836484a4

Matsui H, Ihara Y, Fujio Y, Kunisada K, Akira S, Kishimoto T, Yamauchi-Takihara K (1999) Induction of interleukin (IL)-6 by hypoxia is mediated by nuclear factor (NF)-kappa B and NF-IL6 in cardiac myocytes. Cardiovasc Res 42(1):104–112

Min IM, Pietramaggiori G, Kim FS, Passegue E, Stevenson KE, Wagers AJ (2008) The transcription factor EGR1 controls both the proliferation and localization of hematopoietic stem cells. Cell Stem Cell 2(4):380–391. doi:10.1016/j.stem.2008.01.015

Minamino T, Christou H, Hsieh CM, Liu Y, Dhawan V, Abraham NG, Perrella MA, Mitsialis SA, Kourembanas S (2001) Targeted expression of heme oxygenase-1 prevents the pulmonary inflammatory and vascular responses to hypoxia. Proc Natl Acad Sci U S A 98(15):8798–8803. doi:10.1073/pnas.161272598

Minet E, Michel G, Mottet D, Piret JP, Barbieux A, Raes M, Michiels C (2001) c-JUN gene induction and AP-1 activity is regulated by a JNK-dependent pathway in hypoxic HepG2 cells. Exp Cell Res 265(1):114–124. doi:10.1006/excr.2001.5180

Mohle R, Bautz F, Rafii S, Moore MA, Brugger W, Kanz L (1998) The chemokine receptor CXCR-4 is expressed on CD34+ hematopoietic progenitors and leukemic cells and mediates transendothelial migration induced by stromal cell-derived factor-1. Blood 91(12):4523–4530

Montani D, Perros F, Gambaryan N, Girerd B, Dorfmuller P, Price LC, Huertas A, Hammad H, Lambrecht B, Simonneau G, Launay JM, Cohen-Kaminsky S, Humbert M (2011) C-kit-positive cells accumulate in remodeled vessels of idiopathic pulmonary arterial hypertension. Am J Respir Crit Care Med 184(1):116–123. doi:10.1164/rccm.201006-0905OC

Murphy MJ Jr, Lord BI (1973) Hematopoietic stem cell regulation. I. Acute effects of hypoxic-hypoxia on CFU kinetics. Blood 42(1):81–87

Nakayama T, Mutsuga N, Tosato G (2007) Effect of fibroblast growth factor 2 on stromal cell-derived factor 1 production by bone marrow stromal cells and hematopoiesis. J Natl Cancer Inst 99(3):223–235. doi:10.1093/jnci/djk031

Nallamshetty S, Chan SY, Loscalzo J (2013) Hypoxia: a master regulator of microRNA biogenesis and activity. Free Radic Biol Med 64:20–30. doi:10.1016/j.freeradbiomed.2013.05.022

Newman JH, Holt TN, Hedges LK, Womack B, Memon SS, Willers ED, Wheeler L, Phillips JA 3rd, Hamid R (2011) High-altitude pulmonary hypertension in cattle (brisket disease): candidate genes and gene expression profiling of peripheral blood mononuclear cells. Pulm Circ 1(4):462–469. doi:10.4103/2045-8932.93545

Nishi H, Nishi KH, Johnson AC (2002) Early growth response-1 gene mediates up-regulation of epidermal growth factor receptor expression during hypoxia. Cancer Res 62(3):827–834

Orlic D, Kajstura J, Chimenti S, Limana F, Jakoniuk I, Quaini F, Nadal-Ginard B, Bodine DM, Leri A, Anversa P (2001) Mobilized bone marrow cells repair the infarcted heart, improving function and survival. Proc Natl Acad Sci U S A 98(18):10344–10349. doi:10.1073/pnas.181177898

Pamenter ME, Haddad GG (2014) Do BK channels mediate glioma hypoxia-tolerance? Channels 8(3)

Park YK, Park H (2010) Prevention of CCAAT/enhancer-binding protein beta DNA binding by hypoxia during adipogenesis. J Biol Chem 285(5):3289–3299. doi:10.1074/jbc.M109.059212

Parmar K, Mauch P, Vergilio JA, Sackstein R, Down JD (2007) Distribution of hematopoietic stem cells in the bone marrow according to regional hypoxia. Proc Natl Acad Sci U S A 104(13):5431–5436. doi:10.1073/pnas.0701152104

Peers C, Kemp PJ (2004) Ion channel regulation by chronic hypoxia in models of acute oxygen sensing. Cell Calcium 36(3–4):341–348. doi:10.1016/j.ceca.2004.02.005

Peled A, Grabovsky V, Habler L, Sandbank J, Arenzana-Seisdedos F, Petit I, Ben-Hur H, Lapidot T, Alon R (1999) The chemokine SDF-1 stimulates integrin-mediated arrest of CD34(+) cells on vascular endothelium under shear flow. J Clin Invest 104(9):1199–1211. doi:10.1172/JCI7615

Raoul W, Wagner-Ballon O, Saber G, Hulin A, Marcos E, Giraudier S, Vainchenker W, Adnot S, Eddahibi S, Maitre B (2007) Effects of bone marrow-derived cells on monocrotaline- and hypoxia-induced pulmonary hypertension in mice. Respir Res 8:8. doi:10.1186/1465-9921-8-8

Reeves G (2014) Overview of use of G-CSF and GM-CSF in the treatment of acute radiation injury. Health Phys 106(6):699–703. doi:10.1097/HP.0000000000000090

Ricard N, Tu L, Le Hiress M, Huertas A, Phan C, Thuillet R, Sattler C, Fadel E, Seferian A, Montani D, Dorfmuller P, Humbert M, Guignabert C (2014) Increased pericyte coverage mediated by endothelial-derived fibroblast growth factor-2 and interleukin-6 is a source of smooth muscle-like cells in pulmonary hypertension. Circulation 129(15):1586–1597. doi:10.1161/CIRCULATIONAHA.113.007469

Ryan MA, Nattamai KJ, Xing E, Schleimer D, Daria D, Sengupta A, Kohler A, Liu W, Gunzer M, Jansen M, Ratner N, Le Cras TD, Waterstrat A, Van Zant G, Cancelas JA, Zheng Y, Geiger H (2010) Pharmacological inhibition of EGFR signaling enhances G-CSF-induced hematopoietic stem cell mobilization. Nat Med 16(10):1141–1146. doi:10.1038/nm.2217

Ryan J, Bloch K, Archer SL (2011) Rodent models of pulmonary hypertension: harmonisation with the world health organisation's categorisation of human PH. Int J Clin Pract Suppl 172:15–34. doi:10.1111/j.1742-1241.2011.02710.x

Salnikow K, Kluz T, Costa M, Piquemal D, Demidenko ZN, Xie K, Blagosklonny MV (2002) The regulation of hypoxic genes by calcium involves c-Jun/AP-1, which cooperates with hypoxia-inducible factor 1 in response to hypoxia. Mol Cell Biol 22(6):1734–1741

Sata M, Saiura A, Kunisato A, Tojo A, Okada S, Tokuhisa T, Hirai H, Makuuchi M, Hirata Y, Nagai R (2002) Hematopoietic stem cells differentiate into vascular cells that participate in the pathogenesis of atherosclerosis. Nat Med 8(4):403–409. doi:10.1038/nm0402-403

Schmedtje JF Jr, Ji YS, Liu WL, DuBois RN, Runge MS (1997) Hypoxia induces cyclooxygenase-2 via the NF-kappaB p65 transcription factor in human vascular endothelial cells. J Biol Chem 272(1):601–608

Schoene RB, Swenson ER, Pizzo CJ, Hackett PH, Roach RC, Mills WJ Jr, Henderson WR Jr, Martin TR (1988) The lung at high altitude: bronchoalveolar lavage in acute mountain sickness and pulmonary edema. J Appl Physiol 64(6):2605–2613

Schofield CJ, Ratcliffe PJ (2004) Oxygen sensing by HIF hydroxylases. Nat Rev Mol Cell Biol 5(5):343–354. doi:10.1038/nrm1366

Scholz CC, Taylor CT (2013) Hydroxylase-dependent regulation of the NF-kappaB pathway. Biol Chem 394(4):479–493. doi:10.1515/hsz-2012-0338

Semenza GL (2000) Oxygen-regulated transcription factors and their role in pulmonary disease. Respir Res 1(3):159–162. doi:10.1186/rr27

Serebrovskaya TV, Nikolsky IS, Nikolska VV, Mallet RT, Ishchuk VA (2011) Intermittent hypoxia mobilizes hematopoietic progenitors and augments cellular and humoral elements of innate immunity in adult men. High Alt Med Biol 12(3):243–252. doi:10.1089/ham.2010.1086

Sermeus A, Michiels C (2011) Reciprocal influence of the p53 and the hypoxic pathways. Cell Death Dis 2:e164. doi:10.1038/cddis.2011.48

Stenmark KR, Bouchey D, Nemenoff R, Dempsey EC, Das M (2000) Hypoxia-induced pulmonary vascular remodelling: contribution of the adventitial fibroblasts. Physiol Res 49(5): 503–517

Stenmark KR, Gerasimovskaya E, Nemenoff RA, Das M (2002) Hypoxic activation of adventitial fibroblasts: role in vascular remodelling. Chest 122(6 Suppl):326S–334S

Stenmark KR, Meyrick B, Galie N, Mooi WJ, McMurtry IF (2009) Animal models of pulmonary arterial hypertension: the hope for etiological discovery and pharmacological cure. Am J Physiol Lung Cell Mol Physiol 297(6):L1013–L1032. doi:10.1152/ajplung.00217.2009

Sugiyama T, Kohara H, Noda M, Nagasawa T (2006) Maintenance of the hematopoietic stem cell pool by CXCL12-CXCR4 chemokine signaling in bone marrow stromal cell niches. Immunity 25(6):977–988. doi:10.1016/j.immuni.2006.10.016

Takubo K, Goda N, Yamada W, Iriuchishima H, Ikeda E, Kubota Y, Shima H, Johnson RS, Hirao A, Suematsu M, Suda T (2010) Regulation of the HIF-1alpha level is essential for hematopoietic stem cells. Cell Stem Cell 7(3):391–402. doi:10.1016/j.stem.2010.06.020

Taraseviciene-Stewart L, Kasahara Y, Alger L, Hirth P, Mc Mahon G, Waltenberger J, Voelkel NF, Tuder RM (2001) Inhibition of the VEGF receptor 2 combined with chronic hypoxia causes cell death-dependent pulmonary endothelial cell proliferation and severe pulmonary hypertension. FASEB J 15(2):427–438. doi:10.1096/fj.00-0343com

Taylor CT, Furuta GT, Synnestvedt K, Colgan SP (2000) Phosphorylation-dependent targeting of cAMP response element binding protein to the ubiquitin/proteasome pathway in hypoxia. Proc Natl Acad Sci U S A 97(22):12091–12096. doi:10.1073/pnas.220211797

Tzeng YS, Li H, Kang YL, Chen WC, Cheng WC, Lai DM (2011) Loss of Cxcl12/Sdf-1 in adult mice decreases the quiescent state of hematopoietic stem/progenitor cells and alters the pattern of hematopoietic regeneration after myelosuppression. Blood 117(2):429–439. doi:10.1182/blood-2010-01-266833

Vergadi E, Chang MS, Lee C, Liang OD, Liu X, Fernandez-Gonzalez A, Mitsialis SA, Kourembanas S (2011) Early macrophage recruitment and alternative activation are critical for the later development of hypoxia-induced pulmonary hypertension. Circulation 123(18):1986–1995. doi:10.1161/CIRCULATIONAHA.110.978627

Villarruel SM, Boehm CA, Pennington M, Bryan JA, Powell KA, Muschler GF (2008) The effect of oxygen tension on the in vitro assay of human osteoblastic connective tissue progenitor cells. J Orthop Res 26(10):1390–1397. doi:10.1002/jor.20666

Weir EK, Archer SL (1995) The mechanism of acute hypoxic pulmonary vasoconstriction: the tale of two channels. FASEB J 9(2):183–189

West JB, American College of Physicians, American Physiological Society (2004) The physiologic basis of high-altitude diseases. Ann Intern Med 141(10):789–800

Yan SF, Tritto I, Pinsky D, Liao H, Huang J, Fuller G, Brett J, May L, Stern D (1995) Induction of interleukin 6 (IL-6) by hypoxia in vascular cells. Central role of the binding site for nuclear factor-IL-6. J Biol Chem 270(19):11463–11471

Yan SF, Zou YS, Gao Y, Zhai C, Mackman N, Lee SL, Milbrandt J, Pinsky D, Kisiel W, Stern D (1998) Tissue factor transcription driven by Egr-1 is a critical mechanism of murine pulmonary fibrin deposition in hypoxia. Proc Natl Acad Sci U S A 95(14):8298–8303

Yan SF, Lu J, Zou YS, Soh-Won J, Cohen DM, Buttrick PM, Cooper DR, Steinberg SF, Mackman N, Pinsky DJ, Stern DM (1999) Hypoxia-associated induction of early growth response-1 gene expression. J Biol Chem 274(21):15030–15040

Yang Y, Liu S, Fan Z, Li Z, Liu J, Xing F (2014) Sp1 modification of human endothelial nitric oxide synthase promoter increases the hypoxia-stimulated activity. Microvasc Res 93:80–86. doi:10.1016/j.mvr.2014.03.004

Yin T, Li L (2006) The stem cell niches in bone. J Clin Invest 116(5):1195–1201. doi:10.1172/JCI28568

Yokota T, Oritani K, Butz S, Ewers S, Vestweber D, Kanakura Y (2012) Markers for hematopoietic stem cells: histories and recent achievements. In: Pelayo R (ed) Advances in hematopoietic stem cell research. InTech, Rijeka. doi:10.5772/32381

Zampetaki A, Minamino T, Mitsialis SA, Kourembanas S (2003) Effect of heme oxygenase-1 overexpression in two models of lung inflammation. Exp Biol Med 228(5):442–446

Zepeda AB, Pessoa A Jr, Castillo RL, Figueroa CA, Pulgar VM, Farias JG (2013) Cellular and molecular mechanisms in the hypoxic tissue: role of HIF-1 and ROS. Cell Biochem Funct 31(6):451–459. doi:10.1002/cbf.2985

Chapter 14
Fibrocytes and Pulmonary Vascular Remodeling: The Good, the Bad, and the Progenitors

Kelley L. Colvin, Ozus Lohani, and Michael E. Yeager

Abbreviations

BALT	Bronchus-associated lymphoid tissue
CD11b	Cluster of differentiation factor 11b (aka integrin alpha M (ITGAM))
CD45	Cluster of differentiation factor 45 (aka protein tyrosine phosphatase, receptor type, C (PTPRC))
COPD	Chronic obstructive pulmonary disease
ECM	Extracellular matrix
EGPA	Eosinophilic granulomatosis with polyangiitis
FEV1	Forced expiratory volume exhaled in 1 s
HLA-DR	Human leukocyte antigen-DR (MHC class II cell surface receptor)
ICAM-1	Intracellular adhesion molecule 1 (aka CD54)
IL	Interleukin
IPAH	Idiopathic pulmonary arterial hypertension
IPF	Idiopathic pulmonary fibrosis

K.L. Colvin • M.E. Yeager (✉)
Department of Pediatrics-Critical Care, University of Colorado Denver, Aurora, CO 80045, USA

Cardiovascular Pulmonary Research, University of Colorado Denver, Aurora, CO 80045, USA

Department of Bioengineering, University of Colorado Denver, Aurora, CO 80045, USA

Linda Crnic Institute for Down Syndrome, University of Colorado Denver, Aurora, CO 80045, USA
e-mail: michael.yeager@ucdenver.edu

O. Lohani
Department of Bioengineering, University of Colorado Denver, Aurora, CO 80045, USA

A. Firth, J.X.-J. Yuan (eds.), *Lung Stem Cells in the Epithelium and Vasculature*, Stem Cell Biology and Regenerative Medicine, DOI 10.1007/978-3-319-16232-4_14

MHC	Major histocompatibility complex
MMP	Matrix metalloprotease
PA	Pulmonary artery
PAH	Pulmonary arterial hypertension
PBMC	Peripheral blood mononuclear cell
PG	Proteoglycan
PH	Pulmonary hypertension
PVR	Pulmonary vascular remodeling
RFP	Red fluorescent protein
SAP	Serum amyloid P
SMA	Smooth muscle actin

14.1 Introduction

Pulmonary vascular remodeling (PVR) refers to physiological and often pathobiological changes in the lung vasculature(s). Such changes invariably include elements of vascular cell proliferation and fibrosis. The question of whether PVR may result from stem cell derangement, whether it be local or extrapulmonary in origin, remains unanswered. Recently, the potential contribution of fibrocytes to PVR has received a great deal of investigation. Fibrocytes are bone marrow-derived progenitor-like cells characterized by the expression of both hematopoietic (e.g., CD45, CD11b, CD34) and mesenchymal (e.g., procollagen, smooth muscle actin, vimentin) cell markers. Since the time of their discovery, they have been thought to have a significant influence on pro-fibrotic processes, either directly or in a paracrine fashion. In addition, several investigators have raised the question of whether they have stem-like properties. This chapter will introduce the major concepts of remodeling and the lung-resident cells thought to play key roles in fibrotic processes that contribute to PVR. We will also present the evidence in favor and against whether bone marrow-derived progenitor cells such as fibrocytes participate in PVR and perhaps serve as a source of stem cells [extensive self-renewal (Weiss et al. 2008)] or progenitor cells (limited self-renewal) (Weiss et al. 2008).

14.2 Pulmonary Vascular Remodeling

There are a large number of diseases that either arise from, or are associated with, PVR. These primarily include pulmonary edema and pulmonary hypertension (PH). PH can arise as pulmonary arterial hypertension (PAH), pulmonary veno-occlusive disease in association with left heart disease, in the setting of chronic thromboembolic disease, and by undefined and/or multifactorial mechanisms (Ryan et al. 2011). The lung is the highly complex product of evolutionary pressures that arose to solve several key physiological tasks. First, the lung captures and begins the delivery process of oxygen, needed by every cell in the body. In doing so, it necessarily interfaces with an unpredictably dangerous environment. The lung samples ambient air and largely ignores harmless antigens while retaining the ability to mount aggressive but self-limiting immune and inflammatory responses to dangerous antigens. Due to the combination of unique cell types (surfactant-producing airway epithelium), anatomical structures (ciliation), and an army of resident and influxing immune cells, the airways are well-equipped to deliver oxygen to the alveolar space while defending against pathogens and damage from environmental insults. Second, the lung vasculature delivers blood to the alveolar space in a manner that is matched to the ventilation and distribution of lung oxygen via the airways. The vasculature of the lung is a union of the pulmonary arterial circuit (continuing through the capillary bed and on to the pulmonary venous system to the left atrium) and the bronchial arterial system (systemic circulation). Indeed, the pulmonary and bronchial circulations anastomose with each other along their lengths (Mitzner and Wagner 2004). In addition, a rich plexus of lymphatic vessels accompanies the large bronchovascular structures as well as the periphery of the lung along the pleural surfaces (Brotons et al. 2012). These lymphatics are absolutely essential for lung homeostasis with regard to regulation of interstitial fluid volume, but also required as a leukocyte superhighway facilitating continuous surveillance of the lung–external environment interface (Miserocchi 2009). By virtue of this unique combinatorial vascularity, the lung takes the equivalent of the entire cardiac output and utilizes special physiological adaptations (hypoxic pulmonary vasoconstriction) to deliver blood cells to conditioned air to effectively oxygenate the body within a compound framework of immunological barriers.

As stated, a large number of diseases either directly interrupt, or themselves result from the disruption of the complex physiology outlined above. In this section, we will broadly outline the pathobiology of PVR in a variety of disease settings, arbitrarily divided into discussion of proximal and distal remodeling.

14.2.1 Proximal Vessels

The pulmonary artery (PA) originates at the outflow tract of the right ventricle just distal to the pulmonic valve. After ascending superiorly and then branching to serve the left and right lungs, the PA anatomically adjoins the large airways in the lung parenchyma. After a number of bifurcations (variable but approximately 13 in mice, ~17 in humans) (Pu et al. 2012), the airway-PA bundle simplifies into the terminal bronchiole to open into the cluster of alveoli where gas exchange occurs. At the level of large bronchovascular bundles, the adventitia connecting the airway and the PA is subject to significant remodeling in a number of pathological conditions. For example, asthma, a chronic inflammatory disease characterized by hyperresponsive airways that may be accompanied by vasculitis (as in Churg–Strauss syndrome, now termed eosinophilic granulomatosis with polyangiitis, or EGPA) (Avdalovic 2015). Indeed, PVR is a major complicating factor in asthma and chronic obstructive pulmonary disease (COPD) (Alagappan et al. 2013; Tam and Sin 2012). In COPD, the histopathological changes include vascular medial thickness, as well as extracellular matrix (ECM) deposition of collagen III and fibronectin (Kranenburg et al. 2006). Furthermore, the degree of vascular remodeling in proximal lung vessels is inversely correlated to the extent of airway obstruction measured as forced expiratory volume in one second (FEV_1) (Kranenburg et al. 2002). Similar data have been obtained in the setting of asthma (Li and Wilson 1997; Salvato 2001) and in animal models of asthma (Avdalovic et al. 2006). Such findings led some to propose the concept of the epithelial mesenchymal trophic unit (EMTU) (Evans et al. 1999), which includes vessels and considers the airway and vessel as a single biological unit. In pulmonary artery hypertension, the proximal vessel significantly remodels in humans (Guignabert and Dorfmuller 2013) as well as in animal models (Archer et al. 2010). Proximal pulmonary arteries remodel in all three layers, and this includes intimal hyperplasia, medial hypertrophy, and adventitial fibrosis (Shimoda and Laurie 2013). Among the other four categories of pulmonary hypertension (Simonneau et al. 2009), PVR is also observed. Recently, the contributions of inflammatory and immune cells have been intensely investigated and it is clear that these cells play influential pleiotropic roles in proximal PVR (Cohen-Kaminsky et al. 2014; Price et al. 2012). Furthermore, the presence of bronchus-associated lymphoid tissues (BALT) has been documented in humans with PAH (Perros et al. 2012) and in animal models of PAH (Colvin et al. 2013) and may represent a crossroad linking innate and acquired immunity. Idiopathic pulmonary fibrosis (IPF) is a chronic fibrosing disease of unknown etiology that is typified by fibroblast proliferation, alveolar destruction, and decreased lung function leading to death (Hanumegowda et al. 2012). Varying degrees of proximal PVR and bronchial wall thickening are observed in IPAH (Cosgrove et al. 2004) as well as the bleomycin model of pulmonary fibrosis (Peao et al. 1994). In some disease contexts, there is variability on the extent of proximal vascular remodeling in the lung, and it is unclear whether the amount of remodeling is actually insufficient or overly robust (Farkas et al. 2011).

14.2.2 Distal Vessels

As in the proximal vasculature, the distal vessels of the lung are remodeled in disease states. In PAH, for example, smaller internal diameter resistance vessels undergo extensive remodeling in all three layers (Fig. 14.1) (Heath and Edwards 1958). Although heterogeneous in presentation, PAH lungs exhibit a high degree of distal perivascular inflammation that correlated positively with intimal and medial remodeling (Stacher et al. 2012). In addition, plexiform lesions are variably present but particularly in those treated with prostacyclin or its analogues (Stacher et al. 2012). In COPD, alveolar capillaries undergo destruction and distal vessels become muscularized due, at least in part, by persistent hypoxia. There is also evidence that muscularized bronchial and pulmonary arteries with intimal thickening harbor CD8+ T lymphocytes. The infiltration of these cells is coincident with release of inflammatory mediators and modulators of the ECM such as matrix

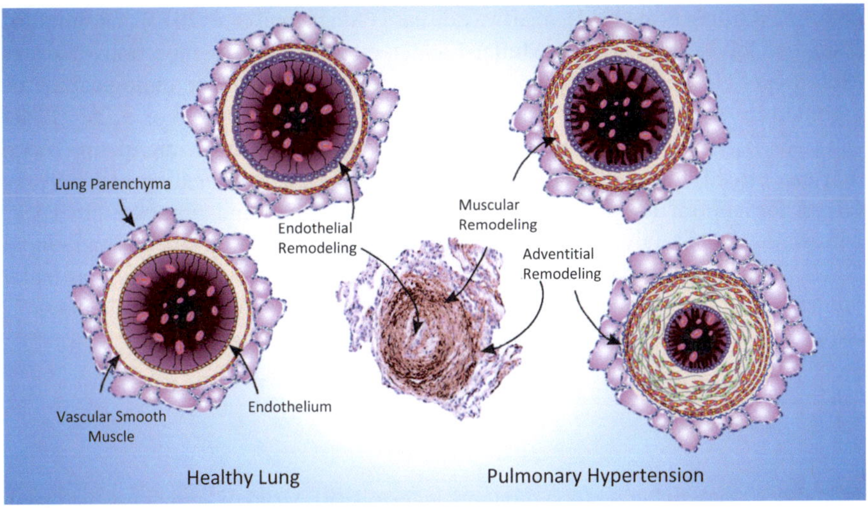

Fig. 14.1 Pathobiology of pulmonary vascular remodeling in pulmonary hypertension. The pulmonary artery is normally characterized by a single endothelial monolayer, a smooth muscle medial layer, and an adventitial layer. As can be seen, the normally patent vessel becomes compromised due to fibrotic remodeling in all three layers (Figure modified from Nicholas Morrell, Cambridge UK, by kind permission)

metalloproteases (MMPs) (Zanini et al. 2008). Recent observations have documented that, even in the face of alveolar loss, there is a progressive distal pulmonary vasculature muscularization that involves increased deposition of collagen type III and fibronectin (Kranenburg et al. 2002, 2005). Such changes may help to explain the development of pulmonary hypertension in the setting of COPD (Wright et al. 1992). Small vessel necrotizing vasculitis caused by hyper-eosinophilia with attendant asthma is characteristic of EGPA/Churg–Strauss syndrome (Churg and Strauss 1951). Interestingly, the presence of anti-neutrophil cytoplasmic antibodies constitutes the principal serologic test to diagnose EGPA (Tervaert et al. 1991). There are a number of autoimmune diseases (systemic sclerosis, lupus) that affect the lung circulation, both proximal and distal, that have been reviewed elsewhere (Muller-Ladner et al. 2009; Carmier et al. 2010). IPF is also associated with small vessel changes in the lung in a broader presentation of heterogeneous vascular change. On one hand, where minimal fibrotic change is observed in IPF, capillary density is increased (Ebina et al. 2004). On the other hand, in regions of pronounced fibrosis, capillary density is decreased (Ebina et al. 2004). Additional studies looking at fibroblastic foci in IPF lend support to the idea that impaired angiogenesis may exist on a continuum (Garantziotis et al. 2008; Cosgrove et al. 2004). Despite its well-known shortcomings, the bleomycin model of rodent IPF presents a similar continuum (Lee et al. 2014; Wang et al. 2010), but the significance of these vascular changes in the pathobiology of IPF is yet unclear (Hanumegowda et al. 2012).

The lung vasculature is subject to change from a large number of sources that originate in the lung or may derive from outside of the lung. These range from air pollution (Grunig et al. 2014), to airway disease (Alagappan et al. 2013), autoimmune disease (Colvin et al. 2013; Muller-Ladner et al. 2009), to ingestion of drugs (Seferian et al. 2013), or infection (Kolosionek et al. 2010). One biological theme common to most of these is lung inflammation, whether chronic or acute (Tuder et al. 2013). As will be discussed below, fibrocytes are cells in the circulating blood that can directly and indirectly influence process of wound healing and fibrosis. Given the central underpinning of inflammatory cells and processes in PVR, controlling the fibrocyte represents a potentially powerful therapeutic paradigm. Next, we summarize the last 20 years of research into fibrocytes, with particular focus on their putative contributions to PVR and the question of their "stemness."

14.3 Fibrocytes

Fibrocytes are peripheral blood mononuclear cells (PBMCs) of hematopoietic origin that have been identified as key players at sites of tissue injury and fibrosis in a wide range of disease settings. Since the original identification of fibrocytes in 1994, there has been a steady growth of findings leading to a better understanding of their functional roles. The subsequent sections temporally map the studies that first established the identity of fibrocytes and the observations that have placed fibrocyte cell subsets as important controllers of a myriad of functions related to fibrosis. Finally, we weigh the evidence for and against the possibility that fibrocytes act as stem or progenitor cells, specifically in the context of PVR.

14.3.1 Discovery and Phenotypic Description

Landmark studies conducted by Bucala et al., using a model system of wound healing, led to the first identification of circulating fibrocytes in 1994 (Bucala et al. 1994). This novel subpopulation of leukocytes emerged after a few weeks in culture with high serum and was characterized as expressing CD34, CD11b, CD18, collagen I, collagen III, and vimentin. Morphologically, fibrocytes appeared as adherent, spindle-shaped cells reminiscent of primary fibroblasts in culture. They were localized to regions of scar formation and were suspected of being important modulators in pathological fibrotic responses (Bucala et al. 1994). This study and other initial observations of fibrocytes included them being referred to as "blood-borne fibroblasts" and "fibroblast-like cells" (Quan et al. 2004). Collectively, these data reinforced earlier studies in 1969 by Stirling and Kakkar, who used cannulation and diffusion chamber techniques to demonstrate that collagen-producing cells were derivatives of circulating blood and not elements dislodged from the walls of blood vessels (Stirling and Kakkar 1969). Interestingly, Bucala et al. speculated that fibrocytes could be required early on in wound healing and/or could contribute to excessive scar formation (Bucala et al. 1994).

It is well established that disturbances (infection, trauma, etc.) in tissues elicit wound repair responses from the host. Injury leads to release of chemokines and the recruitment of diverse cell populations to the site of injury. The wound repair response can be divided into four phases: (1) clot formation, (2) inflammation, (3) cell migration and proliferation, and (4) remodeling and resolution (Cordeiro and Jacinto 2013). Neutrophils along with monocytes are the first key players to appear at the site of injury during the inflammatory phase. Their function is to remove debris, clot, or pathogens that might be present in the wound (Cordeiro and Jacinto 2013). Through specific signal transduction processes, monocytes are able to mature into diverse sets of macrophages as well as fibrocytes, and in turn are able to secrete chemotactic cytokines and growth factors (Aggarwal et al. 2014). Through activation of context-specific gene programs, macrophages work to promote immunity, activate fibroblasts, augment cell proliferation, and to stimulate angiogenesis (Molawi and Sieweke 2013). In skin, the epithelium is responsible for covering the wound in the subsequent proliferative phase and activated fibroblasts produce collagen actively along with other matrix proteins, while new matrix is remodeled in the following maturation phase (Kendall and Feghali-Bostwick 2014). The newly organized matrix, or granulation tissue, provides tensile strength to the site of inflammation and the integrity of the injured tissue is restored more or less to the original state. It should be noted that stem cell niches in the lung tissue itself, while normally low in turnover rate, are activated to proliferate in response to injury (Wansleeben et al. 2013). Although the specific details regarding the identity and locations of putative lung stem cells have been a challenging task to date, it is likely that both lung-resident and non-lung-resident stem and progenitor cells contribute to injury responses (Weiss et al. 2008). Since the first functional studies of circulating fibrocytes that showed their presence in exudate fluid (Quan et al. 2004), and their rapid recruitment to injury sites and production of collagen (Bucala et al. 1994), fibrocytes have been investigated in numerous aspects of tissue repair.

Peripheral blood fibrocytes express CD34 cell surface antigen (Quan et al. 2004), a marker common to hematopoietic stem cells of both myeloid and lymphoid lineage (Bucala et al. 1994; Brown et al. 1991). CD34 expression often decreases with passing time in culture or when hematopoietic cells enter into tissue, a finding confirmed in culture and in vivo (Bucala et al. 1994; Aiba and Tagami 1997). The presence of collagen and its co-expression with unique hematologic markers combined with morphology is usually the minimum criteria to characterize fibrocytes (Bucala et al. 1994; Chesney et al. 1997). Fibrocytes have been found to express a wide variety of cell surface proteins (Pilling et al. 2009) and immune receptors (Crawford et al. 2012), for which a large number of putative functions can be deduced. It was the discovery of the class II major histocompatibility molecule (MHC) HLA-DR being expressed by fibrocytes that led to studies investigating their ability to present antigens (Chesney et al. 1997). Studies in human and mice PBMC led to the affirmation that fibrocytes express the necessary catalog of cell surface proteins requisite for antigen presentation and they readily migrate to lymph nodes and sensitize T lymphocytes (Chesney et al. 1997). Similar results have been found in a model of hepatotoxic injury (Kisseleva et al. 2011) and thus position fibrocytes as key bone marrow-derived cells capable of linking innate and acquired immune systems. Future studies that interrogate other fibrocyte surface markers should provide functional context for their phenotypic expression.

14.3.2 Roles in ECM Biology and Inflammation

Since their initial discovery, more than 300 papers have investigated functional aspects of fibrocyte biology, with a strong emphasis on the ECM remodeling properties of fibrocytes. Fibrocytes differ significantly from the resident tissue fibroblast with respect to the level of expressions of various collagen types (I, III, V, and VI), matrix proteoglycans, and hyaluronan (Bellini and Mattoli 2007; Pilling et al. 2009). Following an initial differentiation from monocytic precursor(s), transforming growth factor beta (Abe et al. 2001) and endothelin-1 (Schmidt et al. 2003) can induce further differentiation into fibroblast and myofibroblast-like cells. Collagen I, III, and V are fibrillar collagens, while collagen VI is non-fibrillar (Myllyharju and Kivirikko 2004). The collagens type I, III, and V can be assembled differently and comprise distinct structural components of the ECM (Ushiki 2002). Collagen VI, on the other hand, forms a network of beaded filaments (Myllyharju and Kivirikko 2004). Comparison of cultured human fibrocytes to cultured human fibroblasts showed significantly lower expressions of mRNA encoding the chains for collagen type I, III, and V in fibrocytes, but expressed comparable levels of mRNA-encoding collagen type VI (Bianchetti et al. 2012). Fibrocytes express lower levels of mRNA-encoding proteoglycans (PGs) such as decorin, but transcribe high levels of mRNA-encoding PGs like perlecan and versican (Bianchetti et al. 2012). Additionally, fibrocytes express higher levels of mRNA encoding for hyaluronan synthase 2. Fibroblast and fibrocytes are both able to incorporate the collagen and PGs actively secreted into the matrix (surrounding as well as underlying), with fibrocytes predominantly producing the proteins associated with the membrane (Bianchetti et al. 2012). However, fibrocytes produce collagens that are deposited into the matrix at a much lower capacity than do fibroblasts (Bianchetti et al. 2012). Collagen VI, hyaluronan, versican, and perlecan modulate cell–to–cell interaction as well as cell–to–matrix interactions (Wu et al. 2005; Hirose et al. 2001; Howell and Doane 1998; Myllyharju and Kivirikko 2001; Iozzo 1998). Perlecan is known to interact with other matrix components, bind growth factors, regulate permeability of the basement membranes, and promote angiogenesis (Zhang et al. 2004). Moreover, the PGs assist in maintaining tissue integrity by regulating stress transfer between the cytoskeleton, cell membrane, and the matrix (Iozzo 1998; Al-Jamal and Ludwig 2001). Fibrocytes, by virtue of their ability to secrete high levels of perlecan, versican, hyaluronan, and collagen type VI (relatively high levels in comparison to other ECM proteins), are primary modulators in responses to inflammation and tissue stabilization during wound healing process (Mattoli et al. 2009; Bellini and Mattoli 2007; Mousavi et al. 2005; Madsen et al. 2007; Wienke et al. 2003).

Fibrocytes adhere to fibronectin through integrin β_1 on their cell surface and to immobilized matrices. Blocking the binding of integrin β_1 decreases the adhesion significantly but not entirely (Bianchetti et al. 2012). Phenotypic analysis led to the revelation that fibrocytes express other collagen receptors, in addition to integrin β_1. Through a series of blocking experiments, one such receptor involved in adhesion to collagen was the mannose receptor Endo180. Endo180 is not only responsible for adhesion of cells to collagen, but also promotes the uptake of collagen and collagen fragments by alternatively activated macrophages and hepatic stellate cells (Engelholm et al. 2003; Behrendt et al. 2000). The bound collagen is transported to the late endosome and then routed to the lysosome for degradation. Fibrocytes were able to internalize fragments of collagen type I more efficiently than non-denatured collagen type I or intact collagen VI, and collagen uptake is inhibited when excess amounts of uPA is present (Bianchetti et al. 2012).

Besides direct effects on ECM protein turnover, fibrocytes play important roles in chronic inflammation that impact the matrix. Fibrocytes secrete interleukin (IL)-6, IL-8, chemokine (C-C) motif ligand 2 (CCL2, also known as monocyte chemoattractant protein-1, or MCP-1), CCL3, as well as intercellular adhesion molecule-1 (ICAM-1) in response to interleukin-1 (Chesney et al. 1998). Thus, fibrocytes, upon entry into injured tissue, amplify and modify subsequent immune responses. Such responses can greatly influence tissue remodeling. For example, fibrocytes from patients with asthma respond to Th2 cytokines by producing collagen (Bellini et al. 2012). IL-33, an IL-1 cytokine family member, enhances proliferation of fibrocytes and augments their output of IL-13 and IL-5 (Hayashi et al. 2014), both of which are Th2 cytokines implicated in asthma pathobiology. IL-33 also induces fibrocyte MMP 9 activity, which in turn may activate latent transforming growth factor beta (TGF-beta), a pleiotropic growth factor and controller of matrix biology and fibrosis (Hayashi et al. 2014). In addition to crosstalk with innate immune cells, fibrocytes communicate with lymphocytes. Regulatory T lymphocytes (Tregs) can inhibit the recruitment of fibrocytes, which decreases the extent of tissue remodeling in the murine model of lung fibrosis by inducible TGF-beta over-expression (Peng et al. 2014). This recent finding is of potentially major importance since deficiency of Tregs appears to mechanistically underpin the development of PVR in both pulmonary hypertension (Tamosiuniene et al. 2011) and IPF (Kotsianidis et al. 2009). Intriguingly, fibrocytes have also been shown to contribute to other (later) stages of the inflammatory response, including resolution. Recently, it was shown that circulating fibrocytes stabilize blood vessels in vitro and in vivo during angiogenesis (Li et al. 2014). The mechanism appeared to be twofold in that fibrocytes inhibited angiogenesis and prevented regression of existing vessels (Li et al. 2014). Thus, the paradigm that fibrocytes are pro-antigenic (Grieb et al. 2011) may now need to be nuanced to include antagonistic functions depending on context. Such a concept may extend to other functions of fibrocytes, as tumor-associated fibrocytes suppress immune cell functions via interferon gamma and tumor necrosis factor (Kraman et al. 2010).

Taken collectively, these data indicate that fibrocytes are endowed with a great degree of phenotypic plasticity that influences, and is influenced, by local cues within the ECM and from cells at sites of injured ECM. Future studies will need to parse these cues and specifically determine what controls fibrocyte differentiation and phenotypic specialization.

14.3.3 Differentiation

Fibrocytes are a bone marrow-derived cell population that express both hematopoietic and stromal cell markers. As discussed above, fibrocytes are key modulators in wound healing, angiogenesis, and pathological fibrosis. Much of what is known about fibrocyte differentiation comes from culture of peripheral blood. The effect of serum on the differentiation and function of fibrocytes is quite pronounced. In the Bucala study, 20 % serum-containing media was used and fibrocytes emerged from such culture conditions in weeks. In contrast, recent studies conducted by Curnow et al., using in vitro conditions where serum is either present or absent, fibrocytes could be cultured in as few as 3 days in the absence of serum (Curnow et al. 2010). The study further indicated that the cells differentiated under serum-free and serum-containing conditions are indeed both fibrocytes, but possess distinct properties that could possibly relate to their proposed functions in vivo (Pilling et al. 2007). These data are in line with studies showing that serum amyloid p (SAP), a pentraxin family member, inhibits fibrocyte differentiation (Pilling et al. 2003). C reactive protein and SAP are liver-derived pattern recognition molecules that help to facilitate clearance of pathogens and cell debris (Hutchinson et al. 2000). Injections of SAP inhibit inflammation and fibrosis in murine models of pulmonary fibrosis (Murray et al. 2010, 2011; Pilling et al. 2007); however, in mice genetically lacking SAP, inflammation and fibrosis are relentless, indicating that a key endogenous function of SAP may be the promotion of fibrocyte differentiation to resolve inflammation and fibrosis (Pilling and Gomer 2014).

Fibrocytes are able to differentiate from CD14+ monocytes as well as from CD14– PBMC (Geissmann et al. 2008). Curnow et al. provide evidence that efficient generation of fibrocytes occurs from PBMC in serum-free conditions, while fibrocytes in the serum-containing conditions differentiate more efficiently from the $CD14^{+}$ monocyte population. Regardless, both types of fibrocytes express the same range of markers (Curnow et al. 2010). This finding raises the possibility that the differentiated cells represent distinct types of fibrocyte that arise from different precursor cells and are present at different frequencies with the subpopulations of CD^{+} monocytes and PBMCs. A number of signaling molecules, including toll-like receptor ligands (Maharjan et al. 2010), semaphorins (Gan et al. 2011), as well as the glycosaminoglycan hyaluronic acid (Maharjan et al. 2011), can affect fibrocyte differentiation. Additionally, it was determined that a high number of gene pathways relating to lipid metabolism were significantly enhanced in serum-containing fibrocyte cultures. This finding led to the deduction that fibrocytes differentiated in serum possess multipotent properties and are able to differentiate to other mesenchymal cell lineages such as adipocytes, as supported by Hong et al. (2005). Whether fibrocytes differentiated in serum-free media retain a similar degree of multipotency is yet to be determined. It is also unclear exactly what micro-environmental cues in the lung constitute the critical paracrine signals and fibrocyte-matrix/fibrocyte-heterotypic cell interactions that might help to explain the putative roles for fibrocytes in PVR.

Fibrocytes have been established as a cell population that can differentiate into myofibroblast-like (Hong et al. 2007) and adipocyte-like cells (Hong et al. 2005).

Fibrocytes possess the potential to differentiate into other mesenchymal lineage cells such as osteoblasts and chondrocytes (Choi et al. 2010). When cultured in osteogenic media in a manner similar to osteo-precursor cells, fibrocytes differentiated into osteoblast-like cells (Choi et al. 2010). Similarly, fibrocytes as well as mesenchymal stem cells (MSCs), when cultured in chondrogenic media, differentiated into highly similar chondrocyte-like cells (Choi et al. 2010). The results suggest that fibrocytes have the capacity to differentiate into different mesenchymal lineage cells. This raises the possibility that fibrocytes could be used for regenerative therapy not only with respect to osteo-related (bone) or chondro-related (articular cartilage) repair, but perhaps to remodel lung tissue compromised by PVR. These results, although tantalizing, have been challenged by some with regard to both terminology and methodology (Weiss et al. 2008). One of the primary reasons for such concerns is the difficulty in the unequivocal identification and fate mapping of fibrocytes, stem cells, and progenitor cells in tissues, especially the lung.

14.3.4 Tracking Fibrocytes

Fibrocytes are defined to be of hematopoietic origin and derived from monocyte lineage cells due to the expression of specific monocyte surface markers CD11b and the leukocyte marker CD45 (Bellini and Mattoli 2007; Reilkoff et al. 2011). The marker CD45 is currently considered the most specific in determining and characterizing cell lines of hematopoietic origin. As mentioned before regarding CD34, the expression of CD45 marker is not a reliable enough condition to distinguish between cells of hematopoietic and non-hematopoietic origin. CD45 is expressed on later stages of hematopoietic cell development (Zambidis et al. 2005), and hence is incapable (used in isolation) of completely identifying these cells. A problem arises when fibrocytes, during the course of differentiation, lose their expression of markers of their hematopoietic origins (Mori et al. 2005; Phillips et al. 2004), and it then becomes challenging to distinguish them from the fibroblasts that are residents in the tissue of interest. The complex cell membrane protein expression pattern, the transient nature of their expression, and the requirement to co-identify a mesenchymal (intracellular) marker, all contribute to the controversial nature of fibrocyte fate tracing. There have been studies that investigated the involvement of bone-marrow-derived cells that are able to produce collagen in response to tissue injury (Mori et al. 2005; Fathke et al. 2004; Barisic-Dujmovic et al. 2010), but have produced results unable to distinguish the true lineage of these cells. This is due to the methods used to track the mesenchymal cells as well as that the hematopoietic cells are similar to resident cells in the models designed (Suga et al. 2014). To effectively be able to determine the presence and the long-term fate of hematopoietic cells, Suga et al. proposed to distinguish hematopoietic cells from non-hematopoietic cells by utilizing Vav 1 gene expression (Suga et al. 2014) and shying away from previous mouse models that targeted either selected lineage cells or specific differential stages (Adams et al. 1999). Vav 1 is a pan-hematopoietic marker that is found expressed by hematopoietic cells and no other tissue (Bustelo et al. 1993; Moores et al. 2000), and whose presence is advantageous in differentiating between cells of hematopoietic origin and non-hematopoietic lineages. Avoiding reliance on the surface markers previously used to identify fibrocytes (Bellini and Mattoli 2007; Reilkoff et al. 2011), Suga et al. used the Vav 1 gene to irreversibly label the cells of hematopoietic origin with Green Fluorescent Protein (GFP) and concomitantly label the non-hematopoietic cells with red fluorescent protein (RFP) (Suga et al. 2014). While examining the influx of different cell types during an injury response, they were able to obtain evidence confirming the existence (presence) as well as the prevalence of the cells over the time span of wound repair. Interestingly, using this novel double transgenic approach, two populations of col I+ cells were found: CD45+/CD11b+ and CD45−/CD11b− cells. The former were found to be fibrocytes and to express alpha-smooth muscle actin (SMA) in early areas of wound healing.

With regard to PVR and fibrocyte tracking, a few studies deserve mention. First, hypoxia-induced pulmonary hypertension in bone marrow-transplanted mice was associated with increased CD45+/GFP+ (donor derived)/collagen I+ lung fibrocytes (Nikam et al. 2010). Use of the clinically approved vasodilatory prostacyclin analog treprostinil (Seferian and Simonneau 2013), dramatically lowered the fibrocyte numbers in the lung, attenuated some of the PVR, but did not improve right heart function (Nikam et al. 2010). In a rat model of irreversible pulmonary hypertension, sex mismatched donor marrow-derived GFP+ cells were observed in areas of PVR and fibrosis (Spees et al. 2008). Many of these cells adopted fibroblast and myofibroblast morphologies, and a small number appeared as bronchial epithelial cells. Whether these cells were fibrocytes or MSCs, as had been earlier reported (Kotton et al. 2001; Ortiz et al. 2003), was not determined from these studies.

14.4 Are Fibrocytes Stem Cells, Progenitor Cells, or Neither and Are they Pro-fibrotic or Pro-resolution?

A bona fide adult stem cell should fulfill important criteria in order to meet the consensus definition (Weiss et al. 2008). They must have the capacity for self-renewal. Generally, adult stem cells have the multipotent capacity to differentiate into cells of the parent tissue. MSCs are stromal-derived cells that self-renew and demonstrate the capacity to differentiate into a variety of cell lineages. The minimal criteria for MSC, as developed by the MSC Committee of the International Society for Cellular Therapy, include (1) plastic adherence in standard tissue culture conditions; (2) expression of CD73, CD90, CD105; (3) no expression of CD11b, CD14, CD19, CD34, CD45, CD79alpha, or HLA-DR; and (4) differentiation in vitro to osteoblasts, adipocytes, and chondroblasts (Dominici et al. 2006). As we have presented above, fibrocytes are a subset of circulating leukocytes that co-express mesenchymal proteins such as collagen or vimentin and home to regions of tissue injury, including the injured lung. Currently, there is no evidence that fibrocytes, despite their ability to differentiate into a number of cell types in response to cytokines, are stem cells.

Although they can proliferate, it remains unclear what capacity for self-renewal they may have. Although they fulfill criteria #1 and #4 above for MSC, they do not express CD90 and do/can express CD11b, CD34, CD45, and HLA-DR; as such, they are not MSC. If fibrocytes do not fulfill the criteria for true stem cells, what are they? We suggest that fibrocytes are a population of bone marrow-derived cells endowed with unique abilities to be mobilized in response to tissue injury, home to that injury, and to participate in complex heterotypic interactions that are context-dependent. They are likely derived from CD14+ monocytes, but may also be derived from CD14– cells. A progenitor cell is defined as any proliferative cell that has the capacity to differentiate into different cell lineages within a given tissue. As discussed above, fibrocytes can differentiate into a variety of lung tissue cells. It is conceivable that fibrocytes may be able to differentiate into cell types, given the variable expression patterns of cell surface molecules. It is appropriate then to term fibrocytes as mesenchymal progenitor cells, as some authors have done (Bellini and Mattoli 2007).

It remains to be seen whether fibrocytes and other bone marrow-derived progenitor-like cells are pro-fibrotic or pro-resolution cells. It seems likely that the term fibrocyte encompasses one or more cell types derived from the bone marrow that are capable of rapid gene expression change and the adoption of a wide variety of unique phenotypes. At sites of PVR, they appear to play both direct and/or paracrine roles in pro-fibrotic or pro-resolution pathways depending on complex and poorly understood contextual cues (Fig. 14.2). Future studies will need to specifically address the following unresolved questions, among others. First, the term fibrocyte appears to mean many things to many investigators. Given the simple two-part requirement of expression of a hematopoietic marker and a mesenchymal marker, a large number of cells

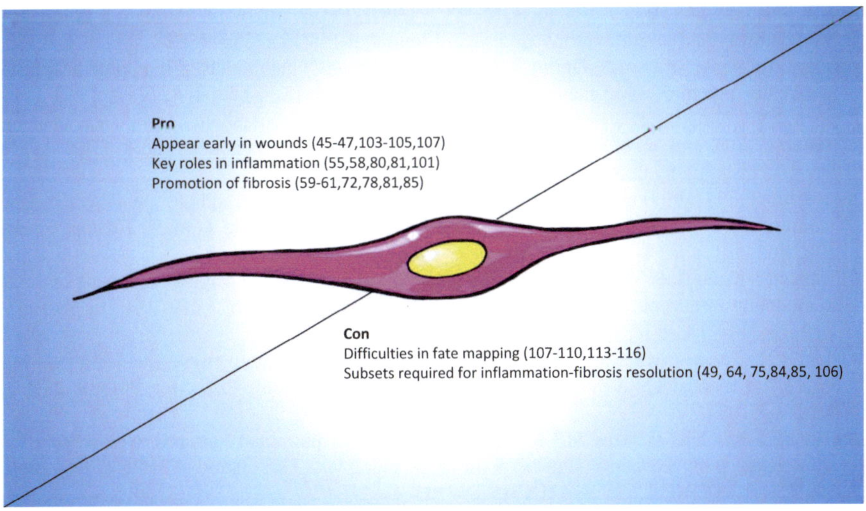

Fig. 14.2 Are fibrocytes pro-fibrotic or pro-resolution of fibrosis? Fibrocyte with classic spindle-shaped morphology, with references supporting roles for and against inflammatory fibrotic remodeling

with distinct expression could be called fibrocytes. For example, a bone marrow-derived cell could express either CD45, CD34, or CD11b and could co-express either procollagen, SMA, or vimentin. This computes to $2^3=8$ combinations and assumes any cell uniquely expresses only one hematopoietic and one mesenchymal marker. It is therefore clear that a fibrocyte could be an accurate term for dozens of potentially unique cell types with specialized functions. Going forward, the field should exercise caution in the reporting and evaluation of studies involving fibrocytes, being clear on exactly what phenotypes (as much as is possible) are under investigation. Second, there is no evidence that fibrocytes serve as stem cells in any tissue injury investigated to date. Rather, it seems their primary role is to influence the tissue injury site in a paracrine manner towards either a pro-fibrotic process or perhaps a pro-resolution mode. Studies are critically needed that detail what types of fibrocytes operate at sites of tissue injury, what signals are they specifically responding to, and what are the kinetics of their arrival, modulation of biological processes, and apoptosis or emigration out of the site. Third, studies in animal models are needed that test the ability of pharmacologic or immunology agents to shape the fibrocyte response to injury. These studies will be extraordinarily difficult in terms of cell tracking and the dissection of micro-environmental cues that contextualize the fibrocyte response. The successful pursuit of these critically important studies should eventually yield new therapies that control the fibrocyte phenotype to achieve the appropriate kinetics of appropriate resolution of inflammation and the amelioration of PVR.

References

Abe R, Donnelly SC, Peng T, Bucala R, Metz CN (2001) Peripheral blood fibrocytes: differentiation pathway and migration to wound sites. J Immunol 166(12):7556–7562

Adams J, Harris A, Strasser A, Ogilvy S, Cory S (1999) Transgenic models of lymphoid neoplasia and development of a pan-hematopoietic vector. Oncogene 18(38):5268–5277

Aggarwal NR, King LS, D'Alessio FR (2014) Diverse macrophage populations mediate acute lung inflammation and resolution. Am J Physiol Lung Cell Mol Physiol 306(8):L709–L725. doi:10.1152/ajplung.00341.2013

Aiba S, Tagami H (1997) Inverse correlation between CD34 expression and proline-4-hydroxylase immunoreactivity on spindle cells noted in hypertrophic scars and keloids. J Cutan Pathol 24(2):65–69

Alagappan VK, de Boer WI, Misra VK, Mooi WJ, Sharma HS (2013) Angiogenesis and vascular remodeling in chronic airway diseases. Cell Biochem Biophys 67(2):219–234. doi:10.1007/s12013-013-9713-6

Al-Jamal R, Ludwig MS (2001) Changes in proteoglycans and lung tissue mechanics during excessive mechanical ventilation in rats. Am J Physiol Lung Cell Mol Physiol 281(5):L1078–L1087

Archer SL, Weir EK, Wilkins MR (2010) Basic science of pulmonary arterial hypertension for clinicians: new concepts and experimental therapies. Circulation 121(18):2045–2066. doi:10.1161/CIRCULATIONAHA.108.847707

Avdalovic M (2015) Pulmonary vasculature and critical asthma syndromes: a comprehensive review. Clin Rev Allergy Immunol 48(1):97–103. doi:10.1007/s12016-014-8420-4

Avdalovic MV, Putney LF, Schelegle ES, Miller L, Usachenko JL, Tyler NK, Plopper CG, Gershwin LJ, Hyde DM (2006) Vascular remodeling is airway generation-specific in a primate model of chronic asthma. Am J Respir Crit Care Med 174(10):1069–1076. doi:10.1164/rccm.200506-848OC

Barisic-Dujmovic T, Boban I, Clark SH (2010) Fibroblasts/myofibroblasts that participate in cutaneous wound healing are not derived from circulating progenitor cells. J Cell Physiol 222(3):703–712

Behrendt N, Jensen ON, Engelholm LH, Mørtz E, Mann M, Danø K (2000) A urokinase receptor-associated protein with specific collagen binding properties. J Biol Chem 275(3):1993–2002

Bellini A, Mattoli S (2007) The role of the fibrocyte, a bone marrow-derived mesenchymal progenitor, in reactive and reparative fibroses. Lab Invest 87(9):858–870

Bellini A, Marini MA, Bianchetti L, Barczyk M, Schmidt M, Mattoli S (2012) Interleukin (IL)-4, IL-13, and IL-17A differentially affect the profibrotic and proinflammatory functions of fibrocytes from asthmatic patients. Mucosal Immunol 5(2):140–149. doi:10.1038/mi.2011.60

Bianchetti L, Barczyk M, Cardoso J, Schmidt M, Bellini A, Mattoli S (2012) Extracellular matrix remodelling properties of human fibrocytes. J Cell Mol Med 16(3):483–495

Brotons ML, Bolca C, Frechette E, Deslauriers J (2012) Anatomy and physiology of the thoracic lymphatic system. Thorac Surg Clin 22(2):139–153. doi:10.1016/j.thorsurg.2011.12.002

Brown J, Greaves M, Molgaard H (1991) The gene encoding the stem cell antigen, CD34, is conserved in mouse and expressed in haemopoietic progenitor cell lines, brain, and embryonic fibroblasts. Int Immunol 3(2):175–184

Bucala R, Spiegel L, Chesney J, Hogan M, Cerami A (1994) Circulating fibrocytes define a new leukocyte subpopulation that mediates tissue repair. Mol Med 1(1):71

Bustelo X, Rubin S, Suen K, Carrasco D, Barbacid M (1993) Developmental expression of the vav protooncogene. Cell Growth Differ 4(4):297–308

Carmier D, Marchand-Adam S, Diot P, Diot E (2010) Respiratory involvement in systemic lupus erythematosus. Rev Mal Respir 27(8):e66–e78. doi:10.1016/j.rmr.2010.01.003

Chesney J, Bacher M, Bender A, Bucala R (1997) The peripheral blood fibrocyte is a potent antigen-presenting cell capable of priming naive T cells in situ. Proc Natl Acad Sci U S A 94(12):6307–6312

Chesney J, Metz C, Stavitsky AB, Bacher M, Bucala R (1998) Regulated production of type I collagen and inflammatory cytokines by peripheral blood fibrocytes. J Immunol 160(1):419–425

Choi YH, Burdick MD, Strieter RM (2010) Human circulating fibrocytes have the capacity to differentiate osteoblasts and chondrocytes. Int J Biochem Cell Biol 42(5):662–671

Churg J, Strauss L (1951) Allergic granulomatosis, allergic angiitis, and periarteritis nodosa. Am J Pathol 27(2):277–301

Cohen-Kaminsky S, Hautefort A, Price L, Humbert M, Perros F (2014) Inflammation in pulmonary hypertension: what we know and what we could logically and safely target first. Drug Discov Today 19(8):1251–1256. doi:10.1016/j.drudis.2014.04.007

Colvin KL, Cripe PJ, Ivy DD, Stenmark KR, Yeager ME (2013) Bronchus-associated lymphoid tissue in pulmonary hypertension produces pathologic autoantibodies. Am J Respir Crit Care Med 188(9):1126–1136. doi:10.1164/rccm.201302-0403OC

Cordeiro JV, Jacinto A (2013) The role of transcription-independent damage signals in the initiation of epithelial wound healing. Nat Rev Mol Cell Biol 14(4):249–262

Cosgrove GP, Brown KK, Schiemann WP, Serls AE, Parr JE, Geraci MW, Schwarz MI, Cool CD, Worthen GS (2004) Pigment epithelium-derived factor in idiopathic pulmonary fibrosis: a role in aberrant angiogenesis. Am J Respir Crit Care Med 170(3):242–251. doi:10.1164/rccm.200308-1151OC

Crawford JR, Pilling D, Gomer RH (2012) FcgammaRI mediates serum amyloid P inhibition of fibrocyte differentiation. J Leukoc Biol 92(4):699–711. doi:10.1189/jlb.0112033

Curnow SJ, Fairclough M, Schmutz C, Kissane S, Denniston AK, Nash K, Buckley CD, Lord JM, Salmon M (2010) Distinct types of fibrocyte can differentiate from mononuclear cells in the presence and absence of serum. PLoS One 5(3):e9730

Dominici M, Le Blanc K, Mueller I, Slaper-Cortenbach I, Marini F, Krause D, Deans R, Keating A, Prockop D, Horwitz E (2006) Minimal criteria for defining multipotent mesenchymal stromal cells. The International Society for Cellular Therapy position statement. Cytotherapy 8(4):315–317. doi:10.1080/14653240600855905

Ebina M, Shimizukawa M, Shibata N, Kimura Y, Suzuki T, Endo M, Sasano H, Kondo T, Nukiwa T (2004) Heterogeneous increase in CD34-positive alveolar capillaries in idiopathic pulmonary fibrosis. Am J Respir Crit Care Med 169(11):1203–1208. doi:10.1164/rccm.200308-1111OC

Engelholm LH, List K, Netzel-Arnett S, Cukierman E, Mitola DJ, Aaronson H, Kjøller L, Larsen JK, Yamada KM, Strickland DK (2003) uPARAP/Endo180 is essential for cellular uptake of collagen and promotes fibroblast collagen adhesion. J Cell Biol 160(7):1009–1015

Evans MJ, Van Winkle LS, Fanucchi MV, Plopper CG (1999) The attenuated fibroblast sheath of the respiratory tract epithelial-mesenchymal trophic unit. Am J Respir Cell Mol Biol 21(6):655–657. doi:10.1165/ajrcmb.21.6.3807

Farkas L, Gauldie J, Voelkel NF, Kolb M (2011) Pulmonary hypertension and idiopathic pulmonary fibrosis: a tale of angiogenesis, apoptosis, and growth factors. Am J Respir Cell Mol Biol 45(1):1–15. doi:10.1165/rcmb.2010-0365TR

Fathke C, Wilson L, Hutter J, Kapoor V, Smith A, Hocking A, Isik F (2004) Contribution of bone marrow-derived cells to skin: collagen deposition and wound repair. Stem Cells 22(5):812–822

Gan Y, Reilkoff R, Peng X, Russell T, Chen Q, Mathai SK, Homer R, Gulati M, Siner J, Elias J, Bucala R, Herzog E (2011) Role of semaphorin 7a signaling in transforming growth factor beta1-induced lung fibrosis and scleroderma-related interstitial lung disease. Arthritis Rheum 63(8):2484–2494. doi:10.1002/art.30386

Garantziotis S, Zudaire E, Trempus CS, Hollingsworth JW, Jiang D, Lancaster LH, Richardson E, Zhuo L, Cuttitta F, Brown KK, Noble PW, Kimata K, Schwartz DA (2008) Serum inter-alpha-trypsin inhibitor and matrix hyaluronan promote angiogenesis in fibrotic lung injury. Am J Respir Crit Care Med 178(9):939–947. doi:10.1164/rccm.200803-386OC

Geissmann F, Auffray C, Palframan R, Wirrig C, Ciocca A, Campisi L, Narni-Mancinelli E, Lauvau G (2008) Blood monocytes: distinct subsets, how they relate to dendritic cells, and their possible roles in the regulation of T-cell responses. Immunol Cell Biol 86(5):398–408

Grieb G, Steffens G, Pallua N, Bernhagen J, Bucala R (2011) Circulating fibrocytes—biology and mechanisms in wound healing and scar formation. Int Rev Cell Mol Biol 291:1–19. doi:10.1016/B978-0-12-386035-4.00001-X

Grunig G, Marsh LM, Esmaeil N, Jackson K, Gordon T, Reibman J, Kwapiszewska G, Park SH (2014) Perspective: ambient air pollution: inflammatory response and effects on the lung's vasculature. Pulm Circ 4(1):25–35. doi:10.1086/674902

Guignabert C, Dorfmuller P (2013) Pathology and pathobiology of pulmonary hypertension. Semin Respir Crit Care Med 34(5):551–559. doi:10.1055/s-0033-1356496

Hanumegowda C, Farkas L, Kolb M (2012) Angiogenesis in pulmonary fibrosis: too much or not enough? Chest 142(1):200–207. doi:10.1378/chest. 11-1962

Hayashi H, Kawakita A, Okazaki S, Murai H, Yasutomi M, Ohshima Y (2014) IL-33 enhanced the proliferation and constitutive production of IL-13 and IL-5 by fibrocytes. Biomed Res Int 2014:738625. doi:10.1155/2014/738625

Heath D, Edwards JE (1958) The pathology of hypertensive pulmonary vascular disease: a description of six grades of structural changes in the pulmonary arteries with special reference to congenital cardiac septal defects. Circulation 18(4 Part 1):533–547

Hirose J, Kawashima H, Yoshie O, Tashiro K, Miyasaka M (2001) Versican interacts with chemokines and modulates cellular responses. J Biol Chem 276(7):5228–5234

Hong KM, Burdick MD, Phillips RJ, Heber D, Strieter RM (2005) Characterization of human fibrocytes as circulating adipocyte progenitors and the formation of human adipose tissue in SCID mice. FASEB J 19(14):2029–2031

Hong KM, Belperio JA, Keane MP, Burdick MD, Strieter RM (2007) Differentiation of human circulating fibrocytes as mediated by transforming growth factor-β and peroxisome proliferator-activated receptor γ. J Biol Chem 282(31):22910–22920

Howell SJ, Doane KJ (1998) Type VI collagen increases cell survival and prevents anti-β1 integrin-mediated apoptosis. Exp Cell Res 241(1):230–241

Hutchinson WL, Hohenester E, Pepys MB (2000) Human serum amyloid P component is a single uncomplexed pentamer in whole serum. Mol Med 6(6):482–493

Iozzo RV (1998) Matrix proteoglycans: from molecular design to cellular function. Annu Rev Biochem 67(1):609–652

Kendall RT, Feghali-Bostwick CA (2014) Fibroblasts in fibrosis: novel roles and mediators. Front Pharmacol 5:123. doi:10.3389/fphar.2014.00123

Kisseleva T, von Kockritz-Blickwede M, Reichart D, McGillvray SM, Wingender G, Kronenberg M, Glass CK, Nizet V, Brenner DA (2011) Fibrocyte-like cells recruited to the spleen support innate and adaptive immune responses to acute injury or infection. J Mol Med 89(10): 997–1013. doi:10.1007/s00109-011-0756-0

Kolosionek E, Crosby A, Harhay MO, Morrell N, Butrous G (2010) Pulmonary vascular disease associated with schistosomiasis. Expert Rev Anti Infect Ther 8(12):1467–1473. doi:10.1586/eri.10.124

Kotsianidis I, Nakou E, Bouchliou I, Tzouvelekis A, Spanoudakis E, Steiropoulos P, Sotiriou I, Aidinis V, Margaritis D, Tsatalas C, Bouros D (2009) Global impairment of CD4+CD25+FOXP3+ regulatory T cells in idiopathic pulmonary fibrosis. Am J Respir Crit Care Med 179(12):1121–1130. doi:10.1164/rccm.200812-1936OC

Kotton DN, Ma BY, Cardoso WV, Sanderson EA, Summer RS, Williams MC, Fine A (2001) Bone marrow-derived cells as progenitors of lung alveolar epithelium. Development 128(24): 5181–5188

Kraman M, Bambrough PJ, Arnold JN, Roberts EW, Magiera L, Jones JO, Gopinathan A, Tuveson DA, Fearon DT (2010) Suppression of antitumor immunity by stromal cells expressing fibroblast activation protein-alpha. Science 330(6005):827–830. doi:10.1126/science.1195300

Kranenburg AR, De Boer WI, Van Krieken JH, Mooi WJ, Walters JE, Saxena PR, Sterk PJ, Sharma HS (2002) Enhanced expression of fibroblast growth factors and receptor FGFR-1 during vascular remodeling in chronic obstructive pulmonary disease. Am J Respir Cell Mol Biol 27(5):517–525. doi:10.1165/rcmb.4474

Kranenburg AR, Willems-Widyastuti A, Mooi WJ, Saxena PR, Sterk PJ, de Boer WI, Sharma HS (2005) Chronic obstructive pulmonary disease is associated with enhanced bronchial expression of FGF-1, FGF-2, and FGFR-1. J Pathol 206(1):28–38. doi:10.1002/path.1748

Kranenburg AR, Willems-Widyastuti A, Moori WJ, Sterk PJ, Alagappan VK, de Boer WI, Sharma HS (2006) Enhanced bronchial expression of extracellular matrix proteins in chronic obstructive pulmonary disease. Am J Clin Pathol 126(5):725–735

Lee AH, Dhaliwal R, Kantores C, Ivanovska J, Gosal K, McNamara PJ, Letarte M, Jankov RP (2014) Rho-kinase inhibitor prevents bleomycin-induced injury in neonatal rats independent of effects on lung inflammation. Am J Respir Cell Mol Biol 50(1):61–73. doi:10.1165/rcmb.2013-0131OC

Li X, Wilson JW (1997) Increased vascularity of the bronchial mucosa in mild asthma. Am J Respir Crit Care Med 156(1):229–233. doi:10.1164/ajrccm.156.1.9607066

Li J, Tan H, Wang X, Li Y, Samuelson L, Li X, Cui C, Gerber DA (2014) Circulating fibrocytes stabilize blood vessels during angiogenesis in a paracrine manner. Am J Pathol 184(2): 556–571. doi:10.1016/j.ajpath.2013.10.021

Madsen DH, Engelholm LH, Ingvarsen S, Hillig T, Wagenaar-Miller RA, Kjøller L, Gårdsvoll H, Høyer-Hansen G, Holmbeck K, Bugge TH (2007) Extracellular collagenases and the endocytic receptor, urokinase plasminogen activator receptor-associated protein/Endo180, cooperate in fibroblast-mediated collagen degradation. J Biol Chem 282(37):27037–27045

Maharjan AS, Pilling D, Gomer RH (2010) Toll-like receptor 2 agonists inhibit human fibrocyte differentiation. Fibrogenesis Tissue Repair 3:23. doi:10.1186/1755-1536-3-23

Maharjan AS, Pilling D, Gomer RH (2011) High and low molecular weight hyaluronic acid differentially regulate human fibrocyte differentiation. PLoS One 6(10):e26078. doi:10.1371/journal.pone.0026078

Mattoli S, Bellini A, Schmidt M (2009) The role of a human hematopoietic mesenchymal progenitor in wound healing and fibrotic diseases and implications for therapy. Curr Stem Cell Res Ther 4(4):266–280

Miserocchi G (2009) Mechanisms controlling the volume of pleural fluid and extravascular lung water. Eur Respir Rev 18(114):244–252. doi:10.1183/09059180.00002709

Mitzner W, Wagner EM (2004) Vascular remodeling in the circulations of the lung. J Appl Physiol 97(5):1999–2004. doi:10.1152/japplphysiol.00473.2004

Molawi K, Sieweke MH (2013) Transcriptional control of macrophage identity, self-renewal, and function. Adv Immunol 120:269–300. doi:10.1016/B978-0-12-417028-5.00010-7

Moores SL, Selfors LM, Fredericks J, Breit T, Fujikawa K, Alt FW, Brugge JS, Swat W (2000) Vav family proteins couple to diverse cell surface receptors. Mol Cell Biol 20(17):6364–6373

Mori L, Bellini A, Stacey MA, Schmidt M, Mattoli S (2005) Fibrocytes contribute to the myofibroblast population in wounded skin and originate from the bone marrow. Exp Cell Res 304(1):81–90

Mousavi S, Sato M, Sporstol M, Smedsrod B, Berg T, Kojima N, Senoo H (2005) Uptake of denatured collagen into hepatic stellate cells: evidence for the involvement of urokinase plasminogen activator receptor-associated protein/Endo180. Biochem J 387:39–46

Muller-Ladner U, Distler O, Ibba-Manneschi L, Neumann E, Gay S (2009) Mechanisms of vascular damage in systemic sclerosis. Autoimmunity 42(7):587–595

Murray LA, Rosada R, Moreira AP, Joshi A, Kramer MS, Hesson DP, Argentieri RL, Mathai S, Gulati M, Herzog EL, Hogaboam CM (2010) Serum amyloid P therapeutically attenuates murine bleomycin-induced pulmonary fibrosis via its effects on macrophages. PLoS One 5(3):e9683. doi:10.1371/journal.pone.0009683

Murray LA, Chen Q, Kramer MS, Hesson DP, Argentieri RL, Peng X, Gulati M, Homer RJ, Russell T, van Rooijen N, Elias JA, Hogaboam CM, Herzog EL (2011) TGF-beta driven lung fibrosis is macrophage dependent and blocked by Serum amyloid P. Int J Biochem Cell Biol 43(1):154–162. doi:10.1016/j.biocel.2010.10.013

Myllyharju J, Kivirikko KI (2001) Collagens and collagen-related diseases. Ann Med 33(1):7–21

Myllyharju J, Kivirikko KI (2004) Collagens, modifying enzymes and their mutations in humans, flies and worms. Trends Genet 20(1):33–43

Nikam VS, Schermuly RT, Dumitrascu R, Weissmann N, Kwapiszewska G, Morrell N, Klepetko W, Fink L, Seeger W, Voswinckel R (2010) Treprostinil inhibits the recruitment of bone marrow-derived circulating fibrocytes in chronic hypoxic pulmonary hypertension. Eur Respir J 36(6):1302–1314. doi:10.1183/09031936.00028009

Ortiz LA, Gambelli F, McBride C, Gaupp D, Baddoo M, Kaminski N, Phinney DG (2003) Mesenchymal stem cell engraftment in lung is enhanced in response to bleomycin exposure and ameliorates its fibrotic effects. Proc Natl Acad Sci U S A 100(14):8407–8411. doi:10.1073/pnas.1432929100

Peao MN, Aguas AP, de Sa CM, Grande NR (1994) Neoformation of blood vessels in association with rat lung fibrosis induced by bleomycin. Anat Rec 238(1):57–67. doi:10.1002/ar.1092380108

Peng X, Moore MW, Peng H, Sun H, Gan Y, Homer RJ, Herzog EL (2014) CD4+CD25+FoxP3+ Regulatory Tregs inhibit fibrocyte recruitment and fibrosis via suppression of FGF-9 production in the TGF-beta1 exposed murine lung. Front Pharmacol 5:80. doi:10.3389/fphar.2014.00080

Perros F, Dorfmuller P, Montani D, Hammad H, Waelput W, Girerd B, Raymond N, Mercier O, Mussot S, Cohen-Kaminsky S, Humbert M, Lambrecht BN (2012) Pulmonary lymphoid neogenesis in idiopathic pulmonary arterial hypertension. Am J Respir Crit Care Med 185(3):311–321. doi:10.1164/rccm.201105-0927OC

Phillips RJ, Burdick MD, Hong K, Lutz MA, Murray LA, Xue YY, Belperio JA, Keane MP, Strieter RM (2004) Circulating fibrocytes traffic to the lungs in response to CXCL12 and mediate fibrosis. J Clin Invest 114(3):438–446

Pilling D, Gomer RH (2014) Persistent lung inflammation and fibrosis in serum amyloid P component (APCs–/–) knockout mice. PLoS One 9(4):e93730. doi:10.1371/journal.pone.0093730

Pilling D, Buckley CD, Salmon M, Gomer RH (2003) Inhibition of fibrocyte differentiation by serum amyloid P. J Immunol 171(10):5537–5546

Pilling D, Roife D, Wang M, Ronkainen SD, Crawford JR, Travis EL, Gomer RH (2007) Reduction of bleomycin-induced pulmonary fibrosis by serum amyloid P. J Immunol 179(6):4035–4044

Pilling D, Fan T, Huang D, Kaul B, Gomer RH (2009) Identification of markers that distinguish monocyte-derived fibrocytes from monocytes, macrophages, and fibroblasts. PLoS One 4(10):e7475. doi:10.1371/journal.pone.0007475

Price LC, Wort SJ, Perros F, Dorfmuller P, Huertas A, Montani D, Cohen-Kaminsky S, Humbert M (2012) Inflammation in pulmonary arterial hypertension. Chest 141(1):210–221. doi:10.1378/chest. 11-0793

Pu J, Gu S, Liu S, Zhu S, Wilson D, Siegfried JM, Gur D (2012) CT based computerized identification and analysis of human airways: a review. Med Phys 39(5):2603–2616. doi:10.1118/1.4703901

Quan TE, Cowper S, Wu S-P, Bockenstedt LK, Bucala R (2004) Circulating fibrocytes: collagen-secreting cells of the peripheral blood. Int J Biochem Cell Biol 36(4):598–606

Reilkoff RA, Bucala R, Herzog EL (2011) Fibrocytes: emerging effector cells in chronic inflammation. Nat Rev Immunol 11(6):427–435

Ryan J, Bloch K, Archer SL (2011) Rodent models of pulmonary hypertension: harmonisation with the world health organisation's categorisation of human PH. Int J Clin Pract Suppl 172:15–34. doi:10.1111/j.1742-1241.2011.02710.x

Salvato G (2001) Quantitative and morphological analysis of the vascular bed in bronchial biopsy specimens from asthmatic and non-asthmatic subjects. Thorax 56(12):902–906

Schmidt M, Sun G, Stacey MA, Mori L, Mattoli S (2003) Identification of circulating fibrocytes as precursors of bronchial myofibroblasts in asthma. J Immunol 171(1):380–389

Seferian A, Simonneau G (2013) Therapies for pulmonary arterial hypertension: where are we today, where do we go tomorrow? Eur Respir Rev 22(129):217–226. doi:10.1183/09059180.00001713

Seferian A, Chaumais MC, Savale L, Gunther S, Tubert-Bitter P, Humbert M, Montani D (2013) Drugs induced pulmonary arterial hypertension. Presse Med 42(9 Pt 2):e303–e310. doi:10.1016/j.lpm.2013.07.005

Shimoda LA, Laurie SS (2013) Vascular remodeling in pulmonary hypertension. J Mol Med 91(3):297–309. doi:10.1007/s00109-013-0998-0

Simonneau G, Robbins IM, Beghetti M, Channick RN, Delcroix M, Denton CP, Elliott CG, Gaine SP, Gladwin MT, Jing ZC, Krowka MJ, Langleben D, Nakanishi N, Souza R (2009) Updated clinical classification of pulmonary hypertension. J Am Coll Cardiol 54(1 Suppl):S43–S54. doi:10.1016/j.jacc.2009.04.012

Spees JL, Whitney MJ, Sullivan DE, Lasky JA, Laboy M, Ylostalo J, Prockop DJ (2008) Bone marrow progenitor cells contribute to repair and remodeling of the lung and heart in a rat model of progressive pulmonary hypertension. FASEB J 22(4):1226–1236. doi:10.1096/fj.07-8076com

Stacher E, Graham BB, Hunt JM, Gandjeva A, Groshong SD, McLaughlin VV, Jessup M, Grizzle WE, Aldred MA, Cool CD, Tuder RM (2012) Modern age pathology of pulmonary arterial hypertension. Am J Respir Crit Care Med 186(3):261–272. doi:10.1164/rccm.201201-0164OC

Stirling G, Kakkar V (1969) Cells in the circulating blood capable of producing connective tissue. Br J Exp Pathol 50(1):51

Suga H, Rennert RC, Rodrigues M, Sorkin M, Glotzbach JP, Januszyk M, Fujiwara T, Longaker MT, Gurtner GC (2014) Tracking the elusive fibrocyte: Identification and characterization of collagen producing hematopoietic lineage cells during murine wound healing. Stem Cells 32(5):1347–1360

Tam A, Sin DD (2012) Pathobiologic mechanisms of chronic obstructive pulmonary disease. Med Clin North Am 96(4):681–698. doi:10.1016/j.mcna.2012.04.012

Tamosiuniene R, Tian W, Dhillon G, Wang L, Sung YK, Gera L, Patterson AJ, Agrawal R, Rabinovitch M, Ambler K, Long CS, Voelkel NF, Nicolls MR (2011) Regulatory T cells limit vascular endothelial injury and prevent pulmonary hypertension. Circ Res 109(8):867–879. doi:10.1161/CIRCRESAHA.110.236927

Tervaert JW, Limburg PC, Elema JD, Huitema MG, Horst G, The TH, Kallenberg CG (1991) Detection of autoantibodies against myeloid lysosomal enzymes: a useful adjunct to classification of patients with biopsy-proven necrotizing arteritis. Am J Med 91(1):59–66

Tuder RM, Archer SL, Dorfmuller P, Erzurum SC, Guignabert C, Michelakis E, Rabinovitch M, Schermuly R, Stenmark KR, Morrell NW (2013) Relevant issues in the pathology and pathobiology of pulmonary hypertension. J Am Coll Cardiol 62(25 Suppl):D4–D12. doi:10.1016/j.jacc.2013.10.025

Ushiki T (2002) Collagen fibers, reticular fibers and elastic fibers. A comprehensive understanding from a morphological viewpoint. Arch Histol Cytol 65(2):109–126

Wang X, Zhu H, Yang X, Bi Y, Cui S (2010) Vasohibin attenuates bleomycin induced pulmonary fibrosis via inhibition of angiogenesis in mice. Pathology 42(5):457–462. doi:10.3109/00313025.2010.493864

Wansleeben C, Barkauskas CE, Rock JR, Hogan BL (2013) Stem cells of the adult lung: their development and role in homeostasis, regeneration, and disease. Wiley Interdiscip Rev Dev Biol 2(1):131–148. doi:10.1002/wdev.58

Weiss DJ, Kolls JK, Ortiz LA, Panoskaltsis-Mortari A, Prockop DJ (2008) Stem cells and cell therapies in lung biology and lung diseases. Proc Am Thorac Soc 5(5):637–667. doi:10.1513/pats.200804-037DW

Wienke D, MacFadyen JR, Isacke CM (2003) Identification and characterization of the endocytic transmembrane glycoprotein Endo180 as a novel collagen receptor. Mol Biol Cell 14(9):3592–3604

Wright JL, Petty T, Thurlbeck WM (1992) Analysis of the structure of the muscular pulmonary arteries in patients with pulmonary hypertension and COPD: National Institutes of Health nocturnal oxygen therapy trial. Lung 170(2):109–124

Wu YJ, La Pierre DP, Jin W, Albert JY, Burton BY (2005) The interaction of versican with its binding partners. Cell Res 15(7):483–494

Zambidis ET, Peault B, Park TS, Bunz F, Civin CI (2005) Hematopoietic differentiation of human embryonic stem cells progresses through sequential hematoendothelial, primitive, and definitive stages resembling human yolk sac development. Blood 106(3):860–870

Zanini A, Chetta A, Olivieri D (2008) Therapeutic perspectives in bronchial vascular remodeling in COPD. Ther Adv Respir Dis 2(3):179–187. doi:10.1177/1753465808092339

Zhang W, Chuang Y-J, Swanson R, Li J, Seo K, Leung L, Lau LF, Olson ST (2004) Antiangiogenic antithrombin down-regulates the expression of the proangiogenic heparan sulfate proteoglycan, perlecan, in endothelial cells. Blood 103(4):1185–1191

Chapter 15
The Role of Stem Cells in Vascular Remodeling in CTEPH

Amy L. Firth and Jason X.-J. Yuan

Abbreviations

BM-MSC	Bone marrow-derived mesenchymal stem cell
CD11b	Cluster of differentiation factor 11b; aka integrin alpha m
CD14	Cluster of differentiation factor 14
CD19	Cluster of differentiation factor 19; aka B-Lymphocyte antigen
CD34	Cluster of differentiation factor 34
CD45	Cluster of differentiation factor 45; aka lymphochyte common antigen
CD73	Cluster of differentiation factor 73; aka 5′-nucleotidase
CD79α	Cluster of differentiation factor 79alpha; aka B-cell antigen receptor complex-associated protein alpha chain and MB-1 membrane glycoprotein
CD90	Cluster of differentiation factor 90; aka Thy-1
CD105	Cluster of differentiation factor 105; aka Endoglin
CFU	Colony-forming unit
CTEPH	Chronic thromboembolic pulmonary hypertension
EnMT	Endothelial to mesenchymal transition
EPC	Endothelial progenitor cell
eNOS	Endothelial nitric oxide synthase
EPC	Endothelial progenitor cells
HLA-DR	MHC class II cell surface receptor encoded by the human leukocyte antigen complex

A.L. Firth (✉)
Laboratory of Genetics, The Salk Institute for Biological Studies, La Jolla, CA 92037, USA
e-mail: afirth@salk.edu

J.X.-J. Yuan
Department of Medicine, Arizona Health Sciences Center, University of Arizona, Tucson, AZ, USA

A. Firth, J.X.-J. Yuan (eds.), *Lung Stem Cells in the Epithelium and Vasculature*, Stem Cell Biology and Regenerative Medicine, DOI 10.1007/978-3-319-16232-4_15

HMGB1	High-mobility group protein B1
HO	Hemeoxygenase
MAPC	Multipotent adult progenitor cells
MCT	Monocrotaline
miRNA	Micro ribonucleic acid
MSC	Mesenchymal stem cell
MSC-MV	Mesenchymal stem cell-derived micro vesicle
PAE	Pulmonary endarterectomy
PAH	Pulmonary arterial hypertension
PE	Pulmonary embolism
PVR	Pulmonary vascular resistance
RAGE	Receptor for advanced glycation end products
SMC	Smooth muscle cells
TGFβ	Transforming growth factor beta
UTR	Untranslated region
VEGF-R2	Vascular endothelial growth factor receptor 2
vWF	von Willebrand factor

15.1 Introduction

Chronic Thromboembolic Pulmonary Hypertension or CTEPH is caused by the failure of a pulmonary embolism to resolve. CTEPH is a notoriously underdiagnosed and severe form of pulmonary hypertension characterized by stenosis or complete obliteration of the pulmonary artery by an intraluminal thrombus that has become fibrotic. Such persistence of pulmonary thromboemboli occurs in 0.1–9.1 % of all cases (Guerin et al. 2014; Becattini et al. 2006; Poli et al. 2010; Pengo et al. 2004). Obstruction of the pulmonary vascular bed by non-resolving thromboemboli leads to increased pulmonary vascular resistance (PVR). Progressive pulmonary hypertension and right heart failure then ensue. In the absence of intervention, CTEPH prognosis is poor. While the clinical and hemodynamic features of the disease are robustly defined (Banks et al. 2014), there is still a lack of understanding of the associated pathophysiology leading to the progression of the disease. Pulmonary endarterectomy (PAE) surgery remains the frontline treatment for patients diagnosed with CTEPH; however, a number of patients are inoperable or develop persistent pulmonary hypertension subsequent to PAE (D'Armini et al. 2014; Hoeper et al. 2014; Hill et al. 2008; Jamieson et al. 2003). Furthermore, lung reperfusion injury and persistent pulmonary hypertension remain serious complications of PAE (Hsu et al. 2007; Lee et al. 2001).

Recent investigations striving to elucidate the mechanisms involved in the formation and resolution of a venous thrombosis have led to an increased understanding of clot interactions with the surrounding vasculature. Genetic and pathophysiological studies have focused on determining the cellular and molecular pathways including those involving regulation of growth factors, cytokines, metabolic signaling, elastases, proteases, and microRNAs (miRNA) (Zabini et al. 2014; Feng et al. 2014; Wynants et al. 2013; Wang et al. 2013; Ogawa et al. 2009). Furthermore, an increased recognition of stem and progenitor properties in cells resulted in identification of cells, both circulating and within the vascular wall that have a high degree of plasticity.

While the vessel is initially occluded by a pulmonary embolism, in patients who develop CTEPH this becomes a fibrotic coagulant material blocking the central artery. Subsequently, this leads to vascular remodeling of the arterial segments that are both surrounding and distal to the occluded regions and the development of elevated PVR. The process of pulmonary arterial remodeling has been associated with a number of changes in vascular function including; (1) increased proliferation and migration of vascular smooth muscle cells; (2) endothelial dysfunction; adventitial thickening due to excessive adventitial fibroblast and myofibroblast; and (3) a build-up of circulating inflammatory cells. Most recently, both resident vascular progenitor cells and circulating progenitor/precursor cells which become encompassed in the fibrotic clot have been identified and proposed to play a role in the pathogenesis of CTEPH. The abnormal microenvironment/niche that the fibrotic clot provides is thus speculated to drive misguided differentiation and proliferation of these cells contributing to stabilization of the clot and associated pulmonary vascular remodeling (Firth et al. 2010).

This chapter will introduce CTEPH as a form of pulmonary hypertension and provide a summary of the pathogenesis of CTEPH and the postulated contribution of progenitor and stem cells. Further detail on myofibroblasts and endothelial progenitor cells can be found in other chapters in this book.

15.2 Pathogenesis of CTEPH

It is established that CTEPH develops subsequent to a pulmonary embolism arising from a deep vein thrombosis. Studies also suggest that in situ lung thrombosis may contribute to disease progression by promotion of the growth and stabilization of the thromboemboli. Why the PE fails to resolve remains a question to be fully elucidated. As mentioned above, a prothrombotic mutation in fibrinogen has been associated with the incidence of CTEPH (Chen et al. 2010; Morris et al. 2009; Laczika et al. 2002). This was discovered after it was reported that the fibrin in patients with CTEPH was relatively resistant to fibrinolysis in vitro (Li et al. 2013; Morris et al. 2006). Studies have investigated the impact of such mutations and found that they are associated with a fibrinogen that, in vitro, is partially resistant to plasmin-mediated lysis when compared to fibrinogen from healthy controls. If such resistance to fibrinolysis occurs in vivo, one could postulate that the fibrin may persist in the thromboemboli and stimulate the remodeling process. Indeed, we observed a significant effect of chronic exposure of fibrin and fibrinogen on the regulation of intracellular calcium in both primary pulmonary artery smooth muscle and endothelial cells (Firth et al. 2009). Intracellular calcium regulation is central to the control of cell migration, proliferation, and contraction, all components of vascular remodeling process. Further studies are necessary to fully understand the impact of mutant fibrinogen on the pathogenesis of CTEPH. Association of mutations in fibrinogen leading to differences in the structure of fibrin in the clot may account for the development of CTEPH in a subpopulation of CTEPH patients; however, it cannot currently account for all cases. Identification of such a polymorphism may be valuable in acting as a biomarker for identifying CTEPH risk.

It has also been proposed that CTEPH could also arise due to small vessel disease leading to a secondary thrombosis. Clinical evidence does suggest that there is endothelial cell damage in the small pulmonary vessel and intimal hyperplasia likely leading to the development of proximal vascular lesions.

15.3 Identification of Stem and Progenitor Cells Associated with CTEPH

Stem and progenitor are terms that are often used interchangeably. By definition, a progenitor cell is like a stem cell with the capacity for both self-renewal and differentiation, but that has already become more committed differentiating into its target cell. It is difficult to categorize the cell types involved in the pathogenesis of CTEPH until we fully understand their origin and differentiation potential. Here we will describe what is currently known with regard to this. Indeed, recently, specific vasodilator compounds are demonstrated to have some clinical efficacy in inoperable PAE patients, indicating some level of vascular disease (Seyfarth et al. 2010).

15.3.1 Mesenchymal Progenitor Cells

The precise definition of a mesenchymal progenitor or stem cells is still difficult to pinpoint due to a lack of specific cell surface markers. Currently, the minimal criteria as defined by the International Society for Cellular Therapy states that a mesenchymal stem cell (MSC) must be; (1) plastic-adherent; (2) express CD105, CD73, and CD90, and lack expression of CD45, CD34, CD14 or CD11b, CD79α or CD19, and HLA-DR; and (3) must have the capacity to differentiate to osteoblasts, adipocytes, and chondroblasts in vitro (Dominici et al. 2006). As further studies are completed, these criteria will become clearer and I refer you to a paper detailing identification of progenitor cells in the pulmonary vasculature (Firth and Yuan 2012).

Mesenchymal stem cells are currently being pursued as a therapeutic approach for other forms of PH. Administration of BM-MSC has been shown to attenuate the development of monocrotaline-induced PH and endothelial dysfunction in rats (Hansmann et al. 2012; Baber et al. 2007). Nanoscale MSC-derived microvesicles (MSC-MV) (Chen et al. 2014) or exosomes (Lee et al. 2012) have also been demonstrated to have pleiotropic effects in PH by inhibiting the hyperproliferative pathways. Furthermore, MSC over-expressing heme-oxygenase (HO) (Liang et al. 2011) or endothelial nitric oxide synthase (eNOS) (Kanki-Horimoto et al. 2006) are able to attenuate the development of chronic hypoxia-induced PAH and MCT-induced PAH, respectively. The role of MSC pathogenically or therapeutically in CTEPH has not currently been investigated. Other cells with a mesenchymal progenitor phenotype have, however, been identified in CTEPH and thought to contribute to its pathogenesis. These cells are discussed below.

Myofibroblasts

Myofibroblasts are cell type exhibiting significant plasticity with characteristics of both fibroblasts and smooth muscle cells. They are able to produce extracellular matrix and have known roles in wound healing and inflammation. A number of precursors or progenitors for the myofibroblast have been identified. Fibroblasts are the primary source, with smooth muscle cells, pericytes, bone marrow-derived circulating fibrocytes, and epithelial to mesenchymal transition all identified as potential sources.

Cells representing a myofibroblastic phenotype were first identified in endarterectomized tissues from patients with CTEPH in our laboratory (Firth et al. 2010) and are currently thought to be a predominant cell type found in the endarterectomized tissues from CTEPH patients (Sakao et al. 2011; Maruoka et al. 2012). After histological examination, our laboratory identified a population of cells expressing intermediate filaments vimentin and smooth muscle alpha actin in endarterectomized tissues from patients with significant neointimal formation. These cells generated single cell colonies resembling CFU-fibroblast, grew in a network like pattern, and had the capacity for at least adipogenic and osteogenic differentiation (Firth et al. 2010). In a subsequent study similar populations of cells with a "cancer-like" phenotype were identified in the endarterectomized tissue; these cells were hyper-proliferative, anchorage-independent, and invasive (Maruoka et al. 2012). Indeed, this cancer-like hypothesis was supported by a study by Jujo and colleagues (Jujo et al. 2012). After a second passage of isolated myofibroblast-like cells, they noted the presence of a pleomorphic cell type similar to those in pulmonary intimal sarcoma. This tumor is thought to originate from sub-endothelial-mesenchymal cells in the pulmonary vascular wall. These cells behaved like the ones in the aforementioned studies being hyper-proliferative, anchorage-independent, invasive, and serum-independent. The authors defined them as sarcoma-like cells (SCLs) and, when injected into C.B-17/lcr-scid/scidJcl mice, they developed solid, undifferentiated tumors at the site of injection (Jujo et al. 2012). Together, these studies would suggest that there are properties to the stabilized clot that generates a unique microenvironment promoting the differentiation/de-differentiation of cells to enhance intimal remodeling.

There is still a poor understanding of the cellular and molecular mechanisms that lead to the development of CTEPH. In a search for biomarkers, Moser and colleagues identified the expression of Receptor for Advanced Glycation Endproducts (RAGE) and High-Mobility Group Protein B1 (HMGB1) in myofibroblasts isolated from endarterectomized tissues (Moser et al. 2014). Additionally, HMGB1 was significantly elevated in serum of patients with CTEPH (Moser et al. 2014). It will be necessary to develop a more detailed understanding of the RAGE-HMGB1 axis in CTEPH and determine its potential to serve as a biomarker.

15.3.2 Endothelial Progenitor Cells

Endothelial progenitor cells (EPC) is a term that is now known to encompass a number of subpopulations of circulating cells with the capacity to differentiate into endothelial cells (Khakoo and Finkel 2005). Such EPC are considered: (1) derivatives of hemangioblasts expressing surface receptors CD34, VEGFR-2, and CD133; (2) subsets of bone marrow-derived multipotent adult progenitor cells (MAPC) expressing CD133 and VEGFR-2 lacking CD34 or vascular endothelial cadherin (VE-Cad); (3) bone marrow-derived myelo/monocytic cells expressing CD14 and forming vWF, VEGF-R2, and $CD45^+$ endothelial cells. All of these EPC populations have the common ability to retain acetylated LDL. For a more detailed review of this cell type, please refer to Chaps. 9 and 17.

Although putative endothelial progenitor cells have been identified in endarterectomized tissues from patients with CTEPH (Yao et al. 2009), they do not appear to be significantly increased in the circulation (Smadja et al. 2010). This is contrary to what is known for PAH where consistently elevated levels of circulating endothelial progenitor cells have led to them being considered biomarkers (Smadja et al. 2009, 2010). Recruitment of EPC requires a cascade of signaling events which include the adhesion and migration and chemoattraction (Urbich and Dimmeler 2004); a final differentiation to endothelial cells drives neo-vascularization and re-endothelialization. Such events were observed by Yao et al. in endarteractomized tissues indicating, at minimum, a functional recruitment of EPC into the clot region (Yao et al. 2009).

Endothelial to Mesenchymal Transition

Endothelial to mesenchymal transition or EnMT describes a phenotypic switch from an endothelial cell to a cell of mesenchymal origin. Indeed, pulmonary EnMT is thought to contribute to the development of pulmonary fibrosis (Almudever et al. 2013; Diez et al. 2010; Arciniegas et al. 2007). While we still have a relatively poor understanding of the contribution of this process in the development and pathogenesis of pulmonary hypertension, in particular, CTEPH studies do confirm the acquisition of mesenchymal cell markers like alpha-smooth muscle actin, sm22-alpha, and myocardin and EnMT-related transcription factors: slug, snail, zeb1, and endothelin-1 in EPC co-cultured with SMC (Diez et al. 2010) and in PAEC

(Almudever et al. 2013). TGFβ and endothelin-1 signaling was key to the EnMT transition (Almudever et al. 2013; Diez et al. 2010). Consequentially, targeting the BH4 synthesis "salvage pathway" has been proposed to be a novel therapeutic strategy to attenuate pulmonary hypertension in idiopathic pulmonary fibrosis (Almudever et al. 2013). It is plausible that such an EnMT could contribute to the pathogenesis of CTEPH; both EPC and cells with a mesenchymal phenotype have been identified in endarterectomized tissues; however, their precise roles in the stabilization of the clot and in pulmonary vascular remodeling are not precisely defined (Firth et al. 2010; Yao et al. 2009; Jujo et al. 2012; Sakao et al. 2011)

15.4 Changes in the Microenvironment

There are currently two trains of thought regarding the role of the microenvironment in the development and pathogenesis of CTEPH. Firstly, the microenvironment created by the fibrotic clot lodged in the main pulmonary artery is obviously different to the one that cells in the pulmonary artery wall and the circulation are normally exposed to. It is thus possible that factors in this unique niche have an impact on cell state. Secondly, it has also been suggested that the thrombotic emboli persist secondary to the development of small vessel disease (Tanabe et al. 2013; Dorfmuller et al. 2014; Galie and Kim 2006).

As mentioned above, it has been shown that development of persistent thromboemboli is associated to a resistance of the fibrin to plasmin-mediated lysis (Morris et al. 2006, 2007). These studies showed that release of the N-terminal fragments from the beta chain of fibrin was retarded in CTEPH patients (Morris et al. 2006). More recently, the same group showed that the clot also had abnormal architecture likely to contribute to the incomplete resolution of the clot in patients with CTEPH-associated dysfibrinogenemia (Marsh et al. 2013). These patients had five confirmed heterozygous mutations in their fibrinogen, which were directly associated with abnormalities in fibrin structure/lysis (Morris et al. 2009). A subsequent study identified a 28 bp polymorphism in the 3' untranslated region (UTR) of the fibrinogen alpha gene, a complementary sequence to microRNA miR-759 which was associated with the susceptibility to CTEPH (Chen et al. 2010). It is also likely that mutations in BMPR2, already associated with the incidence of PH (Humbert et al. 2002; Atkinson et al. 2002; Lane et al. 2000), are correlated with the incidence of CTEPH providing further insight into CTEPH etiology (Feng et al. 2014).

A recent study discovered that in patients with a lack of history of pulmonary embolism, microvascular disease was predominant and fewer proximal lesions were observed (Dorfmuller et al. 2014). This study suggested a high pathogenic contribution of the peripheral pre-capillary and the post-capillary vasculature and such small vessel disease was also observed in their porcine model of CTEPH (Dorfmuller et al. 2014). The potential involvement of stem and progenitor cells in this process remains to be investigated.

15.5 Summary

The persistence of a lysis-resistant clot in the pulmonary artery provides a unique microenvironment and, given the influential role of the microenvironment on cell differentiation, it is likely this provides a unique niche for the recruitment and differentiation of both circulating and resident progenitor cells. Our current understanding of the role of stem and progenitor cells in this process is relatively poorly understood, but may be critical in identifying potential therapeutic options for resolution and prevention of the disease. It will be interesting to follow the field as it evolves.

References

Almudever P, Milara J, De Diego A, Serrano-Mollar A, Xaubet A, Perez-Vizcaino F, Cogolludo A, Cortijo J (2013) Role of tetrahydrobiopterin in pulmonary vascular remodelling associated with pulmonary fibrosis. Thorax 68(10):938–948. doi:10.1136/thoraxjnl-2013-203408

Arciniegas E, Frid MG, Douglas IS, Stenmark KR (2007) Perspectives on endothelial-to-mesenchymal transition: potential contribution to vascular remodeling in chronic pulmonary hypertension. Am J Physiol Lung Cell Mol Physiol 293(1):L1–L8. doi:10.1152/ajplung.00378.2006

Atkinson C, Stewart S, Upton PD, Machado R, Thomson JR, Trembath RC, Morrell NW (2002) Primary pulmonary hypertension is associated with reduced pulmonary vascular expression of type II bone morphogenetic protein receptor. Circulation 105(14):1672–1678

Baber SR, Deng W, Master RG, Bunnell BA, Taylor BK, Murthy SN, Hyman AL, Kadowitz PJ (2007) Intratracheal mesenchymal stem cell administration attenuates monocrotaline-induced pulmonary hypertension and endothelial dysfunction. Am J Physiol Heart Circ Physiol 292(2):H1120–H1128. doi:10.1152/ajpheart.00173.2006

Banks DA, Pretorius GV, Kerr KM, Manecke GR (2014) Pulmonary endarterectomy: part I. Pathophysiology, clinical manifestations, and diagnostic evaluation of chronic thromboembolic pulmonary hypertension. Semin Cardiothorac Vasc Anesth 18(4):319–330. doi:10.1177/1089253214536621

Becattini C, Agnelli G, Pesavento R, Silingardi M, Poggio R, Taliani MR, Ageno W (2006) Incidence of chronic thromboembolic pulmonary hypertension after a first episode of pulmonary embolism. Chest 130(1):172–175. doi:10.1378/chest.130.1.172

Chen Z, Nakajima T, Tanabe N, Hinohara K, Sakao S, Kasahara Y, Tatsumi K, Inoue Y, Kimura A (2010) Susceptibility to chronic thromboembolic pulmonary hypertension may be conferred by miR-759 via its targeted interaction with polymorphic fibrinogen alpha gene. Hum Genet 128(4):443–452. doi:10.1007/s00439-010-0866-8

Chen JY, An R, Liu ZJ, Wang JJ, Chen SZ, Hong MM, Liu JH, Xiao MY, Chen YF (2014) Therapeutic effects of mesenchymal stem cell-derived microvesicles on pulmonary arterial hypertension in rats. Acta Pharmacol Sin 35(9):1121–1128. doi:10.1038/aps.2014.61

D'Armini AM, Morsolini M, Mattiucci G, Grazioli V, Pin M, Valentini A, Silvaggio G, Klersy C, Dore R (2014) Pulmonary endarterectomy for distal chronic thromboembolic pulmonary hypertension. J Thorac Cardiovasc Surg 148(3):1005–1012. doi:10.1016/j.jtcvs.2014.06.052, e1002

Diez M, Musri MM, Ferrer E, Barbera JA, Peinado VI (2010) Endothelial progenitor cells undergo an endothelial-to-mesenchymal transition-like process mediated by TGFbetaRI. Cardiovasc Res 88(3):502–511. doi:10.1093/cvr/cvq236

Dominici M, Le Blanc K, Mueller I, Slaper-Cortenbach I, Marini F, Krause D, Deans R, Keating A, Prockop D, Horwitz E (2006) Minimal criteria for defining multipotent mesenchymal stromal cells. The International Society for Cellular Therapy position statement. Cytotherapy 8(4):315–317. doi:10.1080/14653240600855905

Dorfmuller P, Gunther S, Ghigna MR, Thomas de Montpreville V, Boulate D, Paul JF, Jais X, Decante B, Simonneau G, Dartevelle P, Humbert M, Fadel E, Mercier O (2014) Microvascular disease in chronic thromboembolic pulmonary hypertension: a role for pulmonary veins and systemic vasculature. Eur Respir J. doi:10.1183/09031936.00169113

Feng YX, Liu D, Sun ML, Jiang X, Sun N, Mao YM, Jing ZC (2014) BMPR2 germline mutation in chronic thromboembolic pulmonary hypertension. Lung 192(4):625–627. doi:10.1007/s00408-014-9580-y

Firth AL, Yuan JX (2012) Identification of functional progenitor cells in the pulmonary vasculature. Pulm Circ 2(1):84–100. doi:10.4103/2045-8932.94841

Firth AL, Yau J, White A, Chiles PG, Marsh JJ, Morris TA, Yuan JX (2009) Chronic exposure to fibrin and fibrinogen differentially regulates intracellular Ca^{2+} in human pulmonary arterial smooth muscle and endothelial cells. Am J Physiol Lung Cell Mol Physiol 296(6):L979–L986. doi:10.1152/ajplung.90412.2008

Firth AL, Yao W, Ogawa A, Madani MM, Lin GY, Yuan JX (2010) Multipotent mesenchymal progenitor cells are present in endarterectomized tissues from patients with chronic thromboembolic pulmonary hypertension. Am J Physiol Cell Physiol 298(5):C1217–C1225. doi:10.1152/ajpcell.00416.2009

Galie N, Kim NH (2006) Pulmonary microvascular disease in chronic thromboembolic pulmonary hypertension. Proc Am Thorac Soc 3(7):571–576. doi:10.1513/pats.200605-113LR

Guerin L, Couturaud F, Parent F, Revel MP, Gillaizeau F, Planquette B, Pontal D, Guegan M, Simonneau G, Meyer G, Sanchez O (2014) Prevalence of chronic thromboembolic pulmonary hypertension after acute pulmonary embolism. Prevalence of CTEPH after pulmonary embolism. Thromb Haemost 112(3):598–605. doi:10.1160/TH13-07-0538

Hansmann G, Fernandez-Gonzalez A, Aslam M, Vitali SH, Martin T, Mitsialis SA, Kourembanas S (2012) Mesenchymal stem cell-mediated reversal of bronchopulmonary dysplasia and associated pulmonary hypertension. Pulm Circ 2(2):170–181. doi:10.4103/2045-8932.97603

Hill NS, Preston IR, Roberts KE (2008) Inoperable chronic thromboembolic pulmonary hypertension: treatable with medical therapy. Chest 134(2):221–223. doi:10.1378/chest. 08-0482

Hoeper MM, Madani MM, Nakanishi N, Meyer B, Cebotari S, Rubin LJ (2014) Chronic thromboembolic pulmonary hypertension. Lancet Respir Med 2(7):573–582. doi:10.1016/S2213-2600(14)70089-X

Hsu HH, Chen JS, Chen YS, Ko WJ, Kuo SW, Lee YC (2007) Short-term intravenous Iloprost for treatment of reperfusion lung oedema after pulmonary thromboendarterectomy. Thorax 62(5):459–461. doi:10.1136/thx.2005.051722

Humbert M, Deng Z, Simonneau G, Barst RJ, Sitbon O, Wolf M, Cuervo N, Moore KJ, Hodge SE, Knowles JA, Morse JH (2002) BMPR2 germline mutations in pulmonary hypertension associated with fenfluramine derivatives. Eur Respir J 20(3):518–523

Jamieson SW, Kapelanski DP, Sakakibara N, Manecke GR, Thistlethwaite PA, Kerr KM, Channick RN, Fedullo PF, Auger WR (2003) Pulmonary endarterectomy: experience and lessons learned in 1,500 cases. Ann Thorac Surg 76(5):1457–1462; discussion 1462–1454

Jujo T, Sakao S, Kantake M, Maruoka M, Tanabe N, Kasahara Y, Kurosu K, Masuda M, Harigaya K, Tatsumi K (2012) Characterization of sarcoma-like cells derived from endarterectomized tissues from patients with CTEPH and establishment of a mouse model of pulmonary artery intimal sarcoma. Int J Oncol 41(2):701–711. doi:10.3892/ijo.2012.1493

Kanki-Horimoto S, Horimoto H, Mieno S, Kishida K, Watanabe F, Furuya E, Katsumata T (2006) Implantation of mesenchymal stem cells overexpressing endothelial nitric oxide synthase improves right ventricular impairments caused by pulmonary hypertension. Circulation 114(1 Suppl):I181–I185. doi:10.1161/CIRCULATIONAHA.105.001487

Khakoo AY, Finkel T (2005) Endothelial progenitor cells. Annu Rev Med 56:79–101. doi:10.1146/annurev.med.56.090203.104149

Laczika K, Lang IM, Quehenberger P, Mannhalter C, Muhm M, Klepetko W, Kyrle PA (2002) Unilateral chronic thromboembolic pulmonary disease associated with combined inherited thrombophilia. Chest 121(1):286–289

Lane KB, Machado RD, Pauciulo MW, Thomson JR, Phillips JA 3rd, Loyd JE, Nichols WC, Trembath RC (2000) Heterozygous germline mutations in BMPR2, encoding a TGF-beta receptor, cause familial primary pulmonary hypertension. Nat Genet 26(1):81–84. doi:10.1038/79226

Lee KC, Cho YL, Lee SY (2001) Reperfusion pulmonary edema after pulmonary endarterectomy. Acta Anaesthesiol Sin 39(2):97–101

Lee C, Mitsialis SA, Aslam M, Vitali SH, Vergadi E, Konstantinou G, Sdrimas K, Fernandez-Gonzalez A, Kourembanas S (2012) Exosomes mediate the cytoprotective action of mesenchymal stromal cells on hypoxia-induced pulmonary hypertension. Circulation 126(22):2601–2611. doi:10.1161/CIRCULATIONAHA.112.114173

Li JF, Lin Y, Yang YH, Gan HL, Liang Y, Liu J, Yang SQ, Zhang WJ, Cui N, Zhao L, Zhai ZG, Wang J, Wang C (2013) Fibrinogen alpha Thr312Ala polymorphism specifically contributes to chronic thromboembolic pulmonary hypertension by increasing fibrin resistance. PLoS One 8(7):e69635. doi:10.1371/journal.pone.0069635

Liang OD, Mitsialis SA, Chang MS, Vergadi E, Lee C, Aslam M, Fernandez-Gonzalez A, Liu X, Baveja R, Kourembanas S (2011) Mesenchymal stromal cells expressing heme oxygenase-1 reverse pulmonary hypertension. Stem Cells 29(1):99–107. doi:10.1002/stem.548

Marsh JJ, Chiles PG, Liang NC, Morris TA (2013) Chronic thromboembolic pulmonary hypertension-associated dysfibrinogenemias exhibit disorganized fibrin structure. Thromb Res 132(6):729–734. doi:10.1016/j.thromres.2013.09.024

Maruoka M, Sakao S, Kantake M, Tanabe N, Kasahara Y, Kurosu K, Takiguchi Y, Masuda M, Yoshino I, Voelkel NF, Tatsumi K (2012) Characterization of myofibroblasts in chronic thromboembolic pulmonary hypertension. Int J Cardiol 159(2):119–127. doi:10.1016/j.ijcard.2011.02.037

Morris TA, Marsh JJ, Chiles PG, Auger WR, Fedullo PF, Woods VL Jr (2006) Fibrin derived from patients with chronic thromboembolic pulmonary hypertension is resistant to lysis. Am J Respir Crit Care Med 173(11):1270–1275. doi:10.1164/rccm.200506-916OC

Morris TA, Marsh JJ, Chiles PG, Kim NH, Noskovack KJ, Magana MM, Gruppo RA, Woods VL Jr (2007) Abnormally sialylated fibrinogen gamma-chains in a patient with chronic thromboembolic pulmonary hypertension. Thromb Res 119(2):257–259. doi:10.1016/j.thromres.2006.02.010

Morris TA, Marsh JJ, Chiles PG, Magana MM, Liang NC, Soler X, Desantis DJ, Ngo D, Woods VL Jr (2009) High prevalence of dysfibrinogenemia among patients with chronic thromboembolic pulmonary hypertension. Blood 114(9):1929–1936. doi:10.1182/blood-2009-03-208264

Moser B, Megerle A, Bekos C, Janik S, Szerafin T, Birner P, Schiefer AI, Mildner M, Lang I, Skoro-Sajer N, Sadushi-Kolici R, Taghavi S, Klepetko W, Ankersmit HJ (2014) Local and systemic RAGE axis changes in pulmonary hypertension: CTEPH and iPAH. PLoS One 9(9):e106440. doi:10.1371/journal.pone.0106440

Ogawa A, Firth AL, Yao W, Madani MM, Kerr KM, Auger WR, Jamieson SW, Thistlethwaite PA, Yuan JX (2009) Inhibition of mTOR attenuates store-operated Ca^{2+} entry in cells from endarterectomized tissues of patients with chronic thromboembolic pulmonary hypertension. Am J Physiol Lung Cell Mol Physiol 297(4):L666–L676. doi:10.1152/ajplung.90548.2008

Pengo V, Lensing AW, Prins MH, Marchiori A, Davidson BL, Tiozzo F, Albanese P, Biasiolo A, Pegoraro C, Iliceto S, Prandoni P (2004) Incidence of chronic thromboembolic pulmonary hypertension after pulmonary embolism. N Engl J Med 350(22):2257–2264. doi:10.1056/NEJMoa032274

Poli D, Grifoni E, Antonucci E, Arcangeli C, Prisco D, Abbate R, Miniati M (2010) Incidence of recurrent venous thromboembolism and of chronic thromboembolic pulmonary hypertension in patients after a first episode of pulmonary embolism. J Thromb Thrombolysis 30(3):294–299. doi:10.1007/s11239-010-0452-x

Sakao S, Hao H, Tanabe N, Kasahara Y, Kurosu K, Tatsumi K (2011) Endothelial-like cells in chronic thromboembolic pulmonary hypertension: crosstalk with myofibroblast-like cells. Respir Res 12:109. doi:10.1186/1465-9921-12-109

Seyfarth HJ, Halank M, Wilkens H, Schafers HJ, Ewert R, Riedel M, Schuster E, Pankau H, Hammerschmidt S, Wirtz H (2010) Standard PAH therapy improves long term survival in CTEPH patients. Clin Res Cardiol 99(9):553–556. doi:10.1007/s00392-010-0156-4

Smadja DM, Gaussem P, Mauge L, Israel-Biet D, Dignat-George F, Peyrard S, Agnoletti G, Vouhe PR, Bonnet D, Levy M (2009) Circulating endothelial cells: a new candidate biomarker of irreversible pulmonary hypertension secondary to congenital heart disease. Circulation 119(3):374–381. doi:10.1161/CIRCULATIONAHA.108.808246

Smadja DM, Mauge L, Sanchez O, Silvestre JS, Guerin C, Godier A, Henno P, Gaussem P, Israel-Biet D (2010) Distinct patterns of circulating endothelial cells in pulmonary hypertension. Eur Respir J 36(6):1284–1293. doi:10.1183/09031936.00130809

Tanabe N, Sugiura T, Tatsumi K (2013) Recent progress in the diagnosis and management of chronic thromboembolic pulmonary hypertension. Respir Invest 51(3):134–146. doi:10.1016/j.resinv.2013.02.005

Urbich C, Dimmeler S (2004) Endothelial progenitor cells: characterization and role in vascular biology. Circ Res 95(4):343–353. doi:10.1161/01.RES.0000137877.89448.78

Wang L, Guo LJ, Liu J, Wang W, Yuan JX, Zhao L, Wang J, Wang C (2013) MicroRNA expression profile of pulmonary artery smooth muscle cells and the effect of let-7d in chronic thromboembolic pulmonary hypertension. Pulm Circ 3(3):654–664. doi:10.1086/674310

Wynants M, Vengethasamy L, Ronisz A, Meyns B, Delcroix M, Quarck R (2013) NF-kappaB pathway is involved in CRP-induced effects on pulmonary arterial endothelial cells in chronic thromboembolic pulmonary hypertension. Am J Physiol Lung Cell Mol Physiol 305(12): L934–L942. doi:10.1152/ajplung.00034.2013

Yao W, Firth AL, Sacks RS, Ogawa A, Auger WR, Fedullo PF, Madani MM, Lin GY, Sakakibara N, Thistlethwaite PA, Jamieson SW, Rubin LJ, Yuan JX (2009) Identification of putative endothelial progenitor cells (CD34+CD133+Flk-1+) in endarterectomized tissue of patients with chronic thromboembolic pulmonary hypertension. Am J Physiol Lung Cell Mol Physiol 296(6):L870–L878. doi:10.1152/ajplung.90413.2008

Zabini D, Heinemann A, Foris V, Nagaraj C, Nierlich P, Balint Z, Kwapiszewska G, Lang IM, Klepetko W, Olschewski H, Olschewski A (2014) Comprehensive analysis of inflammatory markers in chronic thromboembolic pulmonary hypertension patients. Eur Respir J 44(4): 951–962. doi:10.1183/09031936.00145013

Chapter 16
Pulmonary Arterial Hypertension: A Stem Cell Hypothesis

Quentin Felty, Seiijiro Sakao, and Norbert F. Voelkel

Abbreviations

ADAMTS1	A disintegrin and metalloproteinase with thrombospondin motifs 1
BMPR2	Bone morphogenic protein receptor 2
CD133	Cluster of differentiation 133 (aka prominin 1)
CD146	Cluster of differentiation 146 (aka melanoma cell adhesion molecule (MCAM))
EnMT	Endothelial mesenchymal transition
HIF	Hypoxia-inducible factor
HOXC6/8	Homeobox C6/8
IPAH	Idiopathic pulmonary hypertension
IPSC	Induced pluripotent stem cells
Muse	Multilineage differentiating stress-enduring cells
NG2	Neural/glial antigen 2
PAH	Pulmonary arterial hypertension
PDGFR	Platelet derived growth factor receptor
Sca-1	Ly-6 A/E
VEGF	Vascular endothelial growth factor

Q. Felty
Department of Environmental and Occupational Health, Florida International University, Miami, FL, USA

S. Sakao
Department of Respirology, Chiba University, Chiba, Japan

N.F. Voelkel (✉)
Department of Biochemistry and Molecular Biology, Virginia Commonwealth University, Richmond, VA, USA
e-mail: nfvoelkel@gmail.com

A. Firth, J.X.-J. Yuan (eds.), *Lung Stem Cells in the Epithelium and Vasculature*, Stem Cell Biology and Regenerative Medicine, DOI 10.1007/978-3-319-16232-4_16

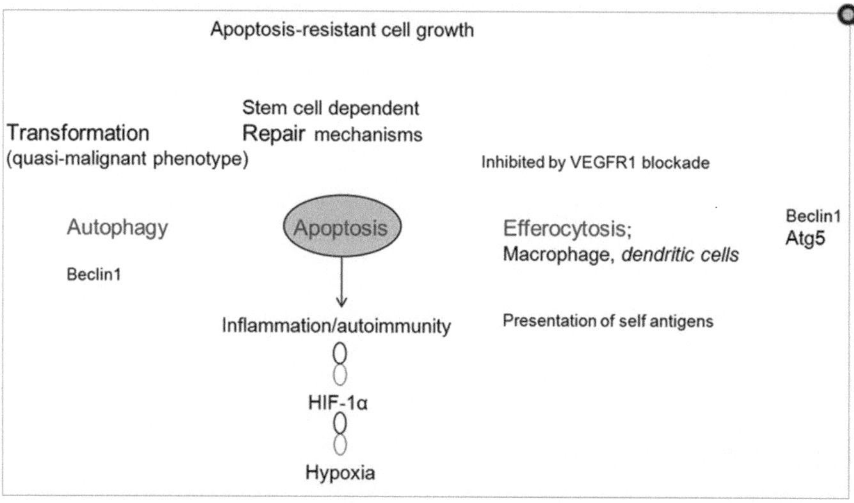

Fig. 16.1 This schematic places pulmonary vascular cell apoptosis, in particular endothelial cell apoptosis, in the center of the lung vascular remodeling process which results in exuberant wound healing. Autophagy is portrayed as a mechanism to protect against apoptosis. Efferocytosis, the removal of apoptotic bodies after engulfment by professional phagocytes or neighboring cells, if inhibited or ineffective, prolongs inflammation and sets the stage for autoimmunity. Chronic hypoxia activates via HIF 1 alpha inflammatory mechanisms, and via VEGF, mobilizes precursor cells from the bone marrow. Within the injured vascular wall, vascular niche precursor cells divide and give rise to cells which obliterate the lumen [With permission from Voelkel (2013)]

16.1 Introduction

All forms of severe and chronic pulmonary arterial hypertension (PAH) are associated with structural alterations of the lung vessels. These structural alterations have been histomorphologically described decades ago (Brenner 1935; Heath and Edwards 1958; Wagenvoort and Wagenvoort 1977; Cool et al. 1999). Generally, all three layers of the vessel wall are affected by the structural changes and they are accompanied by varying degrees of perivascular infiltrates of inflammatory and immune cells (Tuder et al. 1994; Tu et al. 2011; Stacher et al. 2012). The pulmonary vascular resistance rises as a consequence of chronic pulmonary vasoconstriction and/or lumen obliteration caused by proliferating cells, in situ thrombotic events, or vessel loss. Taken together, we propose that most of the structural pulmonary vascular obliterations can be interpreted as a consequence of a wound healing response which follows an injury to the cells of the vessel wall (Voelkel et al. 2012). Both chronic hemodynamic stress and cytotoxic factors can induce endothelial cell apoptosis and loss of pulmonary micro vessels. Predominantly, the increase in the resistance to lung blood flow is due to a combination of lung vessel loss and lung vessel lumen obliteration. In the schematic of Fig. 16.1, apoptosis of endothelial cells takes on a central role; this schematic also illustrates the hypothesis that the failure of clearing apoptotic cells leads to a shift towards altered immunity and autoimmune responses (Jurasz et al. 2010; Sgonc et al. 1996; Erwig and Henson 2007; Nagata et al. 2010; Freire-de-Lima et al. 2006). Thus, as shown in Fig. 16.1, we attempt to build a modern concept of pulmonary vascular remodeling in severe PAH which integrates inflammation/immune response, wound healing, and angiogenesis (Voelkel et al. 2012; Savai et al. 2012) subsequent to vascular cell apoptosis.

Because of the reported phenotype switch of exuberantly proliferating pulmonary vascular endothelial cells in severe PAH (Cool et al. 1999; Tuder et al. 1994; Tu et al. 2011), we had hypothetically applied the concept of the cancer paradigm as outlined by Hanahan and Weinberg (2011) in order to explain the roles of "misguided angiogenesis" and the evolution of quasi-malignant, that is, apoptosis-resistant, proliferating endothelial cell phenotypes in the pathobiology of severe PAH (Voelkel et al. 2012; Rai et al. 2008).

In the following chapter, we will examine a role for stem cells and precursor cells (Majka et al. 2005; Chow et al. 2013; Spees et al. 2008) in the development of exuberant pulmonary vascular endothelial cell growth (Tuder et al. 1994) borrowing from models of wound healing and from cancer stem cell concepts. We will focus on stem cell differentiation and its potentially critical role in the development and maintenance of severe angio-obliterative PAH and will advance the argument for a stem cell-directed therapy of PAH.

16.2 Evidence for a Participation of Stem and Precursor Cells in Severe PAH

Remarkably, several recent reviews covering the topic of the pathology (Stacher et al. 2012) or pathobiology of severe PAH (Rabinovitch 2012; Tuder et al. 2013) have not anticipated or considered a mechanistic role of precursor or stem cells. This is surprising in view of the amount of clinical and experimental data which demonstrate that circulating precursor cells and vessel precursor cells have been described in various forms of PAH (Table 16.1). The majority of these studies report that such cells are present in the blood or in the lungs; however, evidence based on strict mechanistic criteria of stemness is largely lacking. Nevertheless, at this moment in time it is not premature to sketch a stem cell picture of hypertensive pulmonary vascular remodeling and highlight likely mechanistically involved bone marrow-derived and resident lung precursor and stem cells.

We may be biased but believe that the first statement which could be interpreted as a veiled reference to precursor cells in PAH can be found in a study describing the electron microscopic features of plexiform lesions (Smith and Heath 1979) as elaborated in (Tuder et al. 1994): "the lesions appeared to contain cells of an undetermined nature" and the authors go on to suggest that "these cells might represent *primitive angioformative* cells committed to differentiate into endothelial cells" (Smith and Heath 1979). The term "angioformative" was later replaced with "angiogenic," and the cells that had been described as "committed", i.e. not differentiated, are large and have plump nuclei and prominent nucleoli (Fig. 16.2).

Table 16.1 Precursor and stem cells in pulmonary hypertension

Circulating precursor cells in human disease	
Increased endothelial cells in PAH patients	Bull et al. (2003)
Increased bone marrow-derived CD34+/CD133+ cells in IPAH	Aosingh et al. (2008)
Increased CD34+/CD133+ cells in PAH patients	Smadja et al. (2010)
Increased CD34+/CD133+, CD34+/KDR+, CD34+/CD133+/KDR+ cells in PAH	Diller et al. (2008)
Lung tissue precursor cells	
c-Kit+ cells	
In plexiform lesions from IPAH patients	Rai et al. (2008), Montani et al. (2011), Toshner et al. (2009)
Mesenchymal precursor cells	
In thrombendarterectomized tissue	Firth et al. (2010a)
Endothelial cell precursor cells	
In thrombendarterectomized tissue	Yao et al. (2009)
In pulmonary arteries from PAH patients	Farha et al. (2011)
Oct-4+ cells	
In vascular smooth muscle cells from patients with IPAH	Firth et al. (2010b)
CD44+ multipotent stem cells in plexiform lesions in IPAH	Ohta-Ogo et al.(2012)
Precursor cells in animal models of PAH	
Circulating precursor cells	
Decrease in endothelial cell precursors in rat monocrotaline model	Liu et al. (2013)
Lung tissue precursor cells	
Perivascular accumulation of c-kit+ cells in hypoxic mice	Gambaryan et al.(2011), Angelini et al. (2010)
Increased c-kit expression in endothelial and vascular	Farkas et al. (2014)
Smooth muscle cells in the rat Su/Hx model	
Increased number of c-kit+ cells in hypoxic calves	Davie et al. (2004)
Mesenchymal precursors in hypoxic calves	Frid et al. (2006)

Years later, circulating precursor cells were identified in the blood from patients with severe PAH. Bull et al. were the first to describe circulating CD133+ cells and to speculate that these cells might be derived from the PAH patients' bone marrow (Bull et al. 2003). While the original goal of this particular study had been to identify precursor cells which were thought to be released from the "sick lung circulation" (Ghofrani et al. 2010) and can perhaps be used as surrogates for the abnormal, phenotypically altered lung vessel endothelial cells (Cool et al. 1999; Rai et al. 2008), at the time, only a limited number of cell membrane protein markers was known and available for precursor cell characterization.

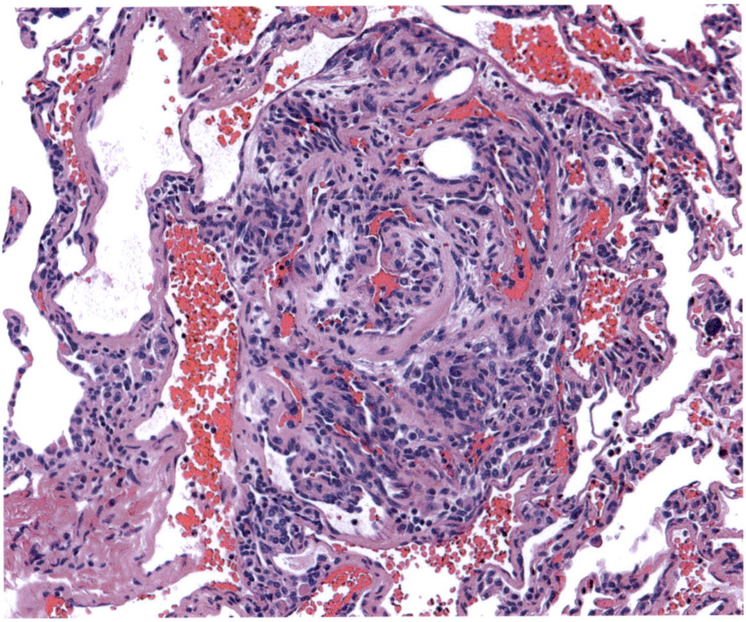

Fig. 16.2 Complex pulmonary vascular lesion in the lung from a patient with severe pulmonary arterial hypertension. The vessel lumen is obliterated by phenotypically altered cells. The cells are large, plump, and have large nuclei and nucleoli. There are many *lumina* within this lesion

16.3 The Cancer Paradigm of PAH

It is now clear and accepted that PAH is not *one* homogeneous disease, but that the term rather defines a *group* of disorders which share a severe pulmonary microangiopathy that is usually progressive and driven by proliferation of apoptosis-resistant vascular cells. This definition does not contradict the presently used WHO nomenclature, but it emphasizes "severe and angio-obliterative" disease features. This microangiopathy defines histologically the abnormal appearance and function of the precapillary arterioles in the lungs of patients with systemic sclerosis, the various forms of hereditary IPAH, HIV infection, and congenital intracardiac shunts. Several, not necessarily mechanistically related, factors and descriptors of PAH prompted us to consider the comparison with malignancies (Hanahan and Weinberg 2011).

First, like many cancers, IPAH is refractory to therapy; that is the vascular lesions do not regress with the treatment of PH by standard pulmonary hypertension drugs or after chemotherapy in the few cases where this has been attempted, as recently documented by the PAH imatinib treatment trial (Ghofrani et al. 2010).

Second, there are several reports of patients developing PAH years after radiation and chemotherapy for Hodgkin's lymphoma, and it is well-appreciated that patients suffering from myelodysplastic disorders can develop PAH (Dingli et al. 2001). However, it remains unclear whether the late occurrence of severe PAH following therapy for Hodgkin's lymphoma is related to the lymphoma or to the chemotherapy. Perhaps the recent occurrence of PAH following treatment of leukemia with dasatinib (Groeneveldt et al. 2013) may suggest a chemotherapy-related mechanism.

In PAH, the complex vascular lesions, like many tumors, are multicellular and the cells are characterized by various degrees of differentiation and proliferation and, again like tumors which contain cancer stem cells, these lesions contain stem cells (Fig. 16.3). Cancer stem cells were initially implicated in hematopoietic malignancies and later also identified in solid tumors (Hanahan and Weinberg 2011). Some of the plexiform lesions in idiopathic PAH are composed of clonally expanded endothelial cells (Lee et al. 1998), which carry mutations (Yeager et al. 2001). In tumors, normal tissue stem cells may undergo oncogenic transformation or progenitor cells may undergo this transformation (Hanahan and Weinberg 2011). Signals that trigger endothelial cell mesenchymal transformation (EnMT) can be released by inflammatory stroma and may also be implicated in stem cell formation and maintenance (Hanahan and Weinberg 2011). Hanahan and Weinberg in their 2011 review of "The Hallmarks of Cancer" describe solid tumors as an assemblage of cells (Hanahan and Weinberg 2011), which contain stem cells, fibroblasts, endothelial cells, immune/inflammatory cells, and pericytes. All of these cell types, prominently including apoptosis-resistant cells, are also found in the vascular lesions of patients with severe PAH. However, one important difference between malignant tumors and the pulmonary vascular lesions in severe PAH is the non-invasiveness of the pulmonary vascular lesions; the lesions do not break through the vessel wall and do not invade alveolar spaces.

16.4 Vascular Stem Cell Niches

Due to the fact that in patients with angioproliferative PAH, there is evidence for circulating, bone marrow-derived precursor cells (see above), and because precursors are found in the vessel wall and in the perivascular spaces, we assume for the purpose of this review that both bone marrow-derived and resident vascular precursor cells participate in the process of pulmonary vascular remodeling (Voelkel et al. 2012; Rai et al. 2008).

There is an increasing number of reports concerned with vascular biology that describe several different precursor and stem cells residing in the vessel wall (Ergun et al. 2011; Pasquinelli et al. 2007; Ingram et al. 2005; Tigges et al. 2013; Majesky et al. 2012; Klein et al. 2011; Zengin et al. 2006). As pointed out above, in human PAH investigations the presence of precursor and stem-like cells has been documented by several groups; however, short of lineage-tracing studies, which cannot be performed in patients, it will be difficult to prove that these cells actually *functio*n

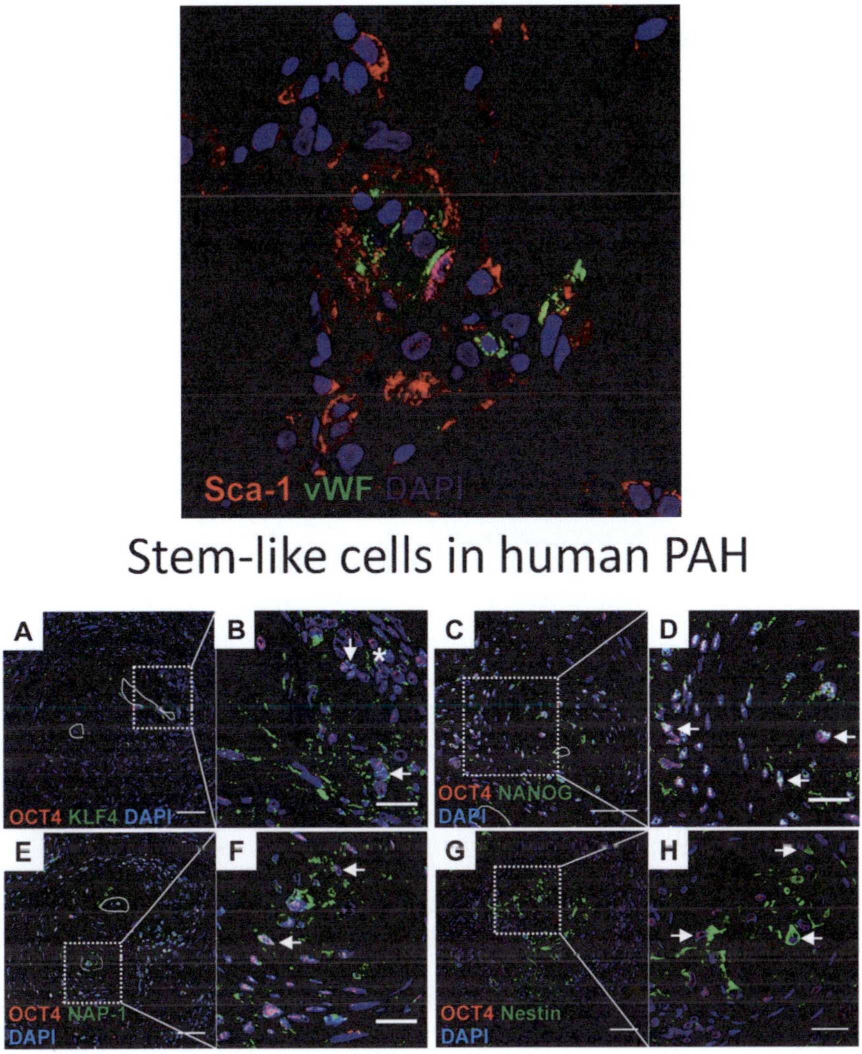

Fig. 16.3 Immunohistochemistry of vascular lesions in a rat lung which develop after a single subcutaneous injection of the VEGF receptor blocker Sugen 5416 when combined with 3 weeks of chronic hypoxia. There are multiple Sca+ cells (**a**) and cells staining positive for the stem cell markers Oct4, Nanog, KLF4, and the marker protein NAP 1 (nucleosome assembly protein 1, a histone chaperone) (**b**). Immunohistochemistry of a human plexiform lesion staining positive for the markers Oct4, Nanog, Nestin, and KLF4 (**c**)

as stem cells, or to demonstrate differentiation or de-differentiation. We are not aware of lineage-tracing studies in any animal model of PAH conducted to demonstrate which of the different precursor cells divide, give rise to apoptosis-resistant cells, and participate in the process of pulmonary vascular remodeling. However, it is likely that in PAH the *adventitial stem cell niche* is a reservoir for precursor cells,

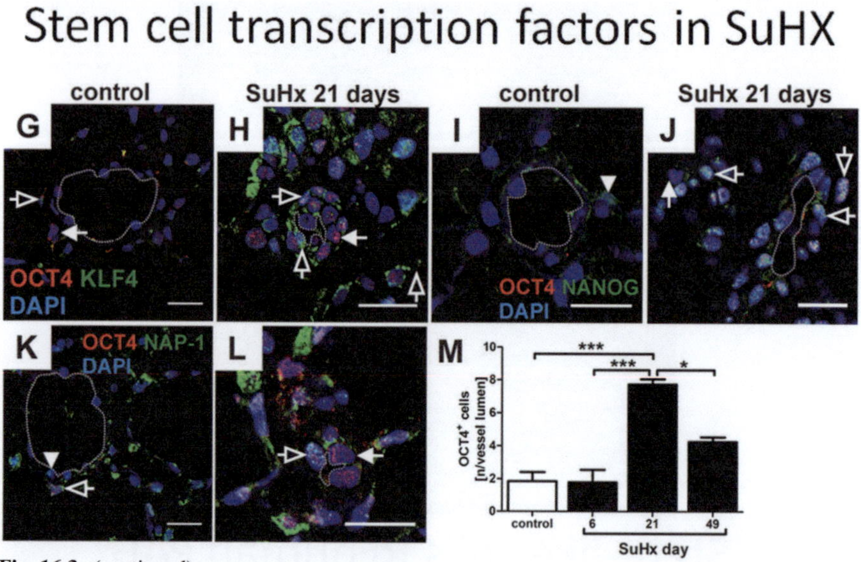

Fig. 16.3 (continued)

as has already been shown in systemic vessels. Rare chondrocytes and adipocytes have been found in the complex vascular lesions from patients with IPAH (Farkas et al. 2013), suggesting the presence of pluri- or multi-potent precursors in these lesions. Of particular interest are pericytes in the adventitia of the remodeled lung vessels (Ergun et al. 2011; Klein et al. 2011).

Klein et al. (2011) showed that CD44+ multipotent stem cells, which are present in plexiform lesions (Ohta-Ogo et al. 2012), can differentiate into pericytes and smooth muscle cells, and that Hox genes are involved in the differentiation into smooth muscle cells. Interestingly, Golpon et al. had shown an altered expression pattern of Hox genes in lungs from idiopathic PAH patients (Golpon et al. 2001); strikingly, the differentially expressed HoxC6 and HoxC8 genes in IPAH lungs are also differentially expressed in the multipotent vessel wall CD44+ stem cells (Klein et al. 2011) and also in cancers (Moon et al. 2012; Du et al. 2014).

There are different subsets of pericytes which can be morphologically distinguished. Lung vessel pericytes can transdifferentiate into contractile smooth muscle cells under chronic hypoxic conditions (Meyrick et al. 1981). Pericytes express NG2, PDGF receptor beta, desmin, and CD146; they can activate ADAMTS1 which cleaves the vessel basement membrane, such that pericytes can directly interact with endothelial cells. Pericytes can express a repertoire of functional phenotypes as shown by culture studies of human pluripotent stem cells (Wanjare et al. 2014); they also synthesize and release VEGF, yet little is known about a precursor role of pericytes in angioproliferative PAH.

As has been pointed out by Ingram et al. (2005), the pulmonary small vessel endothelial cell layer is probably another vessel wall stem cell reservoir or niche, and we speculate that resident endothelial stem cells may account for the EnMT (endothelial cell mesenchymal transition (Medici and Kalluri 2012; Rokavec et al. 2014)) that contributes to hemodynamic stress-induced muscularization of pulmonary arterioles and also to the proliferation of phenotypically altered intima layer cells (Cool et al. 1999).

16.5 The Sugen 5416/Hypoxia Rat Model of Severe Angioproliferative PAH

This rat model of severe PAH was first reported in 2001 (Taraseveciene-Stewart et al. 2001). Briefly, rats are subcutaneously injected with a single dose of the VEGF receptor blocker Sugen 5416 and then exposed to chronic hypoxia for 4 weeks (Taraseveciene-Stewart et al. 2002). After this 4-week period of exposure to hypoxia, the animals are removed from the hypoxic environment and they continue to present with severe angio-obliterative PAH and right heart failure (Oka et al. 2007).

This model is of interest, because the wild-type Sprague Dawley rats used for these experiments do not carry any mutations and the developed disease is refractory to treatment of the animals with pulmonary vasodilator drugs and other drugs (Taraseveciene-Stewart et al. 2006). However, and of mechanistic importance, the disease can be prevented by the inhibition of apoptosis using a pan caspase inhibitor (Taraseveciene-Stewart et al. 2001, 2002). This model also demonstrated for the first time that the degree of pulmonary hypertension was highly correlated with the number of obliterated lung vessels (Oka et al. 2007).

Variants of this model are the combination of Sugen 5416 (VEGFR blockade) and left-sided pneumonectomy, Sugen 5416 treatment of athymic immunocompromised rats, or the immunization of rats with ovalbumin combined with VEGFR blockade. Each of these combinations produces severe and frequently fatal angio-obliterative PAH (Nicolls et al. 2012). Thus, "two hits" are required in this model because Sugen 5416 by itself generates pulmonary emphysema and only a mild degree of PAH (Kasahara et al. 2000). We have recently attempted to elucidate this apparent paradox: that an *anti*angiogenic drug in combination with a second challenge *causes* angio-obliterative PAH (Gomez-Arroyo and Voelkel 2014). As shown by Sakao et al. (2007), induction of lung endothelial cell apoptosis by (Sugen 5416-induced) VEGFR blockade is followed by the emergence of apoptosis-resistant cells, and importantly, a fraction of the initial apoptosis surviving cells are precursor or stem cells. This finding has been recently confirmed by Das et al. (2014); the Sugen 5416/hypoxia rat model of severe PAH is now being interrogated to assess the contribution of precursor/stem cells to pulmonary vascular remodeling (Farkas et al. 2013; Das et al. 2014).

Immunohistochemistry of lung tissue samples from Sugen/chronic hypoxia (Su/Hx) rats with established severe PAH demonstrates the expression of Oct-4, KLF-4, NANOG, Nestin, Ly6A (Sca-1), and c-kit proteins in vascular lesion cells and in the cells occupying the perivascular spaces (Figs. 16.3a, b). Similar aggregates can be shown in the complex pulmonary vascular lesions of patients with severe angio-obliterative PAH (Fig. 16.3c). In the aggregate, these findings suggest that cells expressing precursor and stem cell markers are present in the pulmonary vascular lesion cells. As already pointed out, the presence of cells expressing stem cell markers does not necessarily imply that these cells actually function as stem cells. However, in the context of a vascular stem cell niche and borrowing heavily from David Scadden's "Neighborhood" concept of stem cell niches (Scadden 2014), it is reasonable to consider a stem cell-centric, or stem cell-anchored hypothesis of severe PAH vascular remodeling.

16.6 The Role of the Vascular Stem Cell Niche

For most known multicellular organisms, their relatively constant outward appearance is underscored by an incessant inner transformation in which cells lost to normal wear and tear (turnover) are replaced by the progeny of dividing cells. If the normal fate of cells in adult tissues is plastic and environmental changes such as growth-promoting conditions can switch their development from a program of terminal differentiation to one of active clonogenicity, we need to invest more effort in studying how stemness arises (Sanchez Alvarado and Yamanaka 2014).

In the following we will not depart from the quasi-malignant neoplastic concept of the pathobiology of severe angio-obliterative PAH, but we will redefine this concept by introducing the complementary hypothesis that progenitor/stem-like cells, which are present in the normal vascular wall (Ergun et al. 2011; Pasquinelli et al. 2007; Ingram et al. 2005; Tigges et al. 2013; Majesky et al. 2012; Klein et al. 2011; Zengin et al. 2006), proliferate and transdifferentiate, as the vessel wall responds to vascular injury. The concept of a homeostatic lung vessel structure maintenance program (Taraseveciene-Stewart and Voelkel 2008) accommodates the postulate that in severe PAH, as part of the abnormal wound healing process, the first law of vascular biology, "the law of the endothelial cell monolayer," is being broken (Rai et al. 2008). Thus, if we think about the pathobiology of severe PAH as driven by a wound-healing process (cancer is the wound that never heals) carried out in a cancer-like stem cell niche, we arrive at a concept of innate cell plasticity of the pulmonary vessel wall. For short, in severe PAH, some vessel wall cells, when exposed to severe stress, change their phenotype, and the stress may induce stemness in *normally present* multipotent progenitors (Sanchez Alvarado and Yamanaka 2014).

Such a conceptual model of the pathobiology of severe angioproliferative PAH would explain why endothelial cells can transdifferentiate into smooth muscle cells (Sakao et al. 2007; Frid et al. 2002) (see above), why pericytes may influence the growth of endothelial cells, and why cobblestone (flat) endothelial monolayer cells change their phenotype and turn into large cells that cluster and proliferate (Cool et al. 1999; Tuder et al. 1994; Rai et al. 2008). Such a disease model would also point out the limits of single cell type culture studies and focus our attention on the investigation of multi-cellular niches, where neighboring cells exchange information with the goal of maintaining structural and functional homeostasis. The biological principle of homeostasis and the response to the stress- or injury-induced evolutionary selection pressure, apparently require cellular plasticity.

Apparently, there are hierarchical signaling pathways and topographical blueprints which organize the ongoing conversation between dying and growing cells in all three layers of the vascular wall (Sanchez Alvarado and Yamanaka 2014). It is somewhat intuitive, and this concept suggests how difficult it will be to "reprogram" the altered vessel wall and perivascular cells in the setting of angio-obliterative PAH, to de-remodel by differentiation, in particular when the hemodynamic stress (high pressure or high flow) persists. While in the rodent models of chronic hypoxia-induced PAH there occurs generally successful de-remodeling after several weeks after the chronic hypoxic stress has been withdrawn, this is not the case in the Sugen/hypoxia rat model (see above). In the Sugen/hypoxia model of severe angio-obliterative PAH simvastatin treatment caused a partial reduction of the PAH (Taraseveciene-Stewart et al. 2006), and treatment of rats with established angio-obliterative PAH with a copper chelator reopened most of the obliterated vessels without affecting the muscularization of the pulmonary arterioles. Unfortunately, this Cu^{2+}-chelator-induced de-remodeling is transient as the vessel obliteration returns once the copper chelator treatment has been discontinued (Bogaard et al. 2012). Treatment of the rats with the copper chelator not only reopened vessels and reduced the pulmonary hypertension, but also the number of c-kit+cells in and around the diseased arterioles. If we postulate that progenitor/stem cells drive angioproliferative PAH in the Sugen/hypoxia model, the antiangiogenic copper chelator may have caused pulmonary arteriolar remodeling by influencing the behavior of stem cells, or their crosstalk with the injured vessel wall cells.

There are presently only a few publications which support the concept that copper chelation affects stem cell behavior (Zaker et al. 2013; Jain et al. 2013; Suh et al. 2014). Interestingly, copper chelation does reduce the number of circulating endothelial progenitor cells in women with breast cancer (Jain et al. 2013) and copper chelation-affected neuronal differentiation in induced pluripotent stem cells (iPSC) (Suh et al. 2014).

16.7 The Role of Hypoxia

Chronic hypoxia in rodents causes PAH typically associated with muscularization of the pulmonary arterioles, and this form of vascular remodeling is generally reversible (de Raaf et al. 2014). The putative role of stem cells in this mechanism of hypoxia-induced remodeling is being investigated (see Table 16.1). Because Sugen 5416 alone does not cause significant PAH, we wonder whether chronic hypoxia awakens multilineage differentiating stress enduring (Muse) cells; such mesenchymal Muse cells express Oct-4, Nanog, Sox2, and also CXCL2, a chemokine receptor involved in cell homing (Heneidi et al. 2013).

A growing body of data indicates that most human cancers are hypoxic and that hypoxia-inducible factors (HIFs) aid in the acquisition of more malignant phenotypes of cancer-initiating cells (Mimeault and Batra 2013). For example, it has been suggested that the hypoxic tumor microenvironment maintains glioblastoma stem cells and promotes reprogramming towards a cancer stem cell phenotype (Heddleston et al. 2009). During severe chronic hypoxia, micro-regions of the lung may become severely oxygen-deprived and ischemia-reperfusion may occur (Voelkel et al. 2013); however, exuberant endothelial cell proliferation (Tuder et al. 1994) may not occur without substantial damage to the pulmonary vascular endothelium, as indeed is the case in the Sugen/hypoxia rat model (Taraseveciene-Stewart et al. 2001; de Raaf et al. 2014). As hypoxia, via HIF-1alpha and VEGF, mobilizes bone marrow cells, it is likely that bone marrow-derived precursors home to the lung and participate in pulmonary vascular remodeling; perhaps these cells arrive in the vascular wall via the vasa vasorum (Nijmeh et al. 2014). The process of hypoxia-induced pulmonary arteriolar media hypertrophy may be initiated by EnMT and thus the endothelial cells may be the first responders, quickly providing myofibroblasts, and in doing so, providing the early defense against hemodynamic (shear) stress. Possibly, this EnMT-driven "adequate" adjustment of the resistance vessels to increased shear stress also protects against angio-obliteration (unpublished own data).

16.8 The Role of the Immune System

As stated, the vascular lesions in severe PAH are multicellular and complex and the presence of inflammatory cells in and around these lesions has been appreciated for many years (Wagenvoort and Wagenvoort 1977; Tuder et al. 1994; Hassoun et al. 2009). Here it is important for us to examine whether immune dysregulation can influence progenitor and stem cell behavior during the development of PAH.

In mice, immunization with an aspergillus antigen or ovalbumin followed by repeated antigen challenges causes severe pulmonary arterial muscularization (Daley et al. 2008), while a single injection of the VEGF receptor blocker Sugen 5416 in athymic rats (which are depleted of T lymphocytes, including regulatory T cells) causes severe angio-obliterative PAH (Tamosiuniene et al. 2011; Tian et al. 2013). Thus, both activation of the innate immune system as well as immunoinsufficiency (lack of Treg cells) affect lung vessel remodeling . The role of precursor/stem cells, if any, in the aspergillus and ovalbumin-induced models of pulmonary muscularization still needs to be established. In the VEGF receptor blockade rat models, it is likely that inhibition of VEGF signaling inhibits endothelial cell differentiation (Flamme et al. 1995). The transcription factor Nanog, the safeguard of stem cell pluripotency, promotes in colon cancer a stem-like phenotype and immune evasion (Noh et al. 2012), suggesting that there are immune mechanisms that play a role in stem cell differentiation. Indeed, the immunosuppressant tacrolimus has been shown to influence mesenchymal stem cell behavior in a BMP/Smad-dependent manner (Kugimiya et al. 2005). A recent study demonstrated that tacrolimus activates BMPR2 expression and reverses severe angioproliferative PAH in rat models (Spiekerkoetter et al. 2013). Whether tacrolimus affected pulmonary vascular stem cell behavior (differentiation) is not known.

16.9 Drug Treatment and Effects on Progenitor Cells in Pulmonary Hypertension

We have mentioned the treatment effect of copper chelation on the number of c-kit+cells in the vascular lesions in the Sugen/hypoxia model (Bogaard et al. 2012). Smadja et al. (2011) demonstrated that the prostacyclin analog treprostenil increased the number and angiogenic potential of endothelial progenitor cells in children with pulmonary hypertension and Shenoy et al. (2013) showed that the antitrypanosomal drug, diminazene aceturate, repaired the diminished migratory capacity of angiogenic progenitor cells isolated from patients with PH. Thus, the investigation of drugs used for the treatment of PH in regard to their "off target effects" on progenitor, and stem cells, is in its infancy. So far experimental strategies have included treatment of PAH with mesenchymal stem cells (Liang et al. 2011), transplantation of endothelial progenitor cells in dogs (Takahashi et al. 2004), and of bone marrow-derived endothelial-like progenitor cells in rats (Zhao et al. 2005). Apparently the underlying concept for these strategies is stem cell deficiency or a dysfunctional set of stem cells (Toshner et al. 2009), resulting in a diminished capacity to repair the pulmonary vascular injury. By supplying normally functioning precursor cells, this deficiency can be repaired (Liang et al. 2011; Takahashi et al. 2004; Zhao et al. 2005). In contrast, our hypothesis is that the exuberant wound healing of the injured small lung vessels (Tuder et al. 1994) is initiated and perpetuated by endothelial and pericyte progenitor cells which do not terminally differentiate into mature endothelial cells. It should be intuitive that replacement of dysfunctional stem cells, or the dilution of a dysfunctional precursor pool, is unlikely to accomplish the goal of pulmonary vascular homeostasis and affect the return to the normal pulmonary vascular endothelial cell monolayer if the critically important defect lies in the inability of the cells to differentiate. The investigation of the "quasi malignancy" hypothesis leads to the probing of the concept of a cancer stem cell niche and its translation to the pulmonary angioproliferative pathology. Like modern cancer treatment which strives to turn cancer into a managed chronic disease, by analogy, the treatment goal for severe PAH may become to achieve and maintain an improved level of pulmonary vascular homeostasis by coaxing apoptosis-resistant vessel wall cells towards a terminal state of differentiation. Instead of a cure of PAH by precisely targeting specific gene-dependent molecular mechanisms, such a cellular treatment strategy may turn PAH into a manageable chronic disease.

Acknowledgments Some of the experimental work on which this chapter is based was supported by funding from the Victoria Johnson Center for Lung Research at Virginia Commonwealth University. Excellent immunohistochemistry was performed by Mrs. Daniela Farkas.

References

Angelini DJ, Su Q, Kolosova IA, Skinner JT, Yamaji-Kegan K, Collector M et al (2010) Hypoxia-induced mitogenic factor (HIMF/FIZZ1/RELM alpha) recruits bone marrow-derived cells to the murine pulmonary vasculature. PLoS One 5:11251–11262

Aosingh K, Aldred MA, Vasanji A, Drazba J, Sharp J, Farver C et al (2008) Circulating angiogenic precursors in idiopathic pulmonary arterial hypertension. Am J Pathol 172:615–627

Bogaard HJ, Mizuno S, Guignabert C, Al Hussaini A, Farkas D, Ruiter G, Kraskauskas D, Fadel E, Allegood JC, Humbert M (2012) Copper dependence of angioproliferation in pulmonary hypertension in rats and humans. Am J Respir Cell Mol Biol 46:582–591

Brenner O (1935) Pathology of vessels of pulmonary circulation. Arch Intern Med (Chic) 56:211–457

Bull TM, Golpon H, Hebbel RP, Solovey A, Cool CD, Tuder RM, Gerci MW, Voelkel NF (2003) Circulating endothelial cells in pulmonary hypertension. Thromb Haemost 90:698–703

Chow K, Fessel JP, Ihida-Stansburu K, Schmidt E-P, Gaskill C, Alvarez D et al (2013) Dysfunctional resident lung mesenchymal stem cells contribute to pulmonary microvascular remodeling. Pulm Circ 3:31–49

Cool CD, Stewart JS, Werahera P, Miller GJ, Williams RL, Voelkel NF et al (1999) Three-dimensional reconstruction of pulmonary arteries in plexiform pulmonary hypertension using cell-specific markers: evidence for a dynamic and heterogeneous process of pulmonary endothelial cell growth. Am J Pathol 155(2):411–419

Daley E, Emson C, Giugnabert C, de Waal Malefyt R, Louten J, Kurup VP et al (2008) Pulmonary arterial remodeling induced by a Th2 response. J Exp Med 205:361–372

Das JK, Voelkel NF, Felty Q (2014) Overexpression of ID3 promotes a stem-like molecular signature in human vascular endothelial cells. It's implication in the development of hyper-proliferative endothelial lesions associated with pulmonary arterial hypertension. Microvasc Research 105(2):203–212

Davie NJ, Crossno JT, Frid MG, Hofmeister SE, Reeves JT, Hyde DM et al (2004) Hypoxia-induced pulmonary artery adventitial remodeling and neoveascularization: contribution of progenitor cells. Am J Physiol Lung Cell Mol Physiol 286:L668–L678

de Raaf MA, Schalij I, Gomez-Arroyo JG, Rol N, Happe C, de Man FS et al (2014) SuHx rat model: partly reversible pulmonary hypertension and progressive intima obstruction. Eur Respir J 44(1):160–168

Diller G-P, van Eiji S, Okonko DO, Howard L, Ali O, Thum T, Wort SJ et al (2008) Circulating endothelial progenitor cells in patients with Eisenmenger syndrome and idiopathic pulmonary hypertension. Circulation 117:3020–3030

Dingli D, Utz JP, Krowka MJ, Oberg AL, Tefferi A (2001) Unexplained pulmonary hypertension in chronic myelodysplastic disorders. Chest 120:801–808

Du YB, Dong B, Shen LY, Yan WP et al (2014) The survival predictive significance of HoxC6 and HoxC8 in esophageal squamous cell carcinoma. J Surg Res 188(2):442–450 [Epub 2014 Jan 17]

Ergun S, Tilki D, Klein D (2011) Vascular wall as a reservoir for different types of stem and progenitor cells. Antioxid Redox Signal 15:981–995

Erwig LP, Henson PM (2007) Immunological consequences of apoptotic cell phagocytosis. Am J Pathol 171(1):2–8

Farha S, Aosingh K, Xu W, George D, Comhair S, Park M et al (2011) Hypoxia-inducible factors in human pulmonary arterial hypertension: a link to the intrinsic myeloid abnormalities. Blood 117:3485–3493

Farkas L, Farkas D, Al Hussaini A, Kraskauskas D, Voelkel NF (2013) NfKB inhibition differentially alters accumulation of progenitor and stem-like cell populations and prevents angioobliteration and severe PAH in the Su5416 hypoxia model. Am J Respir Crit Care Med 187 (Abstract)

Farkas D, Kraskauskas D, Drake JI, Alhussaini AA, Kraskauskiene V, Bogaard HJ, Cool CD, Voelkel NF, Farkas L (2014) CXCR4 inhibition ameliorates severe obliterative pulmonary hypertension and accumulation of C-kit(+) cells in rats. PLoS One 9:e89810

Firth AL, Yao W, Ogawa A, Madani MM, Lin GY, Yuan JX (2010a) Multipotent mesenchymal progenitor cells are present in endarterectomized tissues from patients with chronic thromboembolic pulmonary hypertension. Am J Physiol Cell Physiol 298:C1217–C1225

Firth AL, Yao W, Ogawa A, Yuan JX (2010b) Upregulation of Oct-4 isoforms in pulmonary artery smooth muscle cells from patients with pulmonary arterial hypertension. Am J Physiol Lung Cell Mol Physiol 298:L548–L557

Flamme I, Breier G, Risau W (1995) Vascular endothelial growth factor (VEGF) and VEGF receptor 2(flk-1) are expressed during vasculogenesis and vascular differentiation in the quail embryo. Dev Biol 169:699–712

Freire-de-Lima CG, Xiao YQ, Gardai SJ, Bratton DL, Schiemann WP, Henson PM (2006) Apoptotic cells, through transforming growth factor beta, coordinately induce anti-inflammatory and suppress pro-inflammatory eicosanoid and NO synthesis in murine macrophages. J Biol Chem 281(50):38376–38383

Frid MG, Kale VA, Stenmark KR (2002) Mature vascular endothelium can give rise to smooth muscle cells via endothelial-mesenchymal transdifferentiation: in vitro analysis. Circ Res 90:1189–1196

Frid MG, Brunetti JA, Burke DL, Carpenter TC, Davie NJ, Reeves JT, Roedersheimer MT et al (2006) Hypoxia-induced pulmonary vascular remodeling requires recruitment of circulating mesenchymal precursors of monocyte/macrophage lineage. Am J Pathol 168:659–669

Gambaryan N, Perros F, Montani D, Cohen-Kaminsky S, Mazmanian M, Renaud JF et al (2011) Targeting of c-kit + haematopoietic progenitor cells prevents hypoxic pulmonary hypertension. Eur Respir J 37:1392–1399

Ghofrani HA, Morrell NW, Hoeper MM, Olschewski H et al (2010) Imatinib in pulmonary arterial hypertension patients with inadequate response to established therapy. Am J Respir Crit Care Med 182:1171–1177

Golpon HA, Geraci MW, Moore MD, Miller HL, Miller GJ, Tuder RM, Voelkel NF (2001) Hox genes in human lung: altered expression in primary pulmonary hypertension and emphysema. Am J Pathol 158:955–966

Gomez-Arroyo JG, Voelkel NF (2014) The role of vascular endothelial growth factor in pulmonary arterial hypertension. The angiogenesis paradox. Am J Respir Cell Mol Biol 51(4): 474–484

Groeneveldt JA, Gans SJ, Bogaard HJ, Vonk-Noordegraaf A (2013) Eur Respir J 42:869–870

Hanahan D, Weinberg RA (2011) Hallmarks of cancer: the next generation. Cell 144:646–674

Hassoun PM, Mouthon L, Barbera JA, Eddahibi S, Flores SC et al (2009) Inflammation, growth factors and pulmonary vascular remodeling. J Am Coll Cardiol 54:S10–S19

Heath D, Edwards JE (1958) The pathology of hypertensive pulmonary vascular disease: a description of six grades of structural changes in the pulmonary arteries with special reference to congenital cardiac septal defects. Circulation 18:533–547

Heddleston JM, Li Z, McLendon RE et al (2009) The hypoxic microenvironment maintains glioblastoma stem cells and promotes reprogramming towards a cancer stem cell phenotype. Cell Cycle 8:3274–3284

Heneidi S, Simerman AA, Keller E, Singh P et al (2013) Awakened by cellular stress: isolation and characterization of a novel population of pluripotent stem cells derived from human adipose tissue. PLoS One 8:e64752

Ingram DA, Mead LE, Moore DB, Woodard W et al (2005) Vessel wall derived endothelial cells rapidly proliferate because they contain a complete hierarchy of endothelial progenitor cells. Blood 105:2783–2786

Jain S, Cohen J, Ward MM, Kornhauser N, Chuang E, Cigler T, Moore A et al (2013) Tetrathiomolybdate-associated copper chelation decreases circulating endothelial progenitor cells in women with breast cancer at high risk of relapse. Ann Oncol 24:1491–1498

Jurasz P, Courtman D, Babaie S, Stewart DJ (2010) Role of apoptosis in pulmonary hypertension: from experimental models to clinical trials. Pharmacol Ther 126(1):1–8

Kasahara Y, Tuder RM, Taraseveciene-Stewart L, Le Cras TD, Abman S, Hirth PK, Waltenberger J, Voelkel NF, Cool CD, Lynch DA, Flores SC, Voelkel NF (2000) Inhibition of VEGF receptors causes lung cell apoptosis and emphysema. J Clin Invest 106:1311–1319

Klein D et al (2011) Vascular wall resident CD44+ multipotent stem cells give rise to pericytes and smooth muscle cells and contribute to new vessel maturation. PLoS One 6:e20540

Kugimiya F, Yano F, Ohba S, Igawa K, Nakamura K et al (2005) Mechanism of osteogenic induction by FK506 via BMP/Smad pathways. Biochem Biophys Res Commun 338:872–879

Lee SD, Shroyer KR, Markham NE, Cool CD, Voelkel NF, Tuder RM (1998) Monoclonal endothelial cell proliferation is present in primary but not secondary pulmonary hypertension. J Clin Invest 101:927–934

Liang OD, Mitsialis SA, Chang MS, Vergadi E, Lee C, Aslam M et al (2011) Mesenchymal stromal cells expressing heme oxygenase 1 reverse pulmonary hypertension. Stem Cells 29:99–107

Liu JF, Du ZD, Chen Z, Han ZC, He ZX (2013) Granulocyte colony-stimulating factor attenuates monocrotaline-induced pulmonary hypertension by upregulating endothelial progenitor cells via the nitric oxide system. Exp Ther Med 6:1402–1408

Majesky MW, Dong XR, Hoglund V, Daum G, Mahoney WM (2012) The adventitia: a progenitor cell niche for the vessel wall. Cells Tissues Organs 195:73–81

Majka SM, Beutz MA, Hagen M, Izzo AA, Voelkel NF, Helm KM (2005) Identification of novel resident pulmonary stem cells: form and function of side population. Stem Cells 23: 1073–1081

Medici D, Kalluri R (2012) Endothelial-mesenchymal transition and its contribution to the emergence of stem cell phenotype. Semin Cancer Biol 22:379–384

Meyrick B, Fujiwara K, Reid L (1981) Smooth muscle myosin in precursor and mature smooth muscle cells in normal pulmonary arteries and the effect of hypoxia. Exp Lung Res 2:303–313

Mimeault M, Batra SK (2013) Hypoxia-inducing factors as master regulators of stemness properties and altered metabolism of cancer- and metastasis-initiating cells. J Cell Mol Med 17:30–54

Montani D, Perros F, Gambaryan N, Girerd B, Dorfmueller P, Price LC, Huertas A et al (2011) C-kit positive cells accumulate in remodeled vessels of idiopathic pulmonary arterial hypertension. Am J Respir Crit Care Med 183:116–123

Moon SM, Kim SA, Yoon JH, Ahn SG (2012) HoxC6 is deregulated in human head and neck squamous cell carcinoma and modulates Bcl2 expression. J Biol Chem 287:35678–35688

Nagata S, Hanayama R, Kawane K (2010) Autoimmunity and the clearance of dead cells. Cell 140:619–629

Nicolls MR, Mizuno S, Taraseveciene-Stewart L, Farkas L, Drake JI, Al Husseini A, Gomez-Arroyo JG, Voelkel NF, Boggard HJ (2012) New models of pulmonary hypertension based on VEGF receptor blockade-induced endothelial cell apoptosis. Pulm Circ 2:432–442

Nijmeh H, Balasubramanian V, Burns N, Ahmad A, Stenmark KR, Gerasimovaskaya EV (2014) High proliferative potential endothelial colony forming cells contribute to hypoxia-induced pulmonary artery vasa vasorum neovascularization. Am J Physiol Lung Cell Mol Physiol 306(7):L661–L671 [Epub 2014 Feb 7]

Noh K-H, Kim BW, Song K-H, Cho H, Lee Y-H et al (2012) Nanog signaling in cancer promotes stem-like phenotype and immune evasion. J Clin Invest 122:4077–4093

Ohta-Ogo K, Hao H, Ishibashi-Ueda H, Hirota S et al (2012) CD44 expression in plexiform lesions of idiopathic pulmonary arterial hypertension. Pathol Int 62:219–225

Oka M, Homma N, Taraseveciene-Stewart L, Morris KG, Kraskauskas D, Burns N, Voelkel NF, McMurtry IF (2007) Rho kinase-mediated vasoconstriction is important in severe occlusive pulmonary arterial hypertension in rats. Circ Res 100:923–929

Pasquinelli G et al (2007) Thoracic aortas from multiorgan donors are suitable for obtaining resident angiogenic mesenchymal stromal cells. Stem Cells 25:1627–1634

Rabinovitch M (2012) Molecular pathogenesis of pulmonary arterial hypertension. J Clin Invest 122:4306–4313

Rai PR, Cool CD, King JA, Sevens T, Burns N, Winn RA, Kasper M, Voelkel NF (2008) The cancer paradigm of severe pulmonary arterial hypertension. Am J Respir Crit Care Med 178:558–564

Rokavec M, Oner MG, Li H, Jackstadt R, Jiang L, Lodygin D, Kaller M et al (2014) IL-6R/STAT3/miR-34a feedback loop promotes EMT-mediated colorectal cancer invasion and metastasis. J Clin Invest 124(4):1853–1867 [Epub 2014 Mar 18]

Sakao S, Taraseveciene-Stewart L, Cool CD, Tada Y, KasaharaY KK, Tanabe Y, Tatsumi K, Kuriyama T, Voelkel NF (2007) VEGF-R blockade causes endothelial cell apoptosis, expansion of surviving CD34+ precursor cells and transdifferentiation to smooth muscle-like and neuronal cells. FASEB J 21:3640–3652

Sanchez Alvarado A, Yamanaka S (2014) Rethinking differentiation: stem cells, regeneration and plasticity. Cell 157:110–119

Savai R, Pullamsetti SS, Kolbe J, Bienek E, Voswinckel R et al (2012) Immune and inflammatory cell involvement in the pathology of idiopathic pulmonary arterial hypertension. Am J Respir Crit Care Med 186:897–908

Scadden DT (2014) Nice neighbourhood: emerging concepts of the stem cell niche. Cell 157: 41–50

Sgonc R, Gruschwitz MS, Dietrich H, Recheis H, Gershwin ME et al (1996) Endothelial cell apoptosis is a primary pathogenetic event underlying skin lesions in avian and human scleroderma. J Clin Invest 98:785–792

Shenoy V, Gjymishka A, Jarajapu YP, Qi Y, Afzal A et al (2013) Diminazene attenuates pulmonary hypertension and improves progenitor cell function in experimental models. Am J Respir Crit Care Med 187:648–657

Smadja DM, Mauge L, Sanchez O, Silvestre JS, Guerin C, Godier A, Henno P et al (2010) Distinct patterns of circulating endothelial cells in pulmonary hypertension. Eur Respir J 36:1284–1293

Smadja DM, Mauge L, Gaussem P, d'Audigier C, Israel-Biet D et al (2011) Treprostinil increases the number and angiogenic potential of endothelial progenitor cells in children with pulmonary hypertension. Angiogenesis 14:17–27

Smith P, Heath D (1979) Electron microscopy of the plexiform lesion. Thorax 34:177–186

Spees JL, Whitney MJ, Sullivan DE, Lasky JA, Laboy M et al (2008) Bone marrow progenitor cells contribute to repair and remodeling of the lung and heart in a model of progressive pulmonary hypertension. FASEB J 22:1226–1236

Spiekerkoetter E, Tian X, Cai J, Hopper RK, Sudheendra D et al (2013) FK506 activates BMPR2, rescues endothelial dysfunction and reverses pulmonary hypertension. J Clin Invest 123: 3600–3613

Stacher E, Graham BB, Hunt JM, Gangjeva A, Groshong SD et al (2012) Modern age pathology of pulmonary arterial hypertension. Am J Respir Crit Care Med 186:261–272

Suh H, Kim D, Kim H, Helfman DM, Choi JH et al (2014) Modeling of Menkes disease via human induced pluripotent stem cells. Biochem Biophys Res Commun 444:311–318

Takahashi M, Nakamura T, Toba T, Kajiwara N, Kato H et al (2004) Transplantation of endothelial progenitor cells into the lung to alleviate pulmonary hypertension in dogs. Tissue Eng 10: 771–779

Tamosiuniene R, Tian W, Dhillon G, Wang L, Sung YK, Gera L, Patterson A, Agrawal R et al (2011) Regulatory T Cells limit vascular endothelial injury and prevent pulmonary hypertension. Circ Res 109:867–879

Taraseveciene-Stewart L, Voelkel NF (2008) Molecular pathogenesis of emphysema. J Clin Invest 118:394–402

Taraseveciene-Stewart L, Kasahara Y, Alger L, Hirth P, McMahon G, Waltenberger J, Voelkel NF, Tuder RM (2001) Inhibition of the VEGF receptor 2 combined with chronic hypoxia causes cell death-dependent pulmonary endothelial cell proliferation and severe pulmonary hypertension. FASEB J 15:427–438

Taraseveciene-Stewart L, Gera L, Hirth P, Voelkel NF, Tuder RM, Stewart JM (2002) A bradykinin antagonist and a caspase inhibitor prevent severe pulmonary hypertension in a rat model. Can J Physiol Pharmacol 80:269–274

Taraseveciene-Stewart L, Scerbavicius R, Choe KH, Cool CD, Wood K, Tuder RM, Kasper M, Voelkel NF (2006) Simvastatin causes endothelial cell apoptosis and attenuates severe pulmonary hypertension. Am J Physiol Lung Cell Mol Physiol 291:L668–L676

Tian W, Jiang X, Tamosiuniene R, Sung YK, Quian J, Dhillon G et al (2013) Blocking macrophage leukotriene B4 prevents endothelial injury and reverses pulmonary hypertension. Sci Transl Med 5(200):200ra117

Tigges U, Komatsu M, Stallcup WB (2013) Adventitial pericytes progenitor/mesenchymal stem cells participate in the restenotic response to arterial injury. J Vasc Res 50:134–144

Toshner M, Voswinckel R, Southwood M, Al-Lamki R, Howard LS, Marchesan D et al (2009) Evidence of dysfunction of endothelial progenitors in pulmonary arterial hypertension. Am J Respir Crit Care Med 180:780–787

Tu L, Dewachter L, Gore B, Fadel E, Dartevelle P, Simonneau G, Humbert M et al (2011) Autocrine fibroblast growth factor-2 signaling contributes to altered endothelial phenotype in pulmonary hypertension. Am J Respir Crit Care Med 45:311–322

Tuder RM, Groves BM, Badesch DB, Voelkel NF (1994) Exuberant endothelial cell growth and elements of inflammation are present in plexiform lesions of pulmonary hypertension. Am J Pathol 144(2):275–285

Tuder RM, Archer SL, Dorfmueller P, Erzurum SC, Guignabert C, Michelakis E et al (2013) Relevant issues in the pathology and pathobiology of pulmonary hypertension. J Am Coll Cardiol 62:D4–D12

Voelkel NF (2013) Pulmonary vascular diseases: in search of a hub among the spokes-an exercise in hypothesis generation. Pulm Circ 3:723–727

Voelkel NF, Gomez-Arroyo J, Abbate A, Bogaard HJ, Nicolls MR (2012) Pathobiology of pulmonary arterial hypertension and right ventricular failure. Eur Respir J 40(6):1555–1565

Voelkel NF, Mizuno S, Bogaard HJ (2013) The role of hypoxia in pulmonary vascular diseases: a perspective. Am J Physiol Lung Cell Mol Physiol 304:L457–L465

Wagenvoort CA, Wagenvoort N (eds) (1977) Pathology of pulmonary hypertension. Wiley, New York

Wanjare M, Kusama S, Gerecht S (2014) Defining differences among perivascular cells derived from human pluripotent stem cells. Stem Cell Reports 2:561–575

Yao W, Firth AL, Sacks RS, Ogawa A, Auger WR, Fedullo PF, Madani MM et al (2009) Am J Physiol Lung Cell Mol Physiol 296:L870–L878

Yeager ME, Halley GR, Golpon HA, Voelkel NF, Tuder RM (2001) Microsatellite instability of endothelial cell growth and apoptosis genes within plexiform lesions in primary pulmonary hypertension. Circ Res 88:E2–E11

Zaker F, Nasiri N, Oodi A, Amirizadeh N (2013) Evaluation of umbilical cord blood CD34+ hematopoietic stem cell expansion in co-culture with bone marrow mesenchymal stem cells in the presence of TEPA. Hematology 18:39–45

Zengin E et al (2006) Vascular wall resident progenitor cells: a source for postnatal vasculogenesis. Development 133:1543–1551

Zhao YD, Courtman DW, Deng Y, Kugathasan L, Zhang Q, Stewart DJ (2005) Rescue of monocrotaline-induced pulmonary arterial hypertension using bone marrow-derived endothelial-like progenitor cells. Circ Res 96:442–450

Chapter 17
Vascular Repair and Regeneration by Endothelial Progenitor Cells

Glenn Marsboom, Min Zhang, Jalees Rehman, and Asrar B. Malik

Abbreviations

ALI	Acute lung injury
BMPR2	Bone morphogenetic protein receptor 2
BPD	Bronchopulmonary dysplasia
Cdc42	Cell division control protein 42 homolog
COPD	Chronic obstructive pulmonary disease
CXCR4	C-X-C Chemokine receptor type 4
ECFCs	Endothelial colony-forming cells
eNOS	Endothelial nitric oxide synthase
EPC	Endothelial progenitor cell
FoxM1	Forkhead box protein M1
HGF	Hepatocyte growth factor
IGF1	Insulin-like growth factor 1
IL10	Interleukin 10
IPAH	Idiopathic pulmonary arterial hypertension
LPS	Lipopolysaccharide

G. Marsboom • M. Zhang • A.B. Malik, Ph.D. (✉)
Department of Pharmacology, College of Medicine, University of Illinois,
835 S. Wolcott, Chicago, IL 60612, USA

Center for Lung and Vascular Biology, University of Illinois, Chicago, IL, USA
e-mail: abmalik@uic.edu

J. Rehman
Department of Pharmacology, College of Medicine, University of Illinois,
835 S. Wolcott, Chicago, IL 60612, USA

Center for Lung and Vascular Biology, University of Illinois, Chicago, IL, USA

Department of Medicine, Section of Cardiology, University of Illinois, Chicago, IL, USA

A. Firth, J.X.-J. Yuan (eds.), *Lung Stem Cells in the Epithelium and Vasculature*,
Stem Cell Biology and Regenerative Medicine, DOI 10.1007/978-3-319-16232-4_17

MCP1	Monocyte chemoattractant protein-1
MSC	Mesenchymal stem cell
NYHA	New York Heart Association
PAEC	Pulmonary artery endothelial cell
PAH	Pulmonary arterial hypertension
PDE5	Phosphodiesterase 5
Rac1	Ras-related C3 botulinum toxin substrate 1
SDF1	Stromal cell-derived factor 1
VEGF	Vascular endothelial growth factor
VEGFR-2	Vascular endothelial growth factor receptor 2

17.1 EPCs

17.1.1 Discovery of EPCs

Research on vascular grafts paved the way for the description and culture of both early and late outgrowth EPCs. Already in 1963, it was shown that blood contains circulating endothelial cells. In an elegant experiment, young pigs received a special a-cellular Dacron graft consisting of a suspended piece of graft material in the lumen that was kept in place by sutures so that it made no contact with the vascular wall. After 14 days, patches of endothelial cells could be observed on the suspended graft while the sutures holding the suspended graft were not covered with endothelial cells (Stump et al. 1963). Subsequently, in 1998 it was shown that the endothelial cells found on the inside of such grafts are bone marrow derived (Shi et al. 1998). This finding was further supported by the culture of endothelial progenitor cells (EPCs) derived from human peripheral blood (Asahara et al. 1997).

Subsequent experiments demonstrated that there are actually two kinds of EPCs that can be cultured from blood (See Fig. 17.1). The so-called early outgrowth EPCs are present after 7 days of culture express markers of endothelial cells, yet also have monocytic markers and limited proliferation capacity (Rehman et al. 2003). Early outgrowth EPCs can home to ischemic areas, but cannot independently form a vascular network (Sieveking et al. 2008). Their therapeutic potential is believed to depend on the secretion of high levels of growth factors, such as VEGF, SDF1, HGF, and IGF1, that allow them to enhance the network formation of mature endothelial cells (Urbich et al. 2005).

In contrast, late outgrowth EPCs are typically observed after 2–3 weeks of culture and also express endothelial markers. They are more rare than early outgrowth EPCs but are highly proliferative. They can form a vascular network on their own, but do not secrete factors that enhance the network formation of mature endothelial cells (Sieveking et al. 2008). These cells are sometimes also called endothelial colony-forming cells (ECFCs) to highlight their proliferative potential.

	Early outgrowth EPCs	Late outgrowth EPCs (ECFCs)
• Culture time	7-10 days	>7 days
• Endothelial markers	+++	+++
• Monocytic markers	+++	-
• Proliferation	-/+	+++
• Incorporation into vasculature network	-	+++
• Paracrine effects	+++	-/+

Fig. 17.1 Significant differences between early and late outgrowth EPCs. Most importantly, late outgrowth EPCs are highly proliferative and are able to independently form a vascular network. Early outgrowth EPCs lack these characteristics, but secrete a number of growth factors that can help the survival or proliferation

17.1.2 Source of EPCs

While EPCs were originally cultured from peripheral blood, they have subsequently been identified in the bone marrow and even in the endothelium of the intact vessel wall.

In case of the lung, several reports have identified resident EPCs in the lung microcirculation. For example, single cell clonogenic assays of endothelial cells isolated from the pulmonary artery or the pulmonary microvasculature show a hierarchy of proliferative potential. While some cells do not proliferate, others can form large colonies consisting of more than 2,000 cells after 2 weeks of culture, and these colonies were more prevalent in the microvascular endothelial cells. Rare microvascular endothelial cells were able to form very large colonies (>100,000 cells) that show clonogenic potential when reseeded. This highly proliferative population was absent in the main pulmonary artery (Alvarez et al. 2008; Abbas et al. 2003).

When it comes to choosing the ideal source of EPCs for treatment, it has been described that cord blood-derived ECFCs proliferate faster than peripheral blood-derived EPCs (Schwarz et al. 2012) and the implantation of cord blood-derived ECFCs leads to more stable and long lasting blood vessels in a collagen/fibronectin gel (Au et al. 2008). Furthermore, combining different cell types, such as MSCs and EPCs might lead to a greater therapeutic benefit as has been shown in the murine hind limb ischemia model (Schwarz et al. 2012). Moreover, dysfunction of early outgrowth EPCs has been described in mice exposed to chronic hypoxia (Marsboom et al. 2008). In patients with a mutation in BMPR2, the most common mutation in families with hereditary pulmonary arterial hypertension (PAH), late outgrowth

EPCs are dysfunctional (Toshner et al. 2009). Therefore, autologous EPCs might not be very effective in some patients and in vitro manipulations such as treatment with statins might be necessary to improve EPC functionality (Alvarez et al. 2008).

17.1.3 Endothelial Regeneration and EPC Contribution

Natural turnover of endothelial cells is very low (Foteinos et al. 2008), and small lesions can actually be repaired by adjoining cells that can sense the loss of contact and migrate within hours to cover the place occupied by the dying cell. This process is not dependent on endothelial proliferation as proliferation of endothelial cells is usually only detected 1–2 days after a major injury (Reidy and Schwartz 1983, 1984). Nevertheless, the microvascular endothelium can proliferate rapidly, for example, in the endometrium during the luteal phase of the menstrual cycle or in response to exercise. Besides proliferation of local endothelial cells, bone marrow-derived EPCs could potentially contribute to vascular repair. Under baseline conditions, only a small fraction of endothelial cells is derived from the bone marrow. For example, 3 months after a bone marrow transplantation, only 1 % of endothelial cells in the aorta or blood vessels of the skin and brain are from a hematopoietic origin (Crosby et al. 2000). However, insights about bone marrow-derived EPCs contributing to endothelial regeneration in pathological situations with more extensive endothelial damage can be obtained from bone marrow and solid organ transplantation studies. For example, in a subgroup of patients receiving a kidney transplant, endothelial cells derived from the host can be observed in the kidney. The variability between patients does not appear to correlate with time after transplantation, but rather the presence of graft rejection. Most patients without signs of rejection had no recipient-derived endothelial cells in the kidney, while in grafts with vascular rejection, almost all kidneys had at least one third of endothelial cells derived from the recipient (Lagaaij et al. 2001). Furthermore, female patients receiving a hematopoietic stem cell transplant from male donors had significant numbers of pulmonary endothelial cells derived from the donor (approximately 40 %) (Suratt et al. 2003). Interestingly, this occurred exclusively in the alveolar capillaries, while no donor-derived endothelial cells were observed in arterioles or major arteries. Of the three patients studied in this report, one patient did not show endothelial chimerism. While the small sample size precludes a definitive conclusion, possible explanations for the lack of chimerism are the short time between transplant and biopsy in this patient (only 50 days compared to more than 6 months in the two other patients) or the different conditioning treatment prior to transplantation (the patients with high chimerism received total body irradiation compared to Busulfan treatment). These studies highlight that in case of significant endothelial damage, EPCs can contribute to endothelial repair.

Obviously, there is great heterogeneity amongst the endothelial cells of different vascular beds. For example, endothelial cells in the brain have tight junctions to maintain the blood–brain barrier, while endothelial cells in the kidney are fenestrated to allow efficient secretion. Even within a vascular bed, endothelial cells in conduit vessels and microvascular endothelial cells have different characteristics. At this point, it is not clear whether late outgrowth EPCs are able to differentiate in all different phenotypes, but at least one report has demonstrated that EPCs upregulate arterial markers such as ephrin B2 in response to shear stress, suggesting that they can adapt to local conditions (Obi et al. 2009).

17.1.4 Long-Term Engraftment Versus Paracrine Effects of EPCs

Currently, there is a discussion in the field about the importance of EPC long-term engraftment in the vasculature versus transient paracrine effects. Proof that long-term engraftment is necessary for functional improvement comes from a murine hind limb ischemia model where ECFCs were transduced with the ganciclovir-inducible Herpes Simplex 1 thymidine kinase suicide gene. When human ECFCs are injected 24 h after hind limb ischemia induction, they improve perfusion after 14 days. Yet subsequent treatment with ganciclovir completely reverses the functional improvement to what is seen in untreated control animals, suggesting that long-term engraftment and vascular incorporation of injected ECFCs is necessary for improving neovascularization (Schwarz et al. 2012). A similar suicide gene approach showed that the functional improvement of early outgrowth EPCs in a murine myocardial infarction also depends on long-term engraftment. EPC injection 1 day after myocardial infection improves the ejection fraction and capillary density after 2 weeks. Yet these improvements disappear if mice are subsequently treated with ganciclovir (which was not observed in mice receiving EPCs that do not express the suicide gene) (Ziebart et al. 2008). Using high resolution confocal microscopy in a murine hind limb ischemia model, it was shown that bone marrow-derived cells do not actually become endothelial cells, but rather accumulate around nascent blood vessels (Ziegelhoeffer et al. 2004). Similar observations were made in a melanoma model of angiogenesis, where the bone marrow-derived cells were identified as pericytes or monocytes (Rajantie et al. 2004). Other possibilities for EPCs to modify the vasculature are cell fusion and the release of microparticles or microvesicles. For example, in vitro studies have demonstrated the potential of bone marrow-derived cells to fuse with and obtain phenotypic markers of other cell types including embryonic stem cells, although this event appears to be rare (Terada et al. 2002). Endothelial cell-derived microparticles can induce the proliferation of mature endothelial cells (Dignat-George and Boulanger 2011), and microRNAs appear to mediate many of the effects of microparticles (Zhang et al. 2014).

17.1.5 EPCs Can Be Used as a Biomarker or as a Therapeutic Agent

Indirect evidence for the importance of EPCs for vascular health comes from studies demonstrating that patients with coronary artery disease have less and functionally impaired early outgrowth EPCs (Vasa et al. 2001) and even in healthy patients, lower EPC numbers are associated with worse endothelium-dependent vasodilation and a higher cardiovascular risk as assessed with the Framingham risk score (Hill et al. 2003). Therefore, levels of circulating EPCs might be a valuable biomarker for the severity of disease and disease progression.

Moreover, EPC transfer has been tested as a therapy for myocardial infarction, hind limb ischemia, and wound healing. In the next section, we will focus on the work that has been performed using EPCs either as a biomarker or as a therapy specifically for pulmonary diseases.

17.2 Role of EPCs in Pulmonary Vascular Regeneration During Disease

EPCs have been studied in several pulmonary diseases including bronchopulmonary dysplasia (BPD), PAH, acute lung injury (ALI), and chronic obstructive pulmonary disease (COPD). We will summarize the current knowledge for each of these conditions (see Fig. 17.2).

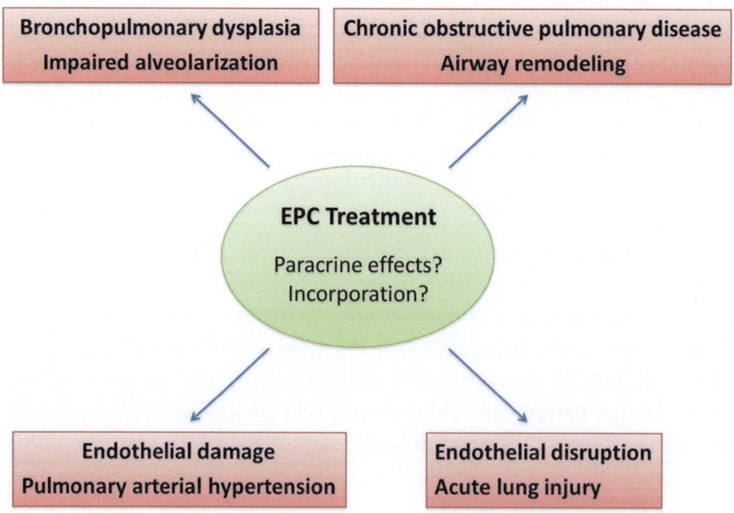

Fig. 17.2 Different pulmonary diseases for which EPC therapies are being considered

17.2.1 Bronchopulmonary Dysplasia

BPD is associated with premature birth and neonatal respiratory distress syndrome. Disruption of the normal lung development results in alveolar simplification and decreased vascular density. Although novel therapies such as prenatal steroids, postnatal surfactant, and improved ventilator strategies have led to a better survival, the prevalence of BPD has not declined.

Pulmonary vascular and alveolar development occur concurrently and recent evidence indicates that disrupted pulmonary blood vessel development can directly lead to impaired alveolarization and BPD (Stenmark and Abman 2005; Thebaud and Abman 2007; Jakkula et al. 2000). Therefore, a current line of thinking is that EPCs could be useful to improve vascularity, which in turn will lead to a normalized alveolar development.

In Vitro Experiments

To determine the paracrine effects of late outgrowth EPCs on pulmonary artery endothelial cells (PAECs), EPCs were isolated from the cord blood of both term and preterm infants and then used to condition medium. Cells were exposed to room air or hyperoxia (50 % oxygen for 24 h). Regardless of whether PAECs were exposed to hyperoxia, conditioned medium from EPCs increased their proliferation and angiogenic network formation. However, if preterm EPCs were exposed to hyperoxia, the conditioned medium became far less effective, while term EPCs were not influenced by hyperoxia. This suggests that EPCs from preterm infants are particularly sensitive to hyperoxia and that autologous EPC transfer might not be an appropriate treatment option for infants with BPD (Baker et al. 2013).

Animal Experiments

Exposure of neonatal mice to hyperoxia (80 % oxygen) for 10 days leads to airway simplification and a 72 % reduction in vascular density (Balasubramaniam et al. 2007). The same treatment also decreased EPC numbers in the lungs, blood, and bone marrow while adult mice actually had increased EPC numbers in the lungs and bone marrow. Interestingly, in vitro exposure of bone marrow-derived EPCs to hyperoxia leads to apoptosis (Balasubramaniam et al. 2007).

Bleomycin administration leads to pulmonary fibrosis in adult rats, yet if neonatal rats are exposed for 14 days to bleomycin, they develop BPD and pulmonary hypertension (Baker et al. 2013). Intravenous injection of either late-outgrowth EPCs or of their conditioned medium reduces right ventricular hypertrophy. EPCs from term infants were equally effective when exposed to hypoxia, yet preterm EPCs exposed to hyperoxia do not significantly decrease right ventricular hypertrophy. In contrast, no improvement of alveolar simplification, vessel density, or pulmonary artery muscularization was observed in response to any of the conditioned media (Baker et al. 2013).

Clinical Studies

In a first study that looked at preterm infants (gestational age <32 weeks and birth weight <1,500 g), there was a reduction of the numbers of late-outgrowth EPCs in the cord blood of infants that would later on develop BPD (Borghesi et al. 2009). Remarkably, levels of late-outgrowth EPCs do not correlate with gestational age or birth weight, suggesting that levels of EPCs would be an independent predictor of the risk to develop BPD (Baker et al. 2012). Not only levels of cord blood EPCs, but also of circulating EPCs in the infant can be used to predict risk: using a similar study population, investigators found that those infants that would later on develop BPD, had similar levels of circulating EPCs at birth, yet had significant lower levels at 7 or 21 days (Qi et al. 2013; Paviotti et al. 2011). The reduction in circulating EPC numbers could be reversed by administration of inhaled nitric oxide (Qi et al. 2013).

17.2.2 Pulmonary Arterial Hypertension

PAH is a disorder characterized by an increase in the pulmonary vascular resistance due to a narrowing of the pulmonary blood vessels. This ultimately leads to right ventricular failure and prognosis remains poor despite the introduction of new therapies including prostacyclin, Phosphodiesterase 5 (PDE 5) inhibitors, and endothelin-1 receptor antagonists. The current 5-year survival rate of patients is only 61 % (Thenappan et al. 2010).

Endothelial damage is considered to be an early event in the onset of PAH (Tuder et al. 2001b) and vascular abnormalities in PAH patients include increased muscularization of the pulmonary arteries and formation of plexiform lesions consisting of disordered proliferating endothelial cells (Tuder et al. 2001a). These lesions are observed in both patients with idiopathic or heritable PAH and in patients that develop PAH in association with other diseases such as scleroderma. Interestingly, in patients with idiopathic PAH (IPAH), these endothelial cells have a monoclonal origin, suggesting that a single endothelial cell gave rise to the plexiform lesion. This is reminiscent of Kaposi's sarcoma and suggests that in the presence of endothelial damage, some endothelial cells can acquire a hyperproliferative phenotype, possibly due to a somatic mutation (Lee et al. 1998).

Further proof for the importance of endothelial damage in the development of PAH comes from the observation that levels of soluble E-selectin, a marker of endothelial inflammation and injury, are increased in PAH patients (Cella et al. 2001). Moreover, novel therapies for PAH are based on supplying endothelial-derived compounds such as prostacyclin and nitric oxide that are reduced in PAH patients.

Also animal models demonstrate that endothelial damage contributes to disease onset, as inhibition of VEGFR-2, an important receptor involved in endothelial survival, leads to an exacerbated disease state (Taraseviciene-Stewart et al. 2001).

A number of factors including increased shear stress, viral infection, or hypoxia sometimes in combination with genetic mutations can lead to PAEC injury. To repair the damaged endothelial cells, either local endothelial cells can proliferate or circulating EPCs can be recruited to repair the damage. In the absence of adequate repair, smooth muscle cells will proliferate and lead to vascular narrowing.

Animal Models

A widely used model to study pulmonary hypertension is the use of the plant pyrrolizidine alkaloid monocrotaline. This compound becomes activated in the liver to monocrotaline pyrrole, a reactive alkylating compound that subsequently causes endothelial damage (Roth and Reindel 1991). Mainly the lungs are affected because this is the first vascular bed encountered by the activated compound after the liver. A single injection of monocrotaline causes a significant increase in the pulmonary artery pressure of rats together with increased muscularization of the pulmonary arteries (Marsboom et al. 2012).

Endothelial damage is an important aspect of monocrotaline-induced pulmonary hypertension and injections of EPCs show beneficial effects in at least two different species. In rats, injection in the jugular vein of bone marrow-derived early outgrowth EPCs 3 days after monocrotaline almost completely prevented the rise in right ventricular systolic pressures and right ventricular hypertrophy. Reversal experiments, where EPCs are administered 3 weeks after monocrotaline injection, also had a positive effect on pressures, hypertrophy, and survival. These effects were modestly improved by transducing EPCs with eNOS (Zhao et al. 2005). Of note, labeled EPCs were present for at least 3 weeks in the distal arterioles of the lungs. In another manuscript, nude rats received monocrotaline followed 7 days later by early outgrowth EPCs derived from human umbilical cord mononuclear cells. EPC transfer lowers the mean pulmonary artery pressure with 14 %, and an even greater effect (−29 %) was observed when EPCs were transduced with the vasodilator adrenomedullin (Nagaya et al. 2003). Similar improvements were observed for muscularization and survival. The beneficial effect of EPC transfer has also been shown in dogs. Dogs receiving an intravenous injection of monocrotaline pyrrole develop severe pulmonary hypertension after 6 weeks (Takahashi et al. 2004). In this model, injection of early outgrowth EPCs derived from peripheral blood (10–14 days of culture) into the lung parenchyma of the lower lobes can improve pulmonary artery pressure and cardiac output and lowers pulmonary vascular resistance (Takahashi et al. 2004).

These positive findings are in contrast with results in the chronic hypoxia model of pulmonary hypertension, where transplantation of early outgrowth EPCs was not able to lower right ventricular pressures or hypertrophy (Marsboom et al. 2008). A possible explanation is that chronic hypoxia leads to EPC dysfunction (reduced adhesion, migration to SDF1α, and incorporation into a vascular network), suggesting that EPC treatment for PAH patients that also experience hypoxemia might have less therapeutic benefit (Marsboom et al. 2008).

Clinical Studies

Based on promising results in the monocrotaline model of pulmonary hypertension, preliminary small-scale clinical studies have been performed to assess the safety and therapeutic efficacy of autologous EPC transfer in IPAH patients (see Table 17.1). The first study used IPAH patients with New York Heart Association (NYHA) functional class II and III (mild to significant symptoms during activity, but comfortable at rest) (Wang et al. 2007). Patients received standard therapy (but no patients on prostacyclin were included) and were randomized to either also receive autologous EPCs ($n=15$) or only continue standard treatment ($n=16$). EPCs were isolated from peripheral blood, cultured for 5 days, and on average 11 million EPCs were subsequently administered intravenously. After 12 weeks, patients receiving EPCs walked on average an additional 50 m during the 6-min walk test, which was the primary end point for this trial. In contrast, the control group's walking distance only improved by less than 6 m on average. The improvement in the EPC group was associated with lower pulmonary artery pressures and vascular resistance and improved cardiac output. No complications related to EPC therapy were observed (Wang et al. 2007). The same group also studied the safety and feasibility of EPC transfer in 13 children with IPAH. Similar improvements in exercise capacity and pulmonary artery pressures were observed, yet this study did not contain a control group (Zhu et al. 2008). Interestingly, at least two other trials with EPCs have been registered on the website of ClinicalTrials.gov. The first study by Canadian researchers is an open-label trial looking at the safety and functional improvement in response to intravenous injection of eNOS transduced EPCs (NCT00469027). The trial was projected to start in 2006, but to the best of our knowledge no results have been published. The other study is based in China and would be the first double-blind clinical trial involving EPCs for the treatment of PAH (NCT00372346). It was also projected to start in 2006, but until now no results have been published. At this point, it is not clear whether recruitment issues, a lack of funding, or possible side-effects are responsible for the delay in publication, but results of these trials will be crucial to assess the therapeutic efficacy of EPCs in PAH.

Prognostic Value of Circulating EPC Numbers in PAH

The development of an EPC capture chip coated with an anti-CD34 antibody, allows measurements on small blood samples (200 μL) without the delays of associated with the isolation and culturing of EPCs from large whole blood samples. Immunostaining for CD31, VEGFR2, and CD45 allows for a rapid (less than 1 h) determination of circulating EPC numbers. This chip therefore could be useful in

Table 17.1 Clinical trials using EPCs for the treatment of PAH

Clinical trial #	Description of trial	Location of trial	Primary goals	Patient population	Number of patients	Trial design
NCT00257413 (2007)	Safety and efficacy study of transplantation of EPCs to treat idiopathic pulmonary arterial hypertension[a]	Zhejiang University	Feasibility, safety, and initial clinical outcome of intravenous infusion of autologous EPCs	Idiopathic PAH	15 NYHA class II–III receiving EPCs vs. 16 controls	Randomized, single-blind
NCT00372346	Safety and efficacy study of transplantation of EPCs to treat idiopathic pulmonary arterial hypertension	Zhejiang University	Feasibility, safety, and initial clinical outcome of intravenous infusion of autologous EPCs	Idiopathic PAH	40 patients, NYHA class II–III	Randomized, double-blind 2006–2007
NCT00641836	Safety and feasibility of autologous endothelial progenitor cells transplantation in patients with idiopathic pulmonary arterial hypertension	Zhejiang University	Ideal quantity of EPCs for therapy, the duration of the therapeutic effect, and moreover, the potential toxicity of such therapy.	Idiopathic PAH	98 patients	Follow-up study, non-randomized and open label. 2005–2007
NCT00469027	Pulmonary hypertension: assessment of cell therapy (PHACeT)	St. Michael's Hospital, Toronto and Sir Mortimer B. Davis–Jewish General Hospital, Montreal, Canada	Safety of autologous progenitor cell-based gene therapy of heNOS	Idiopathic, familial, or anorexigen-induced PAH	18 patients	Non-randomized and open label. 2006–2012. Two centre, phase 1 trial

[a]Wang XX, Zhang FR, Shang YP, Zhu JH, Xie XD, Tao QM, Chen JZ. Transplantation of autologous endothelial progenitor cells may be beneficial in patients with idiopathic pulmonary arterial hypertension: A pilot randomized controlled trial. *J Am Coll Cardiol*. 2007;49:1566–1571

following EPC numbers over time to monitor therapeutic efficacy (Hansmann et al. 2011). However, at this point it is not clear that these are markers truly specific for EPCs, how well they can discriminate between minimally proliferative and highly proliferative EPCs, and whether there is any prognostic value of EPC number determination in PAH. Some researchers have found decreased while other have found increased numbers of circulating EPCs in PAH patients. For example, CD34+/AC133+/VEGFR2+ and AC133+/VEGFR2+ circulating cells are reduced in IPAH patients (Diller et al. 2008; Junhui et al. 2008), while others have shown an increase in CD34+/AC133+/VEGFR2+ and CD34+/AC133+ cells in IPAH patients (Toshner et al. 2009; Asosingh et al. 2008). At this point, it is unclear how this discrepancy can be reconciled, but one possibility is that differences in patient selection in combination with relatively small patient samples lead to a large variability (e.g., the studies showing a decrease in circulating EPC numbers had 20 and 55 patients, while the studies showing an increase had 16 and 17 IPAH patients). When we look at the biggest study so far, using 55 IPAH patients, the reduction in circulating EPCs did not correlate with exercise capacity of NYHA functional class assessment (Diller et al. 2008). Also, incorporation of early outgrowth EPCs into a vascular network was not different between IPAH and control patients (Diller et al. 2008). Highlighting the heterogeneity of findings, another publication has actually found reduced adhesion to the extracellular matrix component fibronectin and decreased migration towards VEGF in EPCs from IPAH (Junhui et al. 2008). Therefore, it is currently not clear whether circulating EPC numbers and their functionality can be used to predict disease outcome and response to therapy.

17.2.3 Acute Lung Injury and Acute Respiratory Distress Syndrome

Damage to the pulmonary endothelium leads to a disruption of the alveolar-capillary membrane in patients with ALI and its most severe form acute respiratory distress syndrome. The endothelial disruption leads to pulmonary edema formation, hypoxemia, and respiratory failure. Despite improved ventilatory strategies, there is a mortality rate of 30–50 %. ALI occurs frequently in association with sepsis.

At the cellular level, endotoxins (lipopolysaccharide or LPS) result in the loss of endothelial barrier function and edema formation, which in turn can lead to hypoxemia. Endothelial repair with EPCs might be a novel mechanism to restore endothelial barrier function and accelerate the recovery.

Animal Experiments

Work from our group has previously shown that LPS-induced injury triggers a small population of lung endothelial cells to proliferate, which is dependent on the FoxM1 transcription factor and is necessary for restoring the endothelial barrier

(Zhao et al. 2006). In addition, we have also shown that bone marrow-derived mononuclear cells can decrease LPS-induced lung microvascular permeability and edema formation by improving the endothelial barrier function (Zhao et al. 2009). Mechanistically, bone marrow-derived mononuclear cells secrete the phospholipid sphingosine-1-phosphate, which upon binding to its receptor on endothelial cells activates the Rho GTPases Cdc42 and Rac1 and leads to the assembly of adherens junctions as well as the restoration of the endothelial barrier (Zhao et al. 2009).

Moreover, 1 week after intranasal delivery of LPS to mice, multiple bone marrow-derived cells can be observed in the lungs (Yamada et al. 2004) of which a subset expresses CD34, a marker observed on hematopoietic and endothelial cells. The number of circulating early-outgrowth EPCs was also increased in response to LPS (Yamada et al. 2004). Indirect evidence for the importance of these circulating EPCs in lung repair was obtained by giving mice a sublethal irradiation followed by LPS administration which would suppress or eliminate bone marrow stem and progenitor cells. This leads to the formation of emphysema-like lesions in the lungs, which were absent when the bone marrow was reconstituted (Yamada et al. 2004).

Rats receiving bone marrow-derived early outgrowth EPCs 30 min after an intravenous injection of LPS have improved survival (Mao et al. 2010). Histological analysis revealed incorporation of EPCs in the vascular wall up to 14 days, and the incorporation is not observed in control rats that did not receive LPS injection. EPCs decreased the typical thickening of the alveolar wall seen after LPS treatment and reduced the number of inflammatory cells present in the lung. The wet-to-dry ratio of the lung also was significantly reduced after EPC treatment, thus indicating improved barrier function. Finally, plasma levels of endothelin-1 were decreased, while the anti-inflammatory cytokine IL10 was increased after EPC injection (Mao et al. 2010).

Clinical Studies

ALI leads to an increase in circulating late-outgrowth EPC numbers and their numbers within 72 h after the onset of ALI is an independent predictor of survival. Patients with high levels of EPCs have a twofold higher survival rate even when other factors such as age, gender, and severity of disease are included in the multivariate analysis (Burnham et al. 2005). While the exact mechanism of the survival benefit is not known, it can be hypothesized that mobilization of EPCs could potentially repair damaged pulmonary endothelium. Also, in humans suffering from pneumonia, an increased number of circulating early-outgrowth EPCs has been observed (Yamada et al. 2005).

17.2.4 Chronic Obstructive Pulmonary Disease

COPD patients have airflow obstruction due to chronic inflammation of the airways. Endothelium-dependent relaxation is impaired in pulmonary arteries of patients with mild COPD, suggesting that endothelial dysfunction or injury is present at the initial stages of COPD (Peinado et al. 1998). Smoking appears to be a major culprit since smokers with a normal lung function have structural abnormalities in the pulmonary arteries and also express less endothelial nitric oxide synthase (Barbera et al. 2001). Endothelium-derived nitric oxide is a potent vasodilator with antiproliferative effects on smooth muscle cells and its downregulation is a hallmark of endothelial dysfunction.

The occurrence of endothelial damage likely explains why COPD patients are at increased risk for developing pulmonary hypertension. Mild pulmonary hypertension defined as mean pulmonary artery pressure higher than 25 mmHg is present in approximately half of COPD patients, with around 4 % of patients having severe pulmonary hypertension (pulmonary artery pressure of more than 35 mmHg). The development of pulmonary hypertension is associated with a significant shorter survival (37 % 5-year survival rate versus 63 % in COPD patients without pulmonary hypertension) (Andersen et al. 2012). Therefore, EPC therapy might be useful to extend the life expectancy of COPD patients.

In a rabbit model of brush injury-induced bronchial denudation, transplantation of tissue-engineered implants containing mature endothelial cells can improve airway regeneration (less airway remodeling and luminal narrowing) (Zani et al. 2008). Although similar experiments have not been performed with EPCs, these findings at least suggest that improving endothelial function by cell transplantation may be beneficial in COPD.

Animal Models

Elastase-induced lung injury mimics the lung pathology seen during COPD. It was shown that an intraperitoneal injection of hepatocyte growth factor leads to a recovery of alveolar structure. While hepatocyte growth factor might have a direct effect on the proliferation of lung epithelial cells (Sakamaki et al. 2002), it is intriguing that an almost sevenfold increase of circulating EPCs was observed and that bone marrow-derived cells were observed in the vascular wall (Ishizawa et al. 2004). Further research is however necessary to confirm the importance of EPCs in improving alveolar structure.

Clinical Studies

Patients with COPD or restrictive lung disease have an approximately twofold reduction of circulating EPC numbers measured as CD34+/CD133+/VEGFR2+ cells/ml blood (Fadini et al. 2006). Interestingly, within a group of COPD patients, those with lower arterial pO_2 levels actually have the highest EPC numbers, suggesting that hypoxemia is inducing mobilization of EPCs. In the patients with restrictive lung disease, patients with lower total lung volumes also have the lowest EPC numbers. As in the COPD patients, those with hypoxemia (in these patients measured as increased hematocrit levels) have higher levels of EPCs. Given the heterogeneous disease background in patients with restrictive lung disease (idiopathic pulmonary fibrosis, obesity-related ventilatory impairment, sarcoidosis, tuberculosis, amyotrophic lateral sclerosis, and kyphoscoliosis) and that some of the patients were smoking, it is difficult to draw definitive conclusions. Nevertheless, it appears that in all of these settings the number of circulating EPCs is reduced.

17.3 Remaining Challenges and Future Directions

Animal experiments have demonstrated the feasibility and efficacy of EPC-mediated vascular repair in the lungs of animals, mainly rodents. Future research should therefore focus on larger animals to establish the numbers of cells needed in patient trials and which cell type (early or late-outgrowth EPCs) is more beneficial in each of the pulmonary vascular diseases. Small animal research would be useful to identify ways to increase the homing of EPCs to the lungs in case insufficient EPCs can be obtained from patients and to better understand any potential side-effects of EPC therapy.

Although initial clinical trials have shown that EPC transfer is safe (as described above), it should be pointed out that under certain conditions, EPCs could actually have detrimental effects. For example, early outgrowth EPCs release a large amount of tissue factor which could contribute to coagulation, while late outgrowth EPCs release the inflammatory chemokine MCP1 (Zhang et al. 2009). Early outgrowth EPCs migrate towards the chemokine stromal cell-derived factor-1 (SDF1) (Marsboom et al. 2008) and blocking CXCR4 can largely prevent (but not reverse) chronic hypoxia-induced vascular remodeling, increased right ventricular pressures, and right ventricular hypertrophy (Gambaryan et al. 2011). This suggests that the homing of progenitor cells (including EPCs) could even contribute to disease progression. Similarly, elimination of circulating monocytes also prevents the development of pulmonary hypertension (Frid et al. 2006). This highlight the complex role that EPCs can play pulmonary vascular disease, having both beneficial and detrimental effects; and that research is needed to identify the optimal EPC type and timing of therapy to maximize the benefits of cell therapy.

Finally, and most importantly, to establish EPC transplantation as a clinical therapy, well designed and double blind clinical trials are necessary. While safety is a serious concern, the lessons learned from larger animal experiments in combination with dose-escalating trials should allow to minimize the risk for patients. At the same time, it is imperative that the results of ongoing clinical trials in PAH patients are published as soon as possible so that other clinical investigators can learn from these initial trials. For example, the numbers of injected cells needed for clinical improvement would benefit trials for other pulmonary vascular disorders as well.

17.4 Conclusion

While animal experiments have shown that EPC transfer can be used to treat several pulmonary vascular diseases, the translation of these findings into clinical practice is clearly lacking. It still needs to be established whether early or late outgrowth EPCs are more effective in the treatment of BPD, PAH, ALI, and COPD. The authors believe that the use of larger animals might be more appropriate to address safety and efficacy of EPC transfer before moving to clinical trials.

Acknowledgements This work was supported by a Parker B. Francis fellowship (G.M.) and grants from the National Institutes of Health HL090152 (A.B.M.), GM094220 (J.R.), and HL118068 (A.B.M. and J.R.).

References

Abbas MM, Evans JJ, Sin IL, Gooneratne A, Hill A, Benny PS (2003) Vascular endothelial growth factor and leptin: regulation in human cumulus cells and in follicles. Acta Obstet Gynecol Scand 82(11):997–1003

Alvarez DF, Huang L, King JA, ElZarrad MK, Yoder MC, Stevens T (2008) Lung microvascular endothelium is enriched with progenitor cells that exhibit vasculogenic capacity. Am J Physiol Lung Cell Mol Physiol 294(3):L419–L430. doi:10.1152/ajplung.00314.2007

Andersen KH, Iversen M, Kjaergaard J, Mortensen J, Nielsen-Kudsk JE, Bendstrup E, Videbaek R, Carlsen J (2012) Prevalence, predictors, and survival in pulmonary hypertension related to end-stage chronic obstructive pulmonary disease. J Heart Lung Transplant 31(4):373–380. doi:10.1016/j.healun.2011.11.020

Asahara T, Murohara T, Sullivan A, Silver M, van der Zee R, Li T, Witzenbichler B, Schatteman G, Isner JM (1997) Isolation of putative progenitor endothelial cells for angiogenesis. Science 275(5302):964–967

Asosingh K, Aldred MA, Vasanji A, Drazba J, Sharp J, Farver C, Comhair SA, Xu W, Licina L, Huang L, Anand-Apte B, Yoder MC, Tuder RM, Erzurum SC (2008) Circulating angiogenic precursors in idiopathic pulmonary arterial hypertension. Am J Pathol 172(3):615–627. doi:10.2353/ajpath.2008.070705

Au P, Daheron LM, Duda DG, Cohen KS, Tyrrell JA, Lanning RM, Fukumura D, Scadden DT, Jain RK (2008) Differential in vivo potential of endothelial progenitor cells from human umbilical cord blood and adult peripheral blood to form functional long-lasting vessels. Blood 111(3):1302–1305. doi:10.1182/blood-2007-06-094318

Baker CD, Balasubramaniam V, Mourani PM, Sontag MK, Black CP, Ryan SL, Abman SH (2012) Cord blood angiogenic progenitor cells are decreased in bronchopulmonary dysplasia. Eur Respir J 40(6):1516–1522. doi:10.1183/09031936.00017312

Baker CD, Seedorf GJ, Wisniewski BL, Black CP, Ryan SL, Balasubramaniam V, Abman SH (2013) Endothelial colony-forming cell conditioned media promote angiogenesis in vitro and prevent pulmonary hypertension in experimental bronchopulmonary dysplasia. Am J Physiol Lung Cell Mol Physiol 305(1):L73–L81. doi:10.1152/ajplung.00400.2012

Balasubramaniam V, Mervis CF, Maxey AM, Markham NE, Abman SH (2007) Hyperoxia reduces bone marrow, circulating, and lung endothelial progenitor cells in the developing lung: implications for the pathogenesis of bronchopulmonary dysplasia. Am J Physiol Lung Cell Mol Physiol 292(5):L1073–L1084. doi:10.1152/ajplung.00347.2006

Barbera JA, Peinado VI, Santos S, Ramirez J, Roca J, Rodriguez-Roisin R (2001) Reduced expression of endothelial nitric oxide synthase in pulmonary arteries of smokers. Am J Respir Crit Care Med 164(4):709–713. doi:10.1164/ajrccm.164.4.2101023

Borghesi A, Massa M, Campanelli R, Bollani L, Tzialla C, Figar TA, Ferrari G, Bonetti E, Chiesa G, de Silvestri A, Spinillo A, Rosti V, Stronati M (2009) Circulating endothelial progenitor cells in preterm infants with bronchopulmonary dysplasia. Am J Respir Crit Care Med 180(6):540–546. doi:10.1164/rccm.200812-1949OC

Burnham EL, Taylor WR, Quyyumi AA, Rojas M, Brigham KL, Moss M (2005) Increased circulating endothelial progenitor cells are associated with survival in acute lung injury. Am J Respir Crit Care Med 172(7):854–860. doi:10.1164/rccm.200410-1325OC

Cella G, Bellotto F, Tona F, Sbarai A, Mazzaro G, Motta G, Fareed J (2001) Plasma markers of endothelial dysfunction in pulmonary hypertension. Chest 120(4):1226–1230

Crosby JR, Kaminski WE, Schatteman G, Martin PJ, Raines EW, Seifert RA, Bowen-Pope DF (2000) Endothelial cells of hematopoietic origin make a significant contribution to adult blood vessel formation. Circ Res 87(9):728–730

Dignat-George F, Boulanger CM (2011) The many faces of endothelial microparticles. Arterioscler Thromb Vasc Biol 31(1):27–33. doi:10.1161/ATVBAHA.110.218123

Diller GP, van Eijl S, Okonko DO, Howard LS, Ali O, Thum T, Wort SJ, Bedard E, Gibbs JS, Bauersachs J, Hobbs AJ, Wilkins MR, Gatzoulis MA, Wharton J (2008) Circulating endothelial progenitor cells in patients with Eisenmenger syndrome and idiopathic pulmonary arterial hypertension. Circulation 117(23):3020–3030. doi:10.1161/CIRCULATIONAHA.108.769646

Fadini GP, Schiavon M, Cantini M, Baesso I, Facco M, Miorin M, Tassinato M, de Kreutzenberg SV, Avogaro A, Agostini C (2006) Circulating progenitor cells are reduced in patients with severe lung disease. Stem Cells 24(7):1806–1813. doi:10.1634/stemcells.2005-0440

Foteinos G, Hu Y, Xiao Q, Metzler B, Xu Q (2008) Rapid endothelial turnover in atherosclerosis-prone areas coincides with stem cell repair in apolipoprotein E-deficient mice. Circulation 117(14):1856–1863. doi:10.1161/CIRCULATIONAHA.107.746008

Frid MG, Brunetti JA, Burke DL, Carpenter TC, Davie NJ, Reeves JT, Roedersheimer MT, van Rooijen N, Stenmark KR (2006) Hypoxia-induced pulmonary vascular remodeling requires recruitment of circulating mesenchymal precursors of a monocyte/macrophage lineage. Am J Pathol 168(2):659–669. doi:10.2353/ajpath.2006.050599

Gambaryan N, Perros F, Montani D, Cohen-Kaminsky S, Mazmanian M, Renaud JF, Simonneau G, Lombet A, Humbert M (2011) Targeting of c-kit+haematopoietic progenitor cells prevents hypoxic pulmonary hypertension. Eur Respir J 37(6):1392–1399. doi:10.1183/09031936.00045710

Hansmann G, Plouffe BD, Hatch A, von Gise A, Sallmon H, Zamanian RT, Murthy SK (2011) Design and validation of an endothelial progenitor cell capture chip and its application in patients with pulmonary arterial hypertension. J Mol Med (Berl) 89(10):971–983. doi:10.1007/s00109-011-0779-6

Hill JM, Zalos G, Halcox JP, Schenke WH, Waclawiw MA, Quyyumi AA, Finkel T (2003) Circulating endothelial progenitor cells, vascular function, and cardiovascular risk. N Engl J Med 348(7):593–600. doi:10.1056/NEJMoa022287

Ishizawa K, Kubo H, Yamada M, Kobayashi S, Suzuki T, Mizuno S, Nakamura T, Sasaki H (2004) Hepatocyte growth factor induces angiogenesis in injured lungs through mobilizing endothelial progenitor cells. Biochem Biophys Res Commun 324(1):276–280. doi:10.1016/j.bbrc.2004.09.049

Jakkula M, Le Cras TD, Gebb S, Hirth KP, Tuder RM, Voelkel NF, Abman SH (2000) Inhibition of angiogenesis decreases alveolarization in the developing rat lung. Am J Physiol Lung Cell Mol Physiol 279(3):L600–L607

Junhui Z, Xingxiang W, Guosheng F, Yunpeng S, Furong Z, Junzhu C (2008) Reduced number and activity of circulating endothelial progenitor cells in patients with idiopathic pulmonary arterial hypertension. Respir Med 102(7):1073–1079. doi:10.1016/j.rmed.2007.12.030

Lagaaij EL, Cramer-Knijnenburg GF, van Kemenade FJ, van Es LA, Bruijn JA, van Krieken JH (2001) Endothelial cell chimerism after renal transplantation and vascular rejection. Lancet 357(9249):33–37. doi:10.1016/S0140-6736(00)03569-8

Lee SD, Shroyer KR, Markham NE, Cool CD, Voelkel NF, Tuder RM (1998) Monoclonal endothelial cell proliferation is present in primary but not secondary pulmonary hypertension. J Clin Invest 101(5):927–934. doi:10.1172/JCI1910

Mao M, Wang SN, Lv XJ, Wang Y, Xu JC (2010) Intravenous delivery of bone marrow-derived endothelial progenitor cells improves survival and attenuates lipopolysaccharide-induced lung injury in rats. Shock 34(2):196–204. doi:10.1097/SHK.0b013e3181d49457

Marsboom G, Pokreisz P, Gheysens O, Vermeersch P, Gillijns H, Pellens M, Liu X, Collen D, Janssens S (2008) Sustained endothelial progenitor cell dysfunction after chronic hypoxia-induced pulmonary hypertension. Stem Cells 26(4):1017–1026. doi:10.1634/stemcells.2007-0562

Marsboom G, Wietholt C, Haney CR, Toth PT, Ryan JJ, Morrow E, Thenappan T, Bache-Wiig P, Piao L, Paul J, Chen CT, Archer SL (2012) Lung (1)(8)F-fluorodeoxyglucose positron emission tomography for diagnosis and monitoring of pulmonary arterial hypertension. Am J Respir Crit Care Med 185(6):670–679. doi:10.1164/rccm.201108-1562OC

Nagaya N, Kangawa K, Kanda M, Uematsu M, Horio T, Fukuyama N, Hino J, Harada-Shiba M, Okumura H, Tabata Y, Mochizuki N, Chiba Y, Nishioka K, Miyatake K, Asahara T, Hara H, Mori H (2003) Hybrid cell-gene therapy for pulmonary hypertension based on phagocytosing action of endothelial progenitor cells. Circulation 108(7):889–895. doi:10.1161/01.CIR.0000079161.56080.22

Obi S, Yamamoto K, Shimizu N, Kumagaya S, Masumura T, Sokabe T, Asahara T, Ando J (2009) Fluid shear stress induces arterial differentiation of endothelial progenitor cells. J Appl Physiol (1985) 106(1):203–211. doi:10.1152/japplphysiol.00197.2008

Paviotti G, Fadini GP, Boscaro E, Agostini C, Avogaro A, Chiandetti L, Baraldi E, Filippone M (2011) Endothelial progenitor cells, bronchopulmonary dysplasia and other short-term outcomes of extremely preterm birth. Early Hum Dev 87(7):461–465. doi:10.1016/j.earlhumdev.2011.03.011

Peinado VI, Barbera JA, Ramirez J, Gomez FP, Roca J, Jover L, Gimferrer JM, Rodriguez-Roisin R (1998) Endothelial dysfunction in pulmonary arteries of patients with mild COPD. Am J Physiol 274(6 Pt 1):L908–L913

Qi Y, Jiang Q, Chen C, Cao Y, Qian L (2013) Circulating endothelial progenitor cells decrease in infants with bronchopulmonary dysplasia and increase after inhaled nitric oxide. PLoS One 8(11):e79060. doi:10.1371/journal.pone.0079060

Rajantie I1, Ilmonen M, Alminaite A, Ozerdem U, Alitalo K, Salven P (2004) Adult bone marrow-derived cells recruited during angiogenesis comprise precursors for periendothelial vascular mural cells. Blood 104(7):2084–2086

Rehman J, Li J, Orschell CM, March KL (2003) Peripheral blood "endothelial progenitor cells" are derived from monocyte/macrophages and secrete angiogenic growth factors. Circulation 107(8):1164–1169

Reidy MA, Schwartz SM (1983) Endothelial injury and regeneration. IV. Endotoxin: a nondenuding injury to aortic endothelium. Lab Invest 48(1):25–34

Reidy MA, Schwartz SM (1984) Recent advances in molecular pathology. Arterial endothelium—assessment of in vivo injury. Exp Mol Pathol 41(3):419–434

Roth RA, Reindel JF (1991) Lung vascular injury from monocrotaline pyrrole, a putative hepatic metabolite. Adv Exp Med Biol 283:477–487

Sakamaki Y, Matsumoto K, Mizuno S, Miyoshi S, Matsuda H, Nakamura T (2002) Hepatocyte growth factor stimulates proliferation of respiratory epithelial cells during postpneumonectomy compensatory lung growth in mice. Am J Respir Cell Mol Biol 26(5):525–533. doi:10.1165/ajrcmb.26.5.4714

Schwarz TM, Leicht SF, Radic T, Rodriguez-Araboalaza I, Hermann PC, Berger F, Saif J, Bocker W, Ellwart JW, Aicher A, Heeschen C (2012) Vascular incorporation of endothelial colony-forming cells is essential for functional recovery of murine ischemic tissue following cell therapy. Arterioscler Thromb Vasc Biol 32(2):e13–e21. doi:10.1161/ATVBAHA.111.239822

Shi Q, Rafii S, Wu MH, Wijelath ES, Yu C, Ishida A, Fujita Y, Kothari S, Mohle R, Sauvage LR, Moore MA, Storb RF, Hammond WP (1998) Evidence for circulating bone marrow-derived endothelial cells. Blood 92(2):362–367

Sieveking DP, Buckle A, Celermajer DS, Ng MK (2008) Strikingly different angiogenic properties of endothelial progenitor cell subpopulations: insights from a novel human angiogenesis assay. J Am Coll Cardiol 51(6):660–668. doi:10.1016/j.jacc.2007.09.059

Stenmark KR, Abman SH (2005) Lung vascular development: implications for the pathogenesis of bronchopulmonary dysplasia. Annu Rev Physiol 67:623–661. doi:10.1146/annurev.physiol.67.040403.102229

Stump MM, Jordan GL Jr, Debakey ME, Halpert B (1963) Endothelium grown from circulating blood on isolated intravascular Dacron hub. Am J Pathol 43:361–367

Suratt BT, Cool CD, Serls AE, Chen L, Varella-Garcia M, Shpall EJ, Brown KK, Worthen GS (2003) Human pulmonary chimerism after hematopoietic stem cell transplantation. Am J Respir Crit Care Med 168(3):318–322. doi:10.1164/rccm.200301-145OC

Takahashi M, Nakamura T, Toba T, Kajiwara N, Kato H, Shimizu Y (2004) Transplantation of endothelial progenitor cells into the lung to alleviate pulmonary hypertension in dogs. Tissue Eng 10(5–6):771–779. doi:10.1089/1076327041348563

Taraseviciene-Stewart L, Kasahara Y, Alger L, Hirth P, Mc Mahon G, Waltenberger J, Voelkel NF, Tuder RM (2001) Inhibition of the VEGF receptor 2 combined with chronic hypoxia causes cell death-dependent pulmonary endothelial cell proliferation and severe pulmonary hypertension. FASEB J 15(2):427–438. doi:10.1096/fj.00-0343com

Terada N, Hamazaki T, Oka M, Hoki M, Mastalerz DM, Nakano Y, Meyer EM, Morel L, Petersen BE, Scott EW (2002) Bone marrow cells adopt the phenotype of other cells by spontaneous cell fusion. Nature 416(6880):542–545. doi:10.1038/nature730

Thebaud B, Abman SH (2007) Bronchopulmonary dysplasia: where have all the vessels gone? Roles of angiogenic growth factors in chronic lung disease. Am J Respir Crit Care Med 175(10):978–985. doi:10.1164/rccm.200611-1660PP

Thenappan T, Shah SJ, Rich S, Tian L, Archer SL, Gomberg-Maitland M (2010) Survival in pulmonary arterial hypertension: a reappraisal of the NIH risk stratification equation. Eur Respir J 35(5):1079–1087. doi:10.1183/09031936.00072709

Toshner M, Voswinckel R, Southwood M, Al-Lamki R, Howard LS, Marchesan D, Yang J, Suntharalingam J, Soon E, Exley A, Stewart S, Hecker M, Zhu Z, Gehling U, Seeger W, Pepke-Zaba J, Morrell NW (2009) Evidence of dysfunction of endothelial progenitors in pulmonary arterial hypertension. Am J Respir Crit Care Med 180(8):780–787. doi:10.1164/rccm.200810-1662OC

Tuder RM, Chacon M, Alger L, Wang J, Taraseviciene-Stewart L, Kasahara Y, Cool CD, Bishop AE, Geraci M, Semenza GL, Yacoub M, Polak JM, Voelkel NF (2001a) Expression of angiogenesis-related molecules in plexiform lesions in severe pulmonary hypertension: evidence for a process of disordered angiogenesis. J Pathol 195(3):367–374. doi:10.1002/path.953

Tuder RM, Cool CD, Yeager M, Taraseviciene-Stewart L, Bull TM, Voelkel NF (2001b) The pathobiology of pulmonary hypertension. Endothelium. Clin Chest Med 22(3):405–418

Urbich C, Aicher A, Heeschen C, Dernbach E, Hofmann WK, Zeiher AM, Dimmeler S (2005) Soluble factors released by endothelial progenitor cells promote migration of endothelial cells and cardiac resident progenitor cells. J Mol Cell Cardiol 39(5):733–742. doi:10.1016/j.yjmcc.2005.07.003

Vasa M, Fichtlscherer S, Aicher A, Adler K, Urbich C, Martin H, Zeiher AM, Dimmeler S (2001) Number and migratory activity of circulating endothelial progenitor cells inversely correlate with risk factors for coronary artery disease. Circ Res 89(1):E1–E7

Wang XX, Zhang FR, Shang YP, Zhu JH, Xie XD, Tao QM, Chen JZ (2007) Transplantation of autologous endothelial progenitor cells may be beneficial in patients with idiopathic pulmonary arterial hypertension: a pilot randomized controlled trial. J Am Coll Cardiol 49(14):1566–1571. doi:10.1016/j.jacc.2006.12.037

Yamada M, Kubo H, Kobayashi S, Ishizawa K, Numasaki M, Ueda S, Suzuki T, Sasaki H (2004) Bone marrow-derived progenitor cells are important for lung repair after lipopolysaccharide-induced lung injury. J Immunol 172(2):1266–1272

Yamada M, Kubo H, Ishizawa K, Kobayashi S, Shinkawa M, Sasaki H (2005) Increased circulating endothelial progenitor cells in patients with bacterial pneumonia: evidence that bone marrow derived cells contribute to lung repair. Thorax 60(5):410–413. doi:10.1136/thx.2004.034058

Zani BG, Kojima K, Vacanti CA, Edelman ER (2008) Tissue-engineered endothelial and epithelial implants differentially and synergistically regulate airway repair. Proc Natl Acad Sci U S A 105(19):7046–7051. doi:10.1073/pnas.0802463105

Zhang Y, Ingram DA, Murphy MP, Saadatzadeh MR, Mead LE, Prater DN, Rehman J (2009) Release of proinflammatory mediators and expression of proinflammatory adhesion molecules by endothelial progenitor cells. Am J Physiol Heart Circ Physiol 296(5):H1675–H1682. doi:10.1152/ajpheart.00665.2008

Zhang M, Malik AB, Rehman J (2014) Endothelial progenitor cells and vascular repair. Curr Opin Hematol 21(3):224–228. doi:10.1097/MOH.0000000000000041

Zhao YD, Courtman DW, Deng Y, Kugathasan L, Zhang Q, Stewart DJ (2005) Rescue of monocrotaline-induced pulmonary arterial hypertension using bone marrow-derived endothelial-like progenitor cells: efficacy of combined cell and eNOS gene therapy in established disease. Circ Res 96(4):442–450. doi:10.1161/01.RES.0000157672.70560.7b

Zhao YY, Gao XP, Zhao YD, Mirza MK, Frey RS, Kalinichenko VV, Wang IC, Costa RH, Malik AB (2006) Endothelial cell-restricted disruption of FoxM1 impairs endothelial repair following LPS-induced vascular injury. J Clin Invest 116(9):2333–2343. doi:10.1172/JCI27154

Zhao YD, Ohkawara H, Rehman J, Wary KK, Vogel SM, Minshall RD, Zhao YY, Malik AB (2009) Bone marrow progenitor cells induce endothelial adherens junction integrity by sphingosine-1-phosphate-mediated Rac1 and Cdc42 signaling. Circ Res 105(7):696–704. doi:10.1161/CIRCRESAHA.109.199778, 698 p following 704

Zhu JH, Wang XX, Zhang FR, Shang YP, Tao QM, Chen JZ (2008) Safety and efficacy of autologous endothelial progenitor cells transplantation in children with idiopathic pulmonary arterial hypertension: open-label pilot study. Pediatr Transplant 12(6):650–655. doi:10.1111/j.1399-3046.2007.00863.x

Ziebart T, Yoon CH, Trepels T, Wietelmann A, Braun T, Kiessling F, Stein S, Grez M, Ihling C, Muhly-Reinholz M, Carmona G, Urbich C, Zeiher AM, Dimmeler S (2008) Sustained persistence of transplanted proangiogenic cells contributes to neovascularization and cardiac function after ischemia. Circ Res 103(11):1327–1334. doi:10.1161/CIRCRESAHA.108.180463

Ziegelhoeffer T, Fernandez B, Kostin S, Heil M, Voswinckel R, Helisch A, Schaper W (2004) Bone marrow-derived cells do not incorporate into the adult growing vasculature. Circ Res 94(2):230–238. doi:10.1161/01.RES.0000110419.50982.1C

Chapter 18
The Use of Embryonic Stem Cells and Induced Pluripotent Stem Cells to Model Pulmonary Arterial Hypertension

Anna R. Hemnes, Eric D. Austin, and Susan Majka

Abbreviations

BMPR2	Bone morphogenic protein (receptor 2)
COPD	Chronic obstructive pulmonary disease
EC	Endothelial cell
ESC	Embryonic stem cells
EUCOMM	European conditional mouse mutagenesis program
FLK-1/KDR	Fetal liver kinase-1/Kinase insert domain receptor
HPAH	Hereditary pulmonary arterial hypertension
IPAH	Idiopathic pulmonary arterial hypertension
iPSC	Induced pluripotent stem cells
KOMP-ES	Knockout mouse project-embryonic stem cell repository
L-EPC	Late outgrowth endothelial progenitor cell
MPSVII	Mucopolysaccharidosis
MSC	Mesenchymal stromal cell
NIH	National Institute of Health

A.R. Hemnes
Department of Medicine, Division of Allergy, Pulmonary and Critical Care Medicine, Vanderbilt University, Nashville, TN, USA

E.D. Austin
Department of Pediatrics, Vanderbilt University, Nashville, TN, USA

S. Majka Ph.D. (✉)
Department of Medicine, Division of Allergy, Pulmonary and Critical Care Medicine, Vanderbilt Center for Stem Cell Biology, Cell and Developmental Biology, Vanderbilt University, P475 MRBIV/Langford, 2213 Garland Ave, Nashville, TN 37232, USA
e-mail: Susan.M.Majka@Vanderbilt.Edu

A. Firth, J.X.-J. Yuan (eds.), *Lung Stem Cells in the Epithelium and Vasculature*, Stem Cell Biology and Regenerative Medicine, DOI 10.1007/978-3-319-16232-4_18

PAH	Pulmonary arterial hypertension
PF	Pulmonary fibrosis
PRDC	Protein related to DAN and cerberus
RV	Right ventricular
SCNT	Somatic cell nuclear transfer
VEGF	Vascular endothelial growth factor

18.1 Pulmonary Arterial Hypertension

Pulmonary arterial hypertension (PAH) is characterized by elevated pulmonary artery pressures and widespread vascular remodeling including endothelial cell (EC) dysfunction and occlusion or rarefaction of the peripheral pulmonary microvasculature (Badesch et al. 2009; Runo and Loyd 2003). Arterial changes occur in all the layers of the vascular wall. In particular, occlusion of the vascular lumen by neointimal remodeling is characterized by the (1) expansion of endothelial cells, (2) microthrombi, and (3) cells with markers of vascular smooth muscle cells and inflammatory cells. A fibroproliferative response characterized by an increase in extracellular matrix components may be seen and vessels may display plexiform lesions, highly characteristic of PAH. Remodeling of the adjacent adventitia is characterized by smooth muscle hypertrophy (Farber and Loscalzo 2004; Pietra et al. 2004). Multipotent mesenchymal stromal cells (MSC) also contribute to the de novo muscularization of microvessels (Chow et al. 2013).

Such progressive vascular remodeling results in right ventricular (RV) dysfunction and subsequent failure (Ghofrani et al. 2009). Because RV function is the primary determinant of mortality in PAH, RV dysfunction exacerbated by dysfunctional BMPR2 signaling may underlie this finding (Hemnes et al. 2013). In support of this theory, BMP signaling is critical to cardiac development (van Wijk et al. 2007), and *BMPR2* mutations are linked to congenital heart disease-associated PAH (Roberts et al. 2004; Rosenzweig et al. 2008). All forms of PAH have a high mortality rate despite current therapeutic options.

18.2 Underlying Causes of Pulmonary Arterial Hypertension

Pulmonary hypertension (PH) is associated with a wide array of comorbid conditions such as chronic obstructive pulmonary disease (COPD) or pulmonary fibrosis (PF); it also occurs as a primary pulmonary vascular disease known as either idiopathic pulmonary arterial hypertension (IPAH) or heritable pulmonary arterial hypertension (HPAH) in which pulmonary vascular disease occurs in the absence of any parenchymal changes or known associated conditions (Simonneau et al. 2013). On average, patients with HPAH die at a younger age than those with IPAH (Sztrymf et al. 2008). About 80 % of HPAH patients have a known BMPR2 mutation characterized as an autosomal dominant disease with low penetrance (approximately 20 %). And approximately 10 % of IPAH patients also have a mutation in BMPR2 (Austin and Loyd 2014). This suggests that additional factors influence the disease onset, progression, and symptoms.

In addition to genetic mutations, dysregulated BMPR2 signaling is strongly associated with the development of idiopathic and other forms of PAH (Atkinson et al. 2002; West et al. 2014). To date, the exact molecular mechanisms through which BMPR2 derangement promotes PAH are unknown. Unfortunately, most genetic rodent models of PAH do not precisely recapitulate the disease pathology, displaying less substantial pulmonary vascular remodeling and inflammation. Alternative animal models have been used such as monocrotaline injection, hypoxia, or the combination of a VEGF receptor antibody and hypoxia. These toxin- or pharmacologically induced rodent models of PAH display substantial remodeling, but are most likely the result of nonspecific activation of signaling networks by mechanisms that are not representative of the underlying causes of PAH. Because of the limitations of these animal models, drug discovery efforts have thus far been of limited success.

18.3 Predicted Benefits of Stem Cell Modeling to Understand Pulmonary Arterial Hypertension

While it remains well-known that dysfunctional BMP signaling results in vascular remodeling and PAH, the precise mechanisms responsible for disease pathogenesis remain elusive (Morrell et al. 2009). To date, the evaluation of early molecular events in the cells of PAH patients has been limited because vascular-specific cells can only be studied at a late stage in the disease process as the risk of lung biopsy is elevated in PAH patients. As a result, specimens are typically obtained at the time of lung transplant or autopsy, when pulmonary vascular changes are end-stage. Further, patient-derived somatic cells also have a limited life span in culture. Moreover, these cells are frequently obtained from patients with end-stage disease exposed to multiple drug therapies, leaving researchers with questions about whether their findings recapitulate disease-initiating processes or are simply the result of pharmacologic intervention or subsequent host compensatory responses. Engineered iPSCs, and subsequently derived endothelium, smooth muscle cells, and cardiac myocytes, are a viable option to address these limitations and to investigate molecular pathways affected by dysregulated BMPR2 signaling. With iPSC technology somatic cells, such as dermal fibroblasts, circulating mononuclear cells or other somatic cells are reprogrammed into pluripotent cells via overexpression of specific transcription factors (Okita et al. 2007, 2008; Wu et al. 2009; Yamanaka 2007, 2008, 2009a, b; Ikonomou et al. 2011; Somers et al. 2010; Sommer et al. 2009, 2010). iPSCs display the critical features which define pluripotent stem cells, including self-renewal, multilineage differentiation, and functional contribution to tissues of all three primary germ layers. With iPSC technology, new paths of scientific inquiry are available to elucidate the underlying mechanisms of PAH. For example, since multiple cell types are likely to contribute to PAH pathogenesis, the ability to differentiate iPSCs into different lineages will allow the analysis of multiple cell types derived from a single patient, including vascular endothelium, smooth muscle cells, and cardiomyocytes (Kattman et al. 2006; Sommer et al. 2009; Geti et al. 2012; West et al. 2014). The use of iPSCs derived from PAH patients also confers the ability to study early, initiating cellular events in the pathogenesis of PAH in relevant vascular and cardiac cell types. Multiple developmental pathways of cardiovascular lineage specification may also be evaluated using embryonic stem cell (ESC) and developmental differentiation protocols as a roadmap (Longmire et al. 2012; Mou et al. 2012). For example, one may evaluate the stages of differentiation and commitment of iPSCs into mesenchymal cells (Bilousova et al. 2010) or endothelium (Choi et al. 1998; West et al. 2014), given that disease processes may represent a re-activation of developmental programs. Therefore, pluripotent cell types such as ESCs and iPSCs will facilitate the study of PAH regardless of its genetic underpinnings, and patient-derived iPSC offer a unique platform for mechanistic studies of, and drug development for, PAH.

18.4 The Use of Stem Cells to Model Pulmonary Arterial Hypertension

18.4.1 Differentiation of Patient Pluripotent Stem cells to Vascular and Cardiac Lineages

BMP signaling plays a necessary and critical role in vascular and cardiac differentiation (van Wijk et al. 2007). Recent publications by West and colleagues have shown molecular and functional differences in iPSCs from PAH patients compared with healthy controls (West et al. 2014). It is therefore conceivable that PAH patient iPSCs, and their differentiated progeny, will also manifest differences in vascular smooth muscle and cardiac differentiation compared to iPSCs from healthy subjects. This approach may also help to determine which types of cells do not act differently from comparable cells found in healthy subjects; such cells could provide resiliency to pathogenesis worth exploring as a novel method to resist PAH in the susceptible individual. Further study of these differences will likely yield new mechanistic insights into PAH disease pathogenesis.

In addition to disease modeling of HPAH, employing iPSC lines with known mutations in receptors that affect BMP signaling via distinct mechanisms will help to advance our understanding of one of the key signaling pathways of cardiovascular development, embryonic patterning, and musculoskeletal formation. On this basis, the novel tool of iPSCs with known *BMPR2* mutations, both activating and inactivating, may present a unique opportunity to address the cellular and molecular basis of cardiovascular diseases more broadly.

18.4.2 *Vascular Lineages: Endothelium, Mesenchyme, and Smooth Muscle and Mesenchyme*

iPSC-derived mesenchymal and endothelial lines may be employed to identify molecular signatures of PAH in the absence of confounding effects including end stage disease, drug therapy, or physiological changes resulting from elevated pulmonary artery pressures. Recent studies have described differentiation of smooth muscle/pericytes, lymphatic, venous, and arterial endothelium from iPSC clones, illustrating the utility of these cells in studying vascular disease pathogenesis (Narazaki et al. 2008; Adams et al. 2013; Choi et al. 2009). PAH iPSC have also been differentiated to endothelium (West et al. 2014; Geti et al. 2012), mesenchymal stromal cells (Fig. 18.1) as well as smooth muscle cells (Fig. 18.2). In addition to characterization of pluripotential ability of the PAH iPSC, further characterization

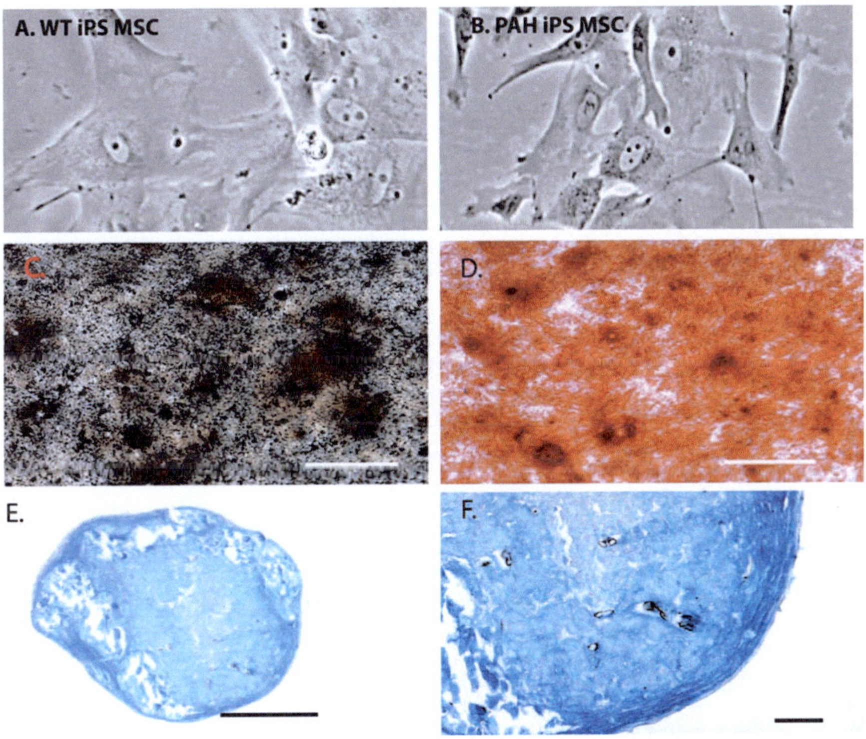

Fig. 18.1 Mesenchymal stromal cell (MSC) differentiation was confirmed via multi-lineage differentiation. (**a**, **b**) WT and BMPR2 mutant iPSC were differentiated to MSC. Subsequent MSC were analyzed for their potential to differentiate into the mesenchymal lineages of bone, fat, and cartilage. (**c**–**f**) Representative images of mesenchymal lineage differentiation. (**c**) Von Kossa stain was used to detect the accumulation of calcium. (**d**) Oil red O localized the appearance of lipid droplets. (**e**–**f**) Alcian blue was used to stain matrix characteristic of chondrocyte microspheres. Scale bar = 500 μM

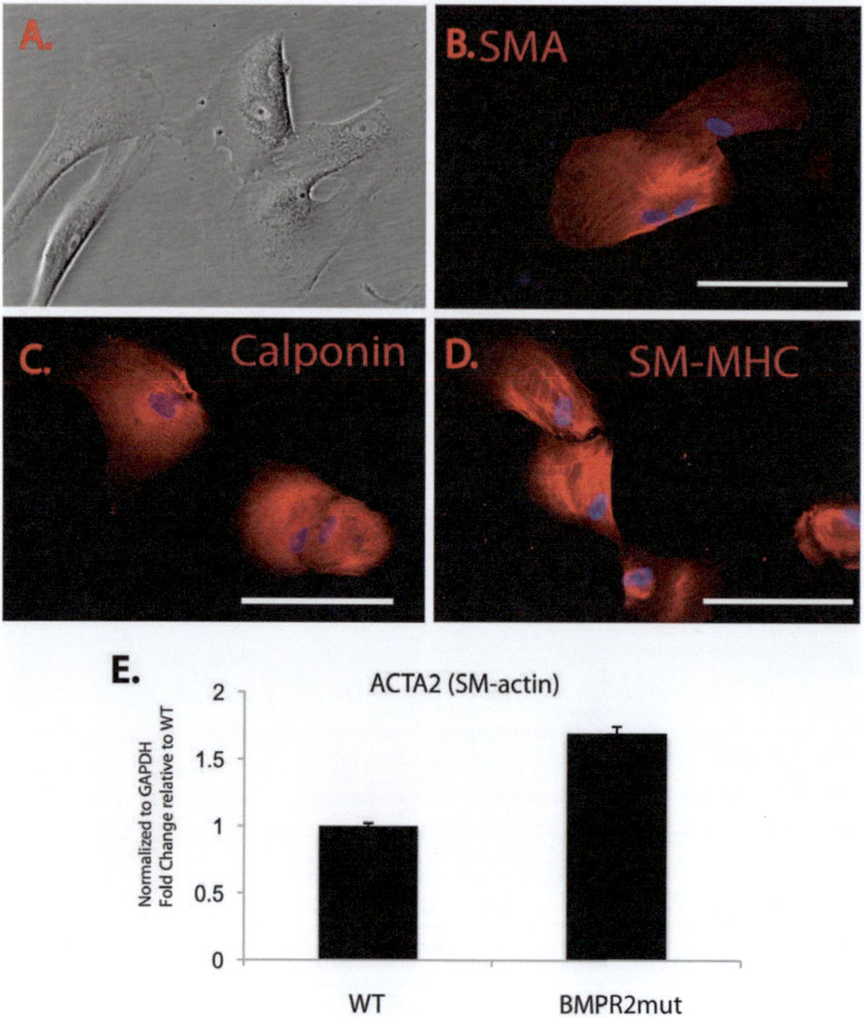

Fig. 18.2 Smooth muscle cell differentiation of PAH iPSC cells. WT and BMPR2 mutant iPSC were differentiated to smooth muscle cells (SMC). (**a**) Representative phase image of PAH SMC. (**b–d**) Successful SMC differentiation was demonstrated by immunofluorescent staining to localize and detect smooth muscle alpha actin (SMA), calponin, and smooth muscle myosin heavy chain (SM-MHC). The proteins localized to the cytoskeletal elements. 200× mag. (**e**) Quantitative PCR was confirmed differential expression of smooth muscle alpha actin message (Acta2) in WT and BMPR2 mutant iPSC-derived SMC

of karyotypic stability and retention of the parent cell mutation of interest was also performed (Fig. 18.3; (West et al. 2014; Geti et al. 2012)).

In first study to report reprogramming of iPSC from PAH patient samples, Geti and colleagues utilized late outgrowth endothelial precursor cells (L-EPCs) isolated from peripheral blood mononuclear cells (Geti et al. 2012). These studies demonstrated the feasibility and high efficiency of reprogramming control and PAH (HPAH and IPAH)

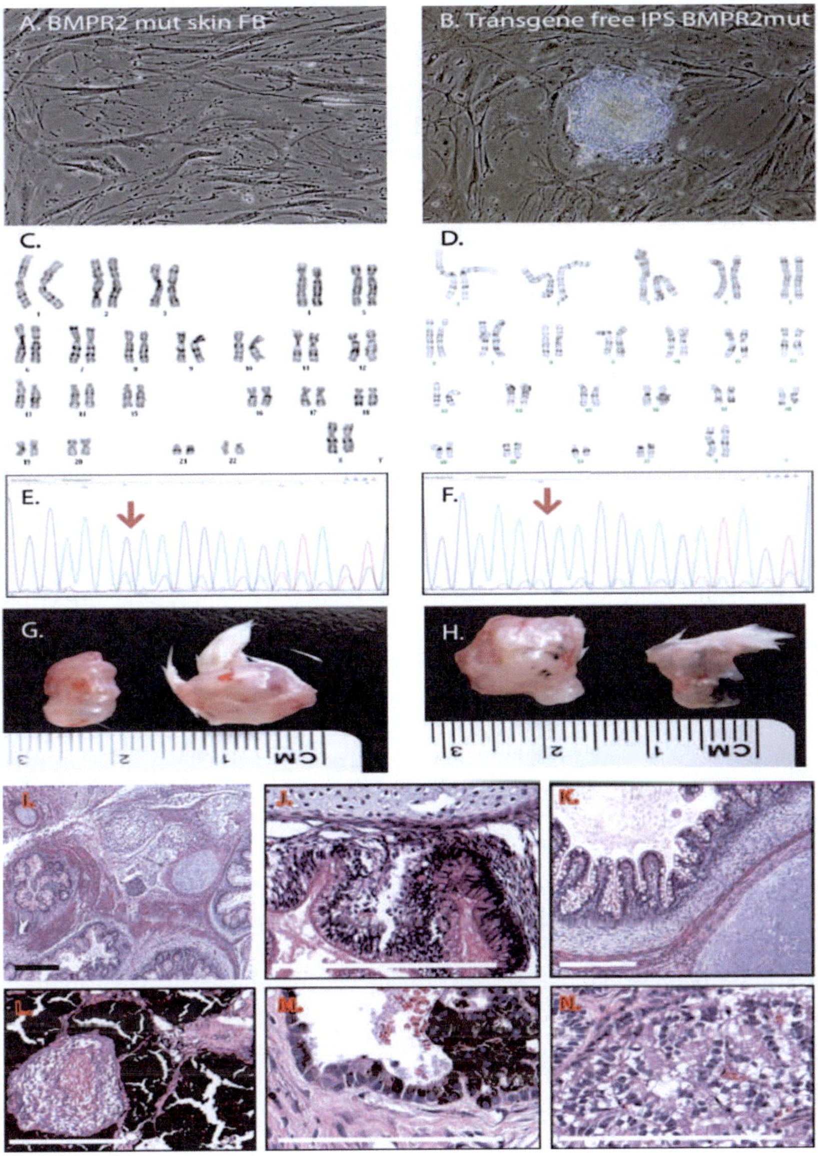

Fig. 18.3 Characterization of PAH iPSC cells from mutant skin fibroblasts. (**a**, **b**) iPSC cell lines were engineered using the STEMCCA vector and an HPAH patient primary skin fibroblast cell line containing a mutation in exon 12 of the BMPR2 gene. (**c**, **d**) Karyotype was confirmed by banding analysis in both the fibroblast (FB) and iPSC cells. (**e**, **f**) The HPAH patient primary skin fibroblast cell lines contained a mutation in exon 12 (nucleotide change 2504delC) of the BMPR2 gene that localized to the cytoplasmic domain of the receptor. Sequencing of BMPR2 mutant iPSC genomic DNA confirmed the retention of this mutation in the BMPR2 mutant iPSC cell clones. In the sequence photo, "C" is denoted by the *blue color*. In the native cell HPAH skin FB, there is a large *blue peak*, and a smaller *green peak*. That smaller *green peak* is actually a shift of the next nucleotide, since the *green* would be "next". *Red arrow* indicates the mutation site. (**g**, **h**) Representative images of teratoma analysis of the wildtype (WT) and BMPR2 mutant iPSC cell lines. (**i–n**) Representative images of (**h**, **e**) stained sections from the BMPR2 mutant iPSC cell-derived teratomas. Visible cell types include: mesodermal proliferation, adipocytes, cyst formation, epithelial cells (gut, respiratory, mucous producing), melanocytes (hypersecreting), and chondrocytes. Scale bar = 100 μM

L-EPC to iPSC using integrating retroviral technology. The resultant iPSC were capable of clonal proliferation and retained normal karyotype, providing evidence that this method is an ideal tool for high-throughput pharmacological screening.

West et al. utilized both excisable and non-integrating reprogramming techniques to engineer control, IPAH, and HPAH patient iPSC lines in a multifaceted approach to evaluate molecular pathways affected by dysregulated BMPR2 signaling (West et al. 2014). While integrating retroviral reprogramming is associated with high efficiency reprogramming, they also have the potential to mask subtle mutations and changes within the genome, hence the choice for excisable or non-integrating approaches to generate iPSC lines. In these experiments, control and PAH iPSCs were differentiated to MSC and endothelium. PAH patient iPSC-derived cells maintained a "PAH" phenotype in vitro, including decreased expression of Tie2, VEGF-A, Flk-1/KDR, and BMP-4, all factors involved in the regulation of vascular stability. Additionally, MSC exhibited decreased rates of apoptosis and increased oxidant stress following exposure to hypoxia (West et al. 2014). Further, gene expression profiles during differentiation across iPSC-derived mesenchymal and endothelial cells were compared to skin fibroblasts from control, HPAH, and IPAH patients, addressing the hypothesis that altered human BMPR2 signaling would result in a genetic signature common across multiple cell types. Common molecular pathways affected by deregulated BMPR2 signaling were identified, the most prominent being Wnt. BMPR2 signaling may regulate both canonical and non-canonical Wnt pathways in endothelium and mesenchymal cells to influence proliferation, survival, and motility during angiogenesis and remodeling of the pulmonary circulation (Laumanns et al. 2009; de Jesus Perez et al. 2011; Alfaro et al. 2010). While the relationship of BMPR2 and Wnt pathways has been defined during development, the regulatory targets of Wnt signaling PAH are largely unknown.

Taken together, these studies exploited iPSC technology and linked deregulated developmental pathways with adult disease, over multiple cell types and differentiation states. Combinations of iPSC and primary patient cell modeling may ultimately enable the identification of cellular defects that initiate the vascular pathology in PAH and provide access to multiple renewable cell types in which to test potential therapies (West et al. 2014; Geti et al. 2012).

18.4.3 Cardiac Myocyte Lineage

Critical differences exist between human and animal cardiomyocytes including mechanical and physiological properties (Sheng and Charles 2013). For example, the mouse heart beats seven times faster and is 1,000-fold smaller than human. While human ESCs may be used to model cardiac diseases, there exist significant obstacles including limited availability of embryonic tissues and ESC lines as well as ethical controversies hindering scientific progress using these experimental techniques. The use of stem cells to model cardiac disease in humans is ongoing especially advancing in those diseases with a pronounced clinical presentation or inherited genetic component (Sheng and Charles 2013; Moretti et al. 2010, 2013). Recent work by our group and others has suggested that BMPR2 may have independent effects on the RV in PAH that cannot be explained by pulmonary vascular disease with evidence of RV lipotoxicity (Hemnes et al. 2013; McGoon et al. 2013). Because of the limitations to animal models and access to human tissue, exploring the mechanisms of this has thus far been challenging. The underlying BMPR2-dependent molecular mechanisms affected in cardiomyocytes during PAH, either heritable or idiopathic, are unknown and ESC as well as iPSC offer an important tool to investigate these mechanisms.

In vitro studies show that the differentiation of iPSCs into cardiovascular progenitors, and ultimately to mature ECs and cardiomyocytes, is enhanced by proper exposure to BMP morphogens at specific time points of the differentiation process (Kattman et al. 2006; Yang et al. 2008). When directing the differentiation of normal ESCs or iPSCs, antagonists of BMP signaling, such as Noggin and PRDC or chemical compounds such as dorsomorphin, influence both the yield and physiologic properties of stem cell-derived cardiomyocytes (Hao et al. 2008). Thus, iPSCs derived from HPAH patients carrying *BMPR2* mutations as well as IPAH patients, with decreased BMP signaling, may manifest differences in cardiac myocyte differentiation compared to iPSCs from healthy individuals without dysfunctional BMP signaling. Thus, iPSCs provide a useful model for comparative studies.

18.5 Murine Embryonic Stem Cell and Induced Pluripotent Stem Cell Models

While significant biological differences exist between rodent and human models of disease, there remains an advantage to using rodent models to study disease processes in an integrative fashion. Starting material is readily available and the ethical concerns associated with human somatic cell nuclear transfer (SCNT) and derivation of ESCs from fertilized embryos are not concerns when studying genetic disease modeled with rodents.

18.5.1 Benefits of Pluripotent Rodent Model Systems

As human iPSC are an invaluable and innovative tool to repeatedly generate multiple cell types from a single clone, murine ESC and iPSC provide the ability to study genetically identical derived cell types indefinitely, from existing, well-characterized rodent models of disease. This ability to re-derive identical cell types from pluripotent lines is in striking contrast to typical primary cell lines isolated from rodent models, which are very heterogeneous in phenotype and proliferate for a limited number of passages. The derivation of ESC or iPSC lines from a rodent model has a few innate differences including the efficiency of generation and degree of manipulation of the starting cell population, which should ideally be minimized to preserve the genomic stability of clones.

Expanding de novo isolated ESC lines is straightforward due to the absence of transfection with vector or infection with virus. If background strain is not a consideration, crossing mutants of interest to an inbred mouse strain (129) can also increase the efficiency of ES cells derivation. Isolation and expansion of ESC requires specific expertise for isolation and cloning of cells from the inner cell mass of a fertilized embryo. In contrast, while iPSC cells require exogenous factors to confer pluripotency, these factors can be administered to any somatic cell type in culture. The specific somatic cell type may influence reprogramming efficiency. While challenging, each technique requires specific expertise.

Both of these model systems in combination with an animal model provide the opportunity to perform gene correction in stem cells and subsequent evaluation of the therapeutic potential of the "autologous" manipulated product in an animal model of disease. Hanna and colleagues performed such a landmark study in 2007 (Hanna et al. 2007). The investigators employed a humanized mouse model of sickle cell anemia to engineer iPSC. Following gene correction of the human sickle hemoglobin allele in iPSC, the cells were differentiated into hematopoietic progenitors and transplanted back into the sickle cell mouse model. Following engraftment of the iPSC-derived hematopoietic cells, the sickle cell defect in iPSC-derived cells was corrected (Hanna et al. 2007).

Another important use of rodent pluripotent stem cells is to further understand the pathogenesis of genetic diseases, which cause embryonic lethality or deformity. One example of such a disease class is lysosomal storage disorders. Using a mouse model of mucopolysaccharidosis (MPSVII), Meng et al. characterized an enzyme deficiency specific to this disease (Meng et al. 2010). Following partial correction of the enzyme deficiency, cell functions and developmental specification of cells improved (Meng et al. 2010).

For the purposes of modeling PAH, there is a paucity of humanized rodent models that would enable study of either adult disease or embryonic lethality, thus demonstrating the benefit of using patient-derived iPSC. To date, such models include the human R899x, delEx4, and delEx2 BMPR2 mutations (Johnson et al. 2012; West et al. 2005, 2008; Majka et al. 2011; Frump et al. 2013). However, ESC may also be stably transfected with variations of mutant human BMPR2 and mechanisms of diseases evaluated in lineage-specific cell derivatives.

18.5.2 KOMP ES Cell Bank

The *k*nock*o*ut *m*ouse *p*roject-*e*mbryonic stem *c*ell repository (KOMP-ES) is an additional resource which may be utilized to study underlying mechanisms of PAH. The KOMP is a sub-portion of an National Institute of health (NIH) initiative in the United States, established with the purpose of creating a comprehensive and public resource comprised of mouse embryonic stem (ES) cells containing a null mutation in every gene in the mouse genome and related materials (mice, sperm, etc.). Programs including but not limited to the NIH-sponsored Knockout Mouse Project (KOMP) (Skarnes et al. 2011), and the European Conditional Mouse Mutagenesis Program (EUCOMM) (Friedel et al. 2007), have combined resources to reach the target goal of murine ES knockout lines representative of each of approximately 20,000 protein-coding genes. Currently, NIH and other agencies have funded the creation and phenotyping of mice from these ESC lines (Ryder et al. 2013). ESC are injected into the blastocoel cavity of 3.5 day old mouse blastocysts. These injected embryos are transferred to the uterine horns of pseudopregnant recipient females. Embryos gestate for approximately 18 days and resulting pups are chimeras, consisting of the ESC and the blastocyst cells. These resources provide valuable tools for the study of heritable mechanisms of human pulmonary and vascular disease in the mouse lines generated as well as in vitro differentiation of primary ES lines to cell types of interest.

18.6 Summary

Both rodent and human pluripotent stem cell technologies, available in ESC and iPSC, offer the potential to investigate a complement of PAH disease spectrum and to characterize genetic and environmental influences on vascular cell phenotype. Moreover, in the years ahead iPSC and their vascular derivatives should promote advancement of PAH research by: (1) providing access to a patient-derived expandable cell population; (2) recapitulating disease pathophysiology in a cell-specific manner; and (3) providing multiple cell types affected in PAH, endothelium, smooth muscle, and cardiac myocytes, with the same genetic makeup. As a caveat to this model system, a single cell type is not likely to recapitulate a complex disease phenotype, thus interaction between multiple cell types may be necessary to recapitulate the disease phenotype. In using pluripotent systems, multiple cell types from a single cell clone can now address complex disease pathogenesis in PAH.

Acknowledgments This work was funded by grants to S.M. Majka from the NIH NHLBI R01HL091105 and R01HL11659701. We would like to extend our gratitude to Dr. Darrell Kotton for the creation of the first BMPR2 mutant iPSC lines submitted to the Vanderbilt University repository (BMP3a and BMP1a).

References

Adams WJ, Zhang Y, Cloutier J, Kuchimanchi P, Newton G, Sehrawat S, Aird WC, Mayadas TN, Luscinskas FW, García-Cardeña G (2013) Functional vascular endothelium derived from human induced pluripotent stem cells. Stem Cell Rep 1(2):105–113, http://dx.doi.org/10.1016/j.stemcr.2013.06.007

Alfaro MP, Vincent A, Saraswati S, Thorne CA, Hong CC, Lee E, Young PP (2010) sFRP2 suppression of bone morphogenic protein (BMP) and Wnt signaling mediates mesenchymal stem cell (MSC) self-renewal promoting engraftment and myocardial repair. J Biol Chem 285(46):35645–35653. doi:10.1074/jbc.M110.135335

Atkinson C, Stewart S, Upton PD, Machado R, Thomson JR, Trembath RC, Morrell NW (2002) Primary pulmonary hypertension is associated with reduced pulmonary vascular expression of type II bone morphogenetic protein receptor. Circulation 105(14):1672–1678

Austin ED, Loyd JE (2014) The genetics of pulmonary arterial hypertension. Circ Res 115(1):189–202

Badesch DB, Champion HC, Sanchez MA, Hoeper MM, Loyd JE, Manes A, McGoon M, Naeije R, Olschewski H, Oudiz RJ, Torbicki A (2009) Diagnosis and assessment of pulmonary arterial hypertension. J Am Coll Cardiol 54(1 Suppl):S55–S66. doi:10.1016/j.jacc.2009.04.011, pii: S0735-1097(09)01214-5

Bilousova G, Hyun Jun D, King KB, DeLanghe S, Chick WS, Torchia EC, Chow KS, Klemm DJ, Roop DR, Majka SM (2010) Osteoblasts derived from induced pluripotent stem cells form calcified structures in Scaffolds both in vitro and in vitro. Stem Cells 29:206–216. doi:10.1002/stem.566

Choi KKM, Kazarov A, Papadimitriou JC, Keller G (1998) A common precursor for hematopoietic and endothelial cells. Development 125:725–732

Choi KD, Yu J, Smuga-Otto K, Salvagiotto G, Rehrauer W, Vodyanik M, Thomson J, Slukvin I (2009) Hematopoietic and endothelial differentiation of human induced pluripotent stem cells. Stem Cells 27(3):559–567. doi:10.1634/stemcells.2008-0922

Chow K, Fessel JP, Kaori Ihida-Stansbury, Schmidt EP, Gaskill C, Alvarez D, Graham B, Harrison DG, Wagner DH Jr, Nozik-Grayck E, West JD, Klemm DJ, Majka SM (2013) Dysfunctional resident lung mesenchymal stem cells contribute to pulmonary microvascular remodeling. Pulm Circ 3(1):31–49

de Jesus Perez VA, Ali Z, Alastalo T-P, Ikeno F, Sawada H, Lai Y-J, Kleisli T, Spiekerkoetter E, Qu X, Rubinos LH, Ashley E, Amieva M, Dedhar S, Rabinovitch M (2011) BMP promotes motility and represses growth of smooth muscle cells by activation of tandem Wnt pathways. J Cell Biol 192(1):171–188. doi:10.1083/jcb.201008060

Farber HW, Loscalzo J (2004) Pulmonary arterial hypertension. N Engl J Med 351(16):1655–1665. doi:10.1056/NEJMra035488, pii: 351/16/1655

Friedel RH, Seisenberger C, Kaloff C, Wurst W (2007) EUCOMM—the European conditional mouse mutagenesis program. Brief Funct Genomic Proteomic 6(3):180–185

Frump AL, Lowery JW, Hamid R, Austin ED, de Caestecker M (2013) Abnormal trafficking of endogenously expressed *BMPR2* mutant allelic products in patients with heritable pulmonary arterial hypertension. PLoS One 8(11):e80319

Geti I, Ormiston ML, Rouhani F, Toshner M, Movassagh M, Nichols J, Mansfield W, Southwood M, Bradley A, Rana AA, Vallier L, Morrell NW (2012) A practical and efficient cellular substrate for the generation of induced pluripotent stem cells from adults: blood-derived endothelial progenitor cells. Stem Cells Transl Med 1(12):855–865. doi:10.5966/sctm.2012-0093

Ghofrani HA, Barst RJ, Benza RL, Champion HC, Fagan KA, Grimminger F, Humbert M, Simonneau G, Stewart DJ, Ventura C, Rubin LJ (2009) Future perspectives for the treatment of pulmonary arterial hypertension. J Am Coll Cardiol 54(1 Suppl):S108–S117. doi:10.1016/j.jacc.2009.04.014, pii: S0735-1097(09)01219-4

Hanna J, Wernig M, Markoulaki S, Sun CW, Meissner A, Cassady JP, Beard C, Brambrink T, Wu LC, Townes TM, Jaenisch R (2007) Treatment of sickle cell anemia mouse model with iPSC cells generated from autologous skin. Science 318(5858):1920–1923. doi:10.1126/science.1152092

Hao J, Daleo MA, Murphy CK, Yu PB, Ho JN, Hu J, Peterson RT, Hatzopoulos AK, Hong CC (2008) Dorsomorphin, a selective small molecule inhibitor of BMP signaling, promotes cardiomyogenesis in embryonic stem cells. PLoS One 3(8):e2904. doi:10.1371/journal.pone.0002904

Hemnes AR, Brittain EL, Trammell AW, Fessel JP, Austin ED, Penner N, Maynard KB, Gleaves L, Talati M, Absi T, DiSalvo T, West J (2013) Evidence for right ventricular lipotoxicity in heritable pulmonary arterial hypertension. Am J Respir Crit Care Med 189(3):325–334. doi:10.1164/rccm.201306-1086OC

Ikonomou L, Hemnes AR, Bilousova G, Hamid R, Loyd JE, Hatzopoulos AK, Kotton DN, Majka SM, Austin ED (2011) Programmatic change: lung disease research in the era of induced pluripotency. Am J Physiol Lung Cell Mol Physiol 301(6):L830–L835

Johnson JA, Hemnes AR, Perrien DS, Schuster M, Robinson LJ, Gladson S, Loibner H, Bai S, Blackwell TR, Tada Y, Harral JW, Talati M, Lane KB, Fagan KA, West J (2012) Cytoskeletal defects in Bmpr2-associated pulmonary arterial hypertension. Am J Physiol Lung Cell Mol Physiol 302(5):L474–L484. doi:10.1152/ajplung.00202.2011

Kattman SJ, Huber TL, Keller GM (2006) Multipotent Flk-1(+) cardiovascular progenitor cells give rise to the cardiomyocyte, endothelial, and vascular smooth muscle lineages. Dev Cell 11(5):723–732. doi:10.1016/j.devcel.2006.10.002

Laumanns IP, Fink L, Wilhelm J, Wolff J-C, Mitnacht-Kraus R, Graef-Hoechst S, Stein MM, Bohle RM, Klepetko W, Hoda MAR, Schermuly RT, Grimminger F, Seeger W, Voswinckel R (2009) The noncanonical WNT pathway is operative in idiopathic pulmonary arterial hypertension. Am J Respir Cell Mol Biol 40(6):683–691. doi:10.1165/rcmb.2008-0153OC

Longmire TA, Ikonomou L, Hawkins F, Christodoulou C, Cao Y, Jean JC, Kwok LW, Mou H, Rajagopal J, Shen SS, Dowton AA, Serra M, Weiss DJ, Green MD, Snoeck H-W, Ramirez MI, Kotton DN (2012) Efficient derivation of purified lung and thyroid progenitors from embryonic stem cells. Cell Stem Cell 10(4):398–411

Majka S, Hagen M, Blackwell T, Harral J, Johnson J, Gendron R, Paradis H, Crona D, Loyd J, Nozik-Grayck E, Stenmark K, West J (2011) Physiologic and molecular consequences of endothelial Bmpr2 mutation. Respir Res 12(1):84

McGoon MD, Benza RL, Escribano-Subias P, Jiang X, Miller DP, Peacock AJ, Pepke-Zaba J, Pulido T, Rich S, Rosenkranz S, Suissa S, Humbert M (2013) Pulmonary arterial hypertension epidemiology and registries. J Am Coll Cardiol 62(25 Suppl):D51–D59. doi:10.1016/j.jacc.2013.10.023

Meng X-L, Shen J-S, Kawagoe S, Ohashi T, Brady RO, Eto Y (2010) Induced pluripotent stem cells derived from mouse models of lysosomal storage disorders. Proc Natl Acad Sci 107(17):7886–7891

Moretti A, Bellin M, Welling A, Jung CB, Lam JT, Bott-Flugel L, Dorn T, Goedel A, Hohnke C, Hofmann F, Seyfarth M, Sinnecker D, Schomig A, Laugwitz KL (2010) Patient-specific induced pluripotent stem-cell models for long-QT syndrome. N Engl J Med 363(15):1397–1409. doi:10.1056/NEJMoa0908679

Moretti A, Laugwitz K-L, Dorn T, Sinnecker D, Mummery C (2013) Pluripotent stem cell models of human heart disease. Cold Spring Harb Perspect Med 3(11)

Morrell NW, Adnot S, Archer SL, Dupuis J, Jones PL, MacLean MR, McMurtry IF, Stenmark KR, Thistlethwaite PA, Weissmann N, Yuan JX, Weir EK (2009) Cellular and molecular basis of pulmonary arterial hypertension. J Am Coll Cardiol 54(1 Suppl):S20–S31. doi:10.1016/j.jacc.2009.04.018, pii: S0735-1097(09)01226-1

Mou H, Zhao R, Sherwood R, Ahfeldt T, Lapey A, Wain J, Sicilian L, Izvolsky K, Lau FH, Musunuru K, Cowan C, Rajagopal J (2012) Generation of multipotent lung and airway progenitors from mouse ESCs and patient-specific cystic fibrosis iPSCs. Cell Stem Cell 10(4):385–397

Narazaki G, Uosaki H, Teranishi M, Okita K, Kim B, Matsuoka S, Yamanaka S, Yamashita JK (2008) Directed and systematic differentiation of cardiovascular cells from mouse induced pluripotent stem cells. Circulation 118(5):498–506. doi:10.1161/circulationaha.108.769562

Okita K, Ichisaka T, Yamanaka S (2007) Generation of germline-competent induced pluripotent stem cells. Nature 448(7151):313–317

Okita K, Nakagawa M, Hyenjong H, Ichisaka T, Yamanaka S (2008) Generation of mouse induced pluripotent stem cells without viral vectors. Science 322(5903):949–953

Pietra GG, Capron F, Stewart S, Leone O, Humbert M, Robbins IM, Reid LM, Tuder RM (2004) Pathologic assessment of vasculopathies in pulmonary hypertension. J Am Coll Cardiol 43(12 Suppl S):25S–32S

Roberts KE, McElroy JJ, Wong WP, Yen E, Widlitz A, Barst RJ, Knowles JA, Morse JH (2004) BMPR2 mutations in pulmonary arterial hypertension with congenital heart disease. Eur Respir J 24(3):371–374

Rosenzweig EB, Morse JH, Knowles JA, Chada KK, Khan AM, Roberts KE, McElroy JJ, Juskiw NK, Mallory NC, Rich S, Diamond B, Barst RJ (2008) Clinical implications of determining BMPR2 mutation status in a large cohort of children and adults with pulmonary arterial hypertension. J Heart Lung Transplant 27(6):668–674. doi:10.1016/j.healun.2008.02.009, pii: S1053-2498(08)00178-2

Runo JR, Loyd JE (2003) Primary pulmonary hypertension. Lancet 361(9368):1533–1544

Ryder E, Gleeson D, Sethi D, Vyas S, Miklejewska E, Dalvi P, Habib B, Cook R, Hardy M, Jhaveri K, Bottomley J, Wardle-Jones H, Bussell J, Houghton R, Salisbury J, Skarnes W, Ramirez-Solis R (2013) Molecular characterization of mutant mouse strains generated from the EUCOMM/KOMP-CSD ES cell resource. Mamm Genome 24(7–8):286–294. doi:10.1007/s00335-013-9467-x

Sheng CC, Charles CH (2013) Pluripotent Stem Cells to Model Human Cardiac Diseases, Pluripotent Stem Cells, Dr. Deepa Bhartiya (Ed.), ISBN: 978-953-51-1192-4, InTech, DOI: 10.5772/54373. Available from: http://www.intechopen.com/books/pluripotent-stem-cells/pluripotent-stem-cells-to-modelhuman-cardiac-diseases

Simonneau G, Gatzoulis MA, Adatia I, Celermajer D, Denton C, Ghofrani A, Gomez Sanchez MA, Krishna Kumar R, Landzberg M, Machado RF, Olschewski H, Robbins IM, Souza R (2013) Updated clinical classification of pulmonary hypertension. J Am Coll Cardiol 62(25 Suppl):D34–D41. doi:10.1016/j.jacc.2013.10.029

Skarnes WC, Rosen B, West AP, Koutsourakis M, Bushell W, Iyer V, Mujica AO, Thomas M, Harrow J, Cox T, Jackson D, Severin J, Biggs P, Fu J, Nefedov M, de Jong PJ, Stewart AF, Bradley A (2011) A conditional knockout resource for the genome-wide study of mouse gene function. Nature 474(7351):337–342. doi:10.1038/nature10163

Somers A, Jean JC, Sommer CA, Omari A, Ford CC, Mills JA, Ying L, Sommer AG, Jean JM, Smith BW, Lafyatis RA, Demierre MF, Weiss DJ, French DL, Gadue P, Murphy GJ, Mostoslavsky G, Kotton DN (2010) Generation of transgene-free lung disease-specific human iPSC cells using a single excisable lentiviral stem cell cassette. Stem Cells 28(10):1728–1740. doi:10.1002/stem.495

Sommer CA, Stadtfeld M, Murphy GJ, Hochedlinger K, Kotton DN, Mostoslavsky G (2009) Induced pluripotent stem cell generation using a single lentiviral stem cell cassette. Stem Cells 27(3):543–549. doi:10.1634/stemcells. 2008-1075, pii: stemcells.2008-1075

Sommer CA, Sommer AG, Longmire TA, Christodoulou C, Thomas DD, Gostissa M, Alt FW, Murphy GJ, Kotton DN, Mostoslavsky G (2010) Excision of reprogramming transgenes improves the differentiation potential of iPSC cells generated with a single excisable vector. Stem Cells 28(1):64–74. doi:10.1002/stem.255

Sztrymf BCF, Girerd B, Yaici A, Jais X, Sitbon O, Montani D, Souza R, Simonneau G, Soubrier F, Humbert M (2008) Clinical outcomes of pulmonary arterial hypertension in carriers of BMPR2 mutation. Am J Respir Crit Care Med 177:1377–1383

van Wijk B, Moorman AF, van den Hoff MJ (2007) Role of bone morphogenetic proteins in cardiac differentiation. Cardiovasc Res 74(2):244–255. doi:10.1016/j.cardiores.2006.11.022, pii: S0008-6363(06)00510-4

West JTY, Fagan KA, Steudel W, Fouty BW, Harral JW, Miller M, Ozimek J, Tuder RM, Rodman DM (2005) Suppression of type II bone morphogenic protein receptor in vascular smooth muscle induces pulmonary arterial hypertension in transgenic mice. Chest 128:553S

West J, Harral J, Lane K, Deng Y, Ickes B, Crona D, Albu S, Stewart D, Fagan K (2008) Mice expressing BMPR2R899X transgene in smooth muscle develop pulmonary vascular lesions. Am J Physiol Lung Cell Mol Physiol 295:L744–L755

West JD, Austin ED, Gaskill C, Marriott S, Baskir R, Bilousova G, Jean J-C, Hemnes AR, Menon S, Bloodworth NC, Fessel JP, Kropski JA, Irwin DC, Ware LB, Wheeler LA, Hong CC, Meyrick BO, Loyd JE, Bowman AB, Ess KC, Klemm DJ, Young PP, Merryman WD, Kotton D, Majka SM (2014) Identification of a common Wnt associated genetic signature across multiple cell types in pulmonary arterial hypertension. Am J Physiol Cell Physiol 307(5): C415–C430

Wu D, Hamilton B, Martin C, Gao Y, Ye M, Yao S (2009) Generation of induced pluripotent stem cells by reprogramming human fibroblasts with the stemgent human TF lentivirus set. J Vis Exp (34). pii: 1553. doi:10.3791/1553

Yamanaka S (2007) Strategies and new developments in the generation of patient-specific pluripotent stem cells. Cell Stem Cell 1(1):39–49

Yamanaka S (2008) Pluripotency and nuclear reprogramming. Philos Trans R Soc Lond B Biol Sci 363(1500):2079–2087. doi:10.1098/rstb.2008.2261, pii: 81R627X1V0591680

Yamanaka S (2009a) Elite and stochastic models for induced pluripotent stem cell generation. Nature 460(7251):49–52. doi:10.1038/nature08180

Yamanaka S (2009b) A fresh look at iPSC cells. Cell 137(1):13–17. doi:10.1016/j.cell.2009.03.034, pii: S0092-8674(09)00333-X

Yang L, Soonpaa MH, Adler ED, Roepke TK, Kattman SJ, Kennedy M, Henckaerts E, Bonham K, Abbott GW, Linden RM, Field LJ, Keller GM (2008) Human cardiovascular progenitor cells develop from a KDR+ embryonic-stem-cell-derived population. Nature 453(7194):524–528. doi:10.1038/nature06894, pii: nature06894

Index

A. Firth, J.X.-J. Yuan (eds.), *Lung Stem Cells in the Epithelium and Vasculature*, Stem Cell Biology and Regenerative Medicine, DOI 10.1007/978-3-319-16232-4

P

R

S

T

V

W

MIX
Papier aus verantwortungsvollen Quellen
Paper from responsible sources
FSC® C105338

If you have any concerns about our products,
you can contact us on
ProductSafety@springernature.com

In case Publisher is established outside the EU,
the EU authorized representative is:
Springer Nature Customer Service Center GmbH
Europaplatz 3, 69115 Heidelberg, Germany

Printed by Libri Plureos GmbH
in Hamburg, Germany